EARTH and ENVIRONMENTS THROUGH TIME:

A Laboratory Manual in HISTORICAL GEOLOGY

EARTH and ENVIRONMENTS THROUGH TIME:
A Laboratory Manual in HISTORICAL GEOLOGY

David R. Hickey
Science Department, Lansing Community College

Additional Illustrations by
Erin O'Brien (Laboratories 5 & 6)

Composition: Randy Miyake/Patti Zeman, Miyake Illustration
Cover image: Erin O'Brien
Interior art: David R. Hickey
Erin O'Brien

A list of registered trademarks found in this text follows the index.

WEST'S COMMITMENT TO THE ENVIRONMENT
In 1906, West Publishing Company began recycling materials left over from the production of books. This began a tradition of efficient and responsible use of resources. Today, up to 95 percent of our legal books and 70 percent of our college texts are printed on recycled, acid-free stock. West also recycles nearly 22 million pounds of scrap paper annually—the equivalent of 181,717 trees. Since the 1960s, West has devised ways to capture and recycle waste inks, solvents, oils, and vapors created in the printing process. We also recycle plastics of all kinds, wood, glass, corrugated cardboard, and batteries, and have eliminated the use of styrofoam book packaging. We at West are proud of the longevity and the scope of our commitment to our environment.

610 Opperman Drive
P. O. Box 64526
St. Paul, MN 55164-1003

Printed in the United States of America

99 98 97 96 95 94 93 92 8 7 6 5 4 3 2 1 0

ISBN 0-314-93920-2

TABLE OF CONTENTS

PREFACE

This manual is intended to provide hands-on experience to students taking a first course in Historical Geology. A variety of exercises have been included to address a broad range of student interests and abilities. While the exercises have been developed for a general audience, many should be appropriate for students majoring in a geology degree program.

The manual is intended to be self-contained; that is, all information necessary for completing each exercise is contained in the discussion. The conceptual background for each laboratory is presented in detail and these discussions contain information that will supplement the presentation in the course text book. The use of terminology has been kept to a minimum, however, as in any discipline, a certain amount of new vocabulary is necessary. When such terms are necessary, the etymology has been provided. Also, a list of vocabulary terms is included at the end of the discussion portion of each laboratory exercise.

Among the variety of exercises included are activities involving the use of hand samples, graphics, photographs and maps. In total there are more than twenty-five global paleogeographic maps and map exercises. Four lab activities address topics related to plate tectonics including ancient climates and oceans, the diversity of life, sedimentation, and geologic structures. Every laboratory exercise has a clear statement of objectives and tips to the student for completing the work.

Available with this lab manual is an Instructor's Guide. This contains preparation suggestions for laboratory exercises, a list of materials needed, pre and post laboratory homework exercises, questions for discussion and exercise solutions. To assist the instructor in preparing the course syllabus, each exercise has been ranked according to level of difficulty, length, and amount of time needed for completion.

ACKNOWLEDGEMENTS

Preparation of a book that will suit a disparate audience at a wide variety of institutions is no easy task and requires the input of many people. The following individuals are thanked for taking the time to read through the manuscript and provide insightful comments and suggestions.

Warren D. Allmon
University of South Florida

Kennard B. Bork
Denison University

Scott Brande
University of Alabama-Birmingham

Barbara Leitner Brande
University of Montevallo

Thomas W. Broadhead
University of Tennessee

Francis H. Brown
University of Utah

James E. Bugh
State University of New York—Cortland

Andrew Cole
Wright State University

Kraig Derstler
University of New Orleans

Charles Wm. Dimmick
Central Connecticut State University

Richard H. Fluegeman, Jr.
Ball State University

R. L. Langenheim, Jr.
University of Illinois-Urbana/Champaign

Joe Ghiold
Louisiana State University

Russell J. Henning
University of Louisville

John A. Howe
Bowling Green State University

Ronald Janowsky
Mohawk Valley Community College

Patricia H. Kelley
The University of Mississippi

Elana L. Leithold
North Carolina State University

James S. Monroe
Central Michigan University

Anne V. Noland
University of Louisville

William C. Parker
Florida State University

James H. Shea
University of Wisconsin-Parkside

Jamie L. Webb
California State University at Dominguez Hills

Reed Wicander
Central Michigan University

Robert Wilson
University of Tennessee

Thanks also go to many people at West Publishing Company for their efforts in bringing this project to fruition.

EARTH and ENVIRONMENTS THROUGH TIME:

A Laboratory Manual in HISTORICAL GEOLOGY

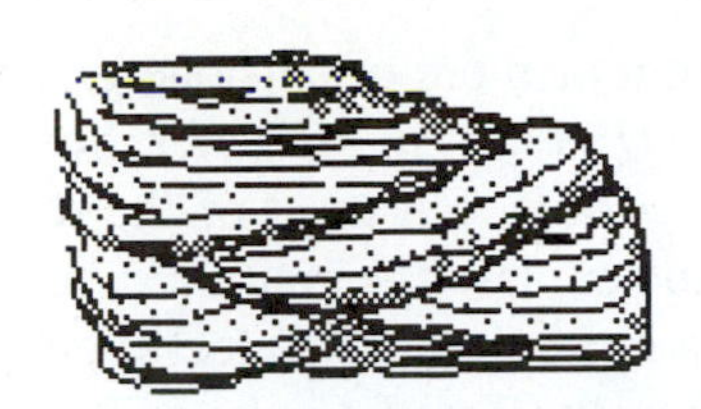

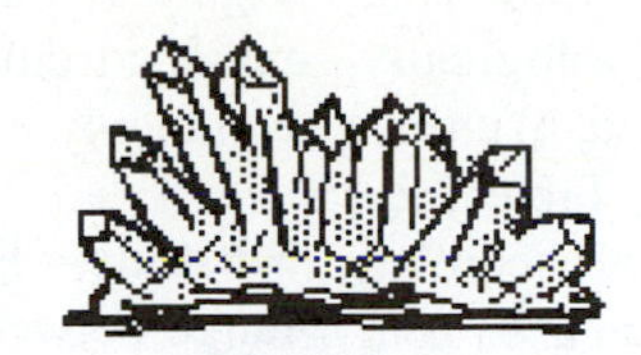

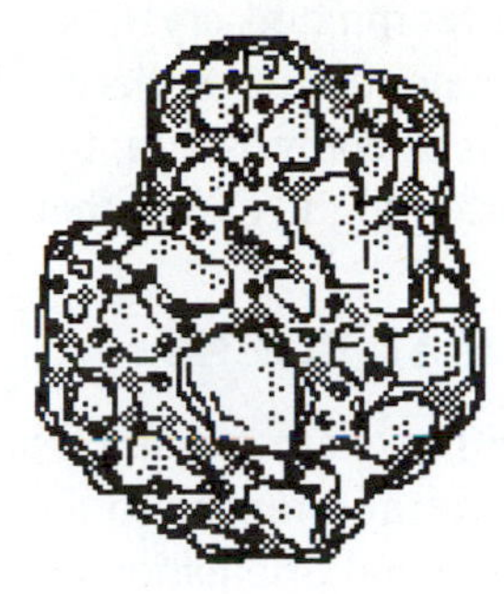

PART I
FIELD AND LABORATORY
MATERIALS AND METHODS

1
Sedimentary Rock Review

Objectives
To demonstrate mastery of the material you should be able to:
1) determine the major mineral constituents and grain sizes of sedimentary rocks; 2) use the classification scheme presented to identify the principal rock types; and, if so instructed, 3) construct an identification scheme based only on rock properties.

Using This Laboratory
This laboratory provides basic references (introduction/review) for the identification and classification of common sedimentary rocks, and thus background for the material to follow. The classification given in Table 1.1 may differ slightly from that (if any) used in your text. Plates are located in Appendix I.

The following apply to every laboratory: 1) worksheets are given the same numbers as the exercise to which they apply; and, 2) terms with which you should be familiar are underlined and explained or defined, and the rock types and fossil groups which you should be able to recognize are underlined.

This laboratory introduces/reviews common sedimentary rock types. Devote your primary attention to the observation and description of rock properties or lithology (*lithos* = rock), not mere identification. Lithologic properties are important not only for descriptive and identification purposes, but also because the broad categories of rock classification are genetic (*genes* = born; or *genia* = origin) and these properties carry information about rock genesis such as conditions of sediment deposition (see also Laboratories 2, 8-10 and 12).

SEDIMENTARY ROCK CLASSIFICATION

Sedimentary rocks are composed either of particles (e.g., the sand-sized grains of sandstone) or precipitated crystals (e.g., halite; NaCl = "rock salt"), and occur in stratified (layered; *stratum* = covering) sequences of beds (strata ≥1cm thick). Laminae are thin layers (<1cm) *within beds*. Grains may consist of rock, mineral, or animal skeleton fragments or plant fibres or mixtures of each. Grains are weathered (disintegrated and/or decomposed) loose and accumulate or are transported (eroded) by water, wind, or ice to a site of deposition, where burial and lithification (transformation to rock) occur. Lithification occurs by cementation of grains by a mineral precipitant (e.g., silica or calcite) and/or compaction and water loss.

Sedimentary rocks are classified in two categories reflecting particulate or crystalline origins: 1) detrital rocks (*attritus* = to rub away) consist of particles of rocks and/or minerals derived from the weathering and erosion of any pre-existing rock type (Table 1.1); and, 2) nondetrital rocks, include: a) biologically formed particles (e.g., shell fragments); and, b) chemically precipitated crystals (e.g., halite; Table 1.2). Because nondetrital rocks can be classified according to origin or composition, the classification used here may differ slightly from that of your text's.

DETRITAL ROCKS

Detrital rocks are classified principally by grain texture—a term referring to the sizes, shapes, sphericity, sorting and orientation of the enclosed grains (see Laboratory 2)—and, secondarily by grain composition or rounding (Table 1.1). Textural classification is based primarily on the dominant *grain size* class (where grain size ≈ diameter (mm)), ranging from boulders (very coarse grained) to microscopic mud (very fine grained). Grains can be composed of minerals—including quartz, feldspar, mica, and clay minerals—and/or of rock fragments. Clastic rocks are those consisting of particles or clasts (*klastos* = broken) of detrital or biological origin.

Conglomerate and Breccia—Conglomerate (Plate I, Figure 1) and breccia (Plate I, Figure 2) are the coarsest grained of detrital rocks, with the largest clasts ranging from pebble-size (>2 mm) to boulders (>256 mm). Conglomerate consists of variously shaped, rounded clasts embedded in a matrix of sand, silt and clay, and cementing minerals. Breccia clasts are little abraded and thus, have "sharp" edges and angular shapes.

Sandstones—Sandstones consist of *sand-sized* particles (1/16-2 mm) of ANY composition cemented by minerals or finer grained particles, and are classified by the bulk mineral composition of the particles bound by cement or finer-grained matrix (Tables 1.1, 1.3). Arkose is a sandstone composed of ≥25% feldspar grains (Plate I, Figure 3; Table 1.1, 1.3). The grains are typically coarse (1.5-2 mm), usually angular, of uniform size and bound by cement. Graywacke is sandstone composed of quartz, feldspar and rock fragments (Plate I, Figure 4; Table 1.1, 1.3). The clasts are bound by a dark colored, fine grained (mud or silt) matrix comprising >20% of the rock and giving it a "dirty" appearance. The sand-sized grains tend to be angular and of mixed sizes in contrast to the more uniformly large and cemented grains of arkose. Subgraywacke consists of abundant quartz and chert, but few feldspar grains, in a mud matrix comprising about 15% of the rock (Table 1.1, 1.3). Quartz sandstone is composed of ≥95% quartz grains (Plate I, Figure 5; Table 1.3). They are usually light in color, ranging from white, yellow and brown to red, and are commonly laminated.

Siltstone—Fifty percent or more of siltstone grains range between 1/256-1/16 mm in diameter. Some grains may be barely visible, but most of the rock resembles hardened mud or clay. The grains are, however, coarse enough to make the surface feel slightly rough and gritty to the teeth (Plate I, Figure 6; Tables 1.1, 1.3). Sandstone grains, by contrast, are easily visible and give the rock a sandpaper-like feel. Thin laminations marked by color differences are common in siltstones.

Shales and Claystones—Shale and claystone are composed of the finest mud-sized grains (< 1/256 mm) of clay minerals too small to be seen with the unaided eye. In contrast to siltstones, these are smooth to the touch and appear to be hardened mud or clay (Plate I, Figure 7; Tables 1.1, 1.3). Claystones are not laminated, whereas those having thin, easily split laminations resembling the pages of a book are called shales.

NONDETRITAL ROCKS

Nondetrital rocks are classified according to origin and composition. They form from the accumulation of biological particles, from chemical precipitates, or both. The classification used here (Table 1.2) includes: 1)carbonate rocks—whether of biological or chemical origin, and of clastic or crystalline texture; 2) crystalline rocks formed by precipitation of dissolved salts during evaporation (hence called evaporites); and, 3) a miscellaneous category of biologic origin, including coals and cherts.

Carbonate Rocks—Carbonate rocks include limestones (Plate I, Figure 8 & 9 & Plate II, Figures 1-4; Table 1.3), composed largely of calcite ($CaCO_3$)—or another crystal shape variant of calcium carbonate known as aragonite— and, dolostone (Plate II, Figure 4), largely of composed of dolomite ($2CaMg(CO_3)_2$). Dolostone originates from chemical process(es), and limestone from disintegrated shell, skeleton and algae, or (less commonly) from marine or fresh water precipitates (Laboratory 2).

Limestone is commonly soft and effervesces readily in a 10% solution of HCl. Dolostone may appear identical to limestone, but it will not react with HCl unless scratched or powdered thus, exposing more reactive surface area.

TABLE 1.1.–CLASSIFICATION of DETRITAL SEDIMENTARY ROCKS.

Texture*	Composition**	Rock Type
Coarse Grained > 2 mm Diam.	Rounded Fragments of any Rock Type or Mineral; Quartz, Quartzite and Chert Dominant	Conglomerate
	Angular Fragments of any Rock Type or Mineral. Quartz, Quartzite and Chert Dominant	Breccia
Medium Grained 1/16 - 2 mm	Quartz with Minor Accessory Minerals	Quartz Sandstone
	Quartz with at least 25% Feldspar	Arkose
	Quartz, Feldspar, Rock Fragments and up to 30% Clay Matrix	Graywacke
	Quartz, Rock Fragments, Chert < 15% Clay Matrix	Subgraywacke
Fine Grained 1/256 - 1/16 mm	Quartz and Clay Minerals	Siltstone
Very Fine Grained < 1/256 mm	Clay Minerals, Quartz, Feldspar, Mica and Chlorite or Organic Carbon	Shale

*Principle Character **Secondary Character

TABLE 1.2. CLASSIFICATION of NONDETRITAL ROCKS.

Composition *	Texture, Other *	Rock Type
	A. CARBONATES	
$CaCO_3$ (Calcite or Aragonite)	Medium - Coarsely Crystalline	Crystalline Limestone
$CaCO_3$ (Calcite or Aragonite)	Microcrystalline (Carbonate Mud)	Micrite,"Lithographic" if very fine
$CaCO_3$ (Calcite or Aragonite)	Fine or Coarsely Crystalline with Fossils	Fossiliferous Limestone
Fossils with Calcite Cement	Largely Sand-sized Fossil Fragments	Coquina
Fossils with Calcite Cement	Microscopic Fossils, Soft	Chalk
$CaMg(CO_3)_2$ Dolomite	Fine to Coarsely Crystalline with/without Fossils	Dolostone
	B. EVAPORITE CHEMICAL PRECIPITATES	
$CaSO_4 2H_2O$ Gypsum	Fine to Coarsely Crystalline 2 Cleavage Planes.	Gypsum
$CaSO_4$ Anhydrite	Fine to Coarsely Crystalline One Cleavage Plane	Anhydrite
NaCl Halite	Fine to Coarsely Crystalline Cubic Cleavage, Salty Taste	Halite
	C. BIOGENIC	
SIO_2 Silica	Microcrystalline, Dense, Smooth Conchoidal Fracture, Hardness = 7	Chert ***
Organic Carbon and < 50% Clay	Fine grained, soft, brown fibrous and unconsolidated to shiny black and blocky to hard and nodular	Coal

**Secondary Character ***Can be of chemical origin as well. *Principal Character

Limestone and dolostone may be composed of: 1) mud-sized crystals (Plate I, Figure 9); 2) large crystals (Plate II, Figure 3); 3) fossils or other clasts bound by a carbonate mud or crystalline matrix (Plate I, Figure 8); 4) loosely-cemented microscopic "shells" (Plate II, Figure 2); 5) chemically precipitated clasts (oöids); or, 6) fossil fragments bound by calcite cement (Plate II, Figure 1).

Chert—Chert is a dense microcrystalline quartz having many of the properties of other quartz minerals (Plate II, Figure 5; Table 1.3). Most bedded chert probably originates from accumulations of the siliceous "shells" of single-celled organisms (Appendix I) and particles of the skeletons of some sponges.

Coal—Coal is composed of ≥50% plant-derived carbon, and silt and/or clay (Plate II, Figure 6). It forms only if abundant plant material accumulates rapidly or within water lacking oxygen because plant fibres decompose readily under other conditions (Laboratories 2 & 7). Various types of coal result from differing degrees of burial temperature and pressure, and differ also in sulfur and other elemental abundances (Table 1.3).

Evaporites—Evaporite rocks form by crystal precipitation during the evaporation of extremely salty water. They are composed of bedded or massive crystals of minerals such as halite ($NaCl$), gypsum ($CaSO_4 \cdot 2H_2O$), and anhydrite ($CaSO_4$; Plate II, Figures 7-10; Table 1.3).

VOCABULARY

Bed, Bedding, Clast, Clastic rocks, Clay, Coarse-grained, Composition, Consolidated, Detrital rocks, Erosion, Fine-grained, Fissile, Grain size (mud, silt, sand, & pebbles-boulders), Intraclast, Lamina (pl., laminae), Lithification, Lithology, Matrix, Mud, Texture, Sand, Stratum (pl., strata), Weathering.

ROCK TYPES

Detrital—Arkose, Breccia, Clastic Rocks, Conglomerate, Graywacke, Quartz sandstone, Sandstone, Shale, Siltstone, Subgraywacke.
Non-detrital—Anhydrite, Aragonite, Carbonate mud, Carbonate rocks, Clastic rocks, Coal, Chert, Coquina (clastic), Crystalline limestone, Dolostone, Evaporites, Fossiliferous limestone (clastic), Rock gypsum, Rock halite.

MINERAL TYPES

Anhydrite, Aragonite, Calcite, Chert, Dolomite, Gypsum, Halite.

TABLE 1.3. MISCELLANEOUS PROPERTIES of SEDIMENTARY ROCKS

ROCK TYPE	COMPOSITION	OTHER COMMON PROPERTIES
Conglomerate	clasts of quartz, chert, quartzite, rock fragments	poorly bedded to unbedded
Breccia	quartz, chert, quartzite, granite	unbedded
Quartz Sandstone	quartz, common accessory minerals: mica, garnet, magnetite	silica/calcite cement. well bedded. light-reddish commonly
Graywacke	quartz, feldspar, clay minerals rock fragments	dark silt-clay matrixmatrix angular grains, poor sorting
Arkose	feldspar, quartz	calcite cement, clay or iron oxide. often coarse, angular grains
Coquina	shell fragments	calcite cement
Siltstone	quartz, mica, clay minerals	angular grains, laminated, various colors
Shale	clay minerals, feldspar, quartz, chlorite, mica, calcite	laminated & fissile. gray, black, green, red, blue, etc.
Claystone	as above	soft-blocky
Limestones	macroscopic & microscopic fossils or unfossiliferous, oöids, pellets, angular plates of lime mud (intraclasts)	Thin to thick-bedded, often light colored, effervesces in dilute HCl
Dolomite	fossil molds & casts, other clasts, above	often pink, hard, poor fossil preservation, often cherty
Chert	microcrystalline quartz -siliceous tests & sponge spicules, altered volcanic ash	hardness = 7, conchoidal fracture ceramic - waxy luster, gray, black, red, brown, white. often occurs as nodules
Coal	plant carbon, clay & silt	brown, fibrous ,unlithified =peat brown, fibrous , lithified =lignite black, oily luster, blocky fracture = bituminous brown-black. hard . nodular =anthracite
Halite	NaCl	transparent-translucent. salty, greasy luster, cubic cleavage
Gypsum	$CaSO_4 \cdot 2H_2O$	satiny-earthy luster, fiberous cleavage 2 directions

LABORATORY EXERCISES

1.1 Specimen Description and Identification—Your instructor will provide a set of sedimentary rock samples as listed on worksheet 1.1.

A) Observe and describe their textures, compositions (where possible), etc.

B) After describing each, determine their origins (i.e., either detrital, chemical or biological) and identity of each rock type.

C) Indicate the reasons for each identification under the heading "Distinguishing Properties."

WORKSHEET EXERCISE 1.1. SEDIMENTARY ROCK DESCRIPTION & IDENTIFICATION.

Composition		Texture	Bedding	Color	Origin	Distinguishing Properties	Rock Type
Indicate: Particle (P) or Crystals (C)	Mineral/ Rock (if poss.)	Grain Size >pebble ≈ sand ≈ silt ≈ mud	Beds (B) or Lamn. (L)	abrevi- ate: Y, R, etc.	Detrital (D) Chemical (C) or Biological (B)		
1.							
2.							
3.							
4.							
5.							
6.							
7.							
8.							
9.							
10.							

LABORATORY EXERCISES

1.2 Lithologic Comparisons—Identify among your set of samples, the rock types listed in worksheet 1.2. Compare and contrast the properties listed for each (where applicable) and note distinguishing and or distinctive properties for each. Note the origins of each as described in exercise 1.1.

WORKSHEET EXERCISE 1.2. COMPARISON & CONTRAST. SEDIMENTARY ROCKS & ORIGINS.

ROCK TYPES	Composition	Particulate/ Crystalline	TEXTURE	Distinguishing Properties	ORIGINS
A. Unfossiliferous Limestone					
Shale					
B. Black Shale					
Coal					
C. Limestone, Unfossiliferous					
Chert					
D. Crystalline Limestone					
Dolostone					
E. Quartz Sandstone					
Coquina					
F. Graywacke					
Conglomerate					
G. Arkose					
Breccia					
H. Fossiliferous Limestone					
Conglomerate					
I. Crystalline Limestone					
Chert					
J. Micaceous Shale					
Siltstone					
K. Graywacke					
Arkose					
L. Gypsum					
Halite					

LABORATORY EXERCISES

1.3. Specimen Description and Identification—Identification and interpretation of rocks requires well developed powers of observation and description—skills which naturally vary among individuals and with experience. This exercise will make you aware of these differences, aid in focusing descriptions and practicing observation by working as a group(s), reinforce the use of lithologic terms, and distinguish observations from inferences and value judgments.

Your instructor will provide samples of each of the sedimentary rocks discussed above. With instructor guidance, each student will describe one lithologic character of each sample and pass it to the next, who will then compare their observation of the same property (noting agreement or disagreement with the prior observation or description), before describing a second property.

As an aid to your efforts, define "observation", "inference", and "value judgement" and avoid or correct the use of such "descriptive" terms during this exercise. Discuss disparate observations and descriptions/terms. After each sample has been described, the group(s) should come to a consensus identification.

Definitions:

Observation—

Inference—

Value Judgement—

1.4. Identification Key—You are the world's first geologist and have set yourself the task of constructing a means of identifying sedimentary rocks. To accomplish this goal you will construct a branching flow chart or binomial key based on observations — not inferences or value judgements. Figure 1.1 illustrates the thought process and diagrams the method for making an ID key. The method involves a sequential subdivision of numbered samples into subsets, in which members of each have, or do not have, the indicated property.

Each sample has been assigned a reference number for use in constructing the key. After observing each numbered sample determine which of those properties (e.g., Property #1:particulate vs. crystalline, Figure 1.1) allows subdivision of the sample into two groups, the members of which exhibit either

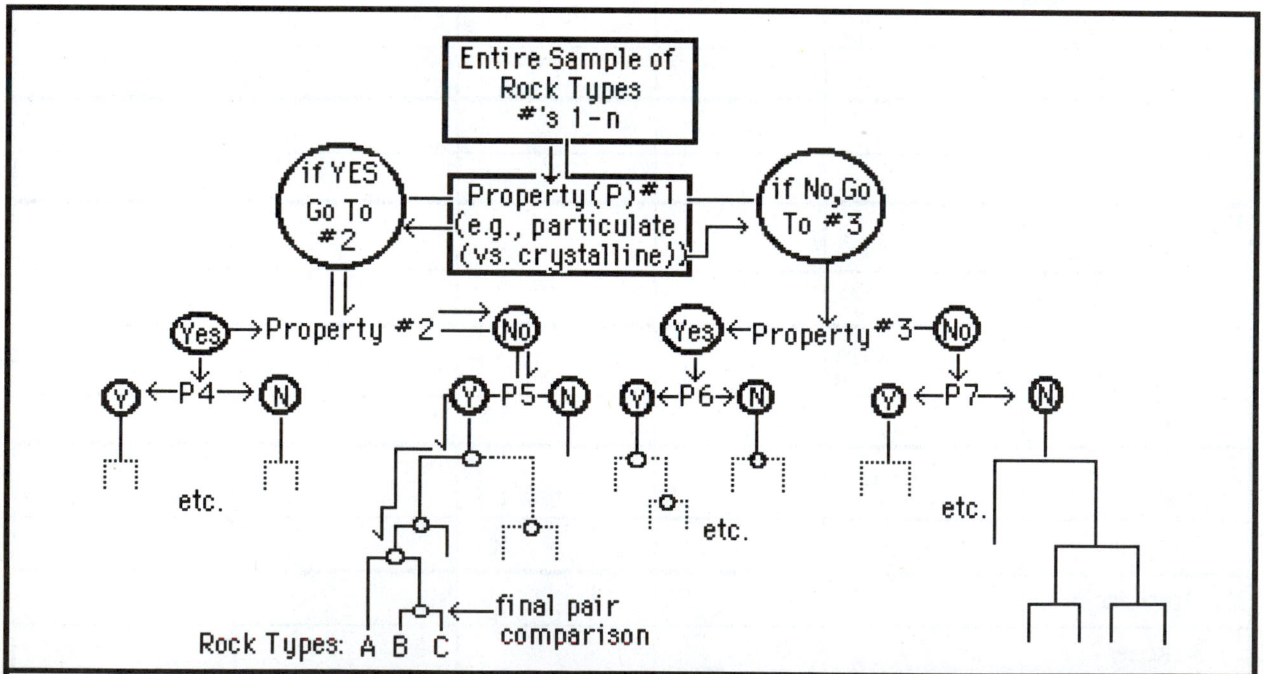

Entire Sample of Rock Types #'s 1-n
if YES Go To #2
Property(P)#1 (e.g., particulate (vs. crystalline))
if No, Go To #3
Yes
Property #2
No
Yes
Property #3
No
Y
P4
N
Y
P5
N
Y
P6
N
Y
P7
N
etc.
etc.
etc.
final pair comparison
Rock Types: A B C

FIGURE 1.1. STRUCTURE of a DICHOTOMOUS IDENTIFICATION KEY.

LABORATORY EXERCISES

1.4. Identification Key (cont.)
property #1 (e.g., = "the rock is comprised of particles") or its alternative ("not comprised of particles"). Alternatively, distinguishing property #1 may be posed as a question: "Are samples # 1, 2, ... n) comprised of particles?" If "Yes", it/they become(s) a member of subset A; if "No", the sample(s) is/are comprised of crystals, and are by default, assigned to subset B. List the sample numbers of specimens belonging to each subset. Repeat this procedure until each sample is separated from all others by a final distinguishing property and give the rock type's name.

A given rock type can occur on both sides of a branch if more than one state of a distinguishing property is possible. Keys can be incorrectly constructed (by use of invalid properties, mis-assignment of rock types, etc.), but there is no single correct answer.

EXERCISE 1.4. WORKSHEET for IDENTIFICATION KEY.

LABORATORY EXERCISES

1.4 Identification Key (cont.)

Figure1.1 presents the beginning steps of the key you are to complete. When completed:

A) Construct a classification outline like that of Tables 1.1 & 1.2 in the space provided on exercise worksheet 1.4;

B) Compare/contrast your results (groupings, types of criteria, types of information, etc.) with the classification used in this manual; and,

C) Answer the following questions:

1) What kind of information is evident from the key?

2) In what ways, or on what bases, are the members of each subset related to one another?

3) Does the key provide a classification as well as a means of identifying sedimentary rocks? Explain.

4) What information is inherent in this manual's classification?

5) Does information which could enhance your (& future geologists') understanding of sedimentary rocks—beyond that of an ordering of properties and convenience for discussion—emerge from your key? Explain.

6) What additional information is needed to enhance the key's explanatory power?

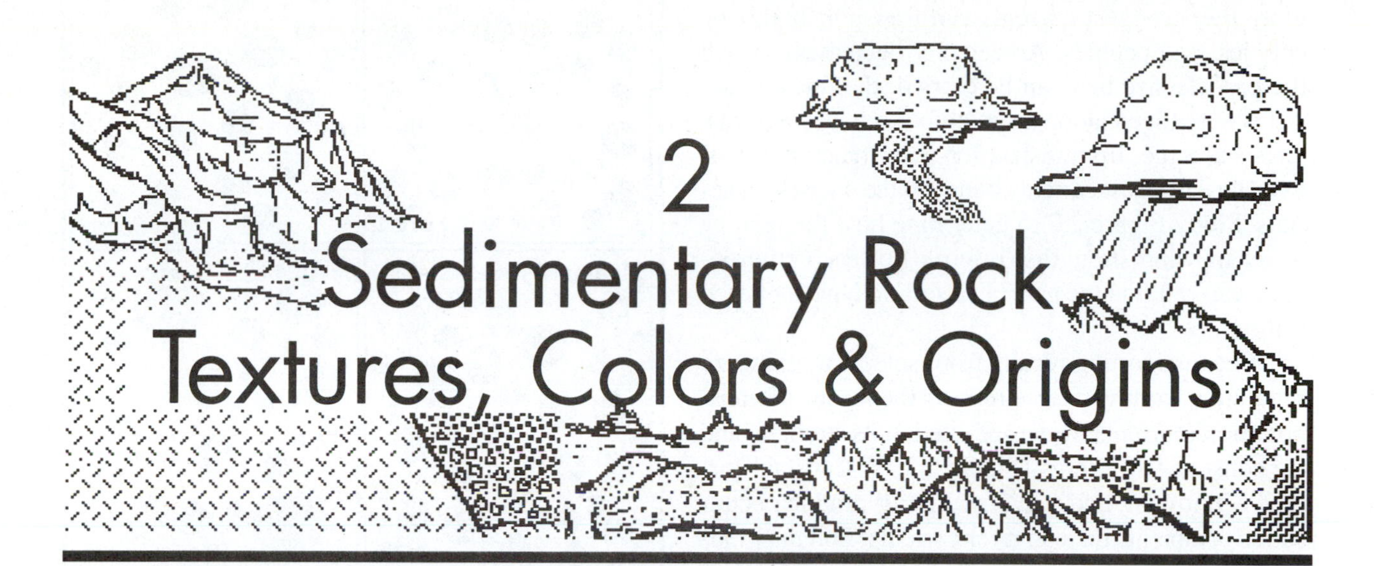

2 Sedimentary Rock Textures, Colors & Origins

Objectives
To demonstrate mastery of the material you should be able to:1) recognize and describe the grain size-sorting, grain roundness and shape, and basic compositions of sedimentary rocks; 2) recognize degrees of lithologic maturity in detrital rocks and understand the factors which affect it; and, 3) use texture and color to determine the general conditions of origin of some sedimentary rocks.

Using This Laboratory
Carefully study the illustrations of environments in which sediment types are deposited and those of sediment textures. Make notes, perhaps in table form, on the factors that influence sediment maturities, and properties which evidence climates, source rock types, and distances and durations of sediment transportation. All information needed to complete the exercises is given below.

Textures, colors and mineral compositions are used in identification and classification of sedimentary rocks, but they can also be used as clues to rock genesis or origins. In this chapter we will learn how to recognize and use these properties to infer very general conditions of weathering, erosion, deposition, and source region climates and rock types.

ORIGINS of TEXTURES

Size, sorting, "shape" (roundness and sphericity), and grain orientations are collectively known as sedimentary texture. These properties provide information about the strength of water agitation, current directions, and conditions of weathering and erosion.

Grain Size and Size Sorting—Particle sizes reflect the strengths of currents, waves or winds that transport grains to depositional sites. Current strength can be inferred from grain size because grains of differing sizes/composition differ in specific gravity/density, and, thus the ease with which they are moved. Thus, grain size sorting (selective separation and deposition according to size classes) reflects the range and variation of the transporting medium's velocity.

Sorting degree refers to the "sameness" of grain sizes in sediment or rock as measured by the per cent abundance of grains belonging to different size classes (Figure 2.1). If most grains in a sample are of similar size, the sample is described as "very well sorted" (Figure 2.1A). If particle size varies greatly among size fractions, the sample is "poorly sorted" (Figure 2.1D). Intermediate states—"well sorted" to "moderately well sorted"—are also recognized (Figure 2.1B, C).

Figure 2.2 shows the dominant grain size and distributions typical of each of several environments. Only the strongest currents can transport boulders, cobbles, and pebbles. As velocity decreases, so will the particle size that can be carried: the finest material is not moved along a stream bottom by weak currents and settles from suspension in the quietest waters. Only the strongest currents transport the coarsest materials. Study Figure 2.2 to determine how the relative water or wind strengths (inferred) of each environment can result in the grain size sorting typical of each setting.

Because sorting results from selective grain size separation, sorting degree reflects the means of transportation (currents, waves or winds), its strength, and the constancy of energy (Figure 2.2, 3). Constant current winnowing and wave-pounding produce better sorted sediments than do river currents. Eolian (windblown) deposits (e.g., desert dunes) tend to be very well sorted, because wind transports only fine sand, silt and dust. Winds sweep the finest dust to high altitudes, ultimately to be deposited in the deep sea. Poorly sorted sediments can indicate limited transportation and/or sudden decreases in water velocity, and thus, its capacity to carry a wide range of sediment-sizes. The size of the coarsest fraction indicates a current's maximum strength, and minimum strength is reflected by the finest fraction. Study Figures 2.2 and 2.3 to determine which depositional sites are likely to have water/wind velocity/agitation that is: 1) high and constant; 2) intermediate and low, but constant; and, 3) fluctuating from moderate or high to low. These environments probably contain very well, moderately well, and poorly sorted sediments, respectively.

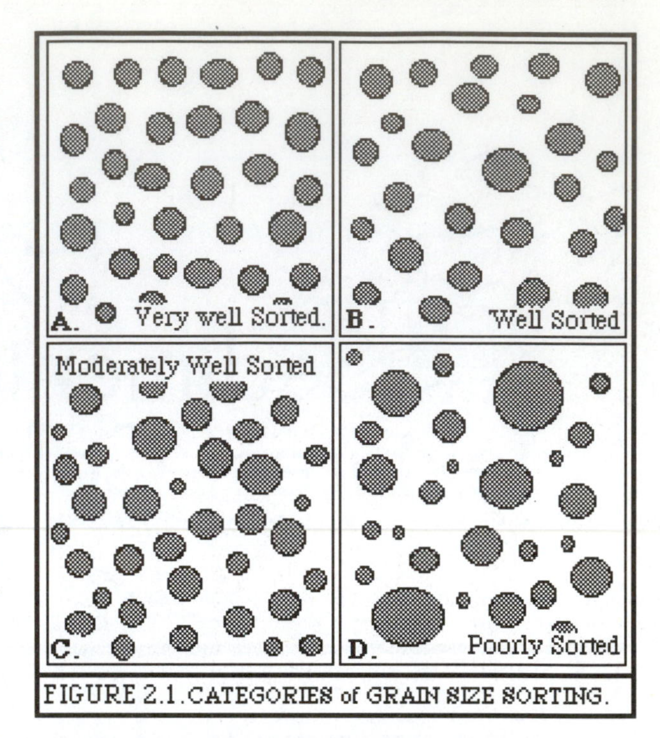

FIGURE 2.1. CATEGORIES of GRAIN SIZE SORTING.

Grain Rounding and Shape—As grains are transported and selectively sorted, collisions slowly change their sphericity and roundness. Grain roundness (Figure 2.4) refers to the angularity (surface roughness due to tiny projections), or lack thereof, not the grain's "shape". The more nearly equal are a grain's long and short axes, the more spherical is its shape. When clasts are heavy, water does not cushion abrasive blows, and thus, large clasts can be of varied sphericity and well

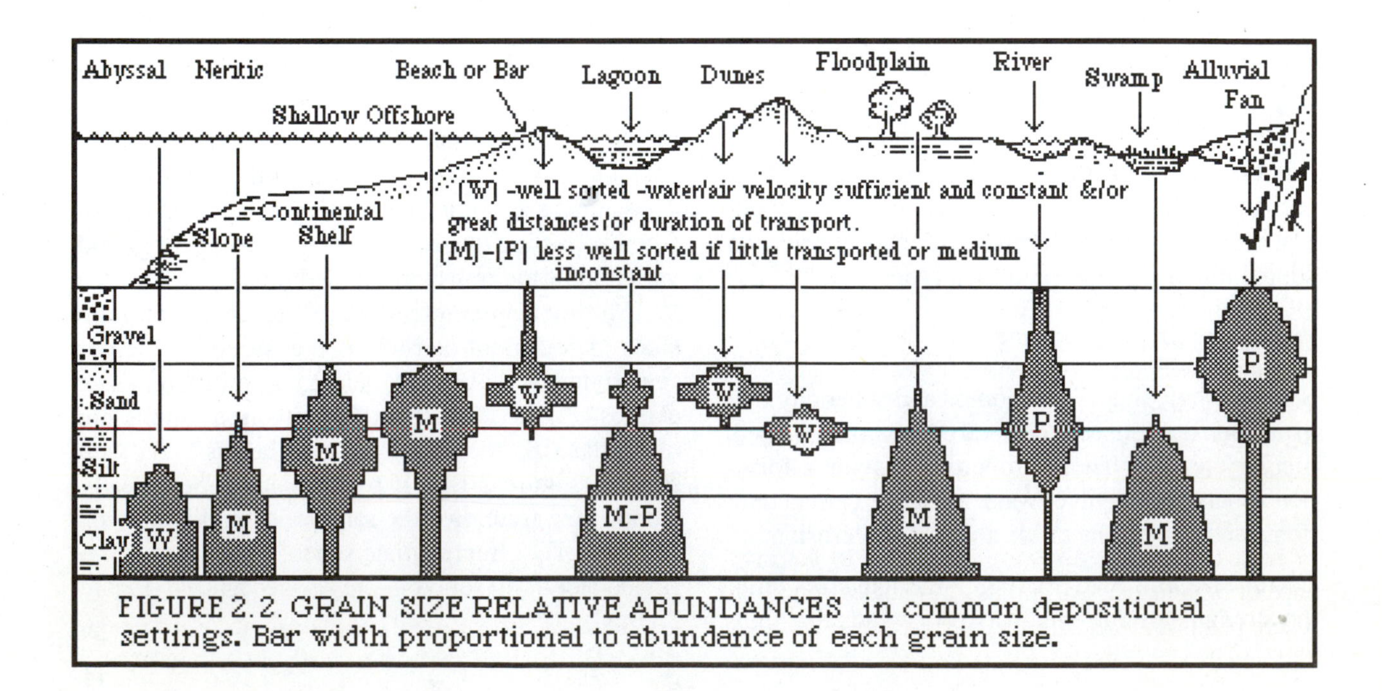

FIGURE 2.2. GRAIN SIZE RELATIVE ABUNDANCES in common depositional settings. Bar width proportional to abundance of each grain size.

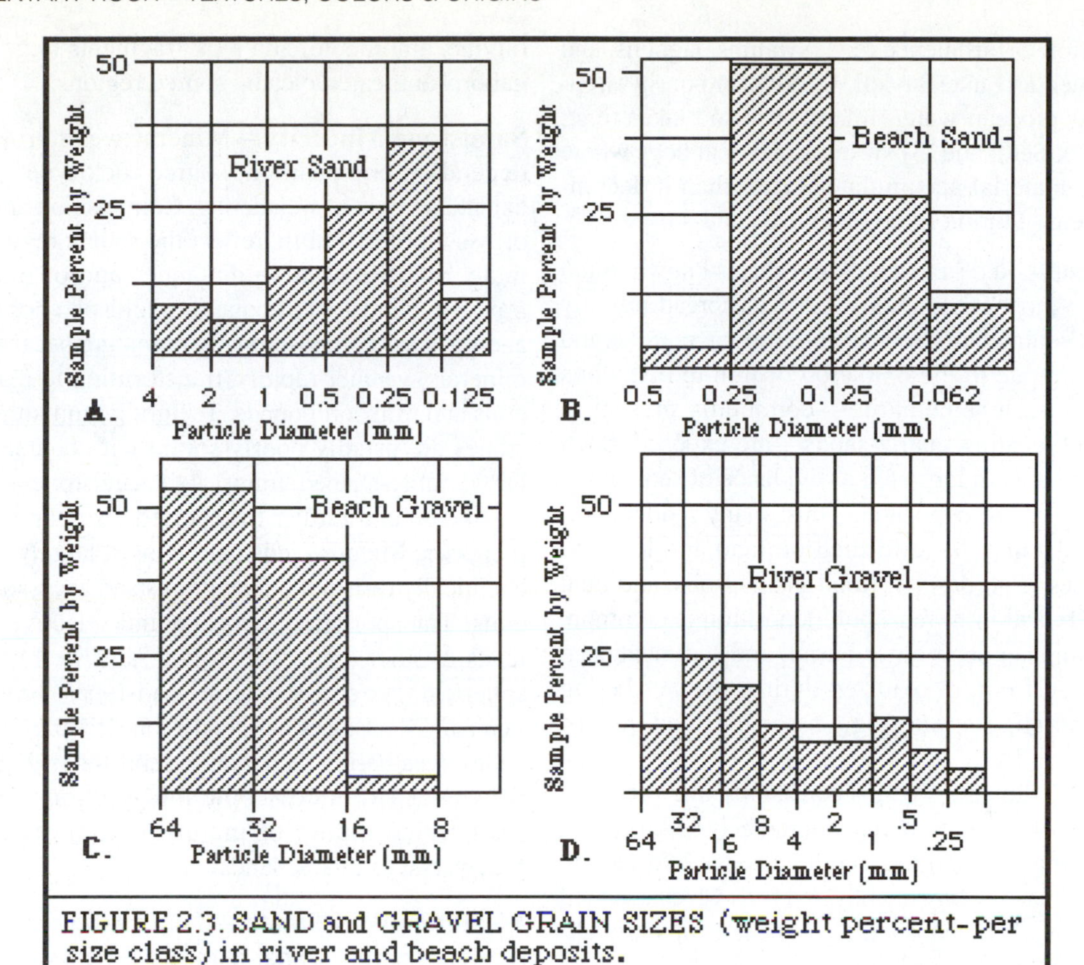

FIGURE 2.3. SAND and GRAVEL GRAIN SIZES (weight percent-per size class) in river and beach deposits.

rounded. As grain size decreases, water cushions grain impacts and abrasion rates decrease. Thus grains become spherical only after extensive transportation. Impacts between wind-blown grains are not well cushioned, and thus produce very spherical, well rounded grains. Water-transported grains tend to be well rounded but less spherical than wind-blown sediments.

ORIGINS of COLORS

Color is a conspicuous property and, although it is not useful for sedimentary rock identification, it can provide clues to conditions of their origin. In fresh rock, colors such as red, green and black may reflect oxidation or reduction (see textbook) of elements such as Fe. Oxidation (e.g., rusting) and reduction, in turn, may indicate (with other evidence) certain environmental conditions during or shortly after, deposition.

Black Sediments—Black coloration of sedimentary rock (e.g., shale) is commonly due to the presence of organic carbon, and thus, indicates conditions where plants and (sometimes) animals were abundant, and which prevented complete tissue decomposition.

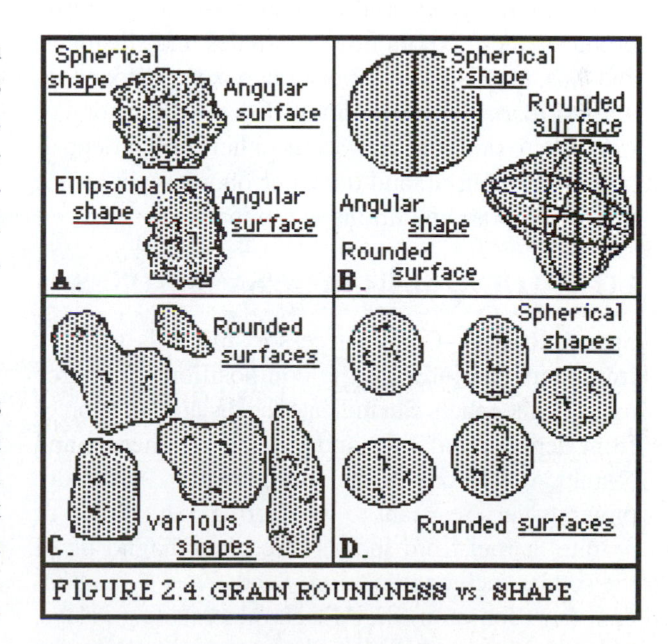

FIGURE 2.4. GRAIN ROUNDNESS vs. SHAPE

Incomplete decomposition typically occurs in oxygen-poor water: most decomposing bacteria cannot exist without it and organic acids from decaying plants may act as antibiotics preventing further decay.

Black sediments may accumulate under three conditions: 1) shallow water without oxygenating circulation or turbulence (e.g., swamps, lagoons, and lakes such as Lake Baykal, USSR); 2) poorly circulated deep ocean water, inland seas and lakes (e.g., the Black Sea); and, 3) swamps and peat bogs where organic material accumulates faster than it decomposes (e.g., Florida Everglades).

Red Beds—Red sandstones and shales known as red beds are most conspicuous and widespread brightly colored sedimentary rocks. Red, brown, maroon and purple are due to the oxidation of iron to rust when minerals such as hematite—commonly present in clays and coating quartz sands—are exposed. Such rocks are often interpreted as evidence of sediments' exposure where deposited. Thus, many reddish colored rocks may have formed on land and lowland shorelines (e.g., deserts, flood plains, soils, and tidal flats). Oxidation to red upon deposition is common, yet it can also occur after burial, and sediments can be derived from, or oxidized during, the weathering of pre-existing strata (e.g., desert-derived dust of abyssal depths).

Red sediments are also commonly interpreted as evidence of arid conditions, but red clay soil forms in tropical regions, and thus, could also have formed in humid climates. Consequently, red beds which acquired their color when deposited are indicators of warm climates—either arid or humid.

The Blue, the Grey, and the Greens—Reduced iron produces these colors in some shales and siltstones, and thus, may indicate deposition in oxygen-poor conditions (marine or nonmarine). However, reduction of iron, like oxidation, can occur either during deposition or after burial, and thus, corroborating evidence of oxygen-poor conditions is needed.

LITHOLOGIC MATURITY: SANDSTONES

Source Rocks—Certain accessory minerals and rock fragments can indicate the compositions of source rocks, while others can indicate the distance of sources from depositional sites and general weathering and climatic conditions of the source region. Accessory minerals very resistant to weathering (e.g., zircon of granite, garnet from schists) are often diagnostic of the source rock composition. Sandstones with abundant (and common) Si, Al-rich minerals (e.g., clays, micas, feldspars, quartz) and rock fragments may be derived from any assortment of continental crust rocks of any rock family (e.g., granite, schist, and quartzite). Very well rounded and sorted quartz sandstones are derived by extensive recycling of other sandstones. Sandstones with abundant Fe, Mg-rich minerals (e.g., olivine, amphibole) and rock fragments (e.g., basalt, gabbro) indicate volcanic source regions.

Sandstone Maturity—Mineral weathering rates depend on composition, source rock type, and the extent and type of weathering. Composition and mode of weathering, in turn, reflect the source region's climate, and the relative distance (and/or duration) traveled. Sandstones containing feldspar, hornblende and mica are compositionally immature because these minerals weather rapidly, transforming to clay minerals and iron compounds. Texturally immature sandstones are usually poorly sorted with coarse, angular, variably-shaped grains. As a sandstone is further weathered and farther transported, its quartz content increases. Mature sandstones consist largely of hard, chemically resistant quartz grains and accessory minerals. Transportation, abrasion and selective sorting leads to finer, very well sorted, and well rounded, spherical, texturally mature sandstones. In general, immature sandstones have undergone less transportation and/or weathering than mature sandstones. Figure 2.5 illustrates (in microscopic thin-section view) the gradational nature of maturity from arkose and graywacke to quartz sandstone.

Weathering & Climate—If only weathering is considered, compositional maturity reflects the source area's climate because chemical and mechanical weathering (see text) predominate in different climates, and each affects maturity differently. Chemical weathering predominates in warm and humid climates, hastening feldspar and Fe, Mg-rich mineral breakdown. Mechanical weathering predominates in arid (hot or cold) environments, while both weathering types occur in temperate climates (Figure 2.5)

ORIGINS of SANDSTONES

Now sandstones can be understood according to conditions of weathering, transportation and deposition. Arkose ("doubly" immature), derives from mechanical weathering of granitic rocks. It is commonly deposited in arid mountain valleys and marine basins and shelves near granitic sources (Figure 2.6).

Graywacke is commonly deposited in deep water basins offshore from volcanically active, mountainous continental margins and volcanic islands (Figure 2.7). It thus consists of mixed igneous (volcanic and otherwise), metamorphic, and sedimentary rock fragments from varied source regions and coastal geographies. Graywacke source rocks are typified by rapid weathering in humid, actively rising highlands.

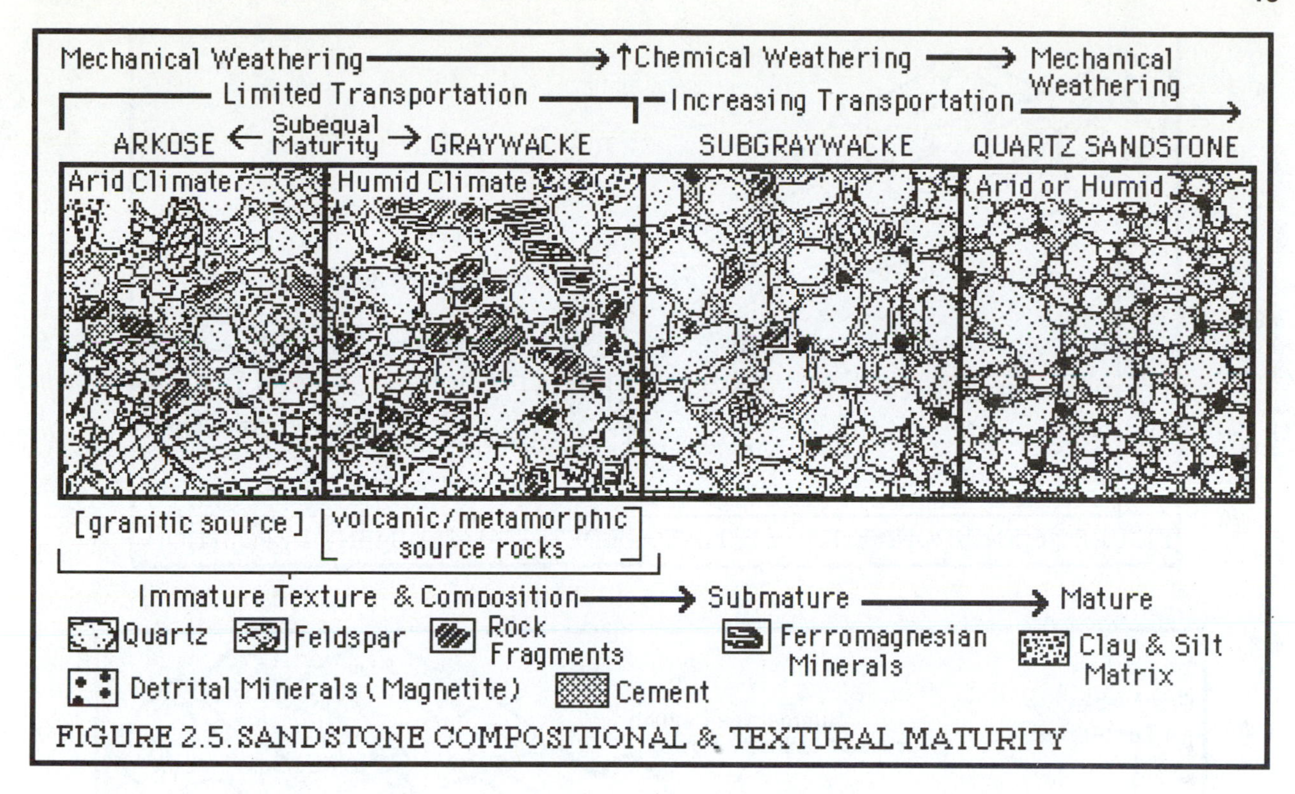

FIGURE 2.5. SANDSTONE COMPOSITIONAL & TEXTURAL MATURITY

However, poor sorting and voluminous matrix indicate rapid and limited transportion.

Many graywackes formed when sediment suspended within turbidity currents,—currents caused by the earthquake or slump-induced sediment suspension (i.e., turbid; hence, turbidity)—rushes down submarine canyon gorges or from otherwise steep slopes, and comes to rest at great depths (Figure 2.7). Couplets of graywacke and shale (deposited during quiet intervals) are known as turbidites (Figure 2.8A, B). These show a gradual decrease in the mean and maximum grain sizes from base (graywacke) to top (shale): a textural change called graded bedding that records the current's passing on rush and wane (Figure 2.8A).

Exceptionally and doubly mature quartz sandstones rarely contain other resistant minerals, and have been recycled many times before (Figure 2.9 on p. 17). Their source rocks were commonly other sandstones or quartzites.

Origins of Shales—Shales occur in each of the environments discussed above, and they vary in clay mineral and silt composition much as do sandstones in adjacent, associated settings (Figures 2.6-2.9).

CARBONATE ROCKS

Limestone Classification—Some of the many types of limestones differ in the kinds of matrix or cement type, and enclosed clasts (Figures 2.10-13 on p. 18). They may consist only of mud-sized crystals of calcite or large visible crystals (crystalline limestone), of either matrix type with fossils or other clasts, or "entirely" of clasts.

There are four clast types: fossils, oöids, intraclasts, and pellets. Limestone may contain fossil clasts (fossiliferous limestone; Figure 2.10). Oöids are tiny spherical grains of concentric aragonite layers enclosing nuclei of shell fragments or other organic material (Figure 2.11A,B). These spheres form in very shallow, constantly agitated water which keeps them in motion, allowing concentric deposition. Intraclasts (*intra* = within) are lumps and plates of partially hardened lime mud enclosed within softer lime mud (Figure 2.13). They commonly form during storms when semi-hardened lime mud plates are torn from the bottom and redeposited forming intramicrite. Pellets are tiny ellipsoidal grains of invertebrate excrement that set within micrite, forms pelmicrite. Pellets resemble oöids but are dark colored, unlayered (Figure 2.12A,B), and being soft, easily disintegrated by currents.

Limestone Origins—Most limestones are of marine origin, having formed in warm, shallow inland or shelf seas by accumulation of animal and algal skeletons. Algae, single-celled animals (Appendix I), reefs (Laboratories 6, 10), shells and skeletons produce enormous volumes of carbonate mud and shell fragments. Unlike detrital rocks, carbonates are deposited at or near their sources, and accumulate only where little detritus dilutes their volume (Figure 2.15).

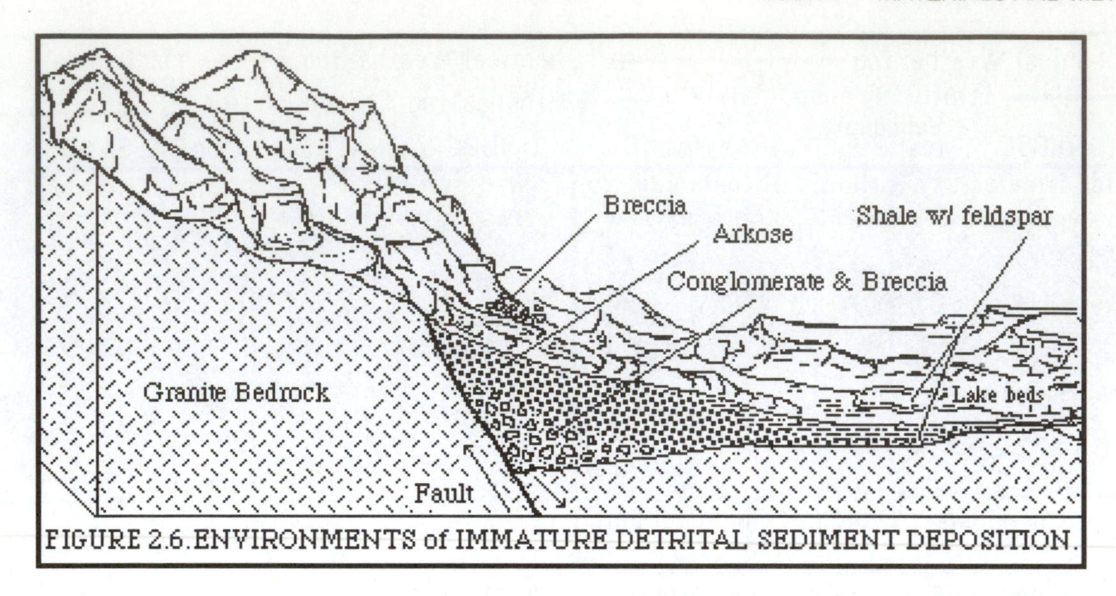

FIGURE 2.6. ENVIRONMENTS of IMMATURE DETRITAL SEDIMENT DEPOSITION.

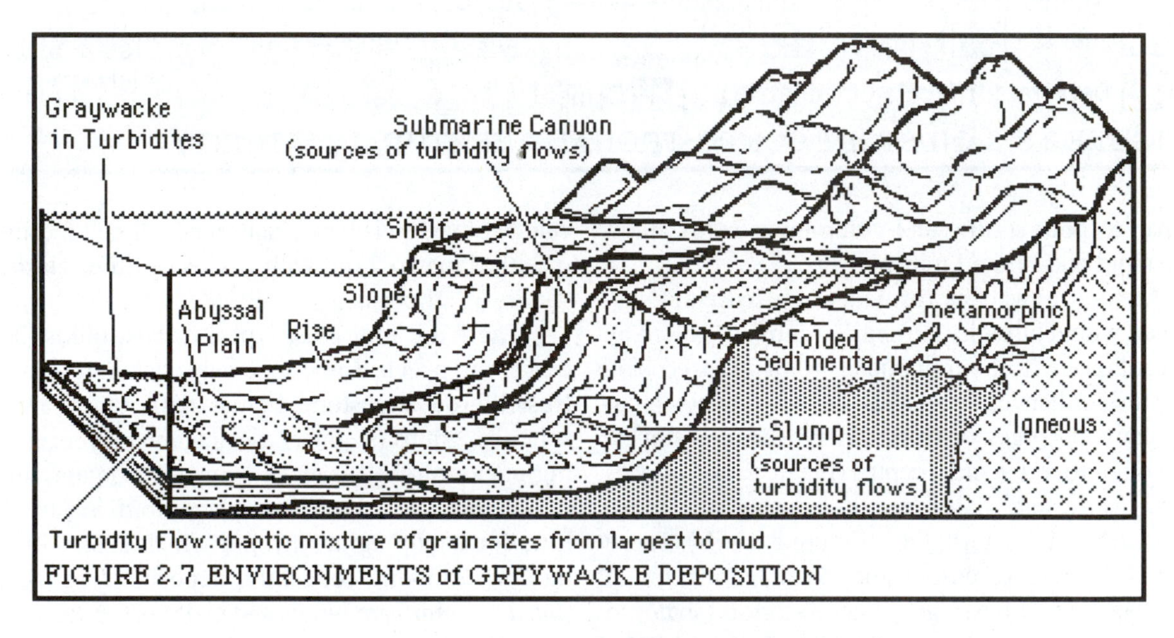

FIGURE 2.7. ENVIRONMENTS of GREYWACKE DEPOSITION

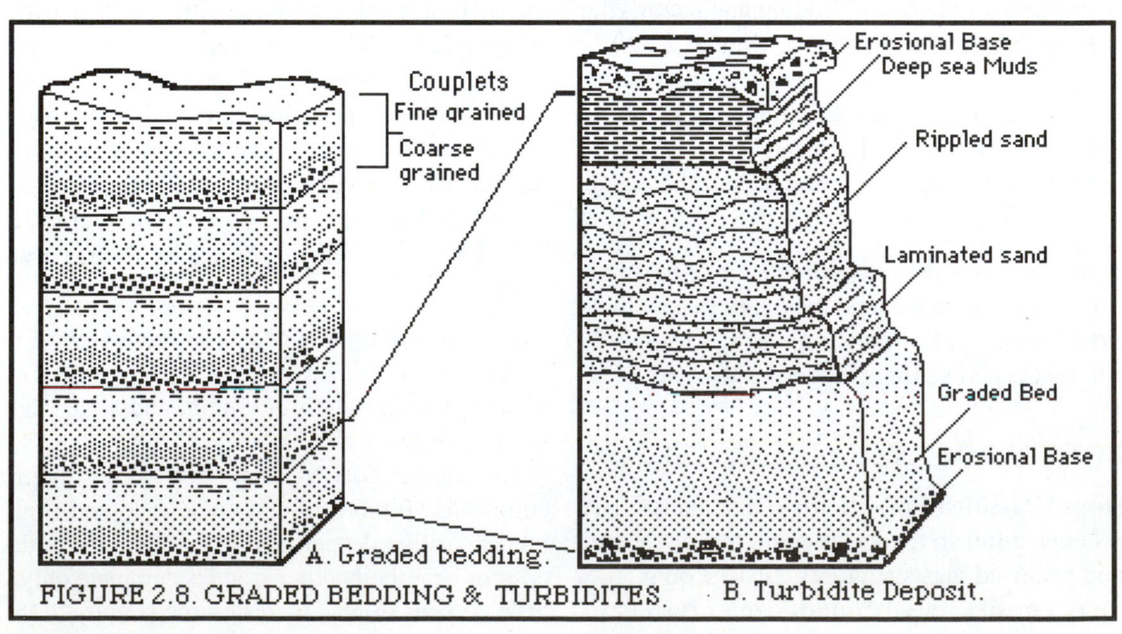

FIGURE 2.8. GRADED BEDDING & TURBIDITES.

Grain size and matrix types of limestones indicate current strength or water agitation. Mud generally accumulates in the absence of strong currents. If washed away, the voids formed between residual clasts are later filled by crystals (spar). Generally, the greater the proportion of micrite to spar, the quieter the water.

Dolostone Origins—Dolostone presently forms in localized portions of regions accumulating limestone, but was widespread in the past. Dolostone forms when magnesium replaces some of the calcium of previously formed calcite. In order for Mg to replace Ca, Mg ions need somehow to be concentrated in highly saline water which is drawn, or flows, through porous carbonate sediment (Figure 2.14A, B).

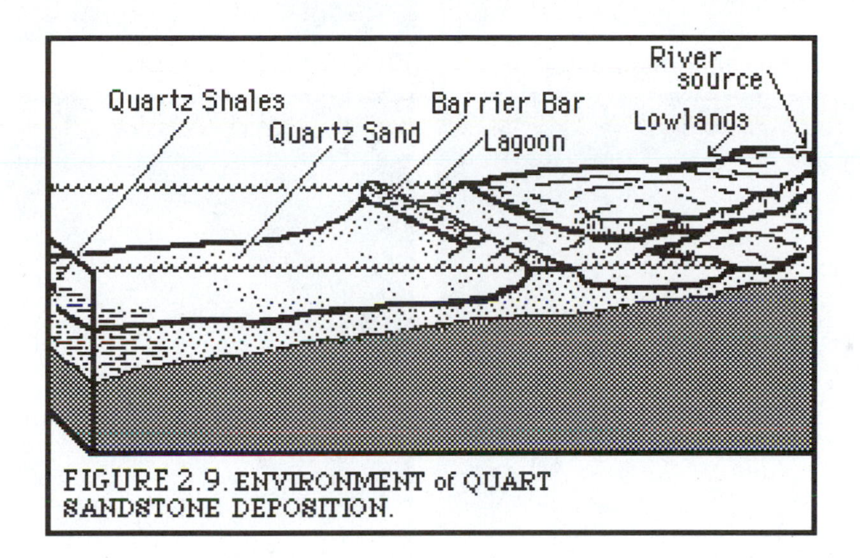

FIGURE 2.9. ENVIRONMENT of QUART SANDSTONE DEPOSITION.

VOCABULARY

Black shale, Coarse/fine-grained, Compositional im/maturity, Graded bedding, Oxidation, Red beds, Reduction, Roundness, Size-Sorting, Sphericity, Texture, Textural im/maturity, Turbidite, Turbidity current.

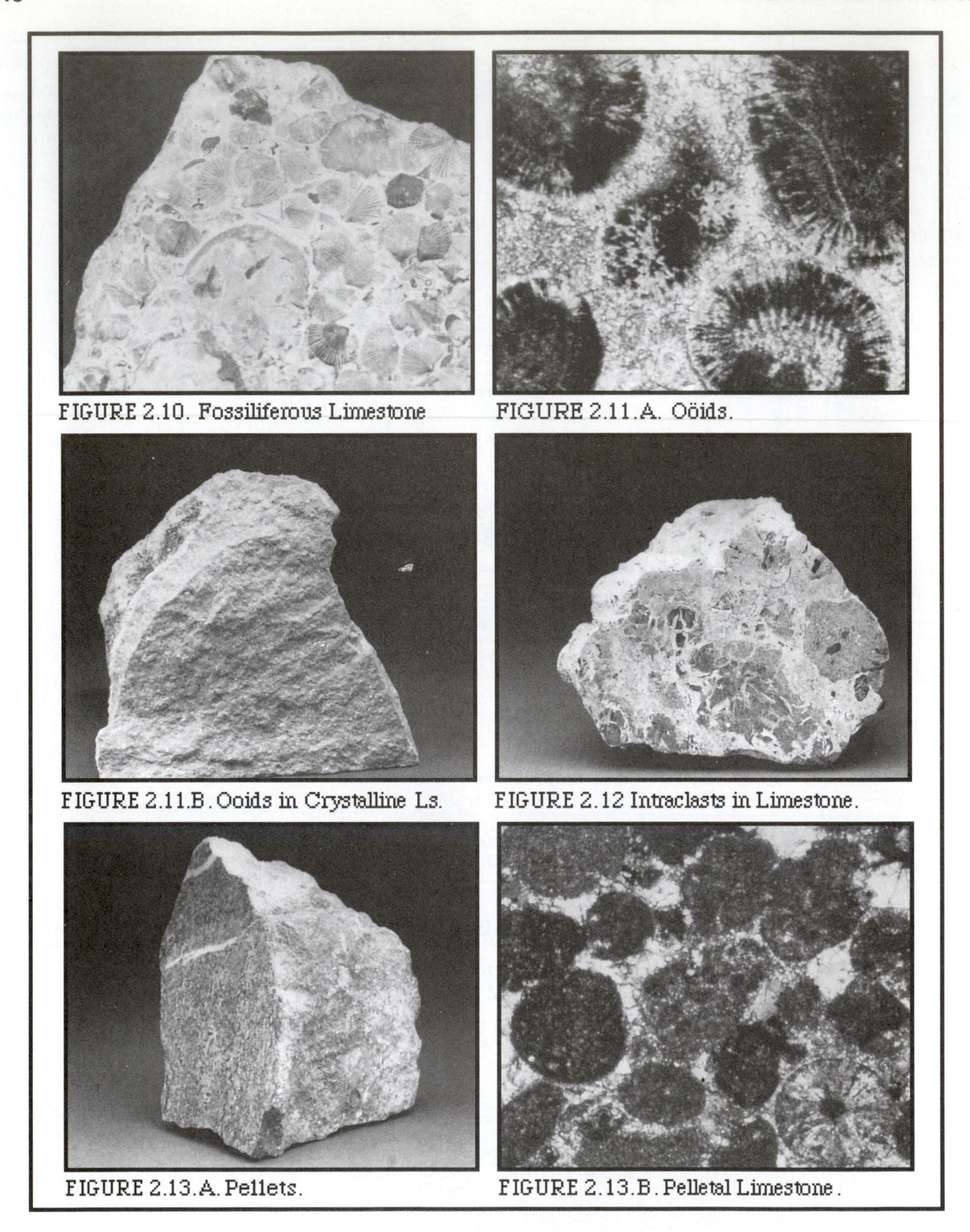

FIGURE 2.10. Fossiliferous Limestone

FIGURE 2.11.A. Oöids.

FIGURE 2.11.B. Ooids in Crystalline Ls.

FIGURE 2.12 Intraclasts in Limestone.

FIGURE 2.13.A. Pellets.

FIGURE 2.13.B. Pelletal Limestone.

LABORATORY EXERCISES

2.1. Group Exercise: Specimen Observation and Description—Samples of sedimentary rocks of the same types used in the group exercise of Laboratory 1 are provided. Form one or several groups. Your instructor will choose one category of properties (e.g., composition) and each group member will observe and describe its state in turn. Announce your descriptions after each student has observed the sample. Discuss any differences and disagreements with the guidance of your instructor. Finally, infer the conditions of the samples' origin, and state the evidence for each.

LABORATORY EXERCISES

2.2. Properties and Origins of Detrital Rocks: Specimens—A) Observe, describe and list the properties of each identified sample in worksheet 2.2. Infer compositional and textural maturities (columns A & B) from these properties.

B) Where possible, and with the aid of your instructor and use of Figure 2.15, indicate by entering the appropriate letters below in each column (C-G) if there was/were:

1) one/many source rocks/areas;
2) one or many source rock types;
3) mechanical (M), chemical (C) weathering or both (B);
4) warm (W)or cold (C) conditions; and,
5) arid (A) or humid (H) conditions; etc..

C) Determine the conditions of origin of each from reference to discussions and figures above. Indicate your conclusions about each rock type's origins by matching the encircled letters "A-P" placed in various cartooned environmental settings in Figure 2.15 with each of the listed rock types in worksheet 2.2 (Column H).

Note: It is likely for any one rock type to form in several environments, and the converse.

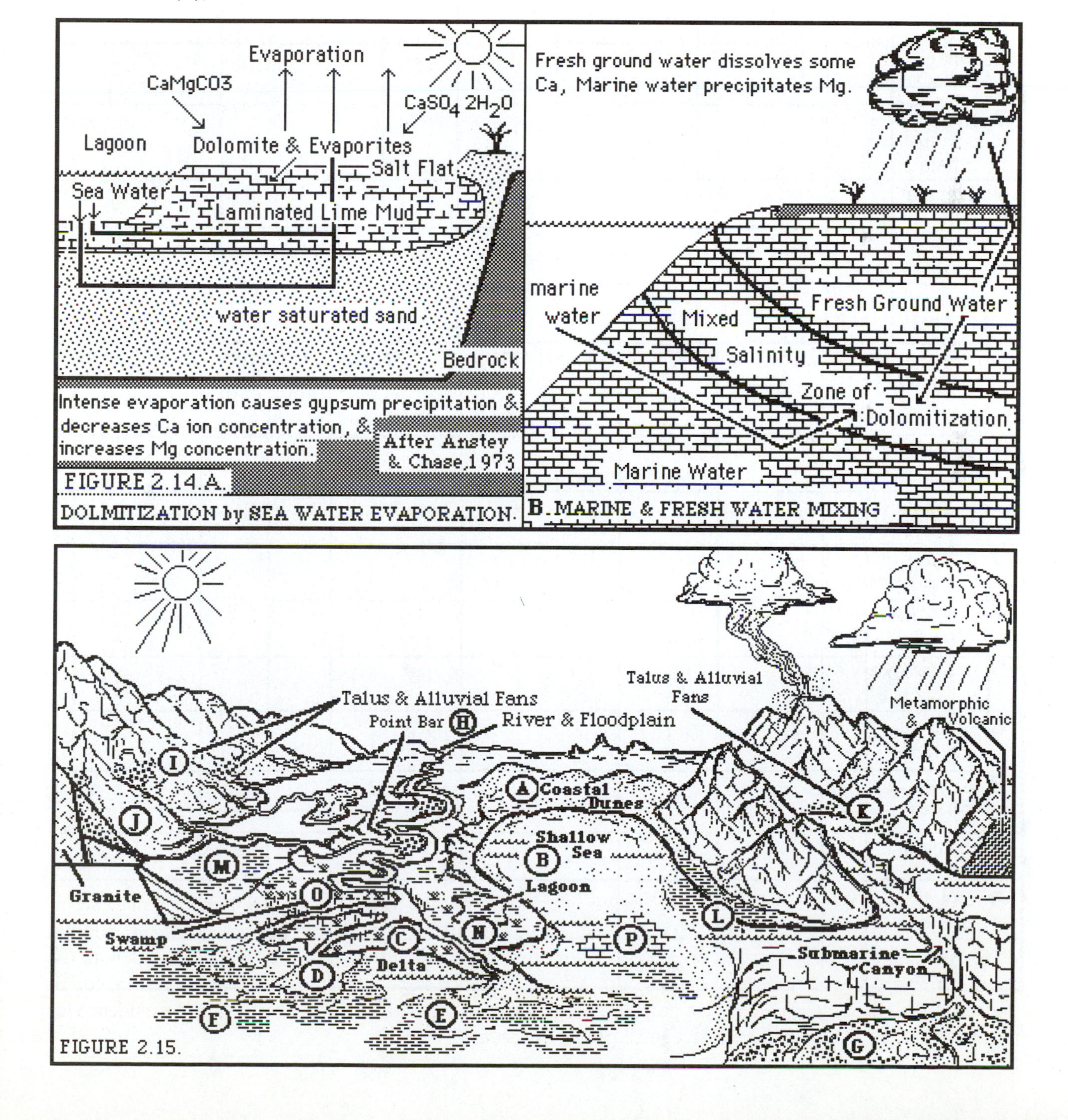

FIGURE 2.14.A. DOLMITIZATION by SEA WATER EVAPORATION. B. MARINE & FRESH WATER MIXING

FIGURE 2.15.

WORKSHEET EXERCISE 2.2. See FIGURE 2.15.

Sample #	Grain Size	Mineral Composition	Grain Size-Sorting	Grain Rounding	Grain Shape	**A.** Hi/Lo Comp. Maturity:	**B.** Textural Maturity: Hi/Lo
1. Arkose							
Red Quartz Sandstone **2.**							
Micaceous Siltstone **3.**							
Conglom-erate **4.**							
Graywacke **5.**							
Subgray-wacke **6.**							
7. Green Micaceous Shale							
Breccia **8.**							
Black **9.** Shale							
Red **10.** Shale							

C. Transportation: Extensive (E) or Limited (L)?		**D.** Source Areas: One or Many?	**E.** Source Rock Type(s) =/> 1?	**F.** Weathering: Mechanical (M) or Chemical? (C)	**G.** Probable Climate			**H.** PROBABLE ORIGIN FIGURE 2.16. Sites A-P
					Warm (W), Arid/Humid	Temp-erate	Cold, Arid	
1. Arkose								
2. Red Quartz Sandstone								
3. Micaceous Shale								
4. Conglomer-ate								
5. Graywacke								
6. Subgray-wacke								
7. Green Shale								
8. Breccia								
9. Black Shale								
10. Red Shale								

LABORATORY EXERCISES

2.3. Carbonate Rock Description and Identification—Numbered, unidentified carbonate rock types are provided. Observe and describe on worksheet 2.3, the matrix type, and clast type, and assess their estimated percent abundance, degree of sorting and rounding.

WORKSHEET EXRCISE 2.3. CARBONATE ROCK DESCRIPTION & IDENTIFICATION.

MINERAL COMPOSITION Calcite, Aragonite or Dolomite	MATRIX Mud or Crystalline	CLAST TYPE	% CLAST	CLAST SORTING/ Roundness	OTHER	ROCK TYPE
1.						
2.						
3.						
4.						
5.						
6.						
7.						
8.						
9.						
10.						

2.4. Comparison and Contrast of Interpretive Properties—Worksheet 2.4 lists several rock types in columns and rows. In each category (A-C), determine whether the rock type listed in the column has greater (G), lesser (L) or equal compositional maturity, etc. In part D determine whether the dominant mode of weathering for the rock type in each of the leftmost rows was mechanical, chemical or a combination of both.

2.5. Carbonate Rock Origins: Specimens—Carbonate rock type samples are provided, numbered and identified by your instructor. Describe the basic properties of each in the spaces provided and indicate probable origin of each from the type(s) of clasts which are/are not present.

WORK SHEET EXERCISE 2.4. DETRITAL ROCK MATURITIES and ORIGINS.

	A. Compositional Maturity							B. Textural Maturity							C. Duration/Distance of Transport							D.
	Breccia	Arkose	Greywacke	Subgreywacke	Qtz. Sandstone	Siltstone	Shale	Breccia	Arkose	Graywacke	Subgraywacke	Qtz. Sandstone	Siltstone	Shale	Breccia	Arkose	Graywacke	Subgraywacke	Qtz. Sandstone	Siltstone	Shale	Dominant Mode of Weathering Chemical (C) Mech. (M)
Conglomerate→	← =	G						=	G						L							
Breccia																						(M)
Arkose																						
Graywacke																						
Subgraywacke																						
Qtz. Sandstone																						
Siltstone																						
Shale																						

WORK SHEET EXERCISE 2.5. * Clasts: F= fossils, I = intraclasts, P = pellets, O = Oöids

Sample	Mineral Composition	Matrix Type Mud (M) vs. Crystalline	Clast Type(s) F, I, P, O *	Clast Rounding? Yes/No ?	Clast Sorting Y/N ?	Water: Quiet/ Agitated?	ORIGIN(S)
1.							
2.							
3.							
4.							
5.							
6.							
7.							
8.							
9.							

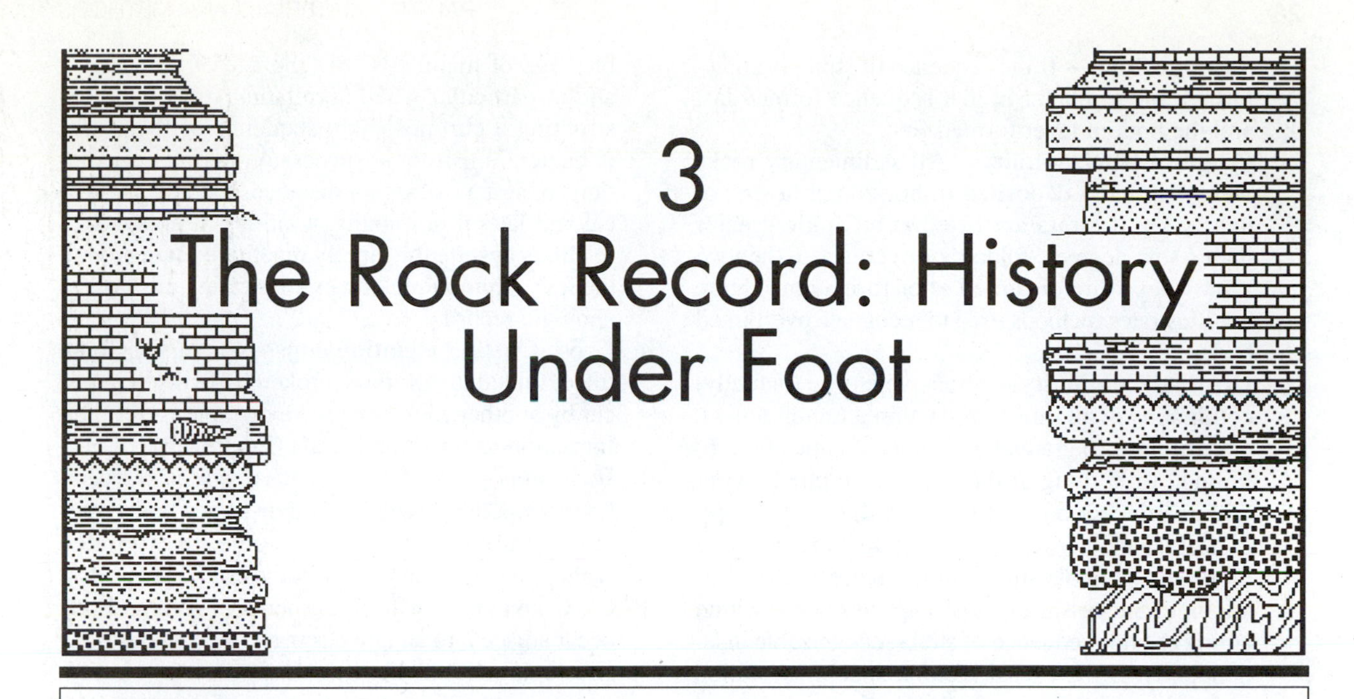

3 The Rock Record: History Under Foot

Objectives

To demonstrate mastery of this material you should: 1) be able to apply the 7 principles of historical geology used to reconstruct sequences of events; 2) recognize the differences among types of unconformities; 3) understand how formations are defined, described, and mapped; and, 4) understand and be able to apply methods of rock correlation. Colored pencils are needed for Exercise 3.9.

Using This Laboratory

Be attentive in this laboratory to illustrations that give a sense of the spatial relations involved in field work: The field is the actual laboratory—with complexities and scales which cannot be literally conveyed by simplified illustrations and hand samples.

In this laboratory you will learn how to use the principles of historical geology to infer the nature and sequence of the geological events of the rock record, and to apply the methods stratigraphers use to describe and correlate strata within and among regions. Although these principles may appear simple—so exceedingly it may seem they could not have been part of the intellectual foundation of a major branch of science—geology's founders nevertheless faced a task no less formidable than reconstructing the Earth's history. And before such an undertaking could be conceived, it was first necessary to recognize that the Earth *had* a history. An understanding of rock origins and historical principles depended on recognizing a depth of time vastly beyond human experience; only then do the few simple principles call from the rocks their story of the Earth's immense journey.

Although the principles of age relations are simple, their application in regions of complex geology can present a daunting challenge. To put the exercises of this lab in proper perspective—that of the field as the true laboratory—we remember that the achievements of geology's founders are truly appreciated only when one stands before a rock exposure, and for the first time asks those questions that will lead to an understanding of history. Being first in the field is a humbling experience!

PRINCIPLES of HISTORY—AGE RELATIONS

A few simple concepts can give us a working knowledge of basic field principles. Given the fundamental principle of uniformitarianism (see text), knowledge that all rocks went through an initial formative period(s), and that sedimentary rocks originated from clastic grains or precipitates, the remaining principles follow directly:

1) **Superposition**—If the sequence of strata is undisturbed, the top layer in that sequence formed last and the bottom layer formed first.
2) **Original Horizontality**—All sedimentary rocks were originally deposited in horizontal layers, or nearly so. If strata are tilted at an angle greater than a few degrees, folded, or overturned, then we know they were disturbed after formation. Figure 3.1 illustrates methods used to recognize overturned beds.
3) **Original Continuity**—Strata were once (laterally) continuous in all directions within a region unless, or until they: a) gradually changed composition; b) thinned to nothing at the surface; c) pinched out within a thicker body of rock; or, d) ended abruptly against another rock type (Figure 3.2B; standard lithologic symbols are given in Figure 3.2A).
4) **Biologic Succession**—Fossil species occur within an irreversible sequence of strata recognizable independently of the causal factors, species evolution and extinction (Figure 3.3). This is the principle that is invoked in a general way whenever we speak of the "age of mammals" or "the age of dinosaurs", and in particular when correlating strata and constructing a chronological sequence of rock units (Chapter 7). Biologic succession permits correlations of strata over great distance, and across vertical and lateral lithologic variations. It allows correlations despite the largely repetitive sequences of the few common rock types comprising most of the geologic record.
5) **Cross-Cutting Relationships**—Strata which are tilted, folded, overturned, broken by faults or cross-cut by another rock type, were disrupted after their formation as horizontal strata (Figure 3.4).
6) **Inclusions**—Smaller fragments of one rock type or body embedded within another larger body must have formed before, and become incorporated within, the latter during its formation (Figure 3.5A, C). Conversely, when fragments of a large body occur adjacent to an igneous mass, the igneous body enveloping the smaller fragments must have intruded the surrounding strata (Figure 3.5B).

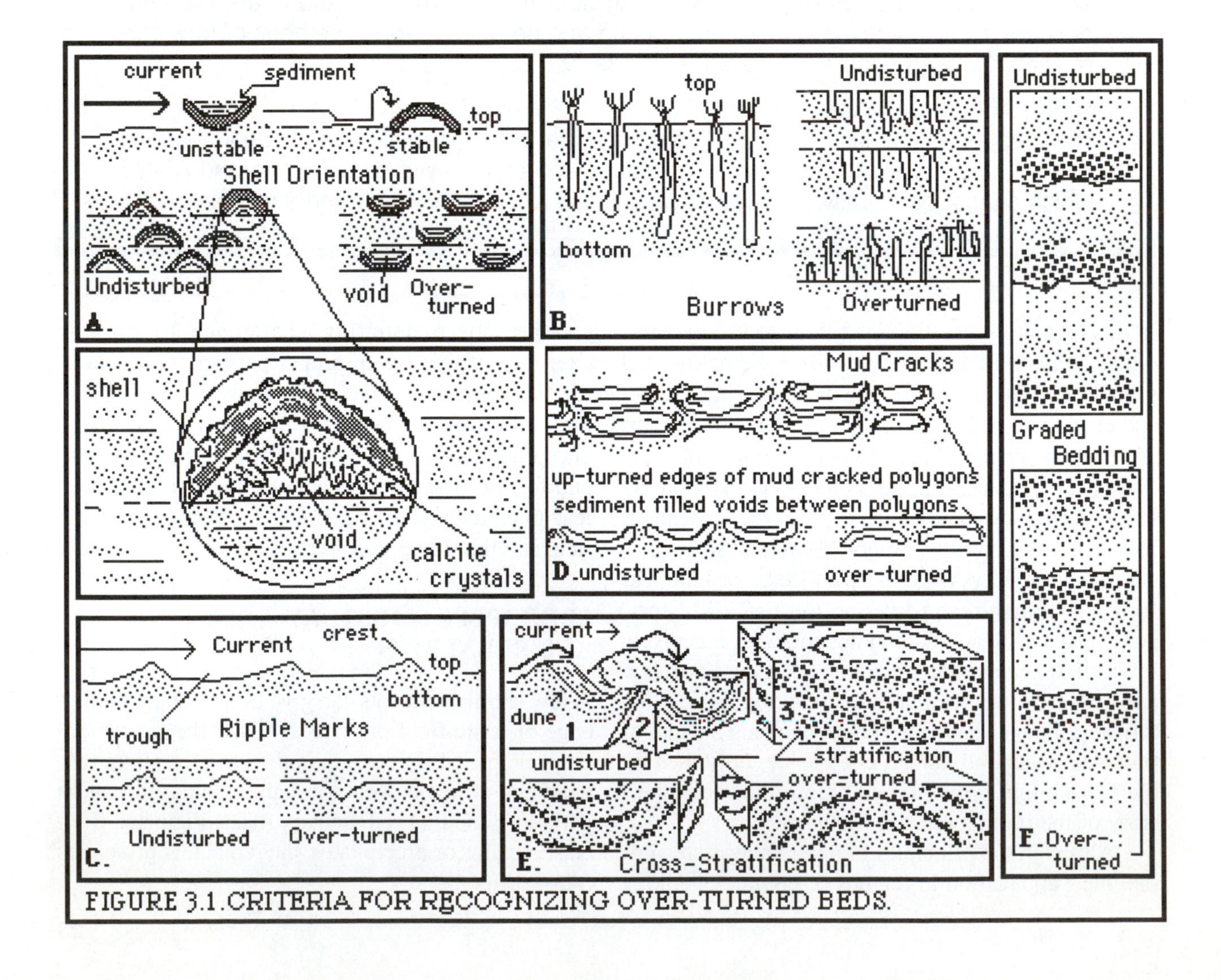

FIGURE 3.1. CRITERIA FOR RECOGNIZING OVER-TURNED BEDS.

EVIDENCE of "MISSING TIME": UNCONFORMITIES

Before the sequence of events in a region can be deciphered, we must discuss one more feature of the rock record: a category of special types of contacts between strata or rock bodies called unconformities. Rocks represent depositional events in time, but oddly enough, much of geologic time in a given region is not represented by rock at all, but rather by regional or even continent-wide surfaces indicating the passage of time for which no regional rock record now exists (Figure 3.8). The duration of unrepresented time is indicated by the ages of the strata above and below.

The various unconformities result from several types and sequences of events:

1) Angular unconformities represent the passage of time during which sedimentary beds were uplifted, tilted and eroded before horizontalstrata formed on the eroded surface (Figure 3.6A). Angular unconformities span long periods of time.

2) Disconformities result from erosion and/or nondeposition and occur between parallel strata. They occur as either: a) an eroded surface between horizontal strata or one containing weathered fragments of the underlying bed (Figure 3.6B); or, b) a simple absence of otherwise predicted fossil species (= paraconformity; Figure 3.6C). Disconformities may embody short or long periods of time.

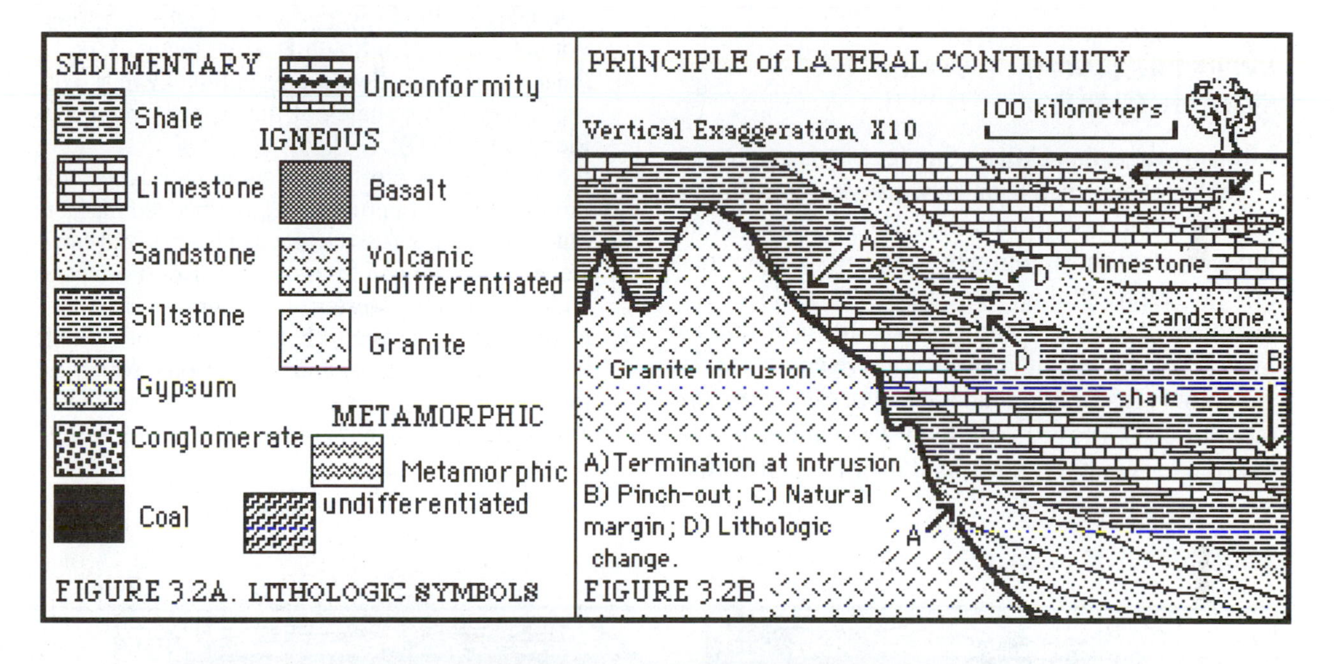

FIGURE 3.2A. LITHOLOGIC SYMBOLS

FIGURE 3.2B.

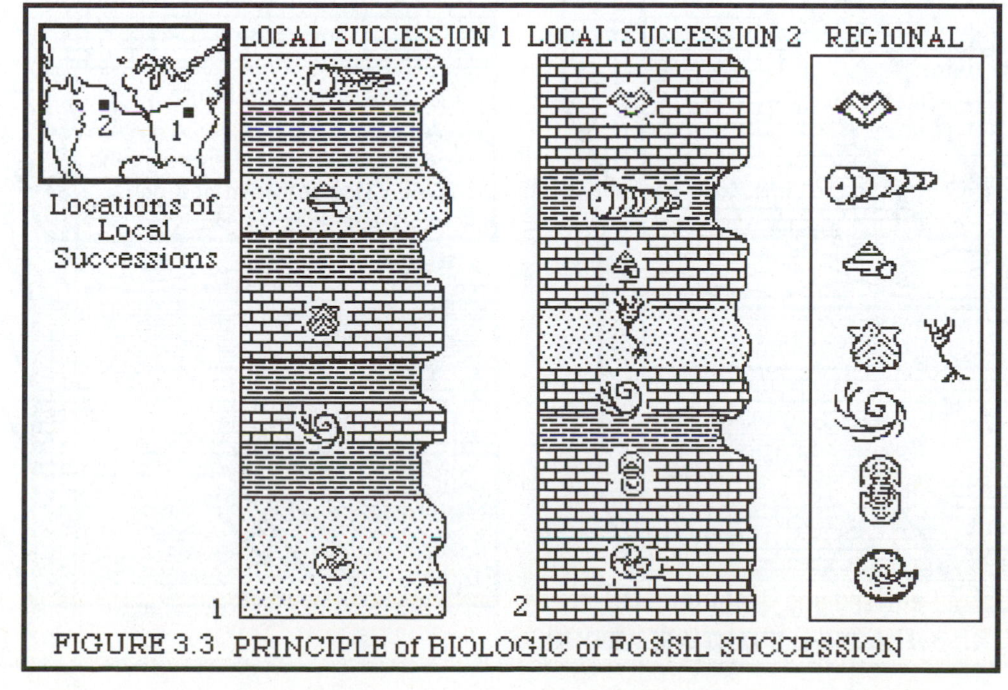

FIGURE 3.3. PRINCIPLE of BIOLOGIC or FOSSIL SUCCESSION

3) A nonconformity results from erosion of intrusive igneous or metamorphic rock surfaces followed by sediment deposition (Figure 3.6D). This type of unconformity formed on the buried landscapes of uplifted, deeply eroded regions exposing igneous or metamorphic rocks, and thus, represents the passage of a long period of time.

All this is simple enough, but geological relationships can quickly become complicated if several disruptive events have occurred (Figure 3.7). Unraveling the sequence of events amounts to working backwards through time to the original horizontality and deposition of each layer in its turn. In this way age relationships of each stratum are established (Figure 3.7, 3.8).

INTERPRETING the LAYERS: STRATIGRAPHY

Lithostratigraphic Units—Stratigraphy is that branch of geology concerned with the sequences, correlations, and interrelationships of strata. The formation is the fundamental lithostratigraphic unit (=rock layer unit) of stratigraphy (Table 3.1; Figure 3.9 -3.11). A formation is defined as a body of rock that is distinctive in its lithology, and thick and extensive enough for its geographic distribution (above and/or below the surface) to be portrayed on a map. Formations can be traced over great distances, and their top and bottom layers are distinguished at their contacts with formations above and below. Here we are concerned only with sedimentary rocks, but the discussion below also applies to igneous and metamorphic bodies.

Because rocks and rock bodies have many properties which commonly vary from place to place, delimiting a body of rock that meets the definition of a formation can be difficult. Formations may be recognized by one or more of the following (e.g., Figures 3.9 and 3.10):

1) distinctive compositions and/or textures (e.g., Ordovician St. Peter Sandstone, widespread in the Mississippi valley region, Figure 3.10);
2) a particular mineralogical composition (e.g., gypsum beds of the Lower Jurassic, Gypsum Springs Formation, western South Dakota, Figure 3.9);
3) the thickest or most widespread lithology (e.g., concretionary black shales of the Upper Cretaceous Pierre Shale, SD);
4) the co-occurrence of two or more rock types, textures or other properties (e.g., interbedded maroon, blue and green shales, white sandstones and limestones of the Upper Jurassic, Morrison Formation widespread in the western U.S.); or,
5) a composite of several lithologies, each of variable thicknesses, positions, and proportions, which are together distinctive (e.g., Upper Cambrian-Lower Ordovician Deadwood Formation, SD, Figure 3.9).

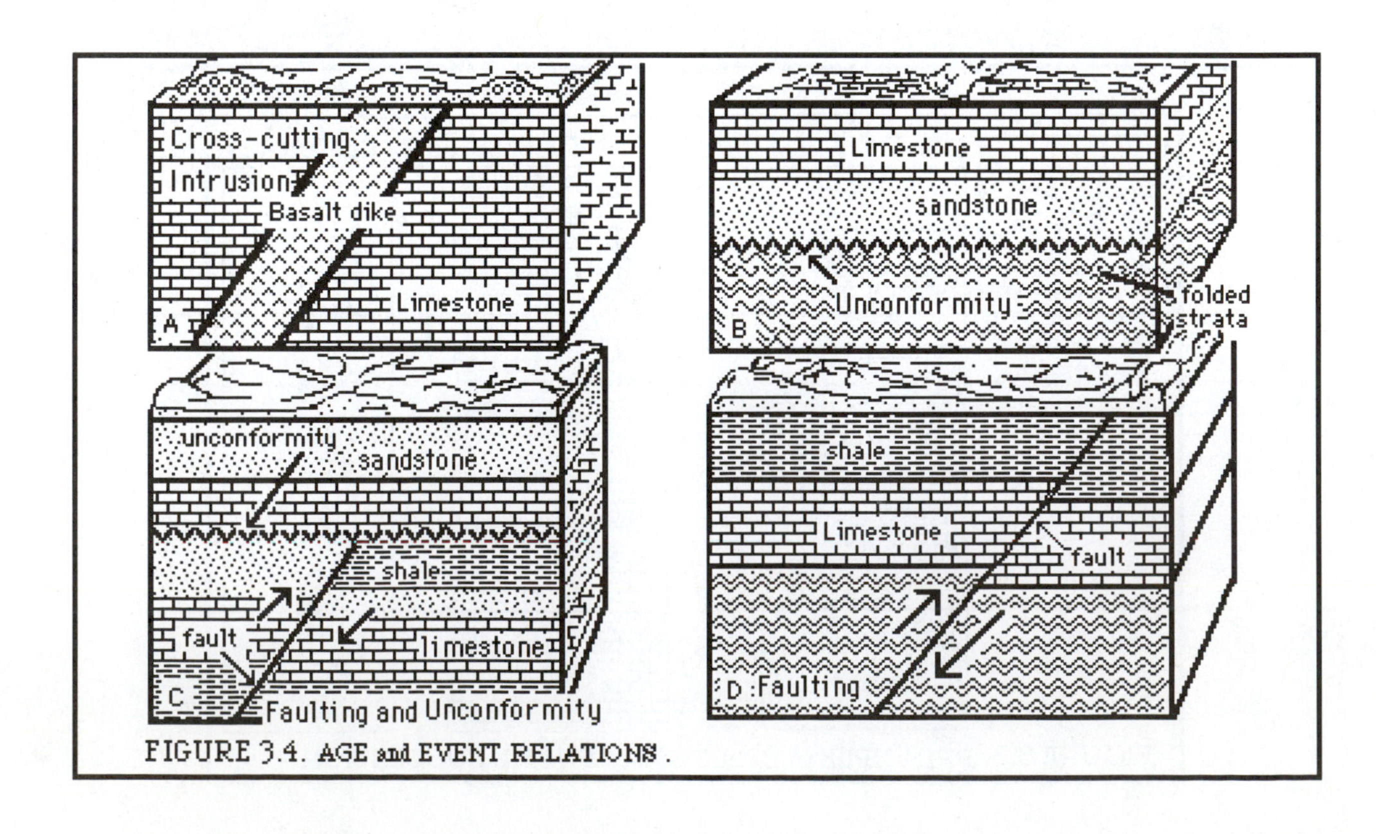

FIGURE 3.4. AGE and EVENT RELATIONS.

Formations consist of members and beds (Table 3.1; Figure 3.9, 3.11). Members are homogeneous portions of formations, usually consisting of a single rock type that may be present throughout the entire region of the formation. Beds are individual layers >1 cm thick. Formations can be subdivisions of groups—litho-stratigraphic units consisting of successive formations sharing features related to their conditions and region of origin. Successive groups form a supergroup. These units are portrayed by stratigraphic columns (standard lithologic symbols, Figure 3.2A) like those of Figure 3.9, and are accompanied by lithologic descriptions.

Correlation—Formations are defined by the characteristics exhibited at specific localities, and are given a geographic place-name, or a place-name combined with a lithologic term. For example, the Pahasapa Limestone in the Black Hills, South Dakota derives its name from the Lakota Indian name meaning "Black Hills" and from its dominant lithology, and the Morrison Formation is named after the Colorado town near the site where the first Late Jurassic sauropod dinosaurs were discovered.

Correlation is the demonstration of lithologic, biologic (fossil content) or temporal equivalency of lithostratigraphic (this chapter), or biostratigraphic or chronostratigraphic units (Table 3.1; Laboratory 8). Exposures are usually correlated by one or more distinctive lithologic characters, the most important of which are rock type, composition, and textures (Figure 3.11).

Six methods are commonly used to correlate formations:

1) Recognition of distinctive compositions and textures (Figure 3.10; St. Peter Sandstone);
2) Physically tracing or "walking out" the unit or a distinctive bed if the exposures are not covered and isolated by soil or sediment;
3) Correlation of exposures adjacent one to another like the links of a chain or sections of a fence (Figure 3.12). [If the first and final columns are found to be correlative, then the rock units in each column can be considered equivalent. Such networks are called fence diagrams.];
4) Recognition of marker beds—distinctive, widespread beds of unusual rock type, composition or mineral content (e.g., chert, volcanic ash, concretions, caves, a certain chemical or mineralogical composition, distinctive color; Figure 3.11));

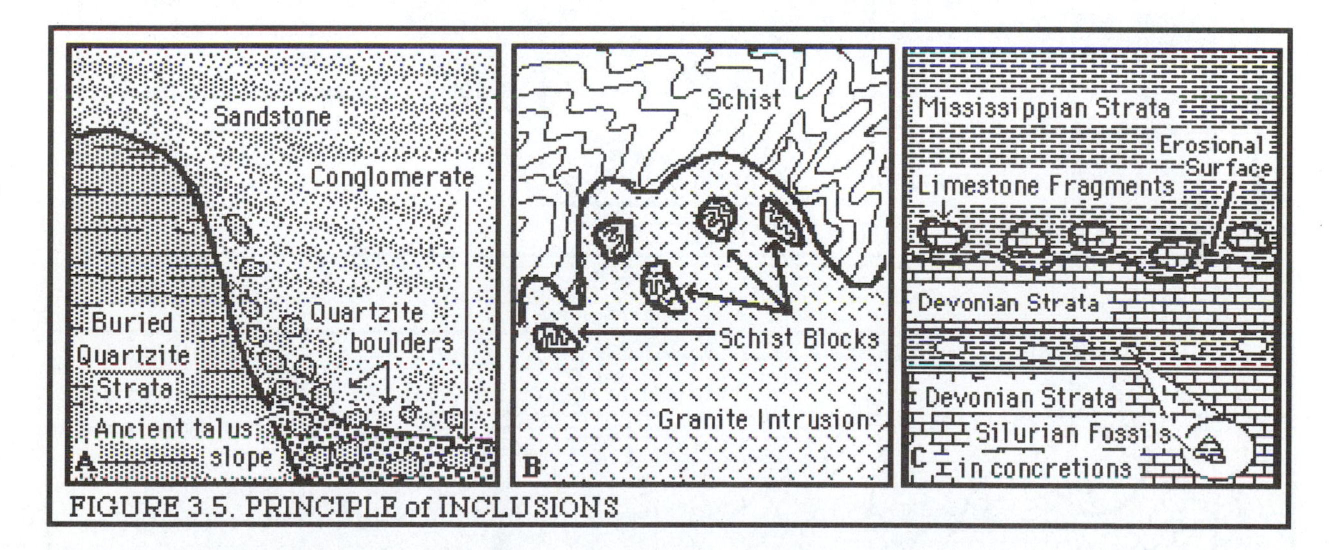

FIGURE 3.5. PRINCIPLE of INCLUSIONS

TABLE 3.1. Units of Stratigraphy and Geological Time.

Lithostratigraphic Units		Chronostratigraphic Units		Geochronologic Units
Supergroup	≠	Erathem	⟷ = ⟷	Era
Group		System	⟷ = ⟷	Period
Formation		Series	⟷ = ⟷	Epoch
Member		Stage	⟷ = ⟷	Age
Bed		Zone		

5) Notation of the positional sequence of beds, members, or formations and the characters of the over or underlying units.

[These may be the only traceable features over a great distance because beds commonly pinchout or intergrade laterally with beds or formations of another rock type (Figure 3.1, 9, 12). Lateral intergradations between sites where stratigraphic columns are described are diagramed as zig-zag lines indicating gradual vertical and lateral changes between sites]; and,

6) Tracing a key horizon—beds deposited in an instant of geologic time (e.g., storm deposit, volcanic ash).

Correlation using key horizons define chronostratigraphic units (Table 3.1)—units all deposited during the same interval of time (Figure 3.13; also Chapter 7). Volcanic ash beds are ideal key horizons because they can be assigned chronometric ages with the techniques discussed in Chapter 4.

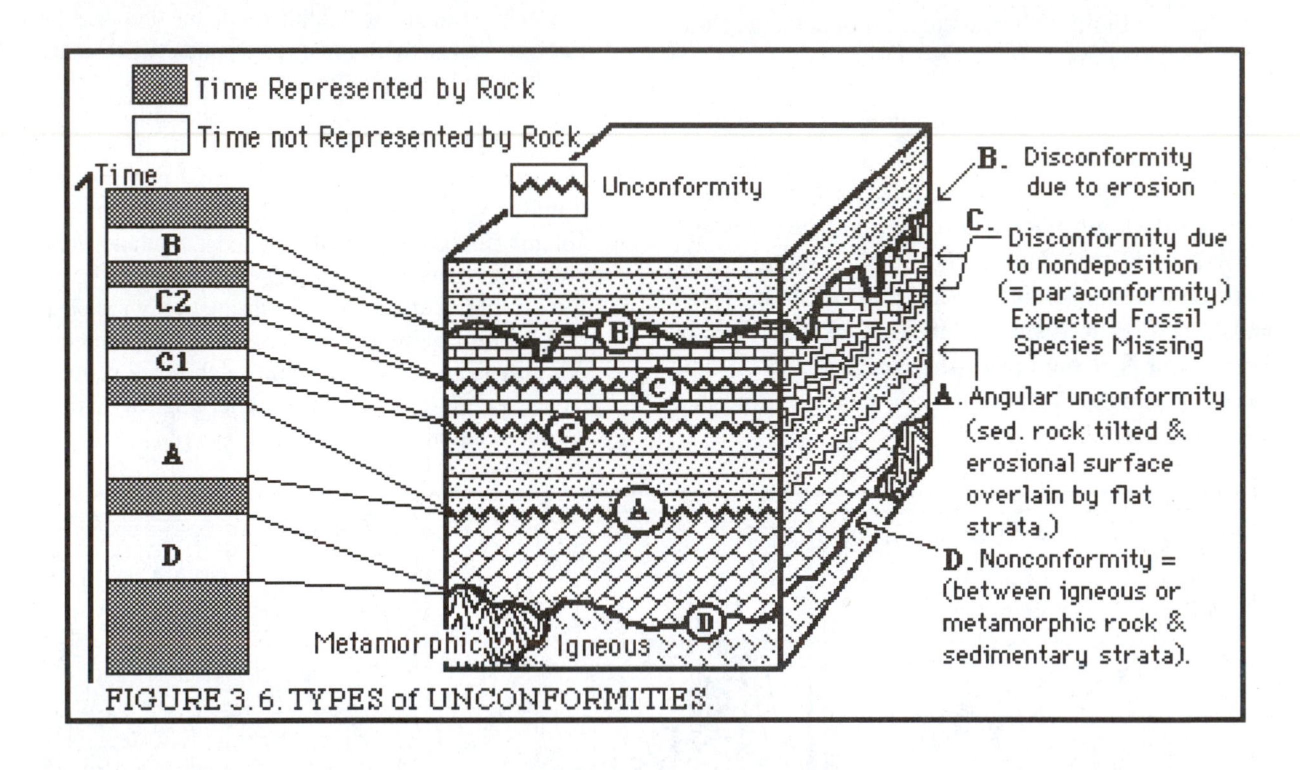

FIGURE 3.6. TYPES of UNCONFORMITIES.

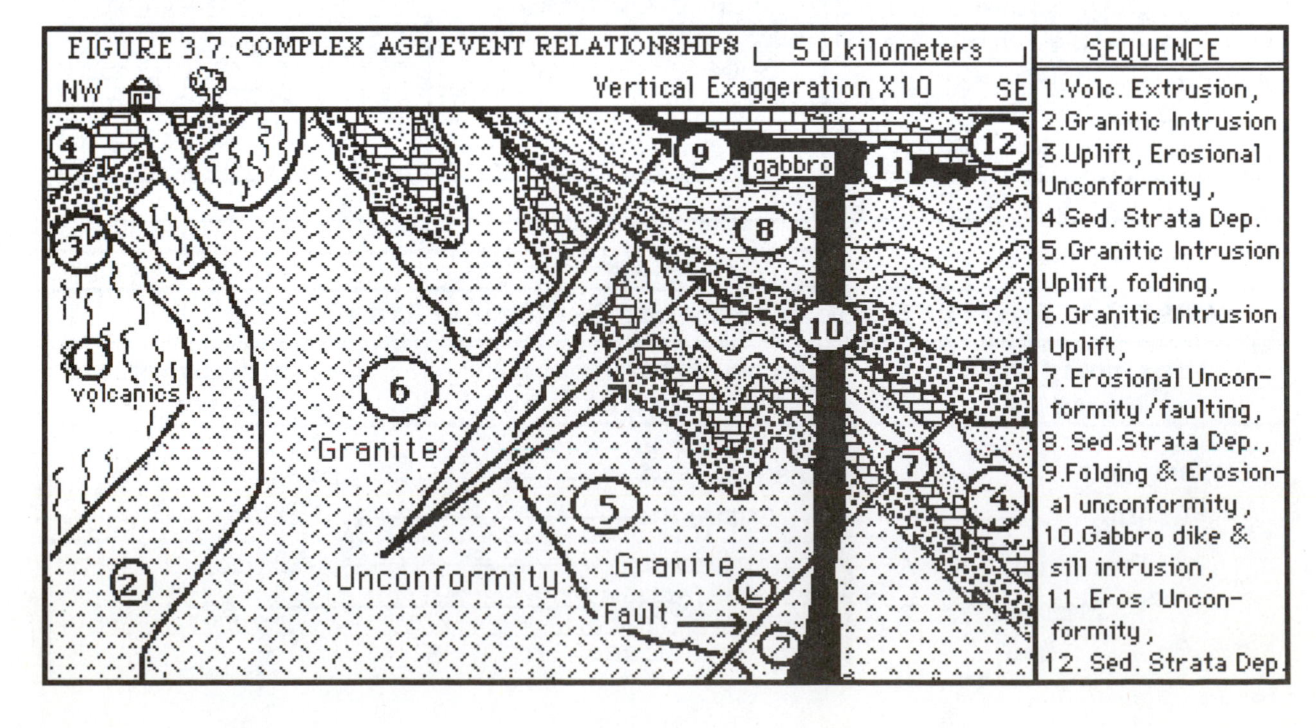

FIGURE 3.7. COMPLEX AGE/EVENT RELATIONSHIPS

MAPPING STRATA

Making a geologic map involves tracing the contacts of formations among exposures and in the subsurface. Each formation's distribution is recorded and color-coded at its appropriate elevation(s) on a topographic map (i.e., a contoured landform elevation map—contour lines represent lines connecting points of equal elevation; see Figure 3.14A-C). To do this as a laboratory exercise, we need to know the approximate thickness of each formation, the elevations of contacts, and the angle at which the rocks are tilted. For purposes of our exercise, imagine the simplest scenario, in which all of the formations shown in the stratigraphic column in Figure 3.14A and B, have no measurable tilt, and thus, the contacts between each describe lines parallel to the imaginary lines of contours on the topographic map (Figure 3.14C). If the rock units recognized are distinctive and mappable, their thicknesses and elevations where the stratigraphic sequence was described can be used to predict the sequence and elevations of strata exposed elsewhere (Figure 3.14C; plotted on topographic map below).

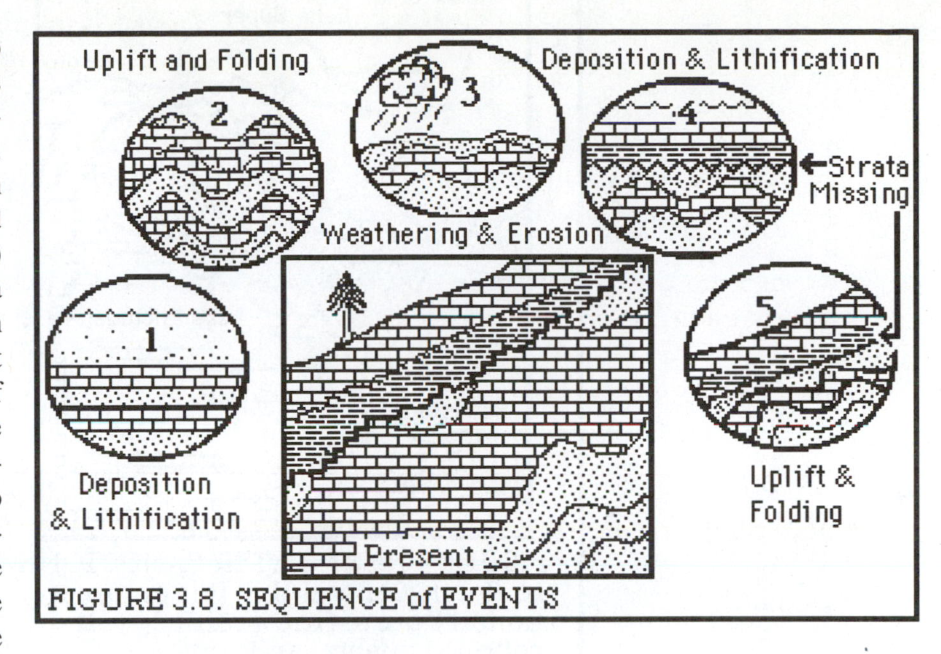

FIGURE 3.8. SEQUENCE of EVENTS

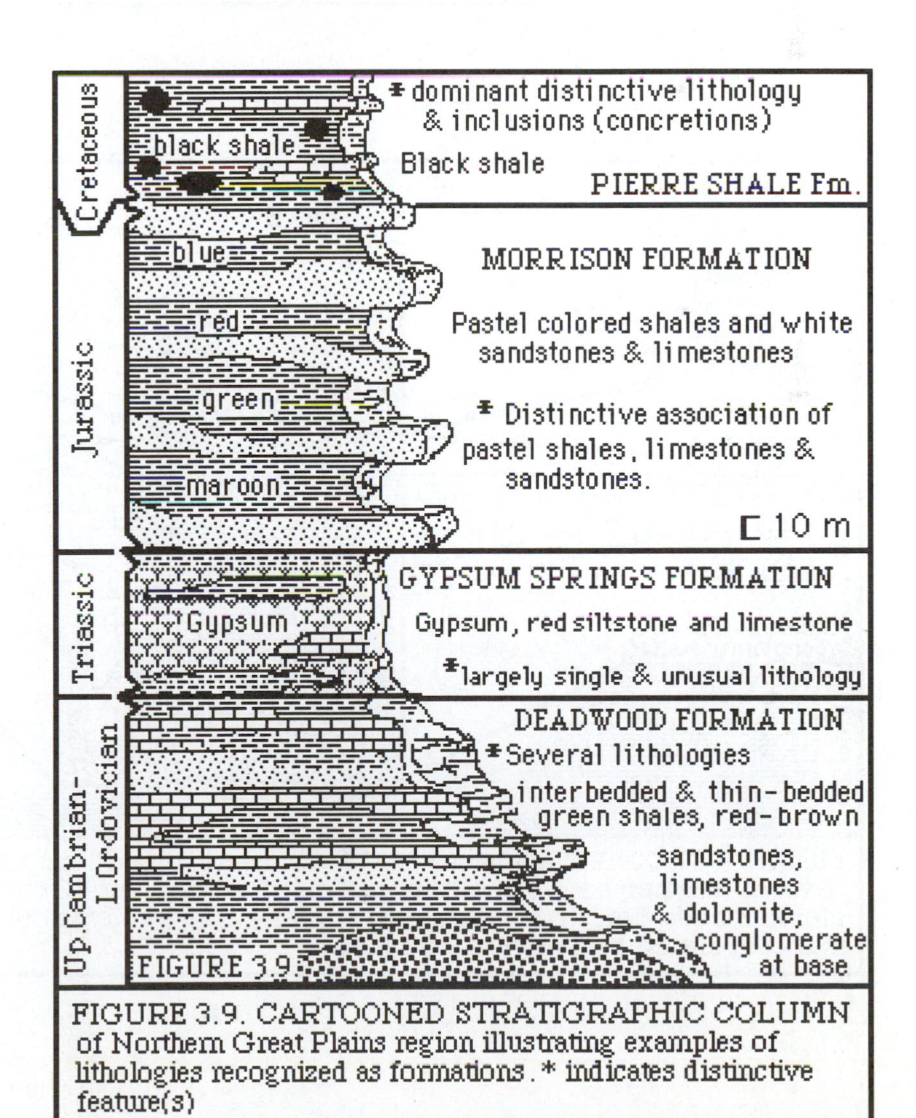

FIGURE 3.9. CARTOONED STRATIGRAPHIC COLUMN of Northern Great Plains region illustrating examples of lithologies recognized as formations. * indicates distinctive feature(s)

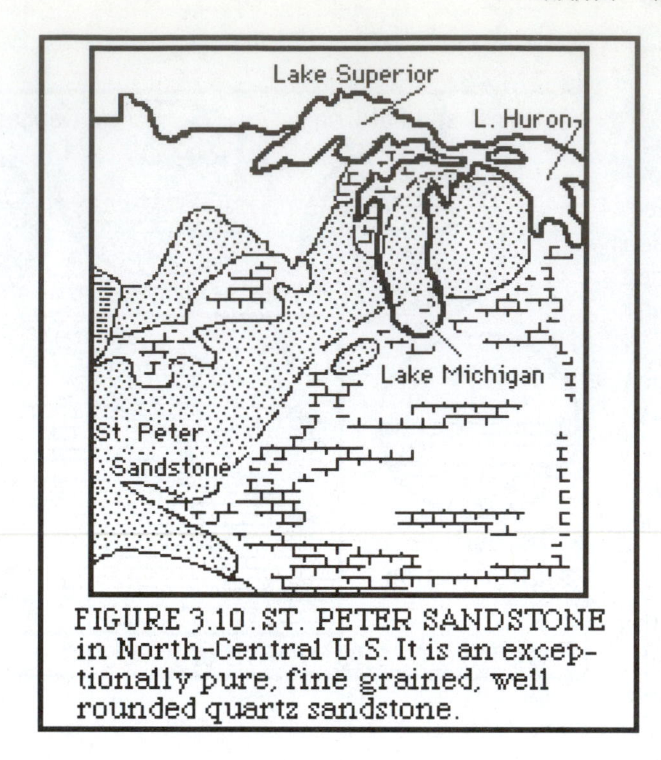

FIGURE 3.10. ST. PETER SANDSTONE in North-Central U.S. It is an exceptionally pure, fine grained, well rounded quartz sandstone.

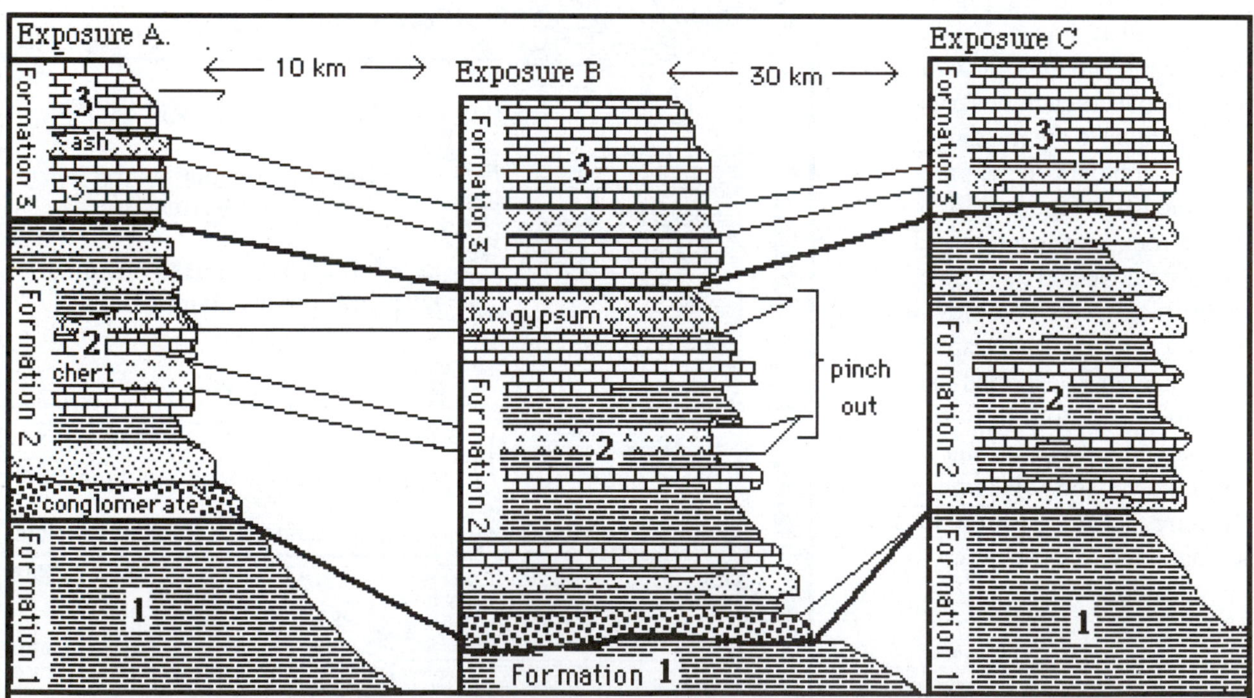

FIGURE 3.11. CORRELATION of STRATA among 3 distant exposures. Fm. #2 contains several marker beds but others not distinctive. Fm. #2 correlated from A to B by marker beds. Fm. #1 & 3 distinctive everywhere. Key horizon (ash) in Fm. 3. Fm. #2 at C correlated by positional sequence between #1 & 2.

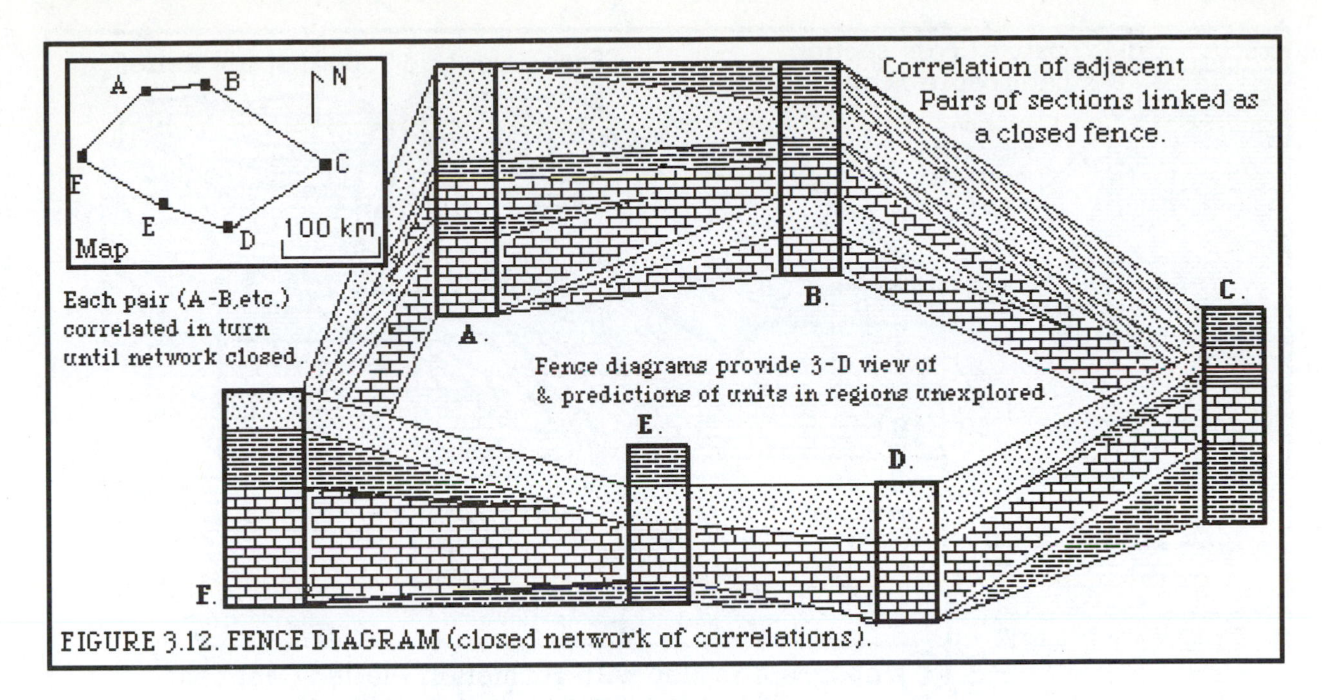

FIGURE 3.12. FENCE DIAGRAM (closed network of correlations).

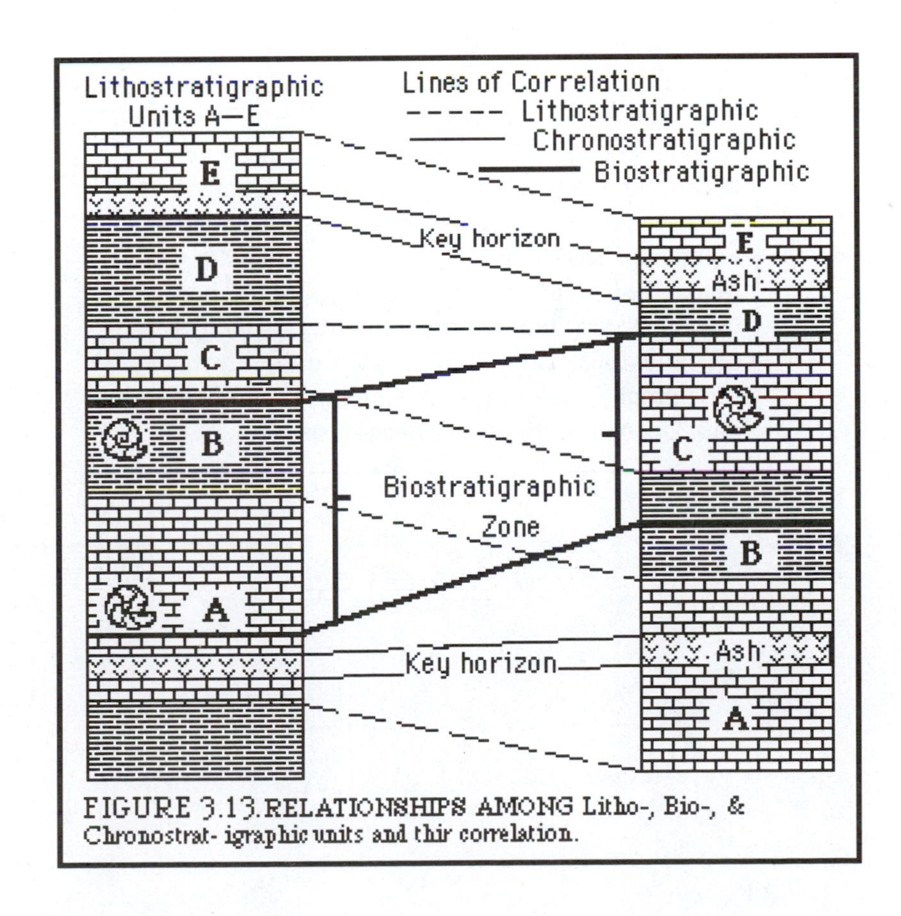

FIGURE 3.13. RELATIONSHIPS AMONG Litho-, Bio-, & Chronostrat- igraphic units and thir correlation.

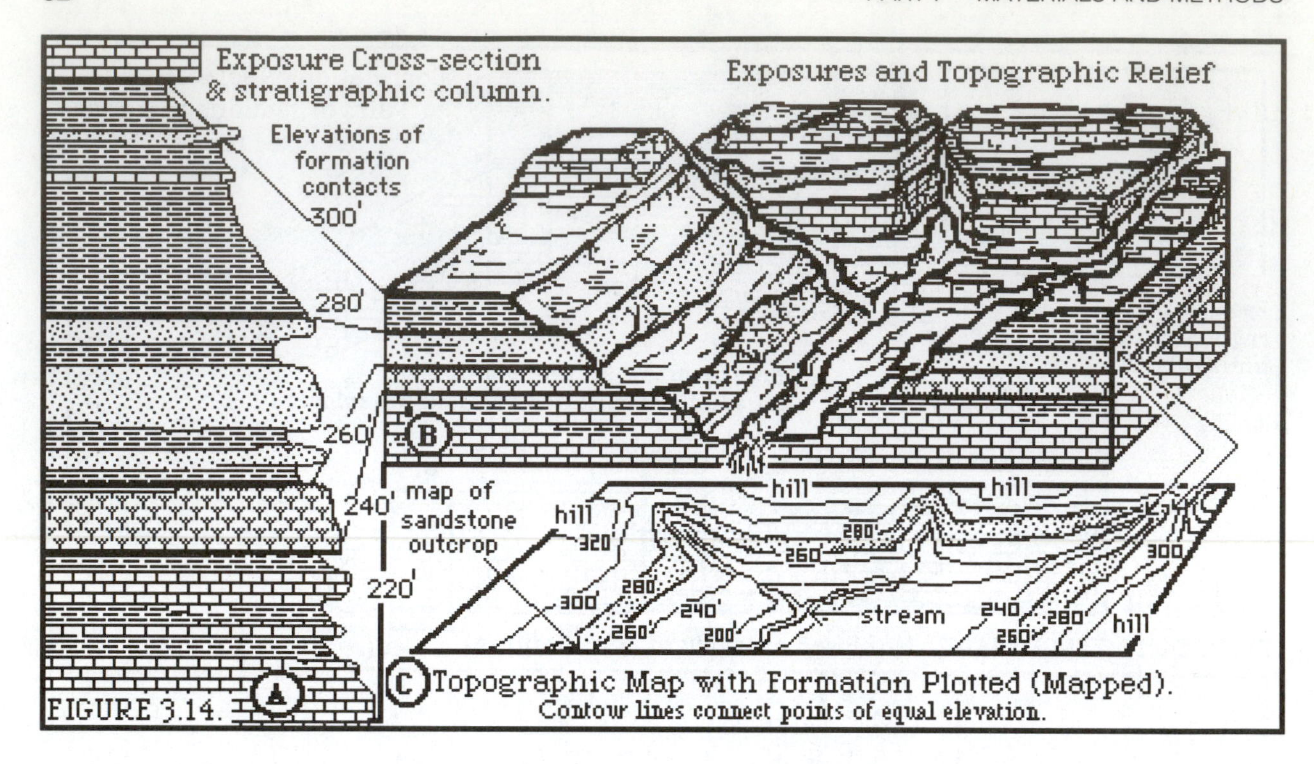

FIGURE 3.14. Topographic Map with Formation Plotted (Mapped). Contour lines connect points of equal elevation.

VOCABULARY

Angular unconformity, Bed, Biologic succession, Biostratigraphy, Biostratigraphic unit, Contact, Cross-cutting or Age relationships, Disconformity, Formation, Fence diagram, Key horizon, Lithology, Marker bed, Member, Nonconformity, Original horizontality, Original lateral continuity, Principle of inclusions, Lithostratigraphic unit, Strata, Stratigraphy, Superposition, Chronostratigraphic unit, Unconformity, Uniformitarianism

LABORATORY EXERCISES

3.1. Correlation—Use the principles of correlation discussed in the text to correlate the strata in columns #1-4 on worksheet 3.1.

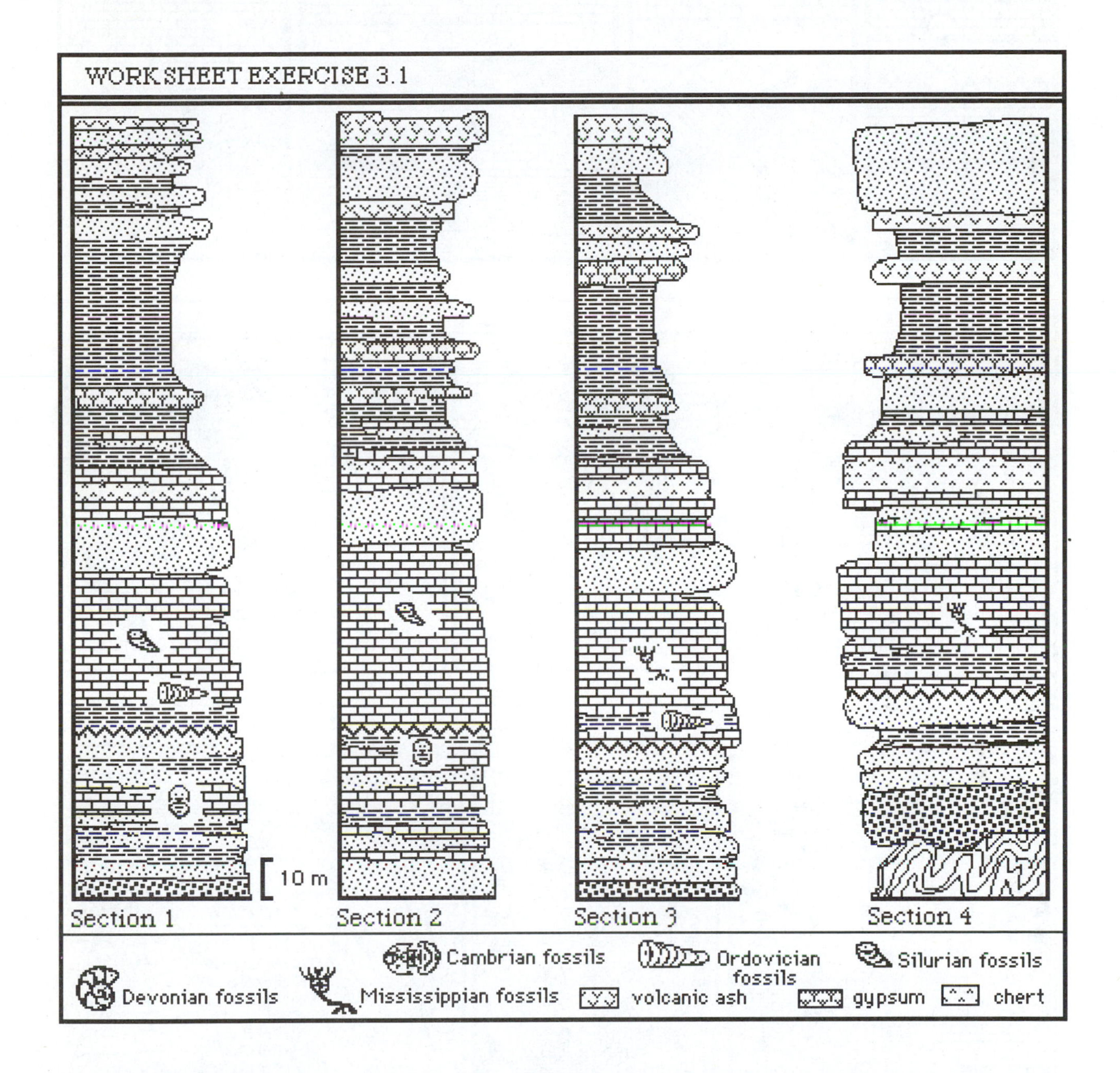

3.2. Principles and Age Relations—Diagrams of relationships between rock units (A—J) are illustrated in worksheet 3.2.

A) Outline the sequence of events in each.

B) Number each different body of rock and faults, folds and unconformities in the correct sequence of events in the circles provided. Note other explanation(s) needed in the spaces provided.

C) What type of unconformity (if any) is present in each? ____________________

D) Apply the principle of biologic succession in "H," and criteria for recognizing original superposition in "I".

E) How can the sequence of fossils in "H" be explained? ____________________

F) Explain the manner in which the upper strata in " I " were disturbed. ____________________

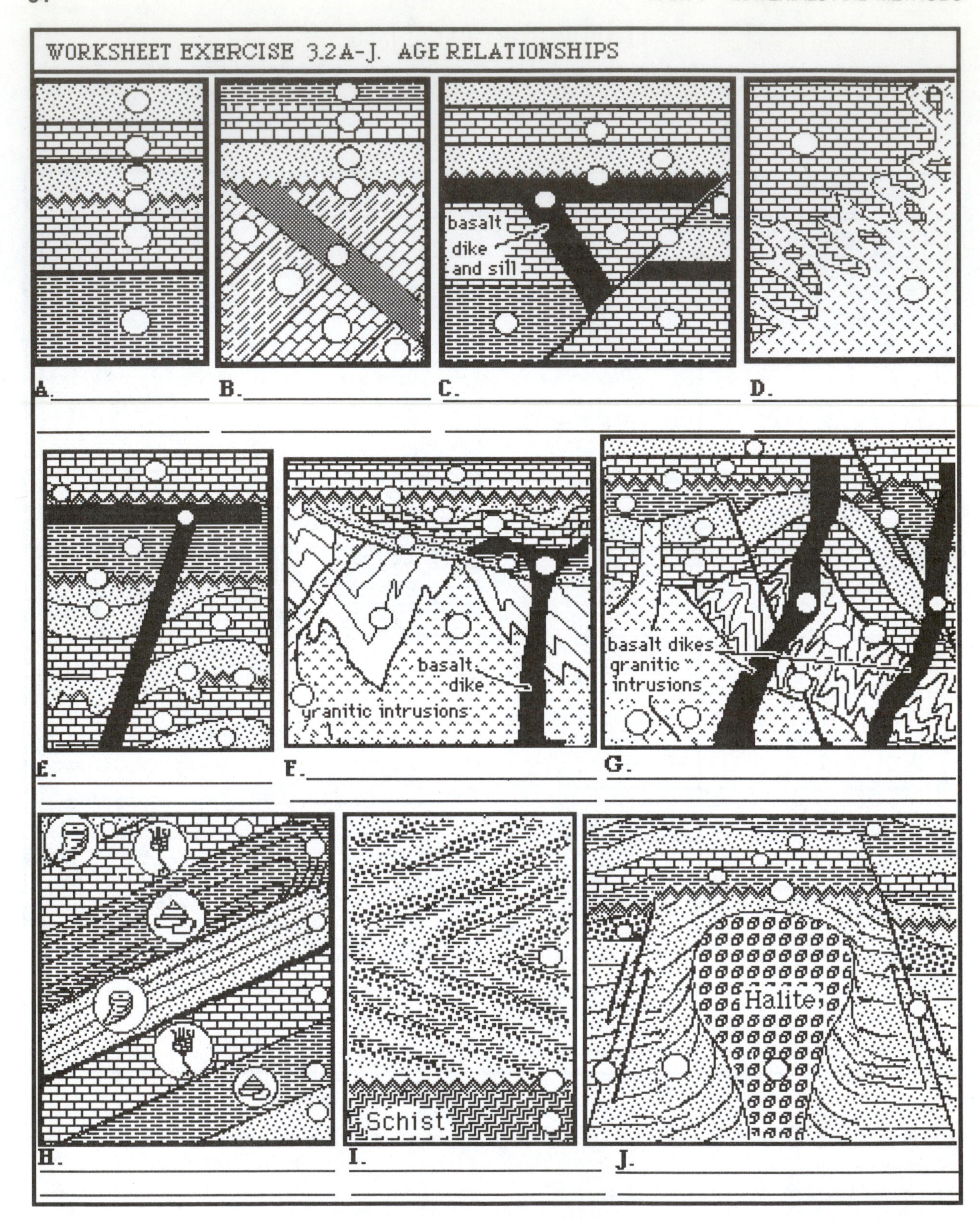
WORKSHEET EXERCISE 3.2A-J. AGE RELATIONSHIPS
basalt
dike
and sill
A.
B.
C.
D.
basalt
dike
granitic intrusions
basalt dikes
granitic
intrusions
E.
F.
G.
Schist
Halite
H.
I.
J.

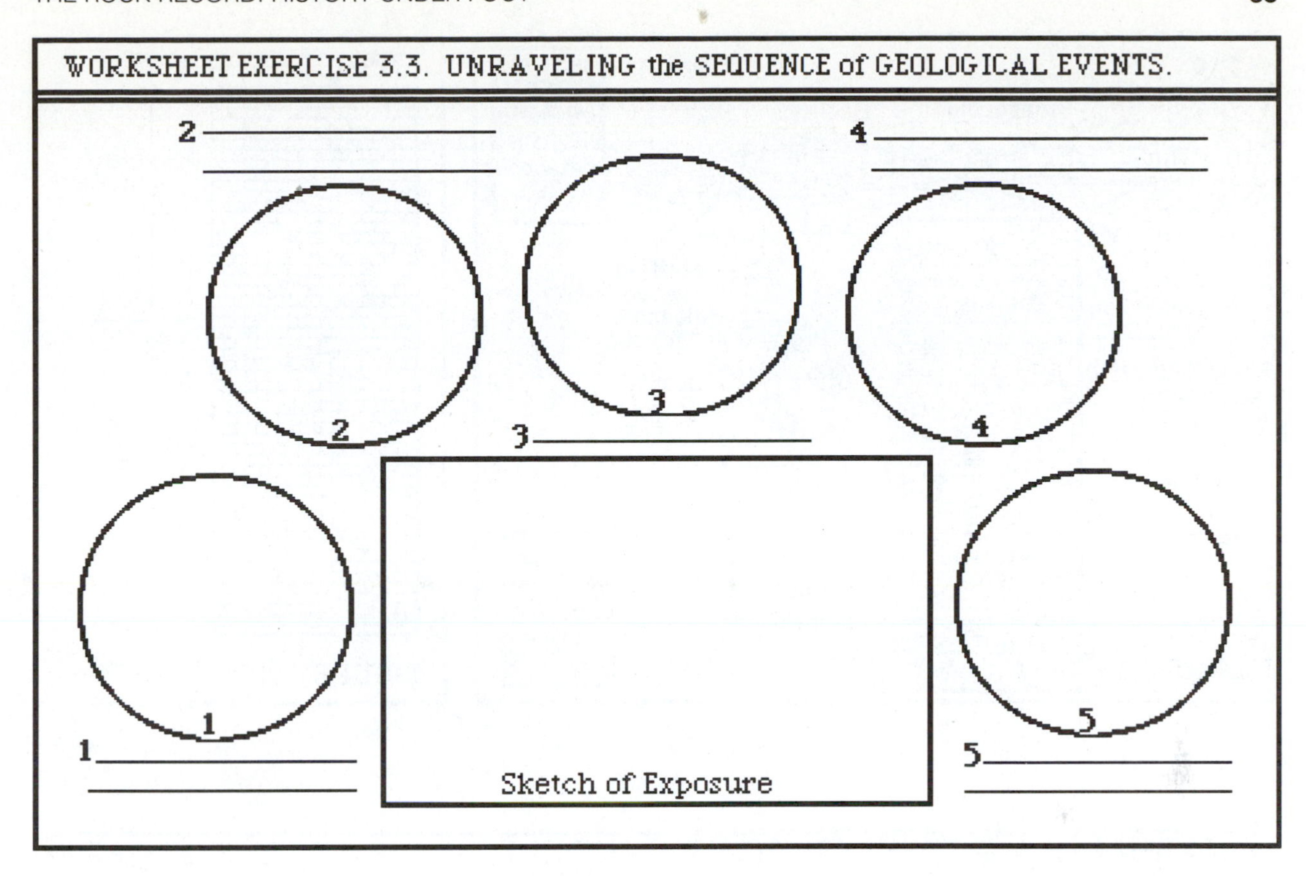

3.3. Sequence of Events—Decipher the sequence of events and age relations of the strata shown in figure B of worksheet 3.2, by drawing each step in their sequence of formation and deformation in the circles provided. First sketch the exposure in the center rectangular space provided.

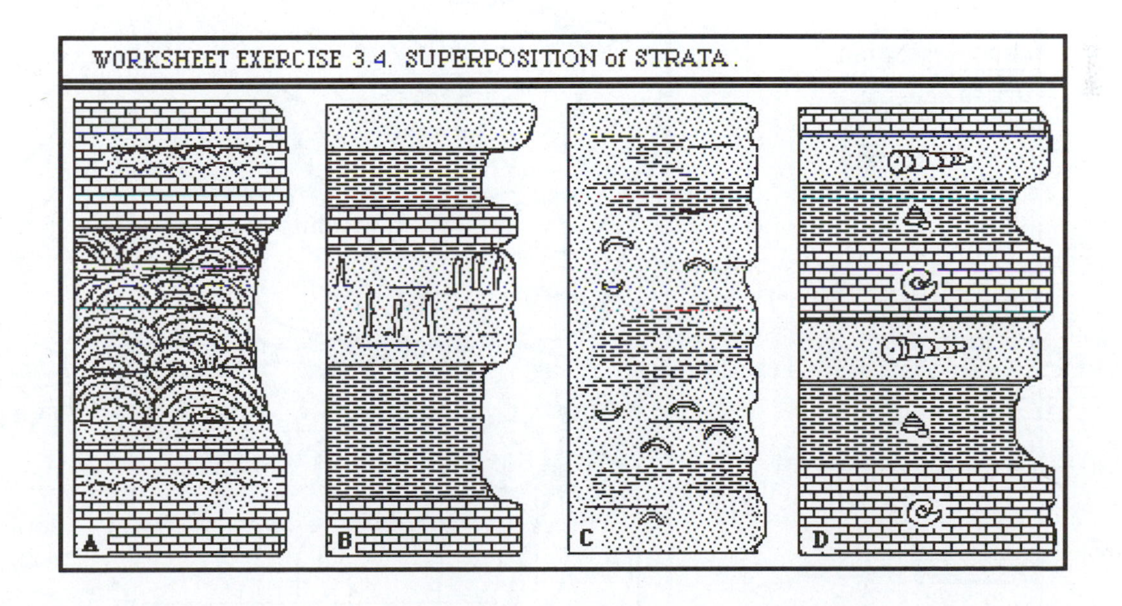

3.4. Principles of Superposition & Biologic Succession—Four exposures are illustrated above.

A) Determine whether the strata in A-C have been overturned and state the reasons for your answers.
A)______________________ B)______________________ C)______________________

B) Exposure D shows a succession of fossil species. Does the succession follow a stratigraphic principle? ___
____

C) How could the observed succession be explained? ______________________________

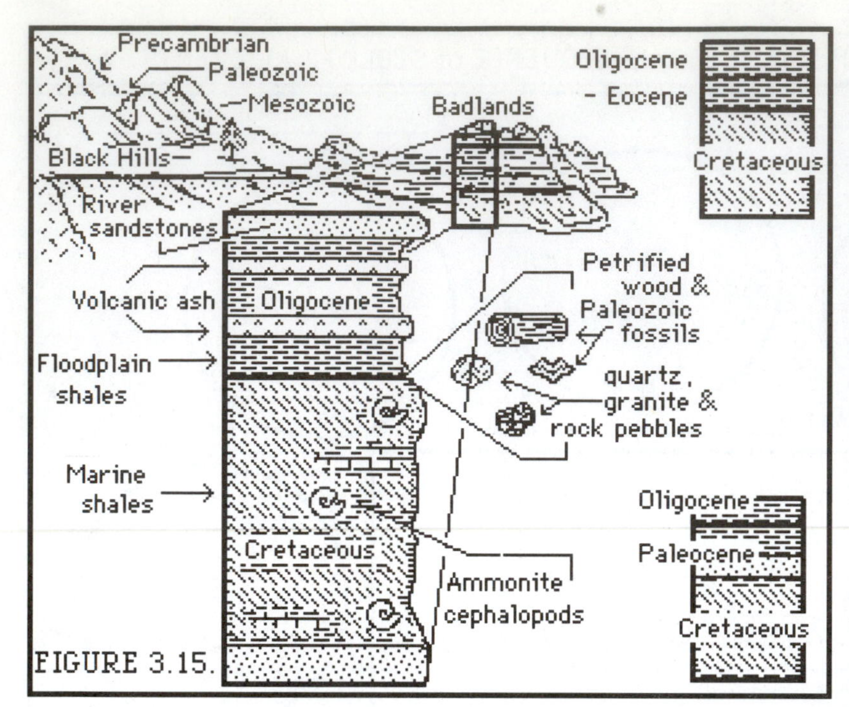

FIGURE 3.15.

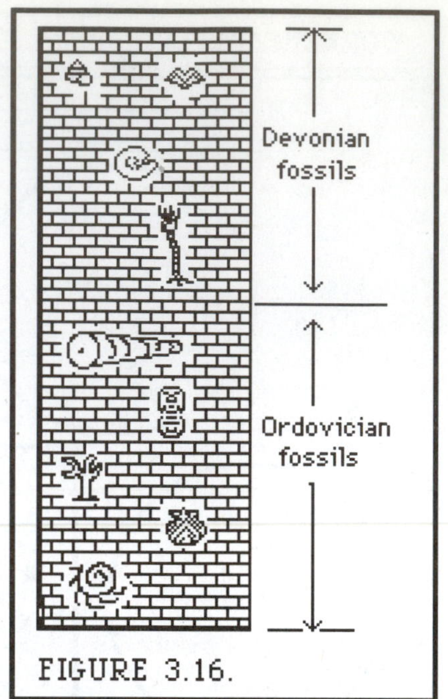

FIGURE 3.16.

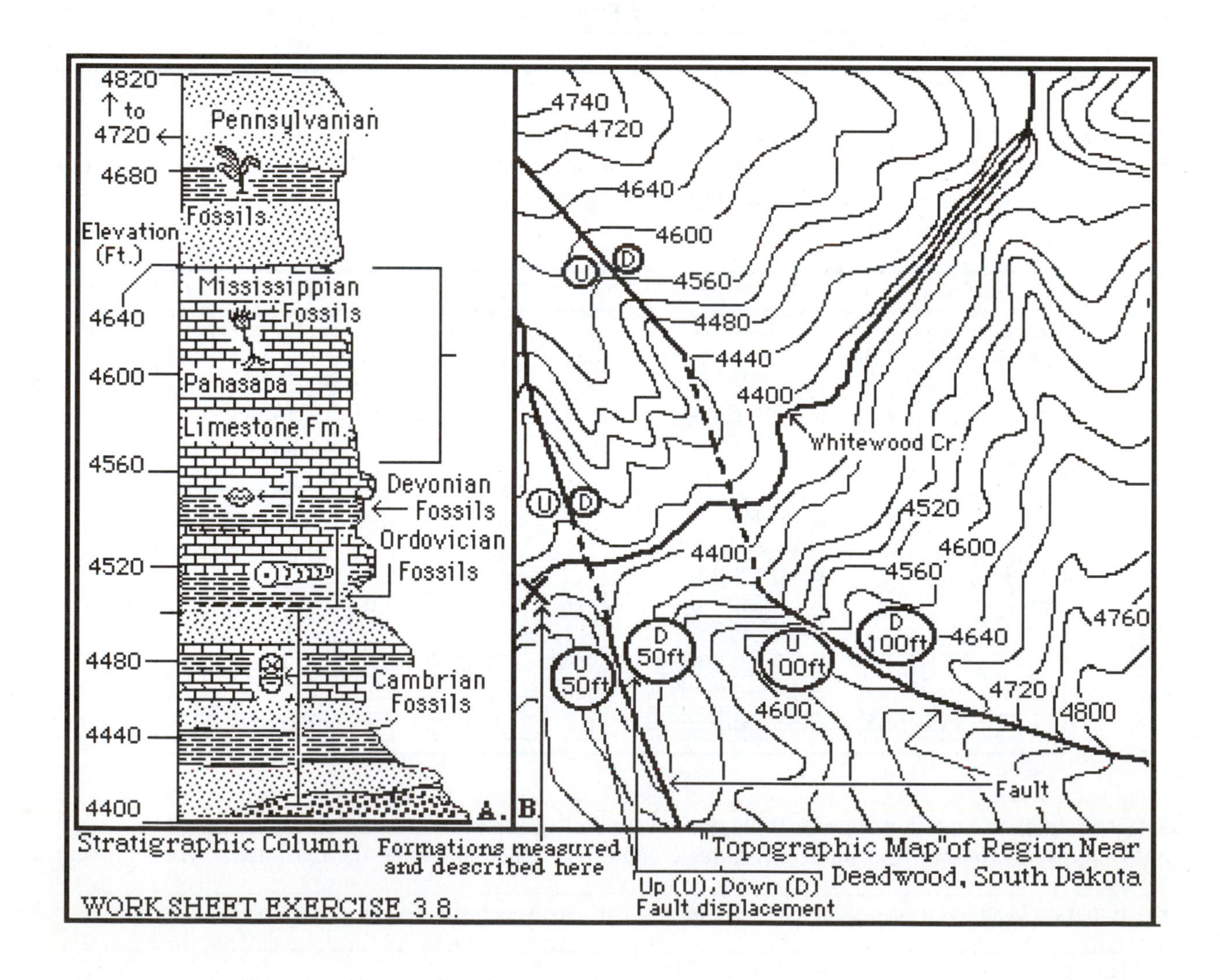

WORKSHEET EXERCISE 3.8.

LABORATORY EXERCISES (Cont.)

3.5. Principles and Events—Refer to Figure 3.15. The lowermost strata exposed in the region of the Badlands, South Dakota, are of Late Cretaceous age. These are overlain by Oligocene strata (Chadron, Brule and Sharps formations, respectively). Petrified wood, quartz and granite pebbles, and pebbles of sedimentary rock containing Paleozoic fossils found in strata of the nearby Black Hills are found at the contact between the Cretaceous and Oligocene strata.

A) What kind of contact is present between the Cretaceous and Oligocene strata? ____________________

B) What sequence of events led to the deposition of these strata? ____________________

____

C) Of what age(s) are the pebbles and petrified wood? ____________________

D) Approximately how many millions of years are missing? ____________________

E) Suppose that elsewhere the same pebbles and fossil wood are found in strata of Eocene age, and that Paleocene strata underlie Oligocene strata in yet another region. Must the strata immediately beneath an erosional disconformity be of the same age everywhere ? Explain. ____________________

____

3.6. Principles and Age Relations—You are meticulously collecting fossils from the bottom of an exposure to the top, in a thick sequence of limestone (Figure 3.16). The strata between the last occurrence of Ordovician fossils and the first Devonian fossils show no erosional contact between the upper and lower beds. Explain your finding. ____________________

____

3.7. Stratigraphic Columns, Descriptions and Correlations: Specimens—Four sets of numbered sedimentary rock samples are provided. For purposes of the exercise, they represent fragments from well cores and are numbered in ascending order (in the stratigraphic column). The thicknesses of the beds represented by each sample are given.

A) Describe each sample fully and draw each of the stratigraphic columns #1 through #4 as in Figures 3.9 and 3.11.

B) Use the appropriate lithologic symbols (Figure 3.2B) and weathering profiles (Figure 3.9) in your drawings.

C) Correlate the beds from column #1 through #4 in turn.

D) Indicate the basis for each correlation.

3.8. Rock-Stratigraphic Units and Mapping—

A) Using the stratigraphic column illustrated in worksheet 3.8, and the accompanying topographic map, make a geologic map in colored pencil of the outcrop of the lowermost formation.

B) Using the descriptions and illustration of the remaining strata in worksheet 3.8, decide which units should be considered formations, members, and marker beds.

C) State the reasons for your decisions and map the formations with colored pencil.

3.9. Correlation by Lithologic Characters—Eight stratigraphic columns (A-H) are illustrated in worksheet 3.9 (next page).

A) Correlate the beds of each adjacent pair and indicate which units might be considered of formation status.

B) Give the reasons for each decision.

3.10. Correlation by Fence Diagrams—Cut worksheet 3.9 into two strips, "A" & "B" as shown. Fold each strip & stand them on end so that each column is arranged according to the map in the upper right corner of the figure. Complete the fence diagram by correlating the units in columns 4 & 5, and 1 & 8 on strips of graph paper (at back of manual), and adding these to the fence. Draw the stratigraphic column that you would expect to find midway between columns 7 and 3.

3.11. Specimens. Tabletop Stratigraphy and Mapping—Several sets of slabs are placed at various numbered locations and heights on the laboratory tables. Each stack of slabs represents the stratigraphic sequence at that location, and each slab, a lithostratigraphic unit.

A) Make a "base map" showing the locations of the laboratory tables, cardinal directions, map scale (1:10,000 actual dimensions of map area), and location of each "exposure".

B) Describe each lithology, measure the thicknesses (cm) and elevations (m) of each above sea level (lab table = 100 m above sea level) with the ruler provided.

C) Draw a stratigraphic column for each exposure showing, thicknesses, elevations, lithologic descriptions and formations (Figure 3.9). Correlate columns #1-3 and #5 using the lithologic characters discussed above (Figure 3.11).

D) Map the distributions of the formations. Infer the location of contacts between "exposures" for the geologic map.

E) Assume that lithologies exposed at approximately the same elevations were deposited at approximately the same time. Given the idealized relationship between grain size of detrital sediments and distance of deposition from shore, and the tendency for carbonates to accumulate in the absence of detrital sediments, determine which column(s) was/were deposited closest to shore and which farthest from shore. Describe the pattern of lithological change from column #1 to columns #4 and 5.

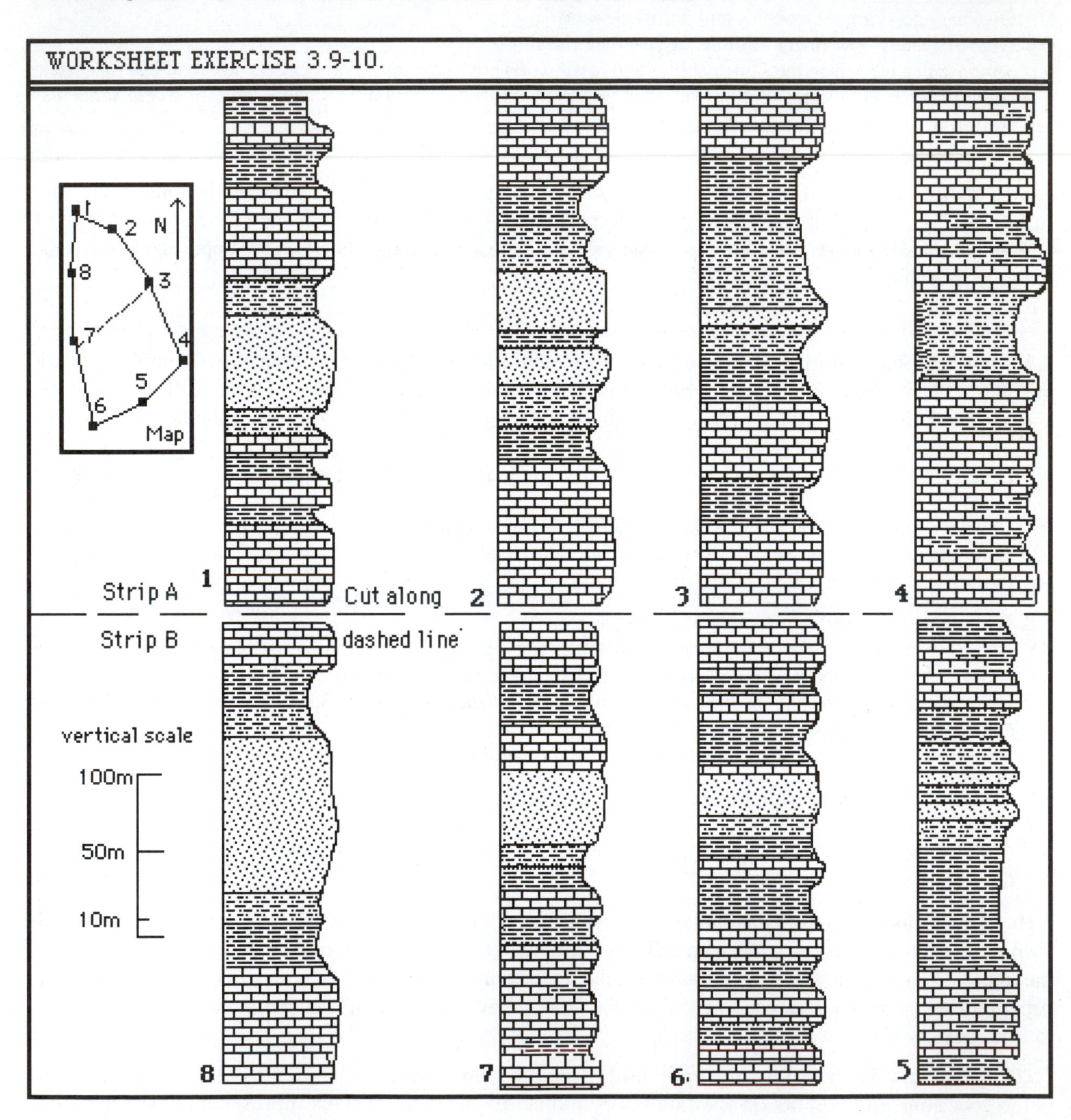

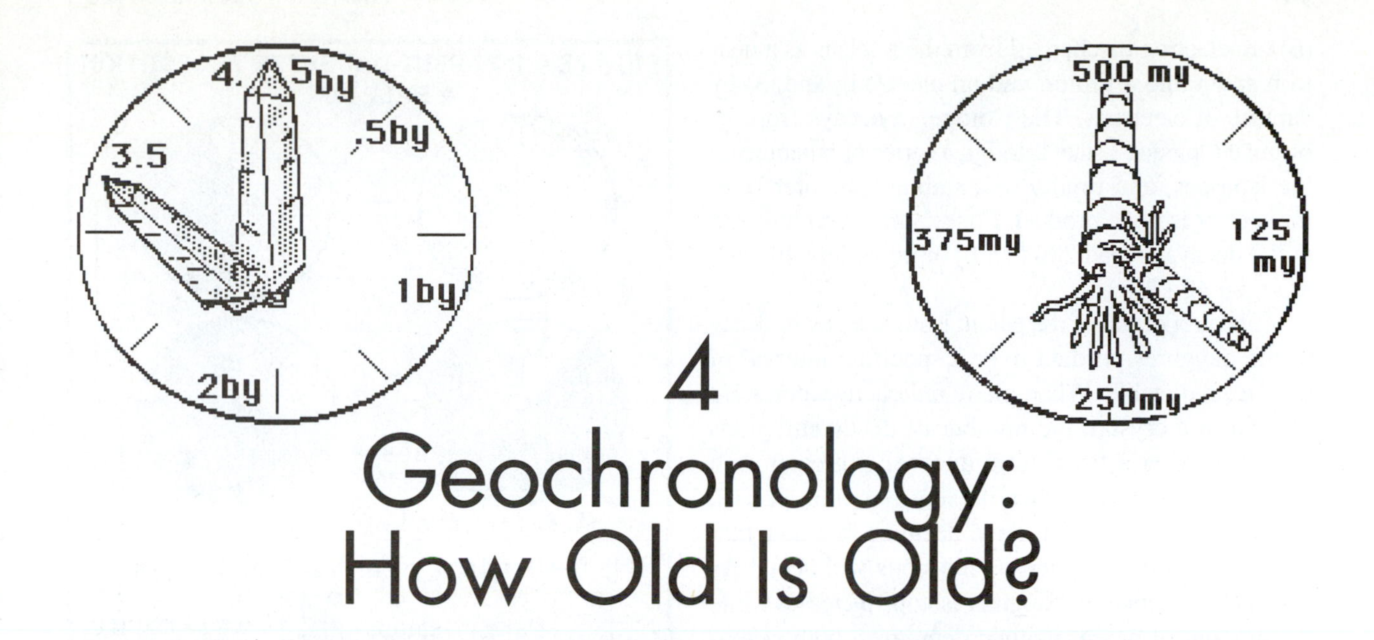

4
Geochronology: How Old Is Old?

Objectives
You should be able to: 1) calculate the age of a sample containing a radiometric isotope if given its isotopic composition and half-life; and, 2) determine which radiometric method should be used to date a sample of a given composition or age.

The geological time scale (inside back cover) is a subdivision of the Earth's history into geologic time (geochronologic) units which encompass, and are represented in the rock record by, chronostratigraphic units (Table 3.1, Laboratory 7): divisions of strata deposited during a single time interval. Such an interval is defined by its strata and tied to an equivalent geochronologic unit by the fossils it contains (the principle of biologic succession), marker beds, or by key horizons (Laboratories 3 and 7). It is geologic time to which we refer when speaking of the Eocene Epoch, Cretaceous Period, or the Paleozoic Era, but the rocks deposited during the Cretaceous Period comprise the Cretaceous System. Thus, geologic time includes both time represented by rock and, that which is not represented due to nondeposition or erosion. References to the physical position of a lithostratigraphic unit (Chapter 3; e.g., the Ordovician System) representing a geochronologic unit (Ordovician Period) use terms such as Lower, Middle and Upper (Ordovician System). References to geochronologic and chronostratigraphic units use "time" terms such as Early and Late Ordovician.

The methods described in Laboratories 3 and 7 allow only placement of strata in sequence and determination of their sequential or "relative", not geochronometric, age relations. Below we will learn how radioactive isotopes in certain minerals are used to determine the"absolute" or geochronometric ages of rocks from a given lithostratigraphic unit; for example, the close of the Cretaceous Period and the extinction of the dinosaurs occurred 65 million years ago and is recorded in rocks of the Upper Cretaceous System.

RADIOMETRIC CLOCKS

We are able to annotate the geologic time scale with a metric (measure) of time that is somewhat related to the human experience of time using a scale divided into units of millions of years before the present (abbr., MYBP where MY =millions of years and BP = before present; also used is BY BP = billions of years BP). Such dates can be affixed to the geologic time scale because several radioactive isotopes are incorporated in the crystals of igneous rock as it cools, each of which spontaneously decays at a constant, experimentally measurable rate. An unstable isotope decays by emissions of: 1) alpha particles (∂) (loss of a positively charged ion of He+); 2) beta particles (ß) (an electron discharged from the nucleus as a neutron splits into a proton and an electron); and, 3) by capture of electrons. The isotope so decays from its original ("parent") state through a series of other unstable isotopes, and finally to a stable "daughter" element or or isotope product. For example uranium 238 (^{238}U) decays to lead 206 (^{206}Pb) or potassium 40 (^{40}K) to argon 40 (^{40}Ar).

The proportion of the parent isotope that will decay to the daughter product over a specified interval of time is predictable. When many radioactive atoms are present in a crystal, the number of decay emissions occurring over a fixed time interval is greater than when few such atoms are present. As time passes, more of the isotope will have decayed to a daughter element, and fewer ∂ and ß emissions will occur. As a result, the amount of daughter isotope increases while the amount of parent isotope decreases, both at regular, statistically predictable rates. The number of atoms which decay over a time interval changes, but the proportion of the remaining isotope which decays in a given time interval remains constant. Thus, as time passes the ratio of parent to daughter isotope will decrease at a known rate. [See your textbook for a complete discussion of these concepts.]

The decay rate of any isotope is most conveniently expressed as a half-life (Figure 4.1). Half-life is the number of years required for one half of the isotope's parent atoms to decay to the daughter product. For example, in the decay series ^{238}U to ^{206}Pb, one half of a known amount of ^{238}U decays to ^{206}Pb after 4.47 BY. One half of the remaining ^{238}U then decays to ^{206}Pb after another 4.47 BY (4.47 by + 4.47 by = 8.94 BY). After two half-lives have passed, 3/4 of the ^{238}U has decayed to ^{206}Pb, and only 1/4 of the parent ^{238}U remains (Figure 4.2). Only 1/8 of the original ^{238}U will remain after three half-lives, (= 13.41 BY bp) have passed, and so on. When the ratio of parent to daughter isotope is plotted against time, the result is an "J"-shaped curve with a negative slope (Figure 4.2).

The relationship between the half-life of any isotope and its decay constant can be determined for any radioactive isotope because the number of atoms of a radioactive element which decay to the daughter product in an interval of time is in constant proportion to the total number of atoms present at that time. Because each isotope has a measurable rate of decay (and thus, half-life) that is constant and unique, it is possible to assign an age to a crystal if its composition and the measured proportion of parent to daughter isotopes are also known.

Determination of a sample's age is possible only if no daughter isotope was present when the sample

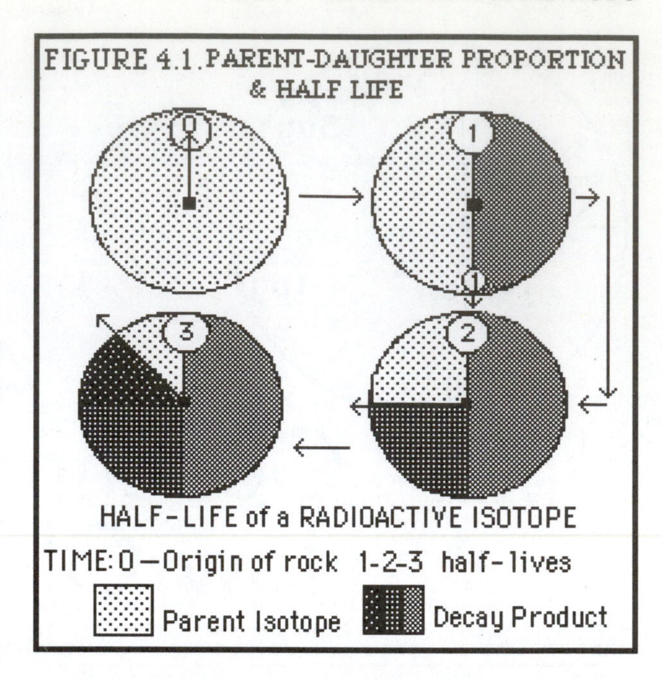

formed and if the parent-daughter ratio has not been altered by external factors. Geologists recognize such alterations, and in the case of metamorphic rocks (e.g., gneiss, metamorphosed from granite), can use them to advantage. Because elements can be gained or lost when rock is heated or metamorphosed, a date for the origin of a rock can be obtained only from those rocks that were not metamorphosed. When metamorphic rock, is "dated", the age obtained is the time of metamorphism rather than that of the parent rock's origin.

There are limits to the number of parent and daughter atoms that can be measured with precision. Isotopes with very long half-lives are not suitable for rocks younger than about 1 million years of age because there are too few daughter atoms to be measured accurately. Furthermore, each radiometric date assigned to a bed is an average age that is calculated from many samples taken from the "same" site. All dates obtained for all samples have an error margin of ± "x" MY. These margins of error reflect inherent variations in the measurements of ages among samples and experimental error for each sample. This error becomes larger with increasing age, but it is generally less than 5% to 10% of the calculated age. As we will see in Laboratory 7, stratigraphic correlations using some fossil species can potentially provide temporal resolution of 1 million year intervals or less!—and thus, in Paleozoic strata, better resolution than that of radiometric methods.

Because isotopes have different half-lives, each is more or less useful for dating rocks of different ages. Experimental error limits measurements to those rocks younger than about seven to eight half-lives of

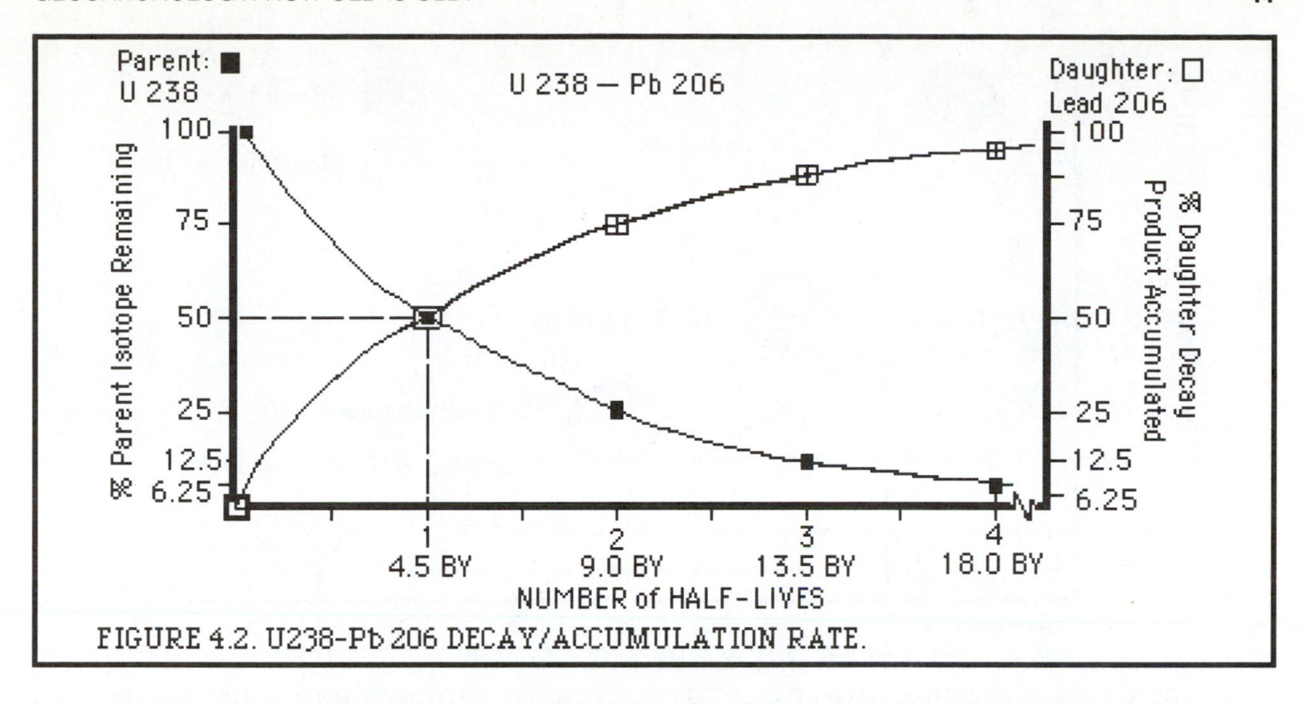

FIGURE 4.2. U238-Pb 206 DECAY/ACCUMULATION RATE.

PARENT	DAUGHTER	HALF-LIFE	ROCK-MINERAL
Uranium 238	Lead 206	4.56 BY	Zircon, Uraninite, Pitchblende
Uranium 235	Lead 207	704 MY	Zircon, Uraninite, Pitchblende
Potassium 40	Argon 40	1.251 BY	Muscovite, Biotite, Hornblende K-feldspar, Volcanic rock, Glauconite, Conodonts
Rubidium 87	Sr 87	48.8 BY	K-mica, K-feldspar, Biotite, Metamorphic rock, Glauconite
Thorium 230	Lead 206	75,KY	Oceanic sediments
Thorium 232	Lead 208	1.39 BY	Zircon, Uraninite, Pitchblende
Carbon 14	Nitrogen14	5,730 yr	Wood, Bone, Shell
Fission Track	---------	1.0 x 1016yr	Mica, K-spar, Zircon,Apatite

TABLE 4.1. COMMONLY USED RADIOACTIVE ISOTOPES.

the isotope used. For example, ^{14}C (carbon 14), which has a half-life of 5,700 years (see below), cannot be used with rocks older than about 40,000–50,000 years. The half-lives of commonly used radiometric isotopes are listed in Table 4.1.

ISOTOPE APPLICATIONS

Minerals and rocks in which the various radioactive isotopes are present are given in Table 4.1. The geological circumstances under which each is most useful are discussed below. Uranium ^{238} and ^{235}U found in the minerals zirconium and uraninite, and pitchblende (uranium oxide) are most often used to date intrusive igneous rocks. (coarse grained igneous rocks which have cooled far below the surface; e.g., granite). Zircon is also found in intrusive igneous rocks and in extrusive (volcanic) igneous rocks like rhyolite (a fine grained compositional "equivalent" of granite). When a magma intrudes overlying sedimentary strata, the age of the intrusion obtained from radiometric dating is the minimum age of the strata that have been intruded.

The ^{40}K -^{40}Ar method can be used on rocks containing muscovite, biotite, potassium feldspar (extrusive igneous rocks such as basalt & rhyolite, and intrusive igneous rocks such as granite), hornblende (many intrusive igneous rocks, including granite and

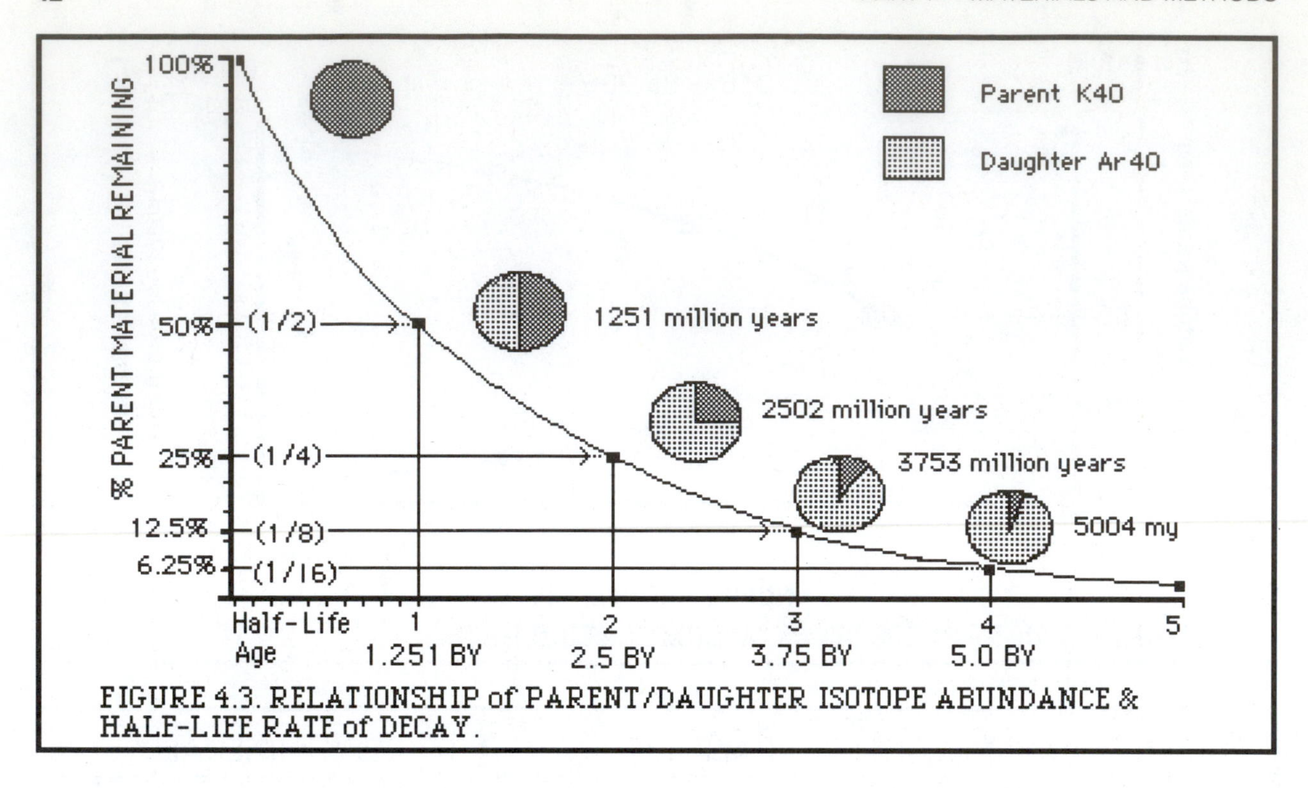

FIGURE 4.3. RELATIONSHIP of PARENT/DAUGHTER ISOTOPE ABUNDANCE & HALF-LIFE RATE of DECAY.

gabbro), and glauconite (a sedimentary mineral). Potassium-argon is a widely used method because ^{40}K-bearing minerals are common in granites and volcanic rocks, and is most often used to date volcanic ashes and basalts. This method is widely used to assign approximate ages to fossiliferous sedimentary strata interbedded with volcanic rocks. The sedimentary strata can then be correlated with regions lacking volcanic strata.

Rubidium 87-strontium 87 are also found in muscovite, biotite, potassium feldspar, glauconite, metamorphic rocks, and in the same types of rocks as ^{40}K. Because the decay product of ^{87}Rb is not as easily affected by reheating as is ^{40}Ar, it is often used to date metamorphic rocks and to provide checks for dates based on ^{40}K—^{40}Ar.

RELATED METHODS

There are many other methods of determining the geochronometric age of rocks, including carbon 14 and fission track dating. Carbon 14 dating is a method used to measure the ages of materials of relatively *recent,* and *organic, origin* only (Figure 4.3). This C isotope forms when cosmic rays strike molecules in the atmosphere, producing stray neutrons which collide with ^{14}N (nitrogen) isotopes causing them to lose a proton. The ^{14}C then becomes part of a carbon dioxide molecule. Although ^{14}C reverts to ^{14}N by ß decay, it is continually taken up by plants during photosynthesis, and becomes part of animal tissues when plants are eaten. When animals and plants die they no longer accumulate CO_2 molecules, and ^{14}C begins to decay to ^{14}N at a steady rate. The age of a sample is determined by measuring the proportion of ^{14}C to total C. Because the half-life of ^{14}C is so short, it is used to provide dates of organic remains from the Holocene (Recent) and Late Pleistocene epochs.

Fission track dating is a relatively new method having some advantages over others. Fission "tracks" are scars in crystals formed when high energy particles escape during the spontaneous fission of uranium. These high energy particles tear electrons away from atoms along the trajectory and leave them with a positive charge so that they repulse one another and widen the original track. The number of original tracks within a crystal are counted and the total amount of radio active isotope present is found by inducing new tracks to form by neutron bombardment. Given a known fission rate, the proportion of natural to induced tracks is used to calculate the sample's age.

VOCABULARY

BY (Billions of years), Carbon 14, Daughter isotope, Fission track, Geochronologic units, Geochronometry, Half-life, MY (Millions of years), Parent isotope, Chronostratigraphic units

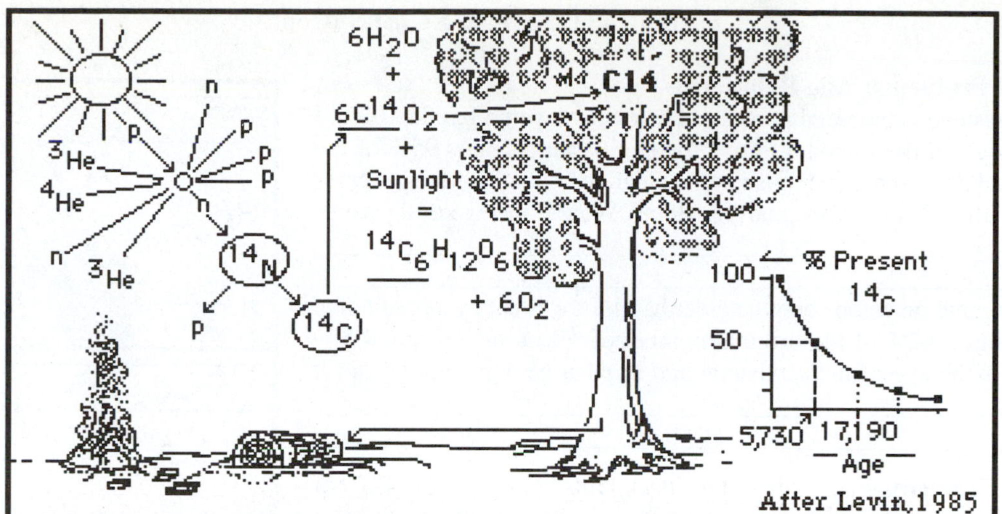

FIGURE 4.4. CARBON 14 FORMATION. Cosmic ray strikes molecule, stray neutron produced strikes N14 atom, which loses proton & becomes C14. C14 in carbon dioxide used by plants becomes part of plant & animal tissues. Decay to N14 begins upon death.

LABORATORY EXERCISES

4.1. Calculations—An isotope has a half-life of 100 days.

A) What fraction of the isotope remains after 10 days? ______________________

After 50 days?______________ After 100 days? ______________

After 200 days?______________

4.2. Methods and Applications—What radiometric method(s) and to what mineral(s) (where applicable) would each method be applied to obtain ages for each of the following:

A) fossil wood in a young glacial till. ______________________

B) a Late Pliocene mammoth bone in strata bracketed by a volcanic ash.______________________

C) a Precambrian granite intrusion. ______________________

D) an Archean gneiss. ______________________

E) an Ordovician shale containing conodonts. ______________________

F) Cenozoic rhyolite? ______________________

G) Precambrian uraninite from South Africa. ______________________

H) an Archean zircon crystal ______________________

I) Mesozoic sea floor spreading ridge basalt. ______________________

4.3. Methods and Applications—You have collected samples of an Archean sandstone, one of the oldest rocks known on the continent of Australia. Your sample contains a single zircon crystal.

A) What data would a radiometric analysis of this crystal provide? ______________________

B) Since most methods of isotope dating require more than one crystal and these are destroyed in analysis, what method would you use to determine its age? ______________________

4.4. Calculations—Use the graph of ^{40}K-^{40}Ar decay in Figure 4.3 to answer the following questions:

A) If 28% of the original number of atoms of ^{40}K in a feldspar crystal have decayed to ^{40}Ar, what is the age of the sample, and in what geologic period was the sample formed? ______________________

____

What is the age and period of formation of a crystal which contains only 10% of the original ^{40}K atoms?

____

B) If 60% of the original number of ^{40}K atoms in a feldspar crystal have decayed to ^{40}Ar, what is the age of the sample and in what geologic period was it formed? ______________________

LABORATORY EXERCISES (Cont.)

4.5. Field Problems: Age Relations—

A) A sandstone is bounded above and below by layered basalts (Figure 4.5A). A feldspar crystal in the younger basalt contains 95% of the original ^{40}K atoms, and those in the older bed contains feldspar crystals with 83% of the original ^{40}K atoms. What is the age of the sandstone? ____________________

____

B) If this same sandstone and the older basalt flow is cut by a basalt dike containing 87% of the original number of ^{40}K atoms (Figure 4.5B) what is the age of the sandstone and in what geologic period did it form? ____________________

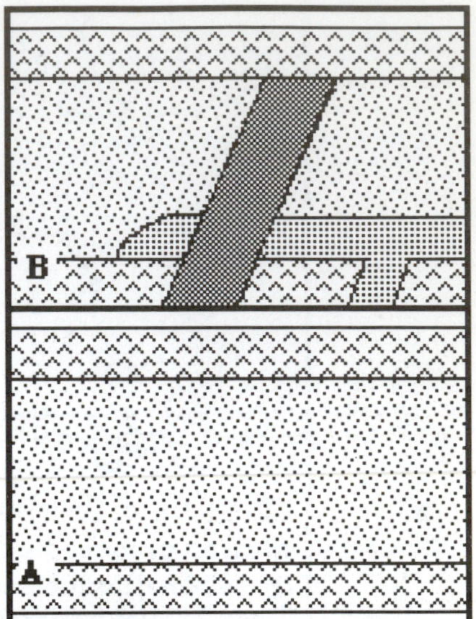

____

FIGURE 4.5.

4.6. Field Problem—The core of the Black Hills uplift in South Dakota is composed of masses of small granitic bodies (Harney Peak Granite) within schists and other deformed metamorphic rocks (Figure 4.6A, B). In one region, a body of gneiss occurs within the metamorphic zone (Bear Mountain). Younger uplifted and tilted Paleozoic—Cenozoic strata (including the latter's Brule Formation), are exposed around the central granitic-metamorphic core (at and around Harney Peak). In the northern portion of the region these strata are strongly deformed and intruded by small bodies of rhyolite and similar igneous rock (e.g., Devil's Tower). See cross-section (4.6A) and sketch map (4.6B) above former.

A) How many episodes of deformation and uplift are indicated? Explain ____________________

____

____

B) Which isotopic methods would you use to obtain the ages of each uplift? Explain.

____

____

____

____

C) You have taken samples of each of the igneous rocks and of the gneiss for radiometric analysis, but your graduate student assistant was careless and lost your records of the sites and rock types on which each analysis is based! This has left you with only the following data samples:

1) –two analyses: 93.38% to 95.5% parent potassium ^{40}K remaining;

2) –two methods applied: a) 99% ^{87}Rb remaining; and, b) 88.5% ^{238}U;

3) –two methods: a) 65.5% parent 238U remaining; and, b) 98.5% parent ^{87}Rb remaining. Use the graphic method, half-life values, and Figure 4.6C to determine the ages of each sample.

1) ____________________

2) ____________________

3) ____________________

D) What are the ages of each uplift? ____________________

____

E) Could the latest uplift have been the source of the volcanic ashes found in the Brule Formation (Cenozoic, Badlands)? ____________________

If not, why, and where might the source(s) have been located? ____________________

____

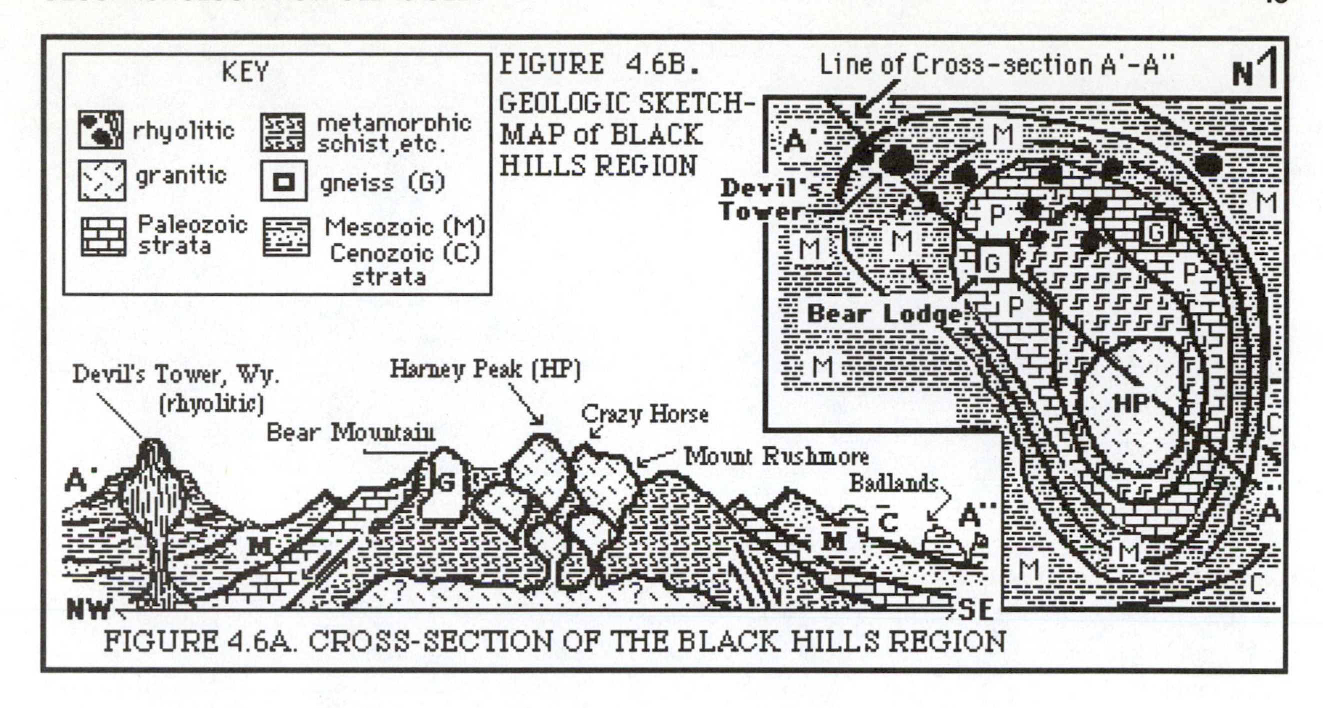

FIGURE 4.6A. CROSS-SECTION OF THE BLACK HILLS REGION

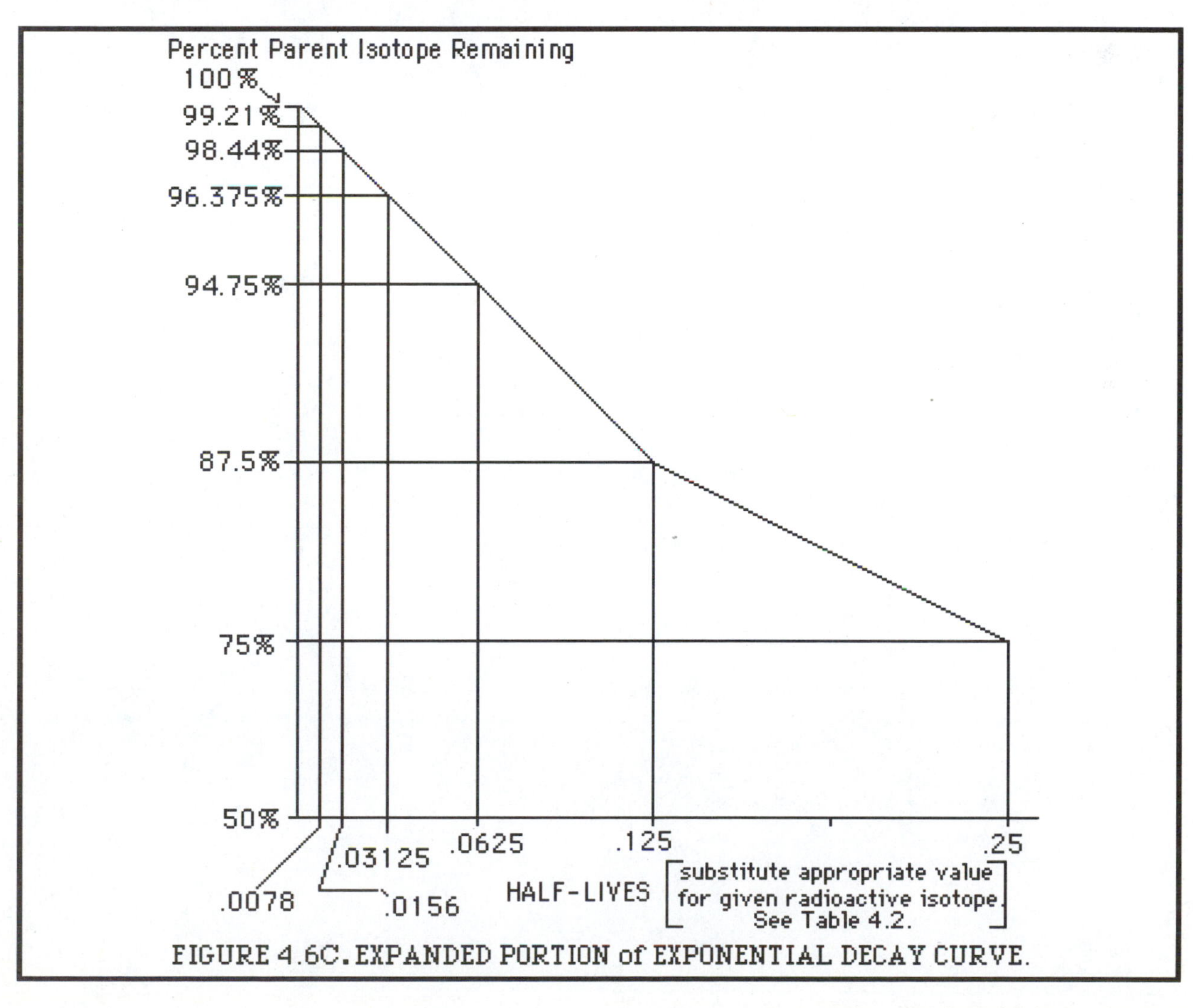

FIGURE 4.6C. EXPANDED PORTION of EXPONENTIAL DECAY CURVE.

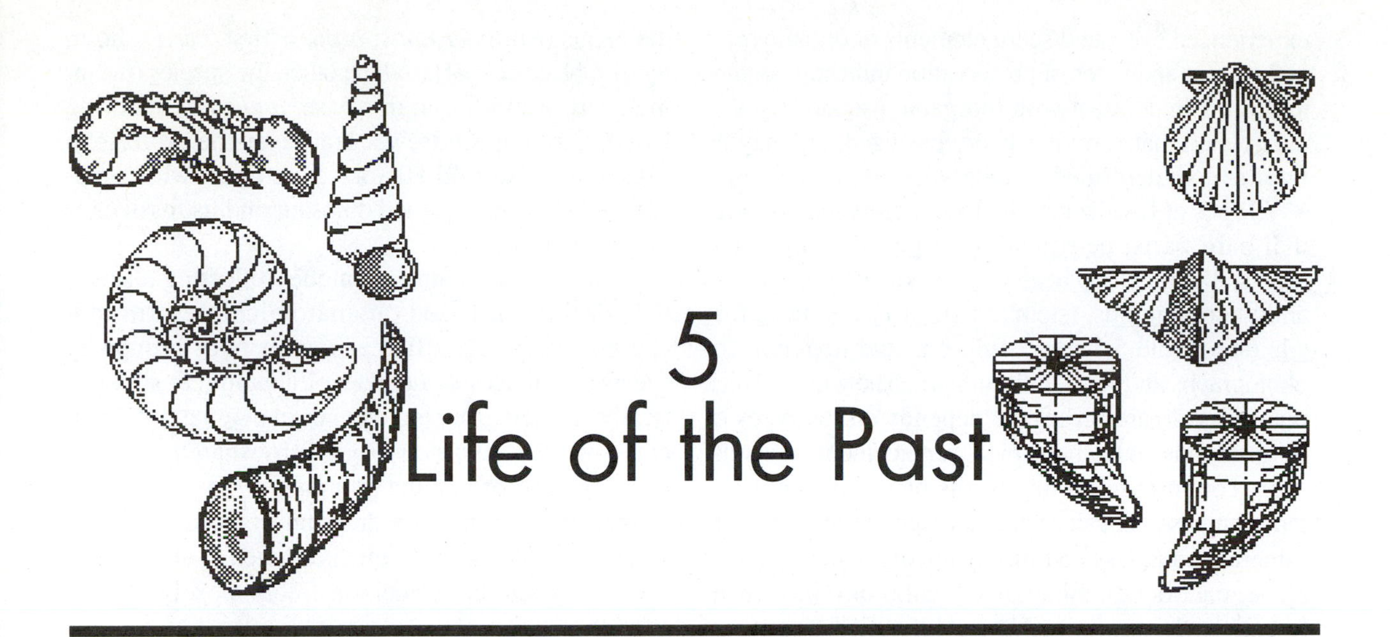

5 Life of the Past

Objectives

To demonstrate mastery of this material, you should: 1) be able to identify specimens representative of several groups and their fundamental distinguishing features; 2) understand that skeletal anatomy reflects fundamental soft anatomy; 3) recognize the various ways in which fossils are preserved; 4) understand that anatomies may be superficially similar and if so, do not reflect descent from a common ancestor nor within-group membership as do homologous features; and, 5) be able to use the Linnaean classification system. Your instructor may also assign the discussion of other "minor"/microscopic groups in Appendix II.

Using This Laboratory

You are not expected to memorize anatomical terms or species names. The discussions of each group include contextual information about soft anatomy, function, shape, etc., for greater understanding of fossil groups as once vibrant, active, evolving animals, not for content memorization. A cursory introduction to the many commonly fossilized, yet unfamiliar animals, requires that you are comfortable with the material before the laboratory begins. Exercise 5.1 is designed as a preparatory self-quiz of Objective 1.

This chapter is an introduction to paleontology—the study of the ancient organisms preserved as fossils (see below). Here we focus on the description, identification and classification of those groups of marine invertebrates (animals without backbones: e.g., clams, crustaceans, insects, corals, snails) common in the fossil record. Although less familiar than vertebrates, they constitute a major part of the biosphere and evolutionary record, are used for correlating strata Laboratory 7), and reconstructing ancient environments, climates and plate movements (Laboratories 6, 8–12).

To understand that recognition and observation of fossils involve the elements of organic form, and to aid in fossil identification and interpretation, we must consider first how fossils are preserved, named, classified, and how paleontologists interpret their anatomies and relationships.

FOSSILIZATION

A fossil is the remains or trace of an organism from the geologic past that has been preserved in sediment or rock. Fossils commonly consist of only hard parts (skeleton, shell, chitin (fingernail-like material), plant fibers, etc.), however, traces of soft tissues can be preserved in exceptional circumstances. An experienced eye can discern elements of organic form and pattern and types of preservation indicating organic origins in contrast with inorganic pseudofossils—even when that form has been fossilized and may be fragmented, deformed or otherwise altered.

Types of fossilization include preservation of original hard parts, petrification or permineralization, recrystallization, replacement, carbonization, external and internal molds, (steinkerns), casts, and trace fossils (see Table 5.1 for definitions and textbook for photographs and further discussion). The way in which many fossils are preserved depends on the types of original materials and typical replacement minerals. Most common fossil organisms had shells or skeletons composed of calcite and/or aragonite, silica (quartz minerals), and less commonly apatite (calcium phosphate; e.g., bone), mineralized chitin or other organic cuticle. Just as minerals are identified by properties such as luster, color and hardness, certain groups of organisms are preserved in characteristic fashions resulting from the involvement of same few original minerals (above) and their common replacements (e.g., calcite, quartz, dolomite, pyrite, and hematite). Thus, many fossils are recognizable in part, by properties of their original or replacement skeletal minerals (Table 5.1).

The majority of fossil invertebrates have some type of external skeleton (exoskeleton; *exo* = outside: e.g., crabs, lobster, many insects, clams, snails), while others have one that is internal—in other words, an endoskeleton (*endo* = within). The exoskeleton may be thought of as the "house" in which the animal lived, but it is more than a simple container: it is a "house" constructed from the inside out and according to a specific genetic code. Likewise, an endoskeleton (e.g., vertebrates, echinoderms (starfish)), may be thought of as a building's suprastructure—to which other structures (soft anatomy) are attached, and one that is also genetically determined. Thus, various aspects of shell and skeletal anatomy reflect soft anatomy, and both hardparts and inference from softparts are used to recognize fossil species.

NAMING and CLASSIFICATION

Naming Organisms—Outside of science, groups of familiar creatures of similar appearance are commonly spoken of as belonging to one or another "family" and a utilitarian or common name is conferred on each "type" (e.g., the Great White Shark). In science every creature is given two formal Latin or latinized Greek names (e.g., *Protoceratops andrewsi*—a primitive Cretaceous horned dinosaur)—one designating the genus (*Protoceratops*; *proto* = first, *cera* = horn, *topos* = place or spot), and the other, the species (meaning "sort or kind"); in this case, *andrewsi* honoring Roy Chapman Andrews, leader of the American Museum of Natural History expedition which discovered this small horned dinosaur and its fossil eggs in Mongolia.).

Each species name is unique, reflecting a scientific definition based on anatomical description. Organisms are classified as members of groups by specific instances or degrees of anatomical similarity. Species are grouped together in a genus (pl., genera; *gen* ="of common origin") by shared similarities—somewhat like the grouping of "brand name " (= species name) foods or drugs under a generic (inclusive) name. The genus name is applied to all anatomically very similar, or demonstrably closely related, species.

Species are recognized by different criteria in paleontology and biology. Biologists define as a species any population(s) which is/are genetically distinct or isolated by effective or actual barriers from breeding with all other populations. Such species commonly do, but need not, differ in their anatomies, skeletal or otherwise. Paleontologists can only infer, on the basis of distinctive hard tissue features or separation in time or geography, that such reflect reproductive isolation, and thus different species.

Just as species are grouped together as members of a genus, genera are grouped together into a family, related families into an order, and so on (Table 5.2). In this laboratory, groups below the rank of subclass are rarely mentioned: recognition and identification of the larger, inclusive categories (usually class and phylum) to which representative species belong is our concern.

Classification, Homology and Analogy—Those classifications most useful in evolutionary studies also reflect evolutionary relationships and descent from a common ancestor. It is on anatomical features which are shared among members of a group by virtue of common origin—not superficial or apparent similarities—that such classifications are (ideally) based. In this laboratory you will learn which features, and thus groups of species, are used to construct and outline respectively, a largely evolutionary classification of marine invertebrates.

Classifications reflecting evolution are based on the presence of homologous features (i.e., having the same relative position, proportion, value or structure due to inheritance—not mere similarity) which are shared among only one group of species. For

TABLE 5.1. STYLES OF FOSSILIZATION

TYPE	ORIGINAL MATERIALS	ORIGINAL MINERALS	MINERAL(S)/ MATERIALS	PROCESS	PRODUCT
Petrification or Permineralization	Shell, bone, other skeletal tissues (plates), chitin, cellulose.	Aragonite, Calcite, SiO2, Apatite.	Silica, calcium carbonate, iron, phosphate.	addition of elements or minerals to pore spaces of original	Original hard parts & added mineral(s)
Replacement	Shell, bone, other skeletal tissues (plates), chitin, cellulose	Aragonite, Calcite, SiO2 Apatite.	Silica, Calcite, Apatite, Hematite, SiO2.	Simultaneous exchange of original & replacement minerals	Good to excellent preservation of hard or soft parts
	Soft tissues		Pyrite, Hematite		
Recrystallization	Shell, bone, other skeletal tissues (plates),	Calcite, SiO2, Aragonite, Apatite.	Calcite, SiO2, Aragonite, Apatite, Dolomite.	Transformation of one mineral to another	Detail lost
Carbonization	Plant fiber, chitin, organic cuticle (hard, unmineralized, soft tissue)	———	———	Compression & loss of volatiles	Thin carbon films
Molds					
Internal (Steinkern)	Hard Parts	Calcite, SiO2 Aragonite, Chitin	Sediments	Sediment fills hard part cavities. Hard part dissolved	Gross shape & internal surface sculpture
External	Hard parts, rarely soft	Calcite, SiO2 Aragonite, Chitin	Sediments	Impressions	Gross shape external surface sculpture
Casts	Hard Parts	Calcite, SiO2 Aragonite, Chitin	Sediments	Sediments fill extrnl/intrnl. molds	Gross Shape
Trace Fossils	Tracks, Trails, Burrows	Hard and soft	Sediments	Impressions, Molds, Casts	Gross Shape

example, whales are grouped with other mammals because all share several homologous features (e.g., mammary glands, hair, lungs, live birth, limb structure, etc.). Furthermore, dolphins and whales are grouped together in the order Cetacea because they, only, share certain anatomical features reflecting the hypothesis that organisms sharing characters do so because they are descendants of a common (mammalian) ancestor.

Features which are merely similar, or analogous, result from convergent evolution (see text): that is, evolution of features resulting from similar function

TABLE 5.2. LINNAEAN HIERACHICAL CLASSIFICATION.

Kingdom	Anamailia
Phylum	Chordata (Subphylum Vertebrata)
Class	Mammalia (mammals)
Order	Carnivora (meat-eating)
Family	Canidae (dog-like carnivores)
Genus	*Canis* (dogs)
species	*familiaris* (domesticated dogs)

and ecological adaptations, not descent from a common ancestor. In an extreme example, *Orca* the "Killer Whale" and *Carcharodon*, the "Great White Shark", although similar (e.g., rapid swimming, marine habitat, dorsal (back) fins, fins and fin-like appendages, similar shapes, etc.), are not classified together because these features are superficial, not homologs. Cetaceans are grouped together by virtue of shared features, and similarly, cetaceans with all other mammals. This classification reflects a common ancestry not shared with sharks. Evolution commonly produces similar, apparently homologous features, muddying the distinction between homologous and analogous features, and thus confounding efforts to classify by criteria of common descent.

A similar phenomenon, parallel evolution (the independent evolution of similar features in groups more or less distantly related by a common ancestor lacking the feature(s)) has also been common in life's history. For example, the mountain goat is not a true goat, but actually a sheep. Both sheep and goats belong to the Bovidae, a group including also cattle, musk ox, bison, gazelles and antelopes,. However, the mountain goat's resemblance to true goats belies the independent evolution of similar features in two groups which share other true homologies.

Thus, frequent, ubiquitous convergent and parallel evolution obscure homology, common ancestry and evolutionary classification. This is an especially common experience for students unfamiliar with marine invertebrates. Similarities abound, and without experience one may understandably group together taxa having merely analogous features. In this laboratory we will discuss homologous features and thereby use the analogy-homology criterion for comparison and contrast of fossil groups. The order in which groups are discussed below was chosen to facilitate (where possible) comparison and contrast of related groups having less than obvious homologies, and unrelated groups with superficially similar features.

Symmetry is repeated regularity of geometric form about one or more planes of mirror image (Figure 5.1). Types of symmetry are very useful for recognizing distantly related inclusive groups (e.g., phylum and class levels) because some type of symmetry is fundamental to the body plan of all animals and thus, (often) evident in their fossilized shells and skeletons.

MAJOR INVERTEBRATE GROUPS SUBKINGDOM I

We will focus on 18 subgroups of 12 animal phyla and protistan groups most common in the fossil record. Classes of the Mollusca and the arthropod class Trilobita (below) are distantly related and are grouped in one subkingdom according to homologous embryological features.

The majority of other phyla important in the fossil record are similarly classified in another subkingdom. The Cnidaria (including corals), and, Porifera (sponges), and (probably) the extinct Archaeocyatha (Appendix I) comprise two or three primitive, and more distantly related groups.

Clams, Snails and Squid: Phylum Mollusca ("Soft-Bodied")

Molluscs are among the most important of fossil invertebrates by virtue of their diversity (number of species), abundance, distribution, and variety of body plans (Figure 5.2). The phylum encompasses disparate groups like squid, clams, chitons (Polyplacophora), snails, tusk shells (Scaphopoda), rostroconchs (extinct), and monoplacophorans. Of seven classes, we are concerned with those abundant as fossils: clams (class Bivalvia), snails (class Gastropoda), and nautiloids, ammonoids and coleoids (class Cephalopoda).

Clams, snails and cephalopods appear very different, yet homologous soft anatomies allie each of these distinctive body plans in one phylum (Figure 5.3-13). The soft anatomy of molluscs includes a large muscle or foot (*poda* = foot) and visceral organs enclosed by a flap of tissue (the mantle) opening outside of, and secreted by, the shell. The body is

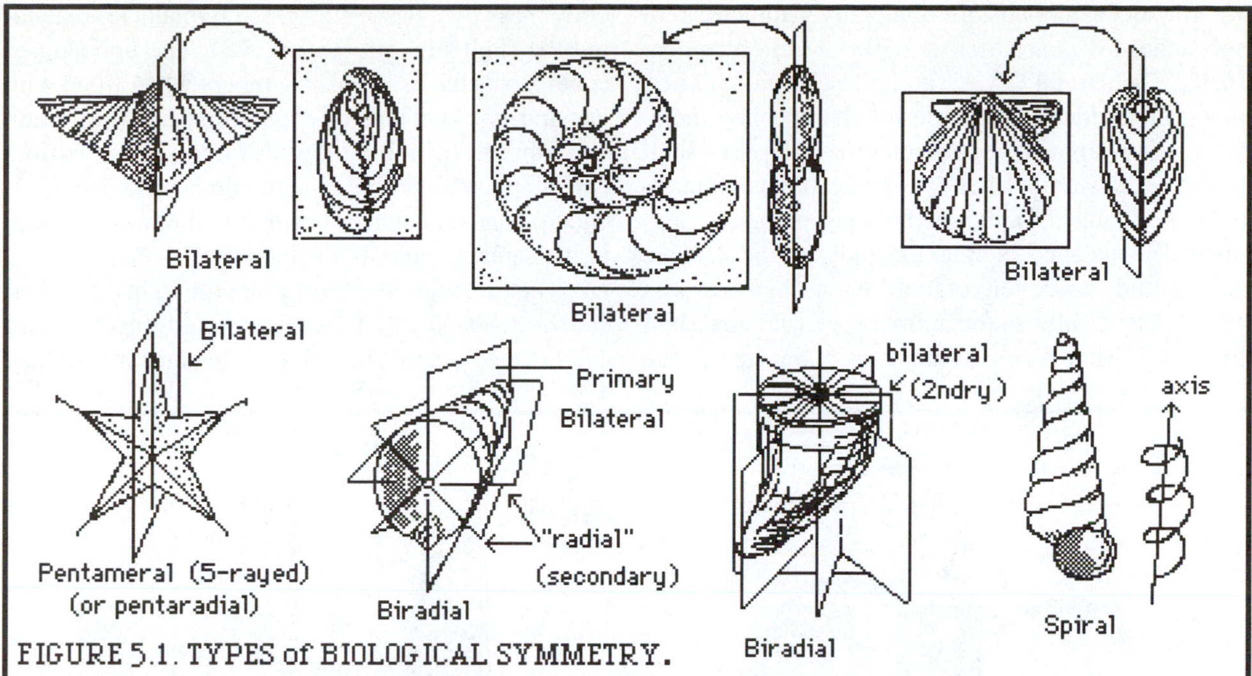

FIGURE 5.1. TYPES of BIOLOGICAL SYMMETRY.
Most groups have either bilateral or radial symmetry. Convergent evolution has produced the same symmetries among unrelated taxa (e/g/. bivalves & brachiopods) Thus additional characters are needed to identify each.

bilaterally symmetrical—or displays remnants thereof—in most species. Molluscs are readily identified at the class level by external and internal features of the shell and its symmetry (Figure 5.1, 3-13).

Nautilus, Squid, and Octopi: (Class Cephalopoda—"Head Foot")

Cephalopods are intelligent, active predatory invertebrates with complex eyes similar to those of vertebrates (Figure 5.3). In cephalopods, the foot forms a tubular siphon (a tube used to draw or expel fluid) used in locomotion (see Laboratory 6), and hook-bearing tentacles or arms extend from the head.

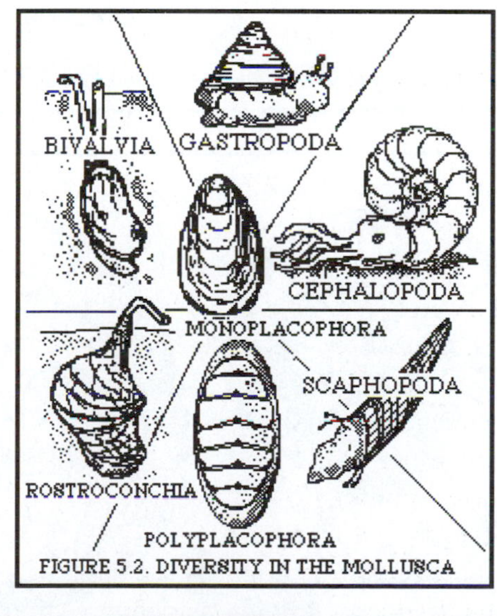

FIGURE 5.2. DIVERSITY IN THE MOLLUSCA

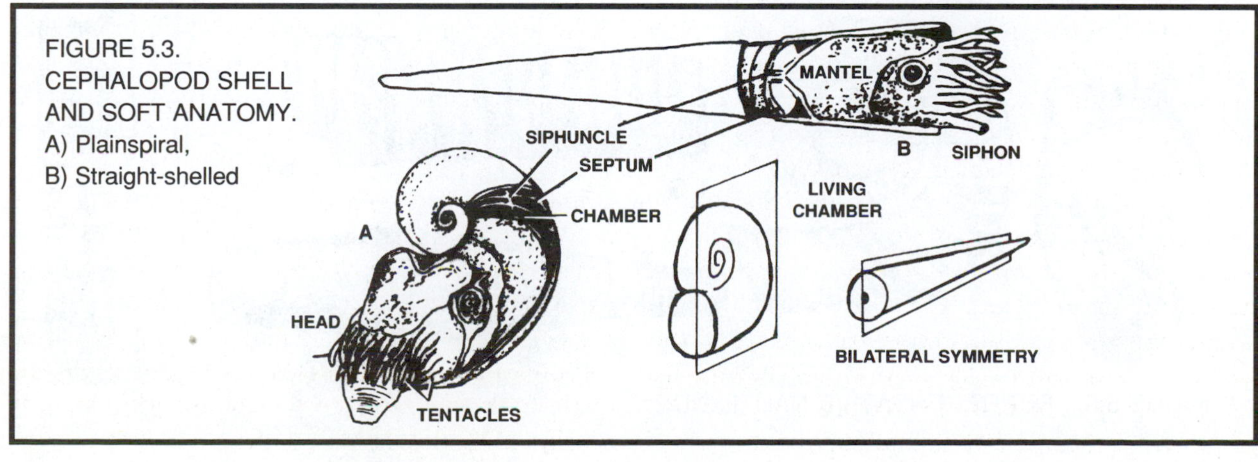

FIGURE 5.3.
CEPHALOPOD SHELL AND SOFT ANATOMY.
A) Plainspiral,
B) Straight-shelled

Most cephalopods are bilaterally symmetrical in soft anatomy as are their straight or coiled conical shells. The animal "lives" in the last formed portion of a shell divided into a series of chambers by skeletal walls or septa. When the outermost layer of shell is stripped away or when it is preserved as an internal mold (Table 5.1), the shell has a superficially segmented appearance. Neither the body nor the shell are segmented, however. Instead each chamber was added periodically as the animal grew and new shell formed. Earlier formed chambers connected to the body mass by a tissue-lined tube (siphuncle) extending the shell's length (Figure 5.3). The unoccupied chambers acted as ballast chambers when filled with gas and liquid. Thus, by regulating the distribution and amount of liquid, *Nautilus* and similar extinct cephalopods were able to maintain a stable head orientation and adjust their position in the water column in the same manner as a submarine!

The same few shell forms are found in all shelled cephalopods (parallel evolution), and thus they are classified by internal homologous features of the shell,

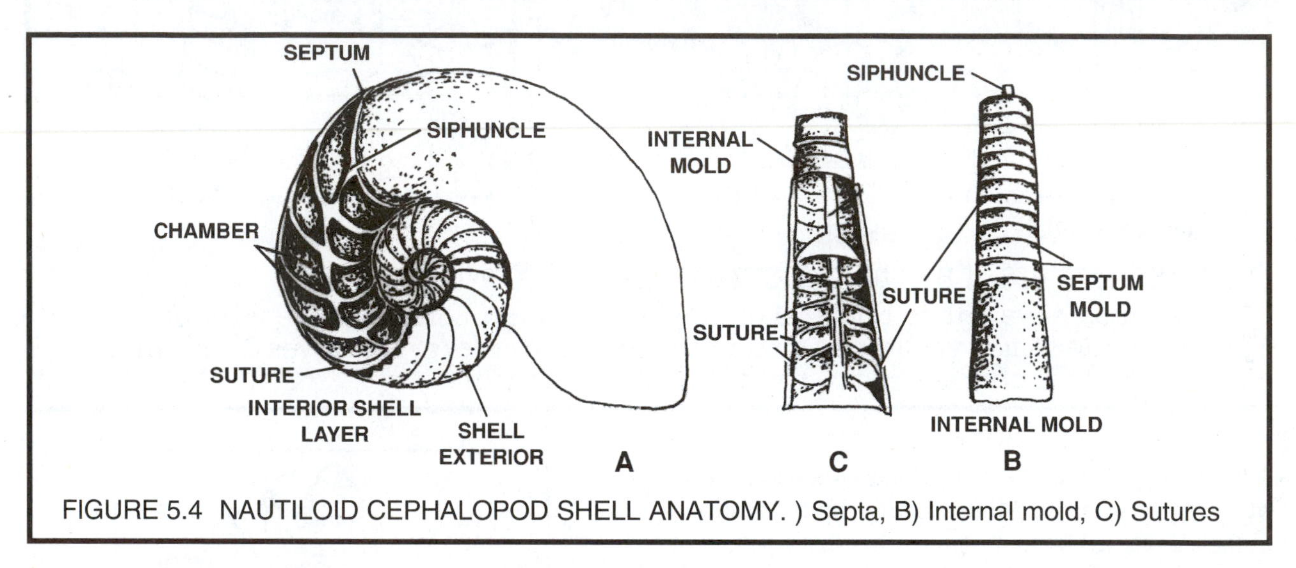

FIGURE 5.4 NAUTILOID CEPHALOPOD SHELL ANATOMY.) Septa, B) Internal mold, C) Sutures

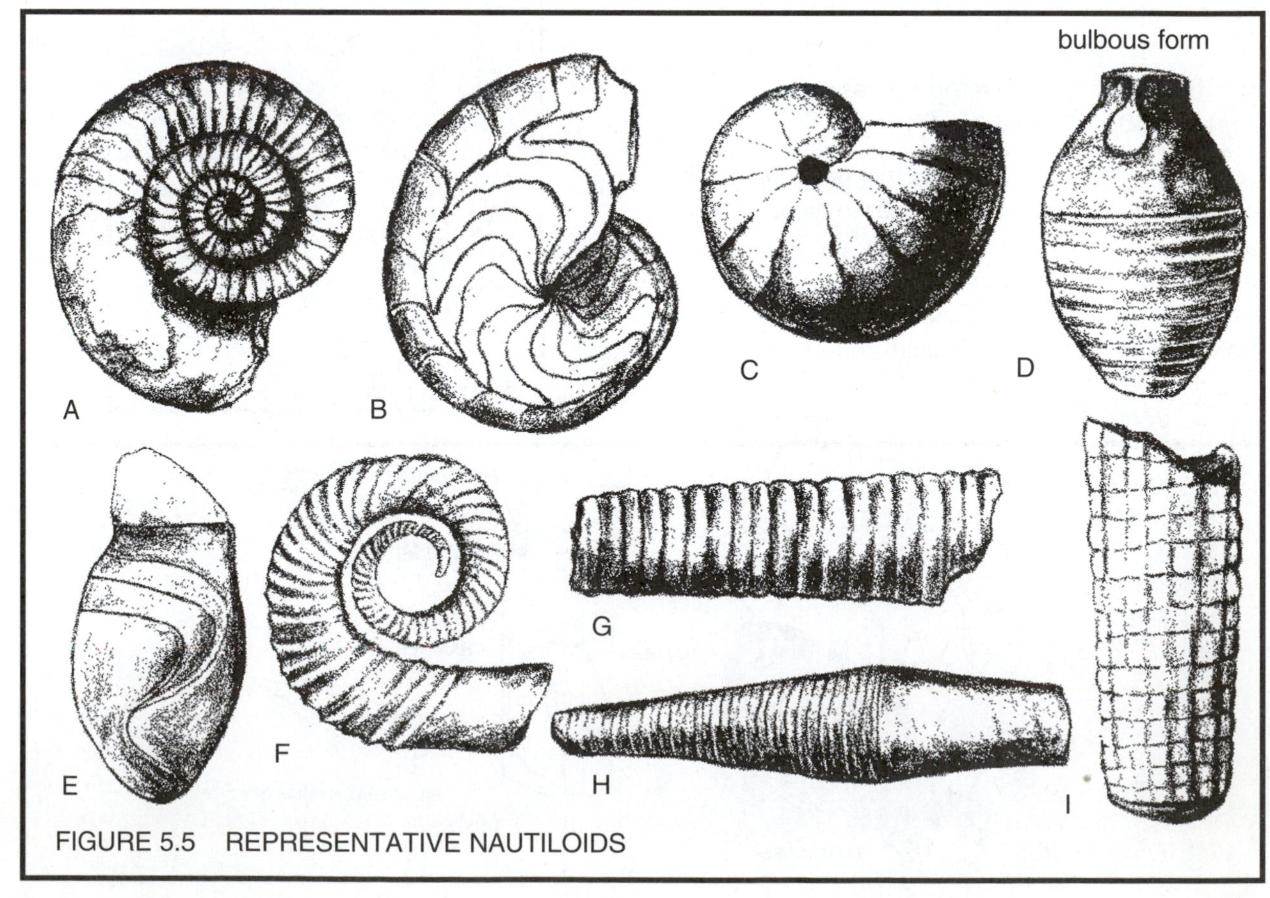

FIGURE 5.5 REPRESENTATIVE NAUTILOIDS

soft parts, and the relationship of the shell to the mantle. Here we recognize three groups of cephalopods: 1) nautiloids (an informal group comprised of the Nautiloidea and two other subclasses); 2) Ammo-noidea, which have external shells; and, 3) Coleoidea having internal (mantle covered) shells or lacking shells.

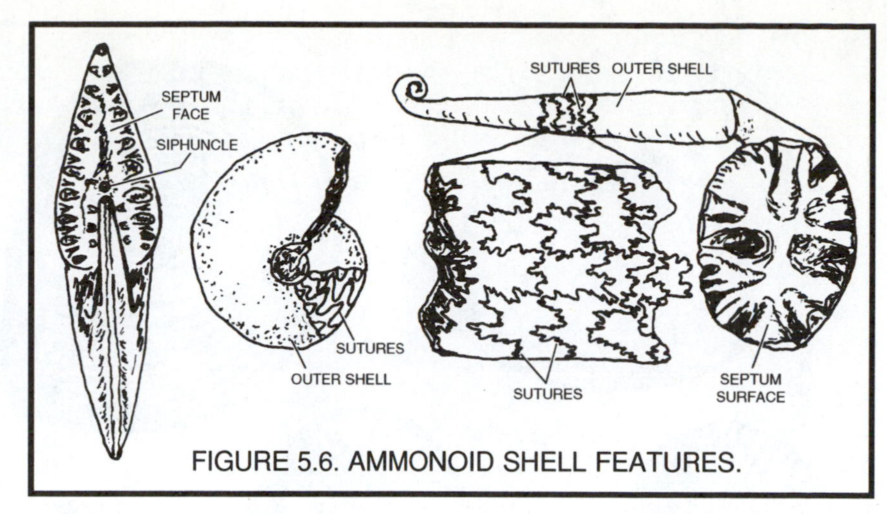

FIGURE 5.6. AMMONOID SHELL FEATURES.

Nautiloids—Much of what is inferred about the anatomy and behavior of cephalopods is based on studies of ("the chambered") *Nautilus* (*nautes*= sailor or sail) —the only surviving nautiloid genus, and one of four diverse subclasses abundant in Early Paleozoic seas (Figure 5.4A). Nautiloids are distinguished from ammonoids (below) by the former's smoothly concave septal surfaces dividing the shell into chambers (Figure 5.4A-C, 5.6). Nautiloid septa form simple concentric or broad, gentle curved sutures where septa join the shell interior when observed as internal molds or on the innermost layer of the exterior shell (Figure 5.4B, C, 5.6). A tubular siphuncle connecting each chamber runs through the center of the septa.

Nautiloid shells are bilaterally symmetrical straight or slightly curved cones, or cones coiled within a plane (Figure 5.5). Some straight and curved forms became bulbous in adulthood, at which time the younger conical shell was discarded. Some straight-shelled Ordovician species reached lengths of 10 m (33 ft) or more!

Subclass Ammonoidea—This extinct subclass differs from nautiloids in that their septa have complexly folded surfaces forming complex, intricate suture patterns (Figure 5.7). Suture pattern complexity ranges from gentle curvature to pronounced zigzag patterns and finally to patterns resembling frost growth on glass (Figure 5.7). Ammonites (*Ammon* = Jupiter; *ite* = stone) are sometimes subdivided according to suture type although distant groups commonly evolved these features in parallel fashion.

The majority of ammonites are coiled within a bilateral symmetry plane of passing through the siphuncle. Some species have straight cones but others are coiled in a spiral like a snail's shell (Figure 5.8). Some bizarre species have shells bent like a saxophone, an unraveled spring (Figure 5.8H, I), and one, "aimlessly" contorted coils of spaghetti (*Nipponites* , Figure 5.8J)! Many are highly ornamented with ribs, knobs and spines. Like their nautiloid Paleozoic precursors, some grew very large—one Mesozoic species reached a diameter of 3 m (10 ft)!

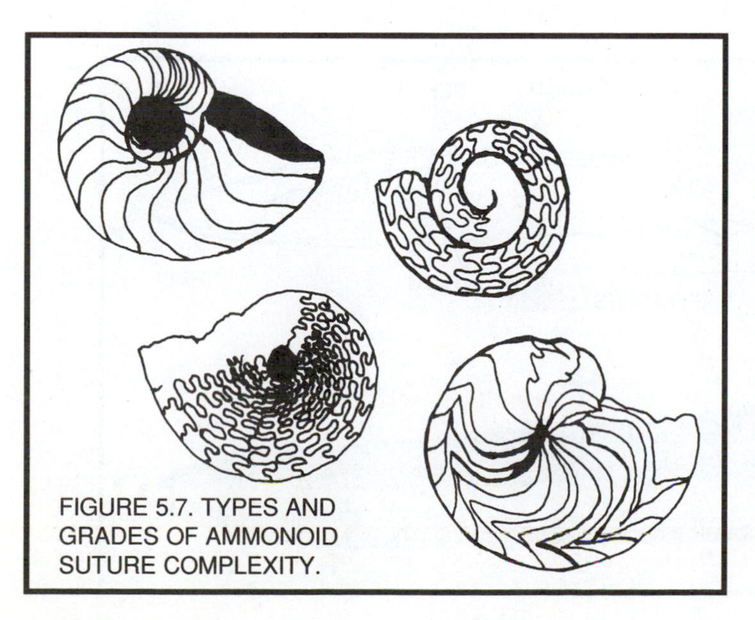
FIGURE 5.7. TYPES AND GRADES OF AMMONOID SUTURE COMPLEXITY.

Squid, Octopi, and Belemnites (Subclass Coleoidea)—Cuttlefish, squid, octopi, the "paper nautilus", and the extinct belemnites are recognized as a separate subclass of cephalopods, Coleoidea (*koleo* = sheath), having only one pair of gills. The shell is small and internal in squid and similar species, but absent in octopi. Mesozoic belemnites ("arrow stones") broadly resemble squid and their shell evokes the image of a cigar or conical arrowhead (Figure 5.9B). Their shell is a narrowly chambered short cone of dense skeleton covered entirely by mantle (Figure 5.9A).

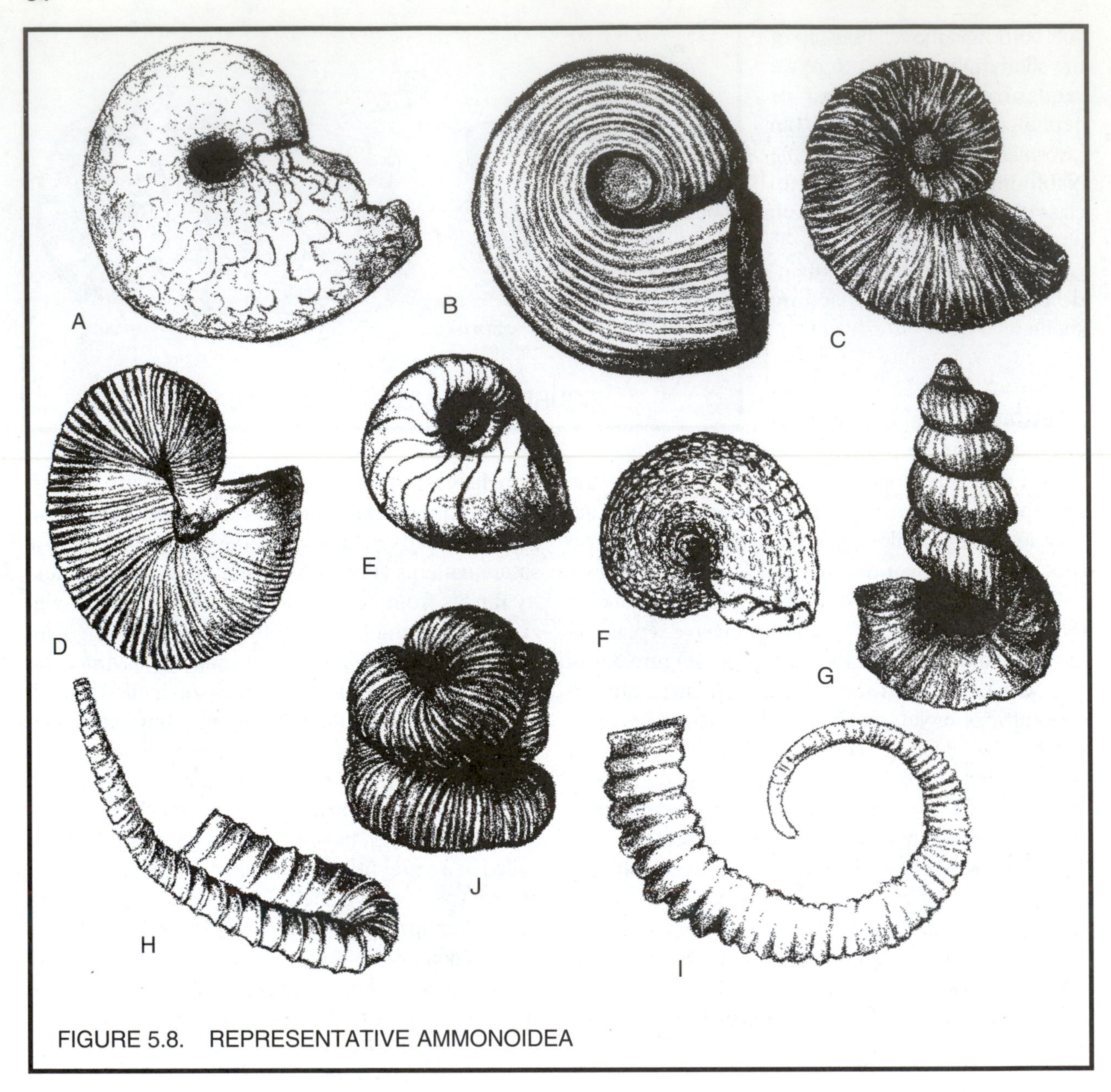

FIGURE 5.8. REPRESENTATIVE AMMONOIDEA

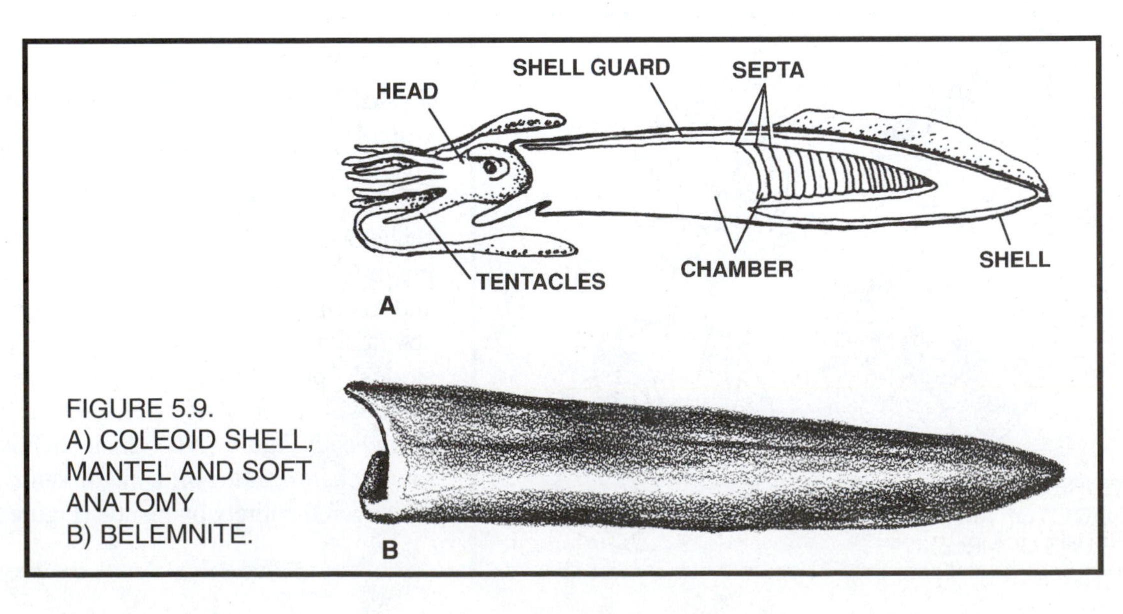

FIGURE 5.9.
A) COLEOID SHELL, MANTEL AND SOFT ANATOMY
B) BELEMNITE.

Snails and Slugs: Class Gastropoda ("Stomach Foot")

Snails or gastropods are the most widespread, diverse, and abundant molluscan class (Figure 5.10A). The long proboscis ends in a mouth and two eyes are commonly mounted on stalked tentacles. Inhalant and exhalant siphons pass water over the gills. The mantle lines the shell's interior and muscles attach the head and foot to the shell's central column. Shelled gastropods have a forward counterclockwise rotation or twisting of an otherwise bilaterally symmetrical body (Figure 5.10A). This rotation brings the rear body forward and above the head so that the shell's spire points backward away from the head. Slugs lack a shell and are bilaterally symmetrical.

Living snails are classified by various soft parts, but fossil species are classified by shell shapes, coiling type, and a few external features like the shape of the notch that held the siphon (Figure 5.10B). The variety of shell shapes and ornamentation is striking (Figure 5.11). Some are limpet-shaped caps, but most are coiled about an axis and lack shell symmetry, or have "spiral" symmetry. A few species coil within a plane of bilateral symmetry (e.g., *Bellerophon*; Figure 5.11C). These species show convergence with coiled cephalopods, but all lack a siphuncle and virtually all, septa.

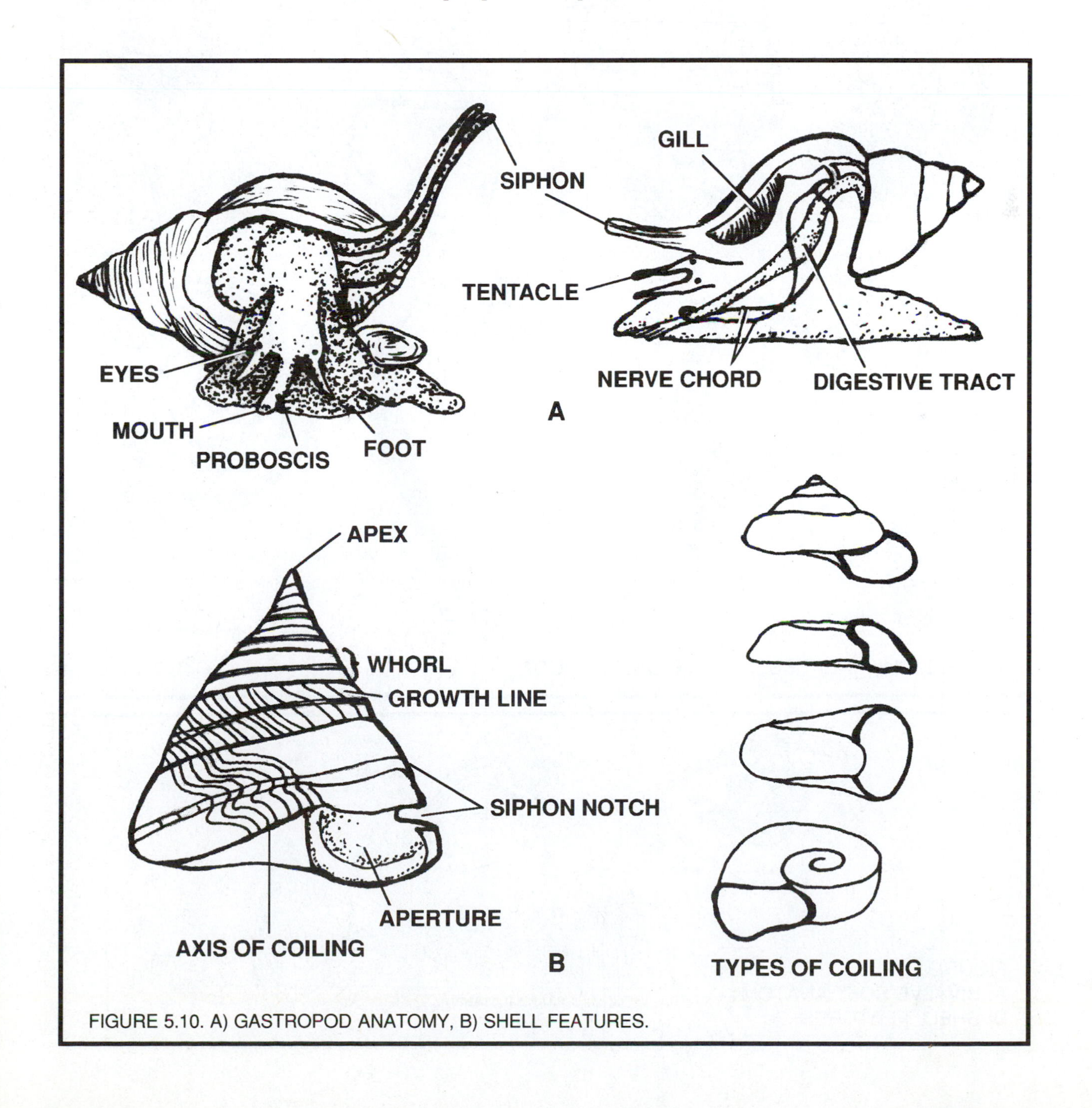

FIGURE 5.10. A) GASTROPOD ANATOMY, B) SHELL FEATURES.

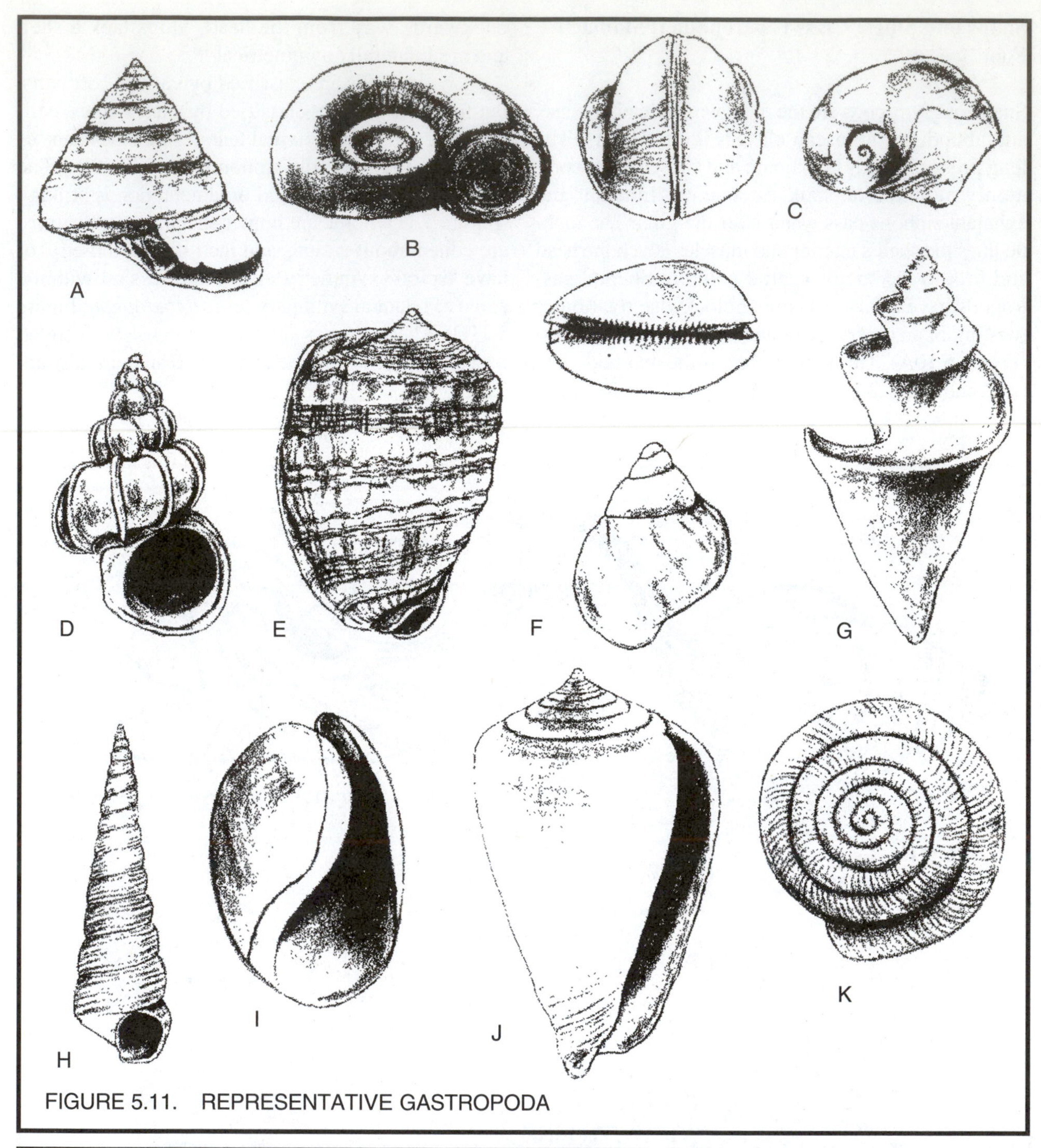

FIGURE 5.11. REPRESENTATIVE GASTROPODA

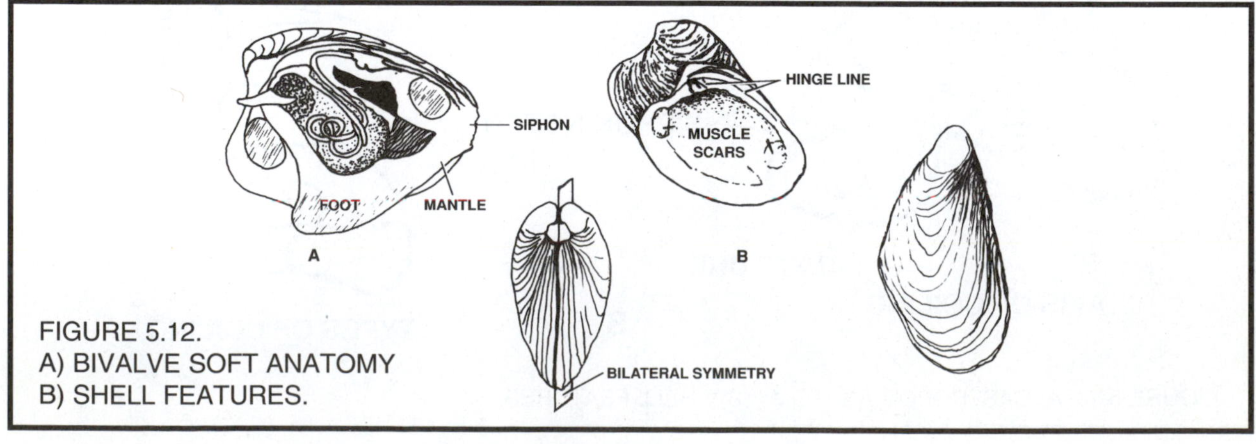

FIGURE 5.12.
A) BIVALVE SOFT ANATOMY
B) SHELL FEATURES.

Clams, Oysters and Mussels: Class Bivalvia ("Two Valves")

Cephalopods and gastropods appear more "complex" than the related Bivalvia (also known as Pelecypoda (*pelecy* = hatchet; *poda* = foot)) but nevertheless, all share very many features of soft anatomy (Figure 5.12A). The name Bivalvia (meaning "two valves" or "two shells") is somewhat misleading because brachiopods (below) and ostracodes (Appendix I) also have two interlocking shells. However, Bivalvia differ from other bi-valved animals in many fundamental ways.

Most bivalves are bilaterally symmetrical about a plane passing between the valves so that each is a mirror image of the other (Figure 5.12B). The shell is secreted by the mantle and the valves are loosely joined along a hinge line bearing "teeth and sockets", or a series of ridges and grooves acting as a hinge. The hinge area is covered by a thin elastic ligament which aids in opening the shell. The shell interior to either side may bear one or more rounded depressions marking the sites of muscle attachments. Muscle contraction pulls the valves together, and relaxation, aided by the spring-action of the ligament, opens the shell. The foot is so-called because many species crawl on, or dig within, the sediment by expanding and contracting this muscle.

The familiar divisions of bivalves into clams, mussels, oysters, and scallops illustrates the diversity of shapes among living species (Figure 5.13) reflecting various life habits and habitats (see Laboratory 6).

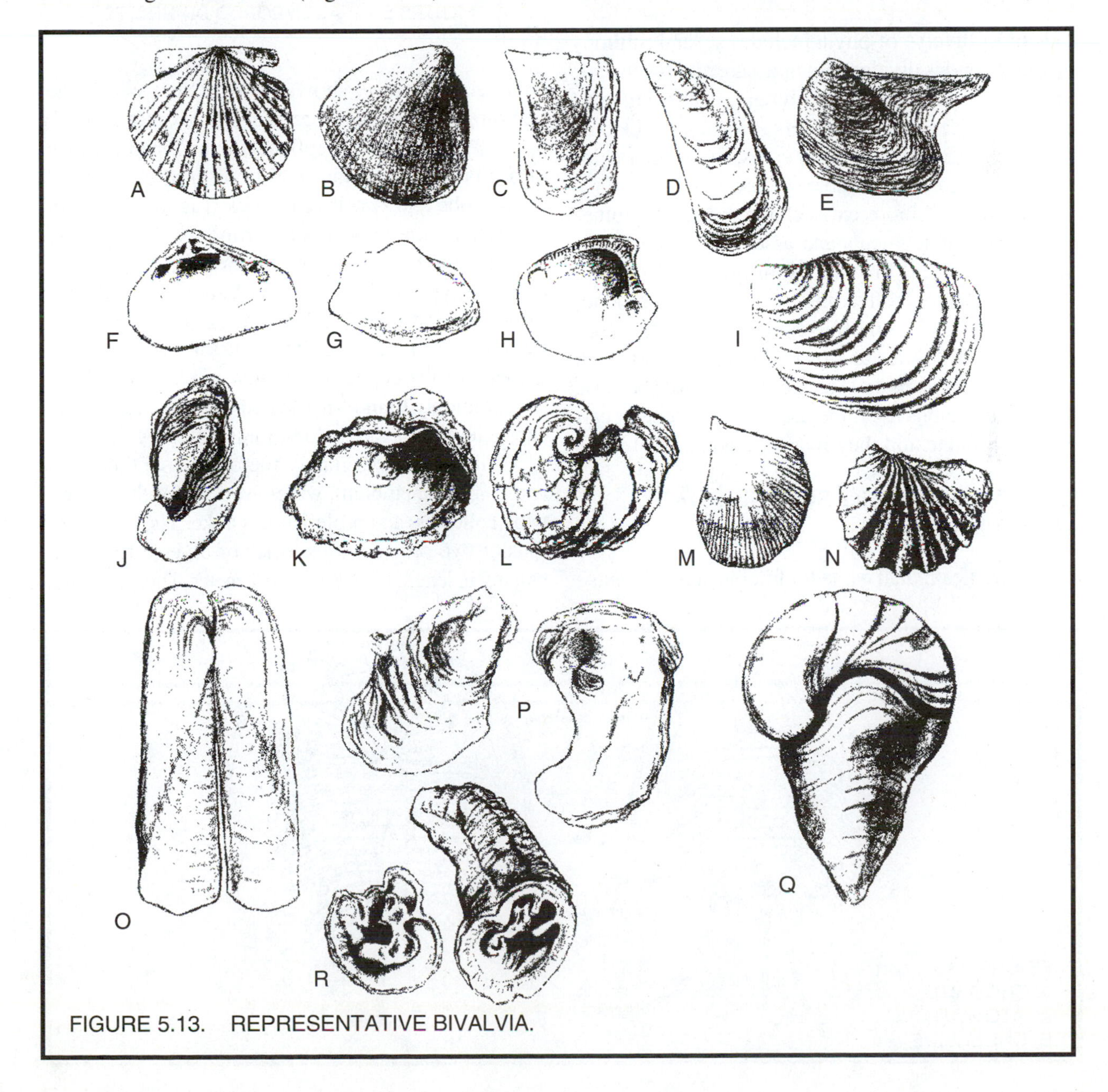

FIGURE 5.13. REPRESENTATIVE BIVALVIA.

The commonness of similar shell shapes among living and extinct groups indicates widespread parallel evolution (e.g., Carboniferous *Myalina* and Recent mussel *Mytilus*; Figure 5.13C,D). Rudists on the other hand, evolved truly bizarre shapes in which a smaller cap or lid-like valve fit on a large, conical one (Figure 5.13R), giving superficial resemblance to solitary corals (below). Gryphaeids (Figure 5.13J), and some rudistids have one large coiled valve and a smaller lid-like valve convergent with gastropods and coiled cephalopods (5.13Q). Other bivalves (e.g., oysters; Figure 5.13K,L,P) had extremely variable, irregular shell shapes.

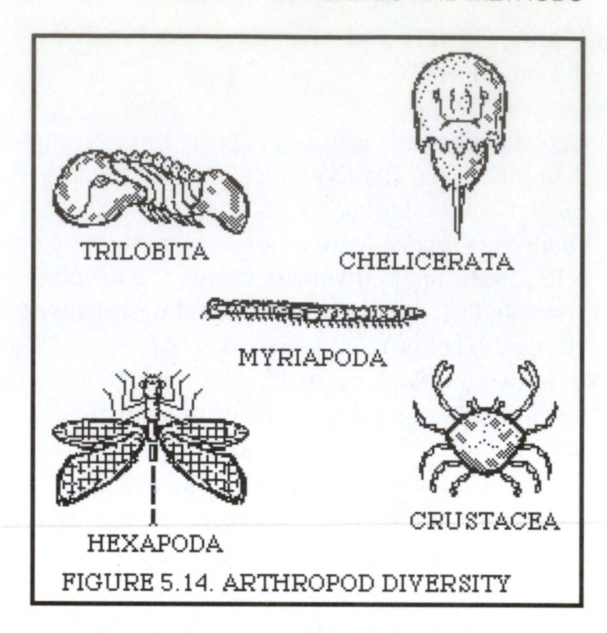

FIGURE 5.14. ARTHROPOD DIVERSITY

Crustaceans, Insects and Trilobites: Phylum Arthropoda ("Joint-Foot")

This most diverse of phyla includes several millions of species and is divided into nine superclasses including the familiar insects, lobsters, crabs, shrimp, millipedes, centipedes, spiders, scorpions, ticks, as well as barnacles, horseshoe crabs, the extinct trilobites, sea scorpions and more (Figure 5.14). All arthropods share a chitinous exoskeleton, bilateral symmetry, a series of segments and associated multifarious appendages, and growth by molting. Arthropod appendages evolved feeding, sensing, and reproductive as well as locomotive functions. Unlike other arthropods, trilobites and ostracodes (small bivalved crustaceans; Appendix I) are common fossils because their exoskeletons are mineralized with calcite or calcium phosphate and they live(d) in aquatic habitats.

"Little Water Bug Like Stone House In": Class Trilobita

The above Ute Indian name for trilobite amulets, *timpe khanitza pachavee*, shows how distinctive are the homologous features of arthropods, extinct and living. The name trilobite reflects the tri-lobed division of a bilaterally symmetrical body into one central (axial) lobe and two lateral lobes (Figure 5.15A,B). Like many other arthropods, trilobites have a body with three primary sections: 1) a head or cephalon; 2) a tail or pygidium comprised of smaller fused segments; and 3) a jointed midsection or thorax. Compound eyes are located on either side of the nose-like glabella of the cephalon in most species. Trilobites are commonly found fragmented because the molting exoskeleton separated along joints (sutures).

Trilobites are arguably the quintessential fossil and Paleozoic emblem, widely sought for their beauty by collectors and professionals alike. The class consists of two orders with species of diverse form and ranging in length from 1 mm to 1 meter. The Agnostida

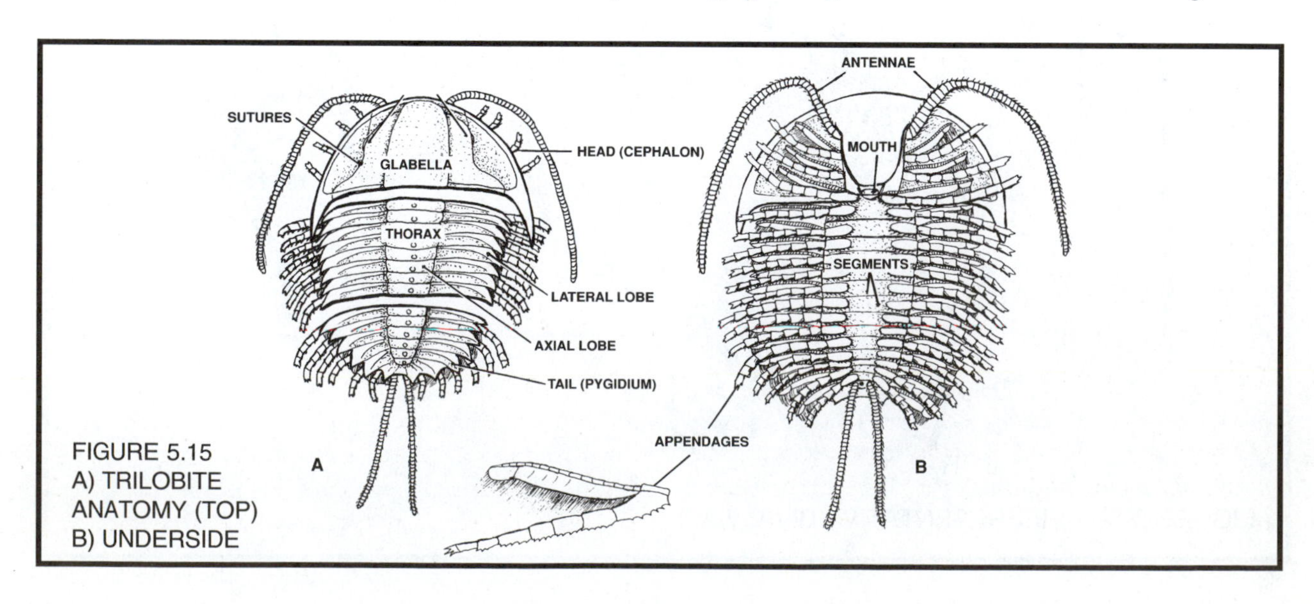

FIGURE 5.15
A) TRILOBITE ANATOMY (TOP)
B) UNDERSIDE

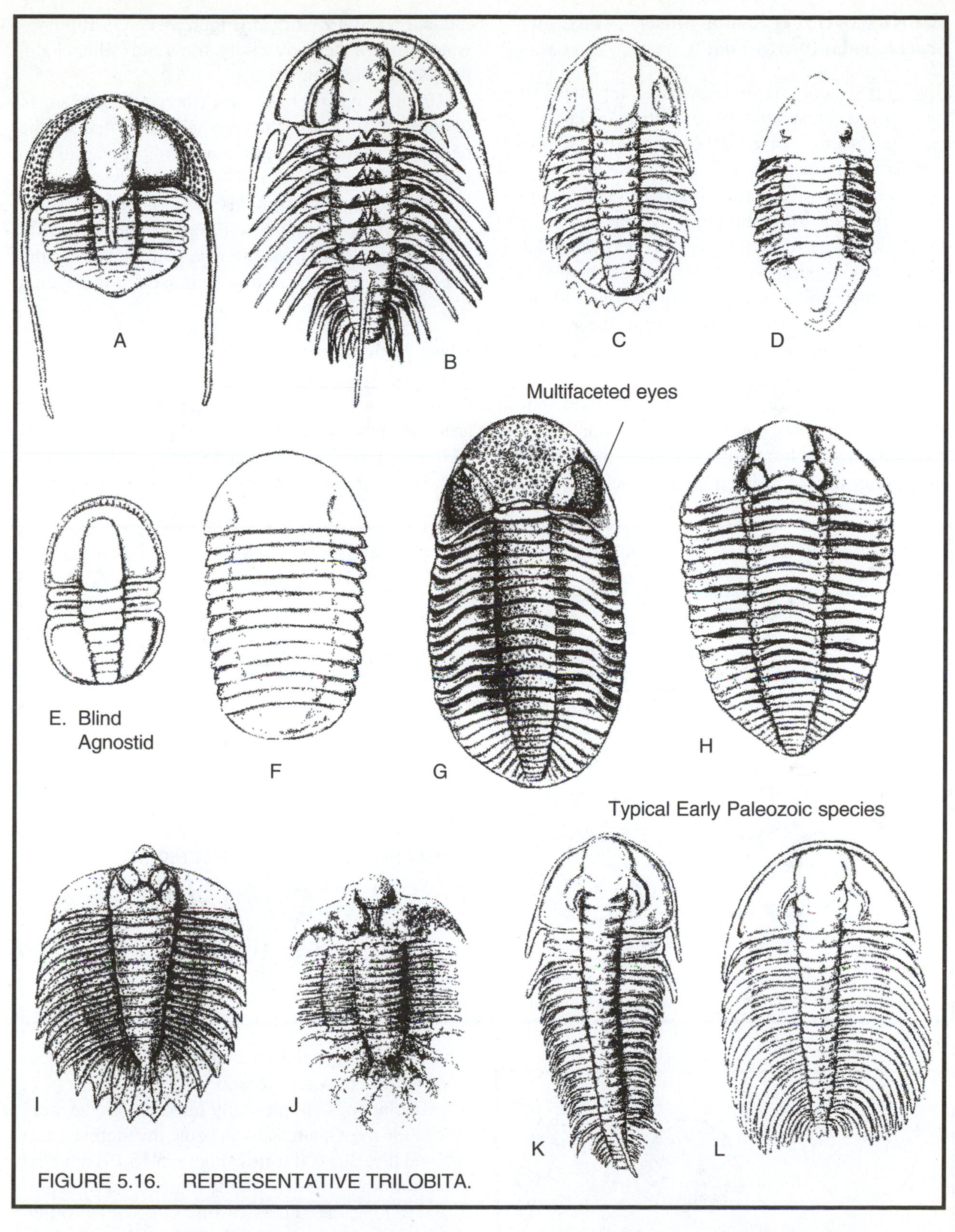

FIGURE 5.16. REPRESENTATIVE TRILOBITA.

are a group of tiny (less than 2 mm in length) species lacking all but two thoracic segments and most cephalic sutures (Figure 5.16E); many also lacked eyes. The Polymerida include all other trilobites, except those rare Cambrian "soft-bodied" species lacking a thoracic segment. Some species had smooth featureless exoskeletons, but others had hundreds of spines. Some had large multifaceted eyes while others were blind. Many early Paleozoic species had very many pygidial and thoracic segments; pygidial spines were also common, especially in Early Paleozoic species.

SUBKINGDOM II "Lamp Shells"— Phylum Brachiopoda: ("Arm-Foot")

Brachiopods, commonly known as lamp shells from some species' resemblance to Arabian oil lamps, are easily mistaken for bivalves because, although the soft parts are fundamentally different, these are enclosed by two hinged valves in both groups (Figure 5.17A, B). While both have bilateral symmetry, the brachiopods' plane of symmetry cuts perpendicular to the plane of separation and thus, across both valves through their apices and dividing each into identical halves (Figure 5.17B,18). In contrast, the symmetry plane of bivalves passes between valves, separating each complete valve.

Brachiopods are not related to bivalves, but to bryozoa (see below), coral-like colonial animals! The two share several homologous features including the lophophore (*lopho* = ridge; *phora* = bearing, a feeding and respiratory organ (Figure 5.17A)). It pumps water through the body cavity, traps and filters food, brings it to the mouth.

In most species one valve (the pedicle) bears an opening at its apex, through which protrudes the pedicle or stalk-like organ anchoring the animal to stable surfaces (Figure 5.17A,B).

The other valve contains the lophophore and its support structure, the brachidium (hence, brachial valve; *brachia* = arm). Interior structures near the valves' apices mark points of muscle attachments (Figure 5.17C).

Class Articulata—There are two classes of brachiopods: Articulata and Inarticulata. The Articulata (*artus* = joint) have valves composed of calcite that articulate along a hinge line and/or by internal projections ("teeth") on one, and corresponding "sockets" on the other (Figure 5.17A,C). Brachiopod shells tend to remain closed after death because the muscles close them when relaxed.

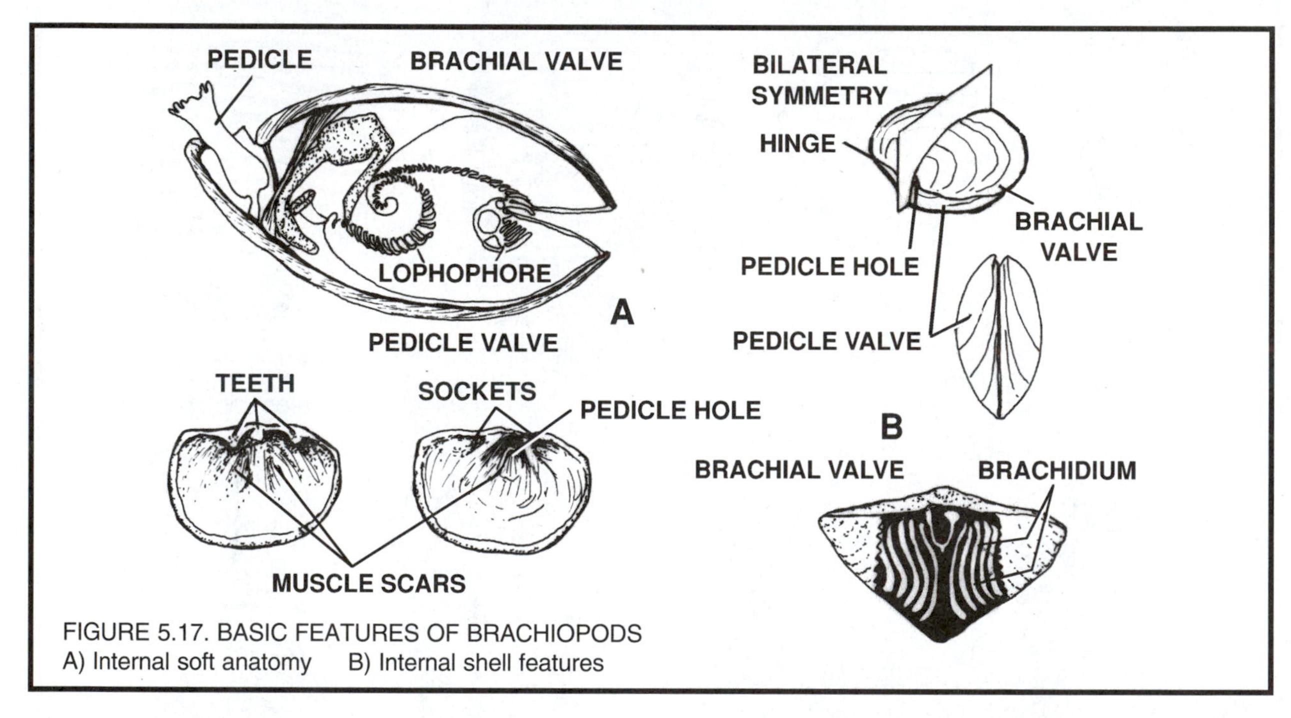

FIGURE 5.17. BASIC FEATURES OF BRACHIOPODS
A) Internal soft anatomy B) Internal shell features

Brachiopods are presently uncommon, but were among the most abundant Paleozoic invertebrates and evolved diverse shell forms (Figure 5.18,19). Parallel evolution was common among brachiopods and produced among the genera of one Ordovician order, nearly the entire range of shell shapes (as broadly defined; Figure 5.19B). One unusual group, the richthofenids evolved shells of one deep conical valve and one flat lid (e.g., *Proricthofenia*, Figure 5.19O) broadly convergent with conical rudistid bivalves and "horn" corals (see below).

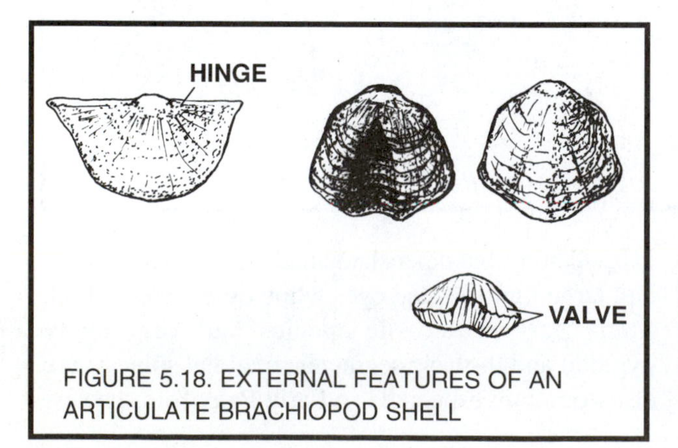

FIGURE 5.18. EXTERNAL FEATURES OF AN ARTICULATE BRACHIOPOD SHELL

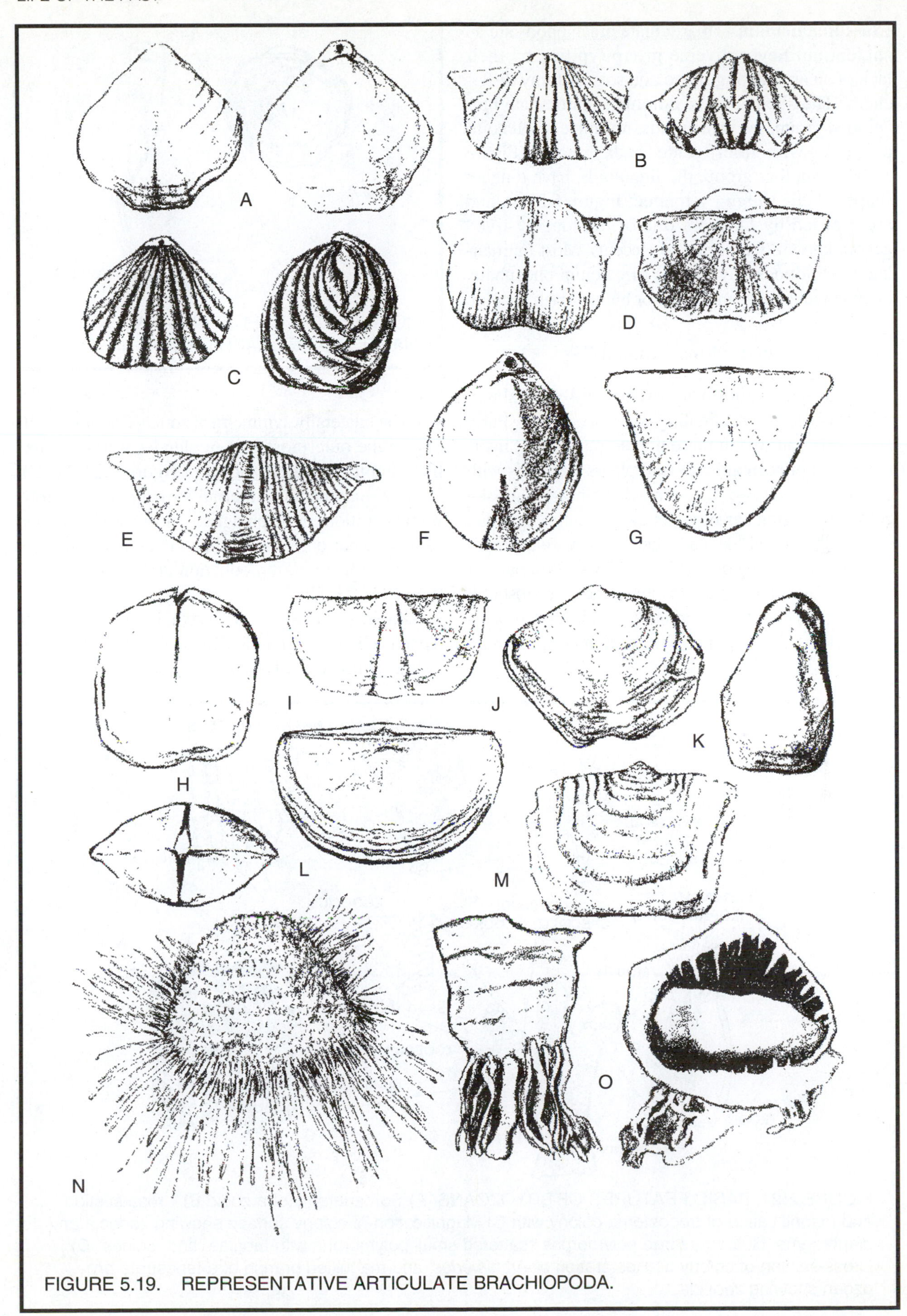

FIGURE 5.19. REPRESENTATIVE ARTICULATE BRACHIOPODA.

Class Inarticulata—Inarticulate brachiopods are an ancient but never diverse group typified by their lack of an articulating hinge and soft anatomy differences. The shells of one common inarticulate brachiopod group, the orbiculoids, resemble small compact discs with an off-center pedicle notch (Figure 5.20A). Another group, the linguloids, have tongue-shaped shells (*linguo* = tongue; Figure 5.20B), and are typified by *Lingula*, one of the longest-lived genera known (Cambrian to Recent). Most inarticulate brachiopods have shells of calcium phosphate, often porcelainous, and white or bluish-black in color.

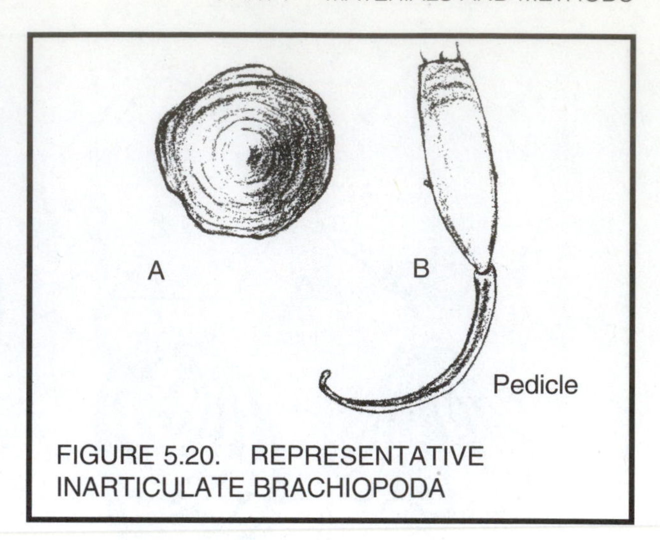

FIGURE 5.20. REPRESENTATIVE INARTICULATE BRACHIOPODA

Phylum Bryozoa: ("Moss Animals")

Bryozoa (*bryo* = moss) are exclusively colonial animals superficially resembling coral (see below). Early zoologists considered these tiny animals to be "intermediate" between animals and plants, or corals with extremely small zooids (a tentacled "individual" of a colony or aggregates of genetically identical zooids). Despite their coral-like endoskeleton, bryozoan zooids have a complex anatomy (Figure 5.21A). Without soft parts or microscopic skeletal study, they are most easily distinguished from corals by specialized zooecia (see below) and small (≤ 1 mm diameter) zooecial openings.

The bilaterally symmetrical zooids (Figure 5.21B) live in the outermost part of slender tubes or box-like chambers called zooecia (Figure 5.21C,D,E). Zooecia may contain internal structures (e.g., horizontal partitions or diaphragms). Most zooids are food gatherers, but others specialize in colonial functions such as protection, reproduction and waste removal (Figure 5.21B,C).

Zooids of a colony are asexually budded clones (genetically identical individuals) of the founding individual (Figure 5.21B,E). Colonies are produced

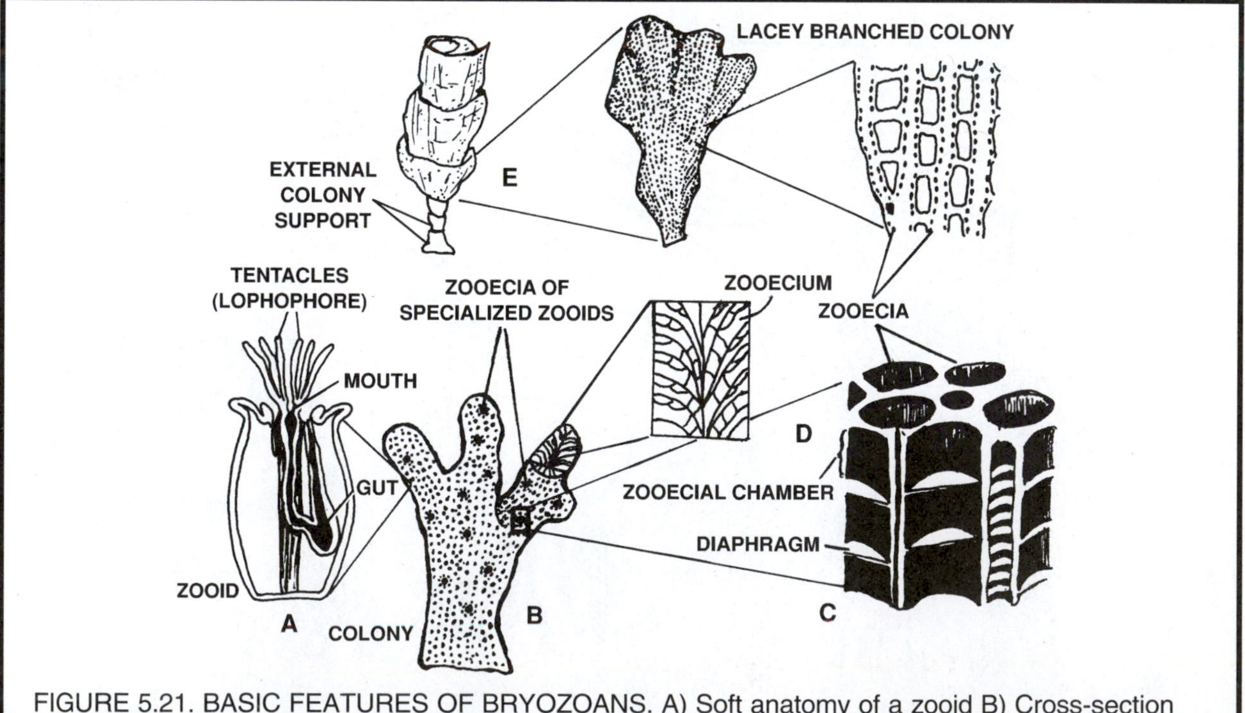

FIGURE 5.21. BASIC FEATURES OF BRYOZOANS. A) Soft anatomy of a zooid B) Cross-section and magnification of trepostome colony with C) Magnification of colony surface showing zooecia and diaphragms, cluster of large polymorphs scattered small polymorphs with tabulae, and "spines" D) Cross-section of colony E) Restoration of *Archimedes*, and magnified branch of a fenestrate bryozoan showing zooecia.

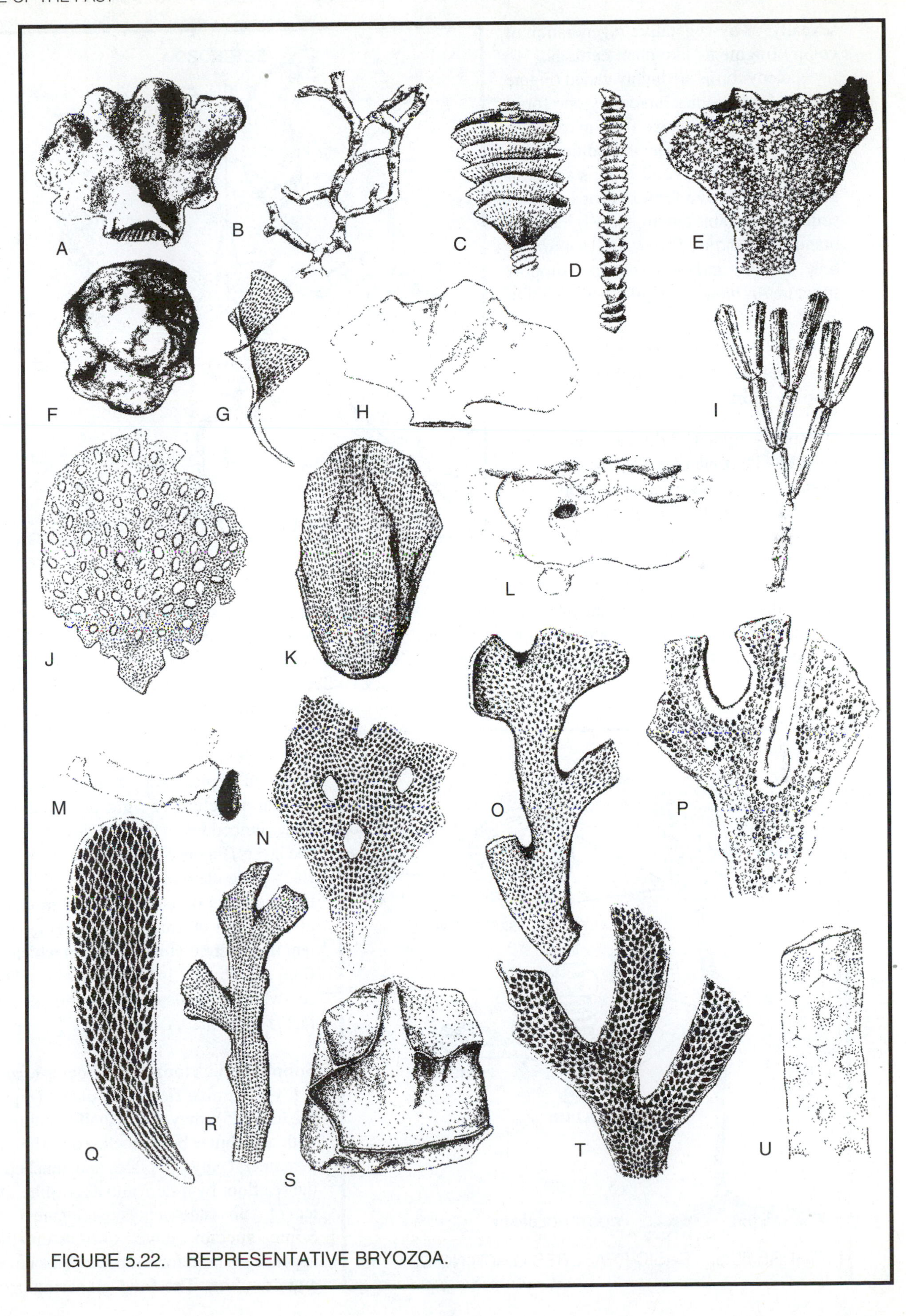

FIGURE 5.22. REPRESENTATIVE BRYOZOA.

sexually, or by vegetative regeneration of colony fragments like plant cuttings.

Colony forms are highly varied (Figure 5.22). Many species have only one form, but some species change form in response to particular environmental conditions. The colony forms of several groups are built around supportive (external or internal) structures. Notable among these is the lacy-branched frond of *Fenestella* (*fenestra* = hole; 5.22C) and its screw-like support structure *Archimedes* (Figure 5.5E, 5.22D).

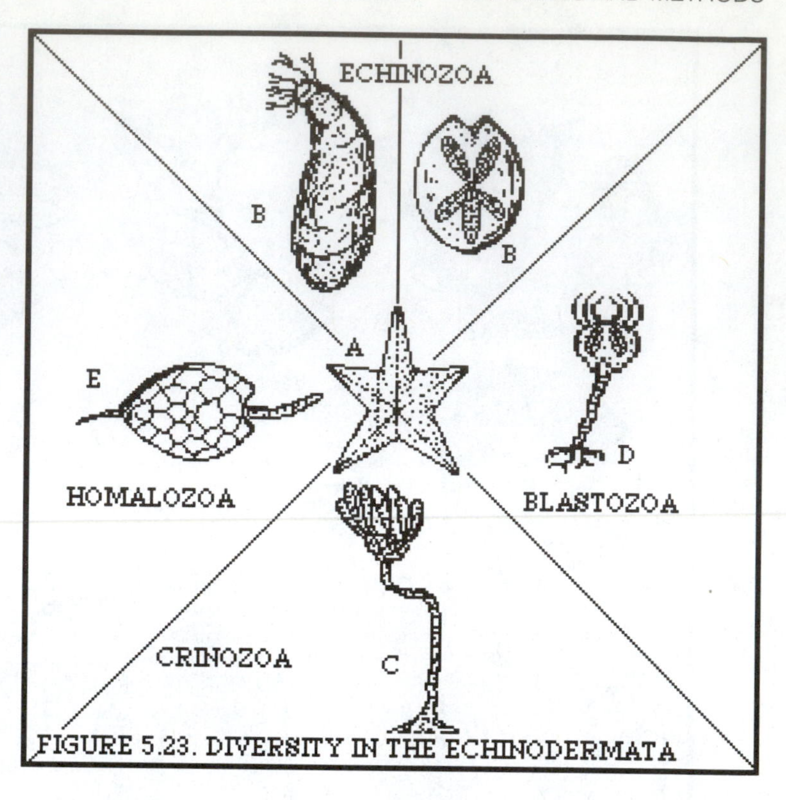

FIGURE 5.23. DIVERSITY IN THE ECHINODERMATA

Crinoids, Starfish, Sea Urchins, and Sand Dollars: Phylum Echinodermata ("Spiny Skin")

The spiny-skinned phylum are truly weird animals: with pentameral patterns impressed on intricate, delicate 'armor-plated flowers', an exceptional fossil evokes an otherworldly quality. Starfish, sea urchins and sand dollars are familiar echinoderms (Figure 5.23A-C) but crinoids, most abundant in the fossil record, are uncommon today. Only 5 of the 20 classes are living, and most species belong to one of 3 subphyla: the Crinozoa, Blastozoa and Echinozoa.

Echinoderms are easily recognized by their unique, primary five-fold (pentameral) symmetry, a character most readily seen in five-armed starfish (Figure 5.23A). Secondary (later evolved) bilateral symmetry is most apparent in mobile groups such as some urchins (Figure 5.23B). Other shared features, include a body cavity enclosed (typically) by a skeletal calyx or similar structure of rigid or articulating plates embedded in the middle of three tissue layers (Figure 5.23C, D, 5.24A, 5.29A.) Each plate cleaves like, and has the distinctive luster of, calcite. All echinoderms also share a unique water vascular system (i.e., a circulatory system in which the fluid salt content is balanced with that of sea water) connected to suckers on tube feet (Figure 5.29A).

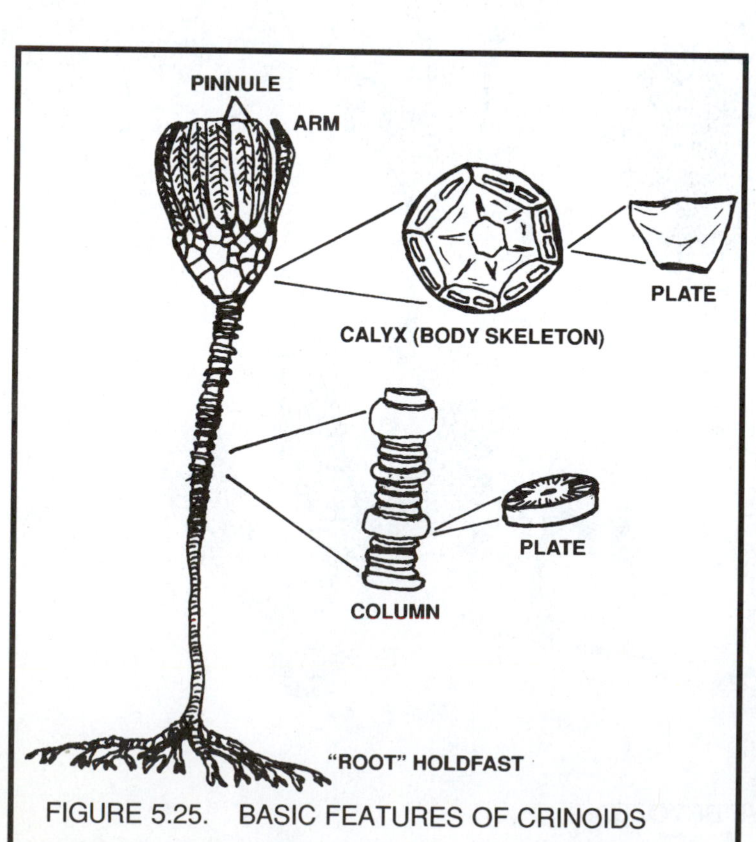

FIGURE 5.25. BASIC FEATURES OF CRINOIDS

Subphylum Crinozoa—Crinoids (*crinus* = lily) resemble flowering plants (Figure 5.23C, 5.25): they are (usually) stemmed with a globular body enclosed within the pentameral calyx of plates and attached to the sea floor by a stem and a root-like system of stems called a holdfast (Figure 5.25). Some species have only stem-like appendages used for temporary attachment and crawling. The food-gathering arms

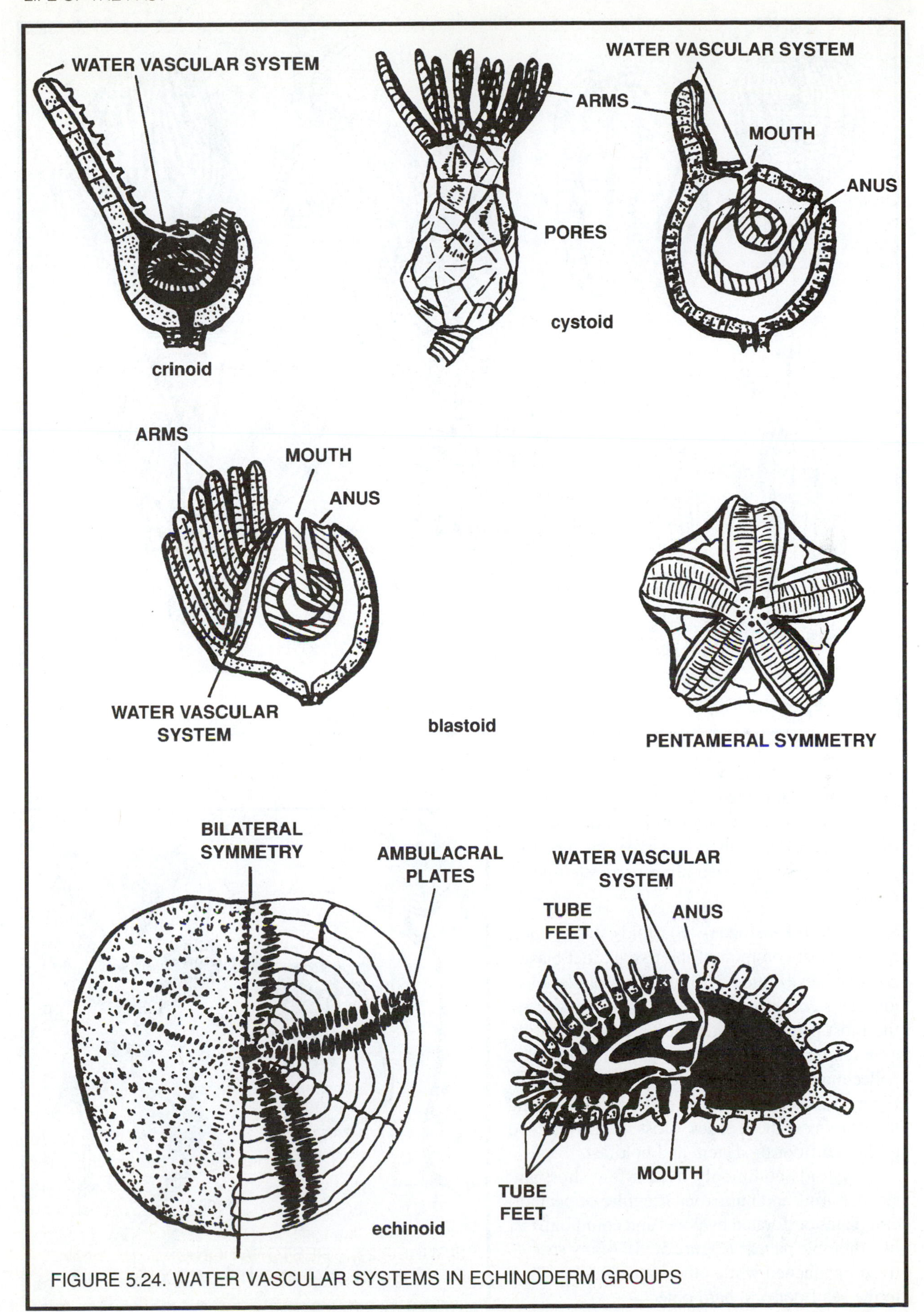

FIGURE 5.24. WATER VASCULAR SYSTEMS IN ECHINODERM GROUPS

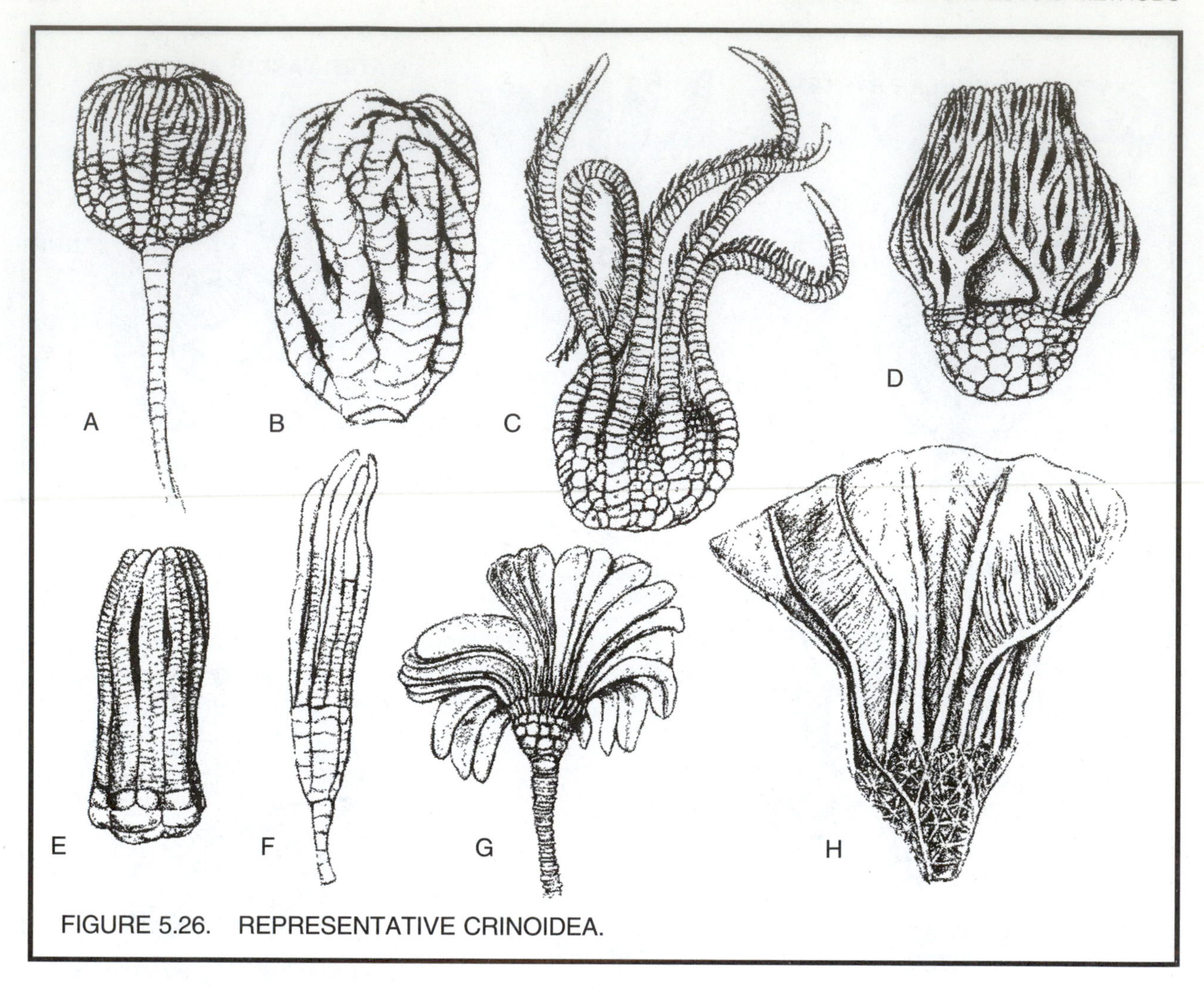

FIGURE 5.26. REPRESENTATIVE CRINOIDEA.

have tube feet and may be finely divided into pinnules resembling a feather or petal (Figure 5.26). The petal-like arms, stem and holdfast consist of circular to pentameral or irregular disc-like plates (columnals) resembling perforated poker chips.

Subphylum Blastozoa—Blastoids and cystoids are the most common of the five extinct blastozoan classes (Figure 5.23D). The globular blastoid calyx contains only a few tightly bound plates, the most prominent of which are pentamerally arranged ambulacral plates holding the food collecting organ and tubed-feet (e.g., *Pentremites*, Figure 5.27). The small "arms" are located along the margins of these plates. Most species attached to the sea floor by a stem and holdfast.

Cystoids commonly have a few short and slender arms and numerous irregular or pentameral plates perforated by pores and commonly set in a rhombic pattern (Figure 5.28). Many species lived unattached while others attached by stems to the sea floor or a hard object.

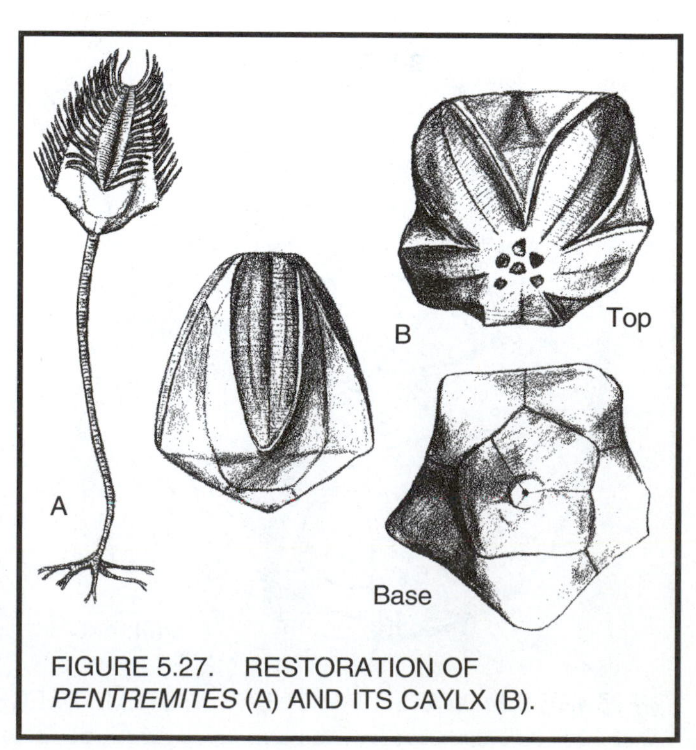

FIGURE 5.27. RESTORATION OF *PENTREMITES* (A) AND ITS CAYLX (B).

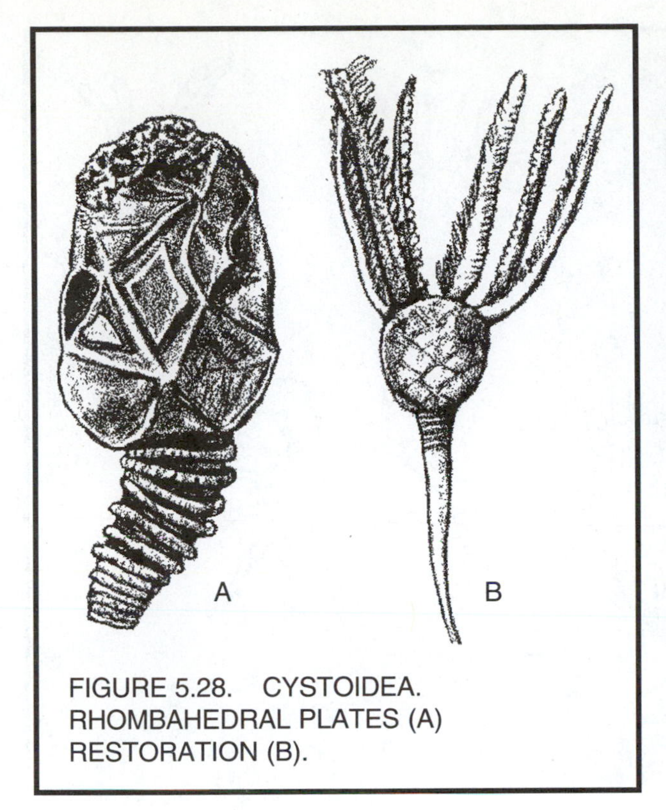

FIGURE 5.28. CYSTOIDEA. RHOMBAHEDRAL PLATES (A) RESTORATION (B).

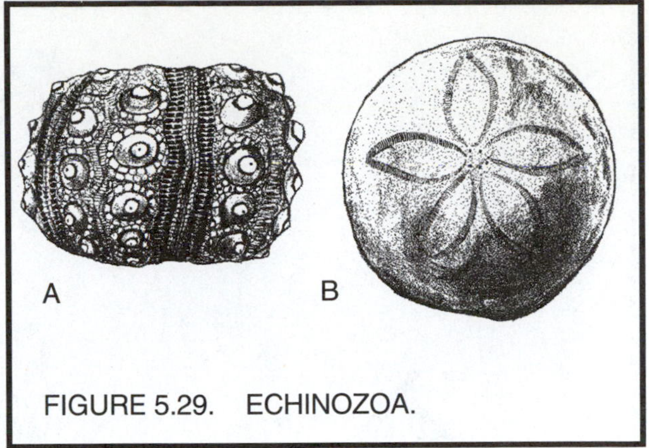

FIGURE 5.29. ECHINOZOA.

Subphylum Echinozoa—Of this group, only the Mesozoic and Cenozoic echinoids (sea urchins and sand dollars) are important in the fossil record (Figure 5.29, 5.23B). Their globular or flattened bilaterally symmetrical skeletons are made of many tightly bound plates. Secondary pentameral symmetry is obvious in the star-shaped ambulacra of some urchins. Most echinoids bear many spines and lack appendages and feeding arms.

Phylum Hemichordata, Class Graptolithina ("Engraved Stones")

Graptolites are an important, entirely extinct group of colonial animals classified with some minor living groups in the phylum Hemichordata (*hemi* = half; animals having a notochord for only part of their life cycle), close relatives of echinoderms and distantly related to chordates (vertebrates and relatives).

Graptolite skeletons, and by inference their soft parts, are very similar to the living pterobranch hemichordates, and probably held zooids with tentacles (Figure 5.30A). The exoskeleton consisted of tubular branches each supporting one or two rows of thecae (zooid chambers; Figure 5.30B). The thin graptolite skeletons (Figure 5.31.) are most commonly preserved on slabs as shiny carbonized films resembling engravings and hence, the name graptolite (grapto = written/ engraved; lithos = stone).

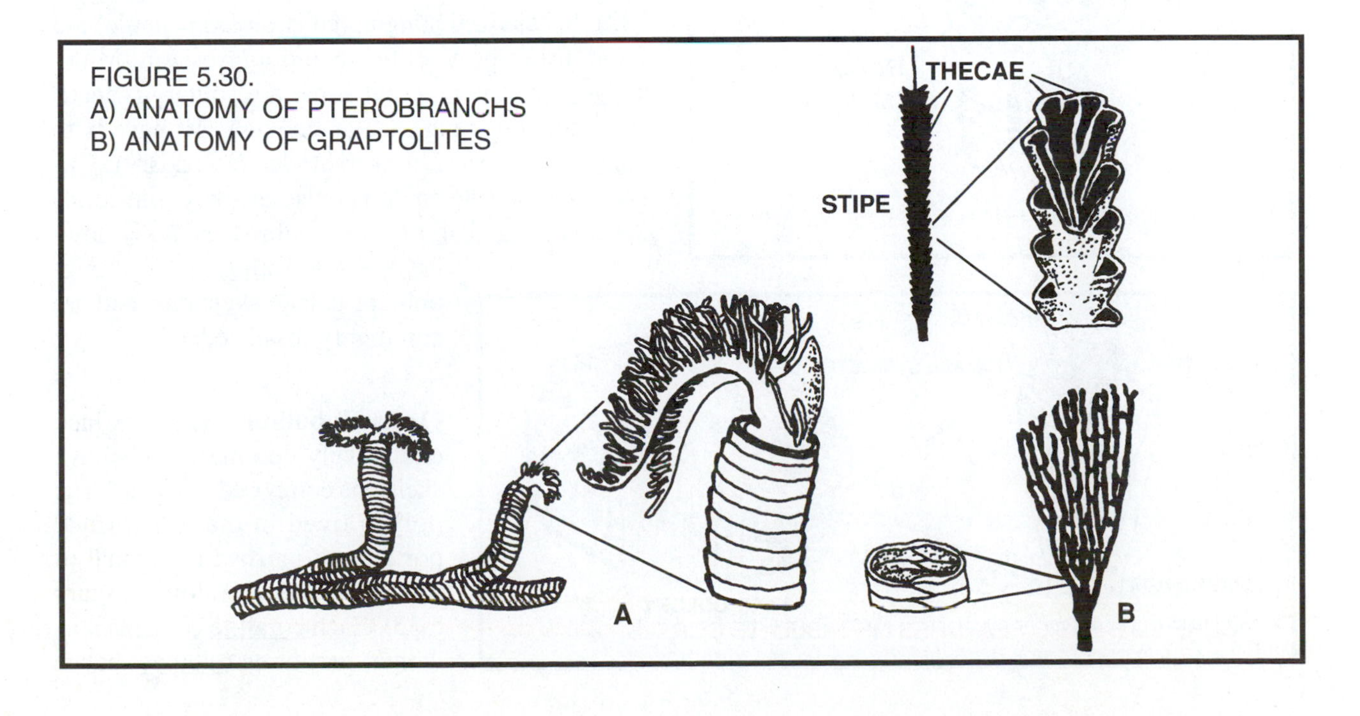

FIGURE 5.30.
A) ANATOMY OF PTEROBRANCHS
B) ANATOMY OF GRAPTOLITES

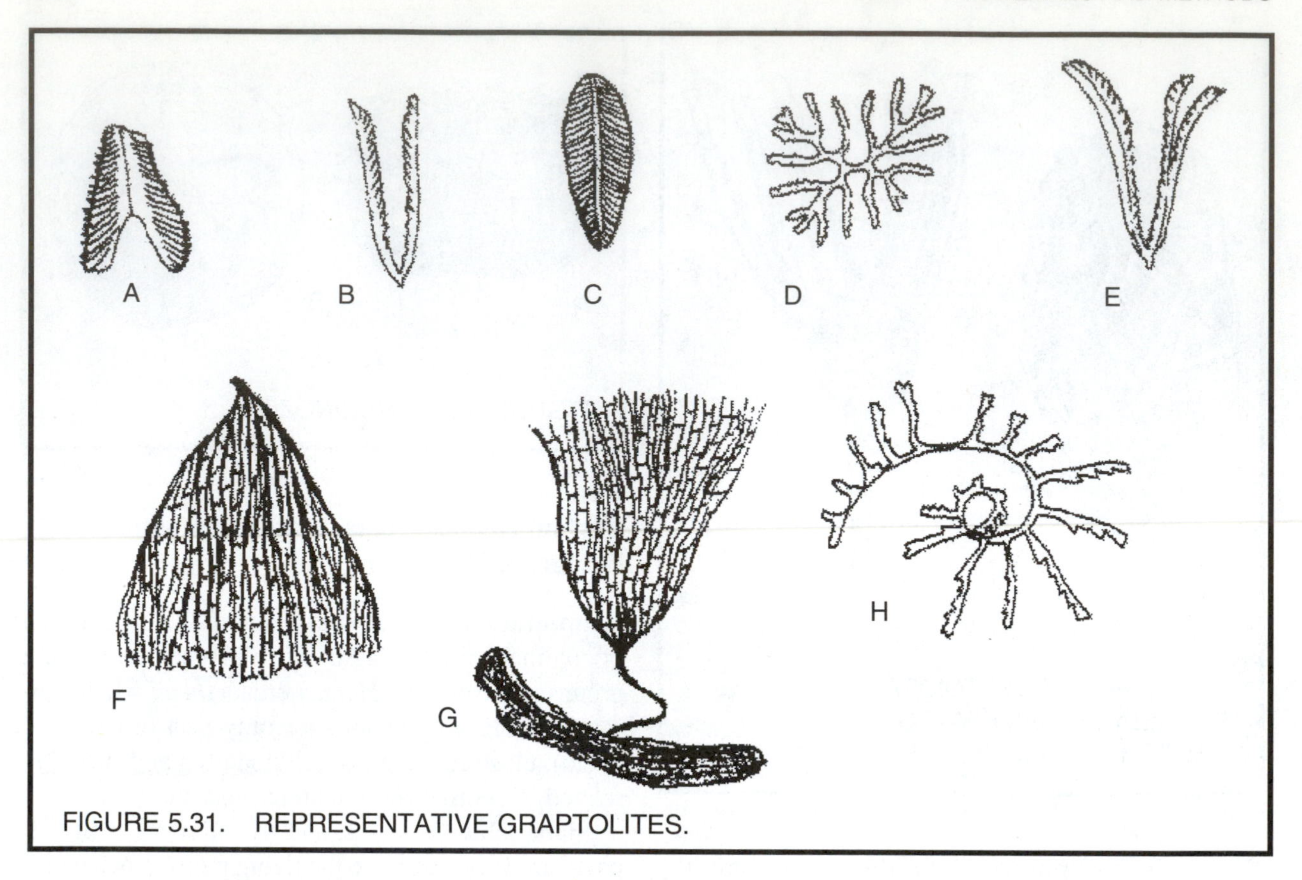

FIGURE 5.31. REPRESENTATIVE GRAPTOLITES.

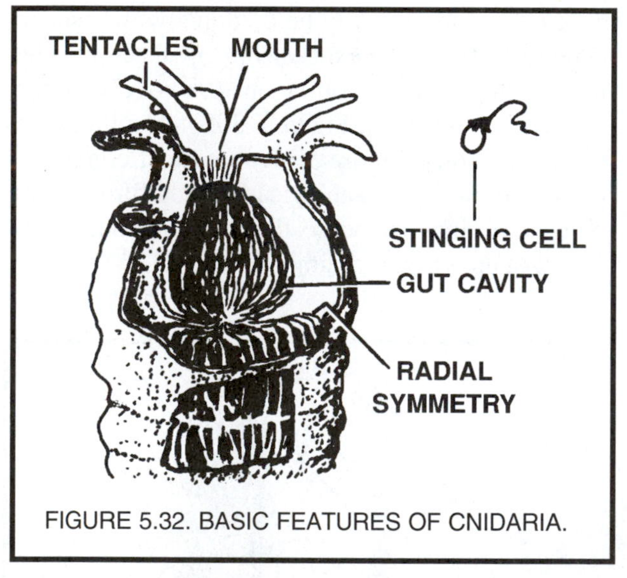

FIGURE 5.32. BASIC FEATURES OF CNIDARIA.

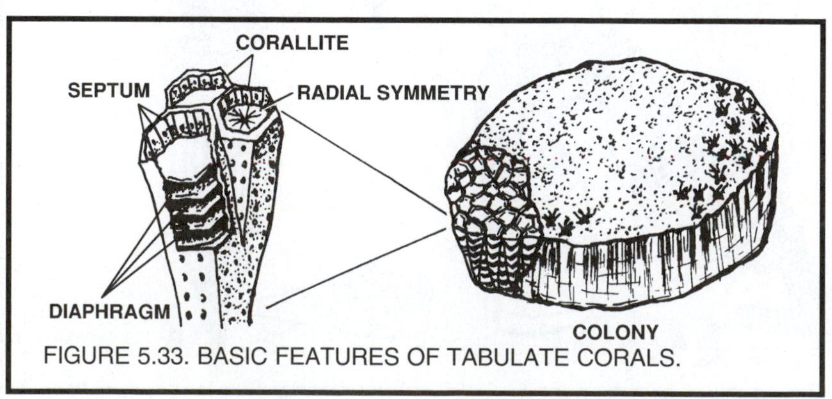

FIGURE 5.33. BASIC FEATURES OF TABULATE CORALS.

Corals, Sea Anemones, Hydra, and Jellyfish: Phylum Cnidaria ("Nettles-bearers")

Corals, hydroids, sea anenomes, and jellyfish belong to this phylum, also known as Coelenterata. These simple animals have only two tissue layers and no organs; just a mouth surrounded by tentacles, a "gut" cavity, simple muscles and a nerve network (Figure 5.32). Specialized stinging cells (*cnida* = nettle) aid in capturing prey, defense and aggression. Many cnidaria are colonial, but some are solitary (genetically individual; e.g., sea anemones). All have both radial (primary) and bi-lateral/radial (secondary) symmetry. Of the five cnidarian classes, only true corals (subclass Zoantharia ("*zoo* = animal; *antho* = flower")) of the Anthozoa, have aragonite or calcite skeletons, and are commonly fossilized.

Order Tabulata—These extinct, exclusively colonial corals have skeletons composed of calcite. The polyps lived in the last formed portions of narrow, polygonal or rounded tubes (corallites; Figure 5.33). Each corallite contains horizontal partitions (tabulae; hence

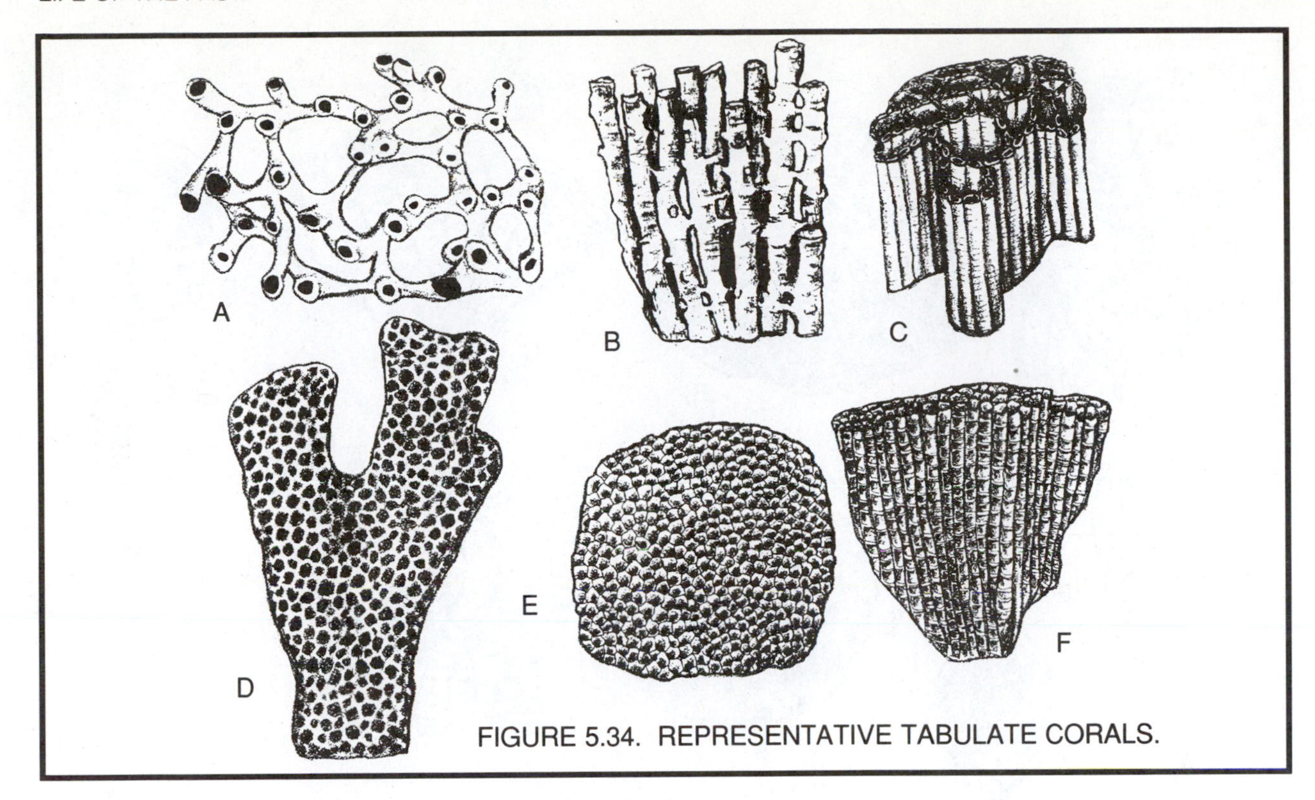

FIGURE 5.34. REPRESENTATIVE TABULATE CORALS.

Tabulata). Radially arranged, short, low ridges or spines along the interior walls of corallites are inconspicuous remnants of septa and thus, radial symmetry. Some colony form varieties are shown in Figure 5.34.

Order Rugosa—Like tabulate corals, Rugosa corals are extinct, have skeletons composed of calcite, and have colonial species resembling one another (Figure 5.36). Other species were large, solitary polyps secreting horn-shaped skeletons, and are thus called "horn corals"(Figure 5.35B). Rugosa corallites have exteriors roughened by daily, monthly and annual growth lines known as rugae ("wrinkles"), hence, the name Rugosa.

The polyp lived within the corallites' cup-like depression, and its gut cavity was partitioned by radially arranged skeletal septa of different lengths (Figure 5.35). Tabulae are usually present in the center of the corallite, often with blister-like partitions between them.

Various colony forms are shown in Figure 5.36. Corallites of some are contiguous and polygonal like

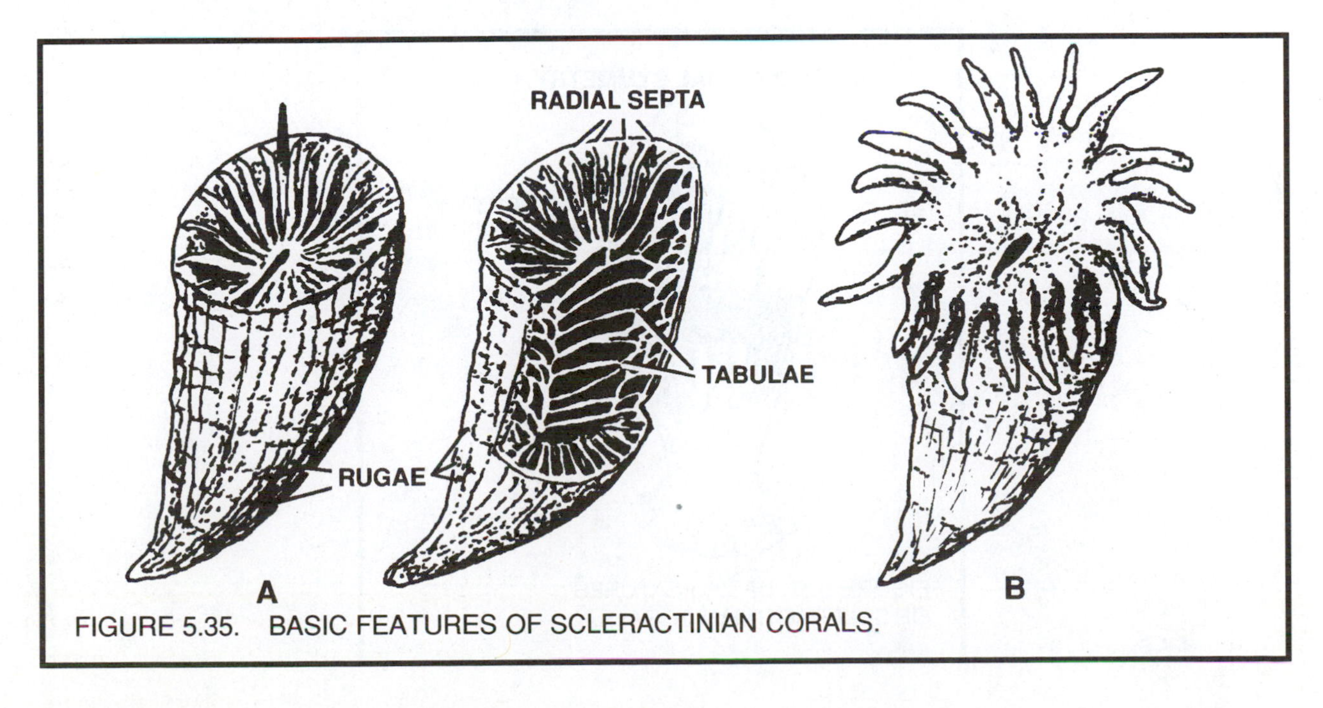

FIGURE 5.35. BASIC FEATURES OF SCLERACTINIAN CORALS.

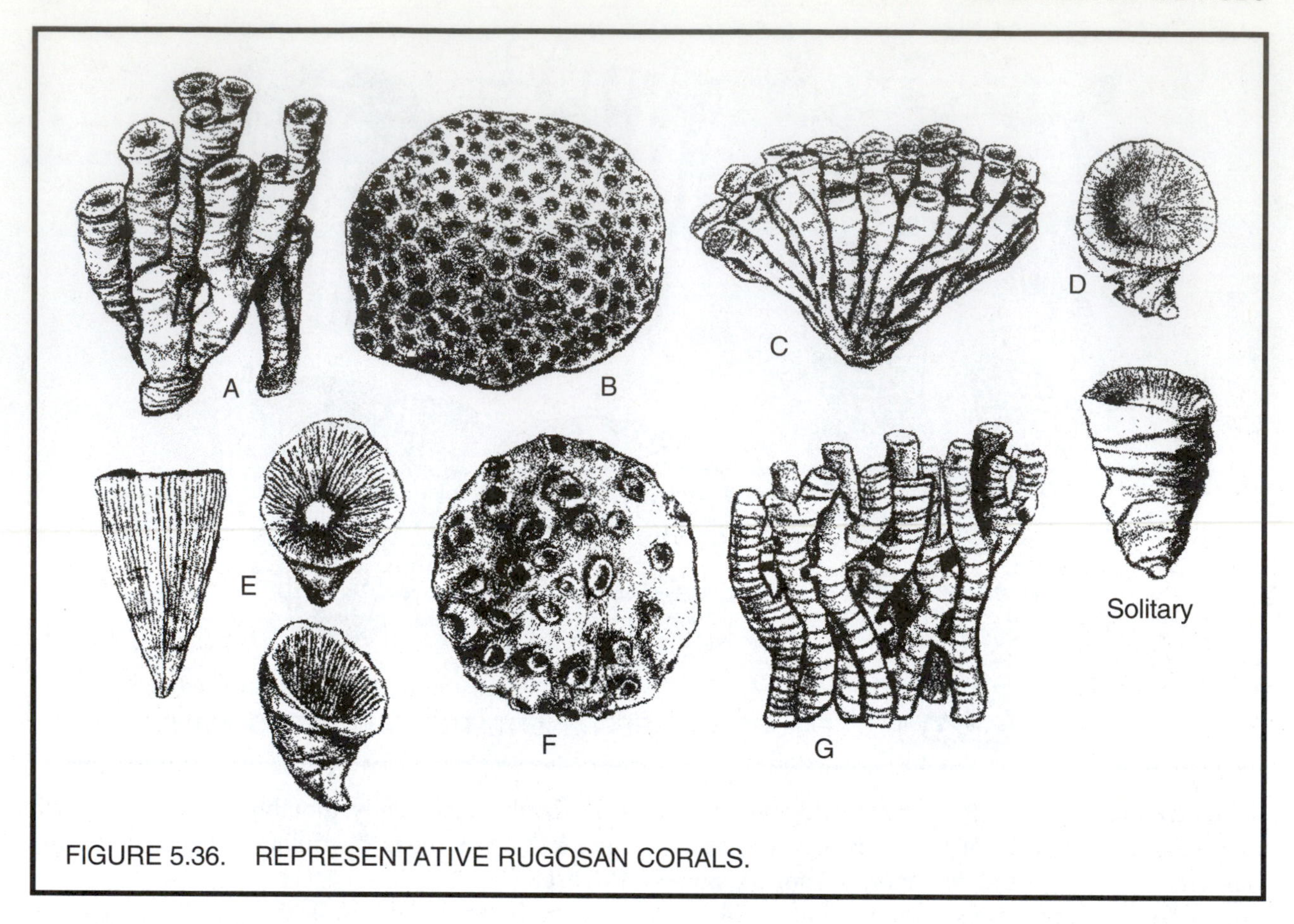

FIGURE 5.36. REPRESENTATIVE RUGOSAN CORALS.

that of *Hexagonaria*,—the fossil, which when cut and polished, is sold as the decorative "Petosky stone" Figure 5.36B.

Order Scleractinia—Scleractinia (*sclera* = hard) are Cenozoic reef-building corals. Rapid, profuse reef-forming growth is possible because their tissues harbor symbiotic algae that provide extra nutrients (Laboratory 6). Scleractinians have aragonite skeletons, dispersed corallites, and numerous septa (5.37). Six major septa give the corallites a six-fold radial symmetry. Most scleractinian corals are colonial and their forms (including the familiar brain coral and

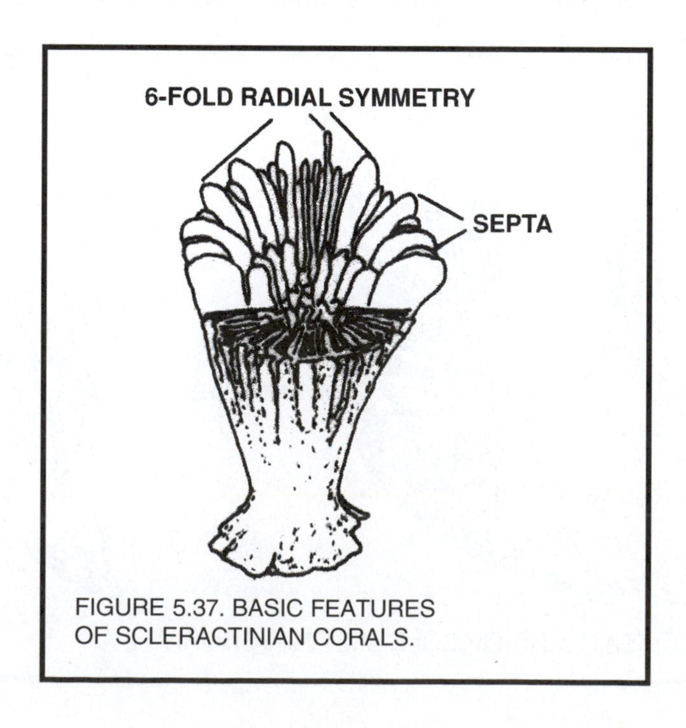

FIGURE 5.37. BASIC FEATURES OF SCLERACTINIAN CORALS.

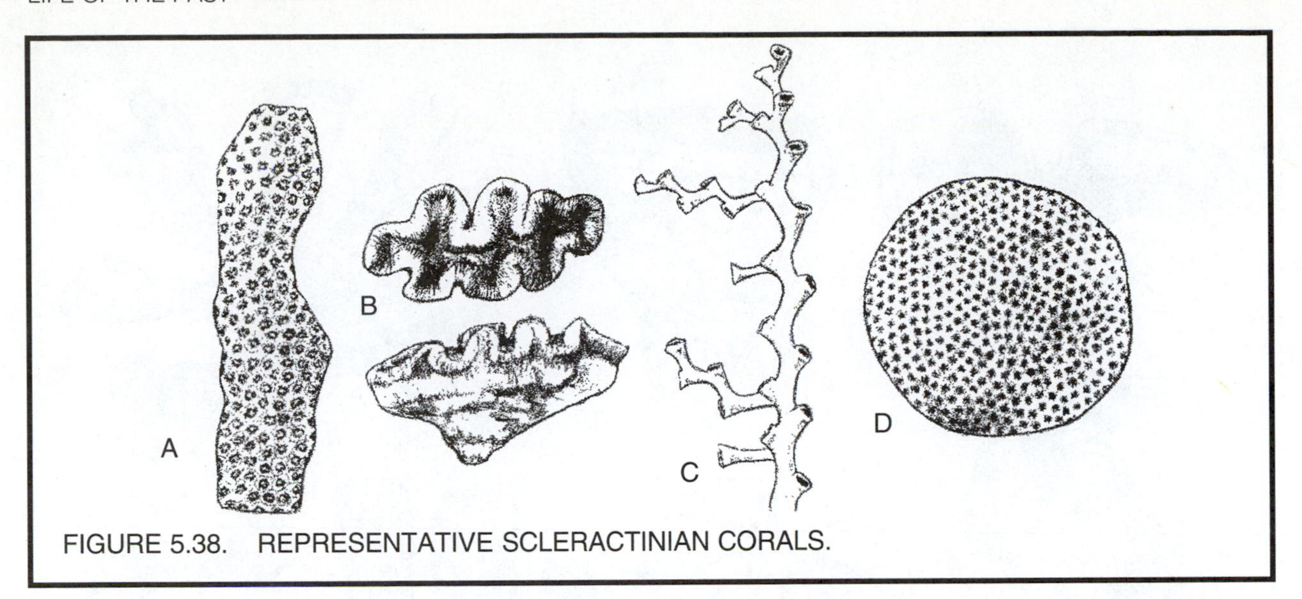

FIGURE 5.38. REPRESENTATIVE SCLERACTINIAN CORALS.

branching staghorn coral) are more varied than those of the Rugosa or Tabulata (Figure 5.38).

Sponges—Phylum Porifera: ("Pore-bearers")

These simplest of multicellular animals are little more than aggregates of slightly differentiated cells organized in poorly defined layers and only very distantly related to other animal phyla. There are no organs —only a system of pore-like canals (hence Porifera) within the skeleton, through which food-bearing water is drawn and expelled from the central cavity (Figure 5.40A). Sponge skeletons may consist of layers of calcite, or needle-like spicules of silica or organic fibers that serve as girder-like supports calcite (Figure 5.40B).

Of the sponges shown in Figure 5.40 J & L, are two common groups of calcareous sponges once classified as corals: the extinct tabulate coral-like chaetetids, and the Stromatoporata (*stroma* = layered; *pora* = pore), a particularly important extinct group of reef-builders. Stromatoporoids lack spicules and features more distinctive than laminations linked by calcareous pillars, cysts and tubes.

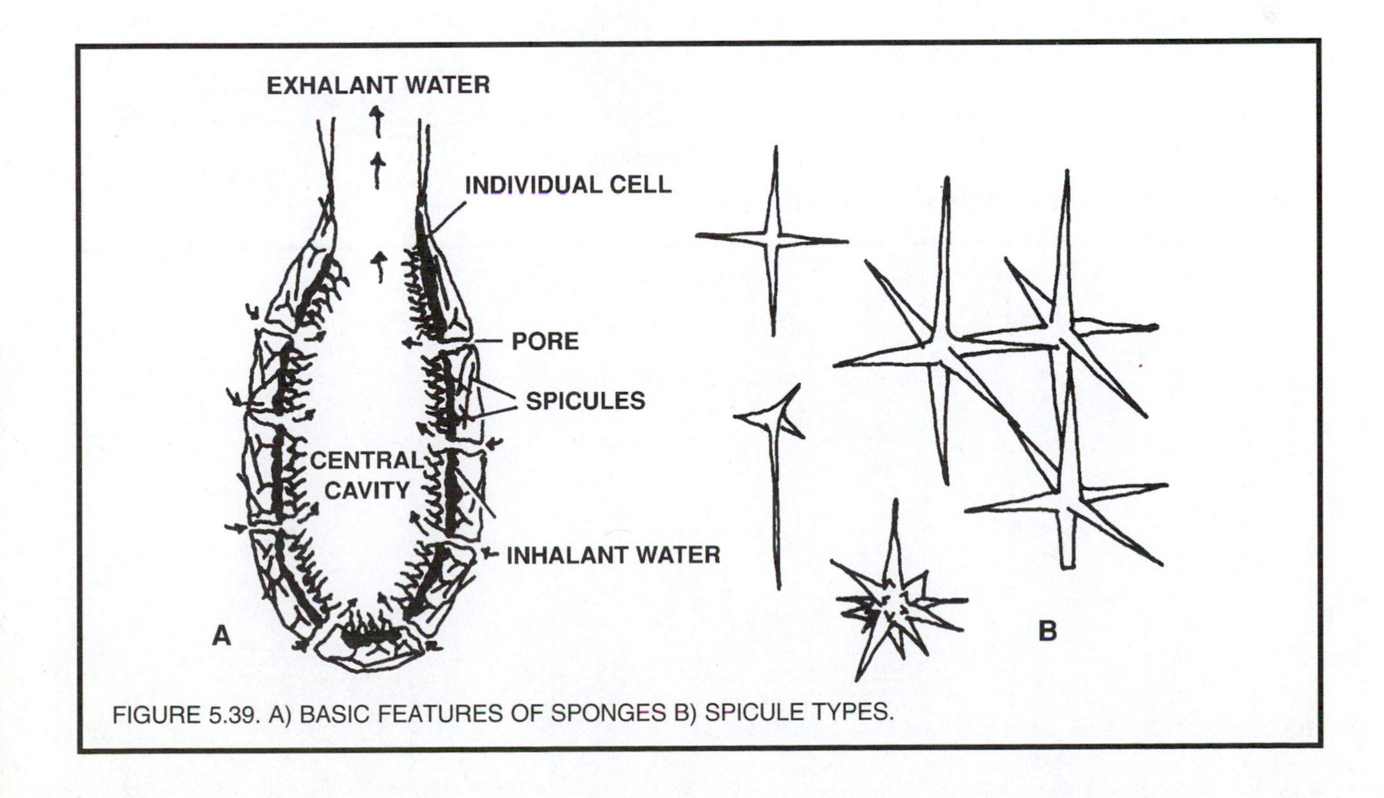

FIGURE 5.39. A) BASIC FEATURES OF SPONGES B) SPICULE TYPES.

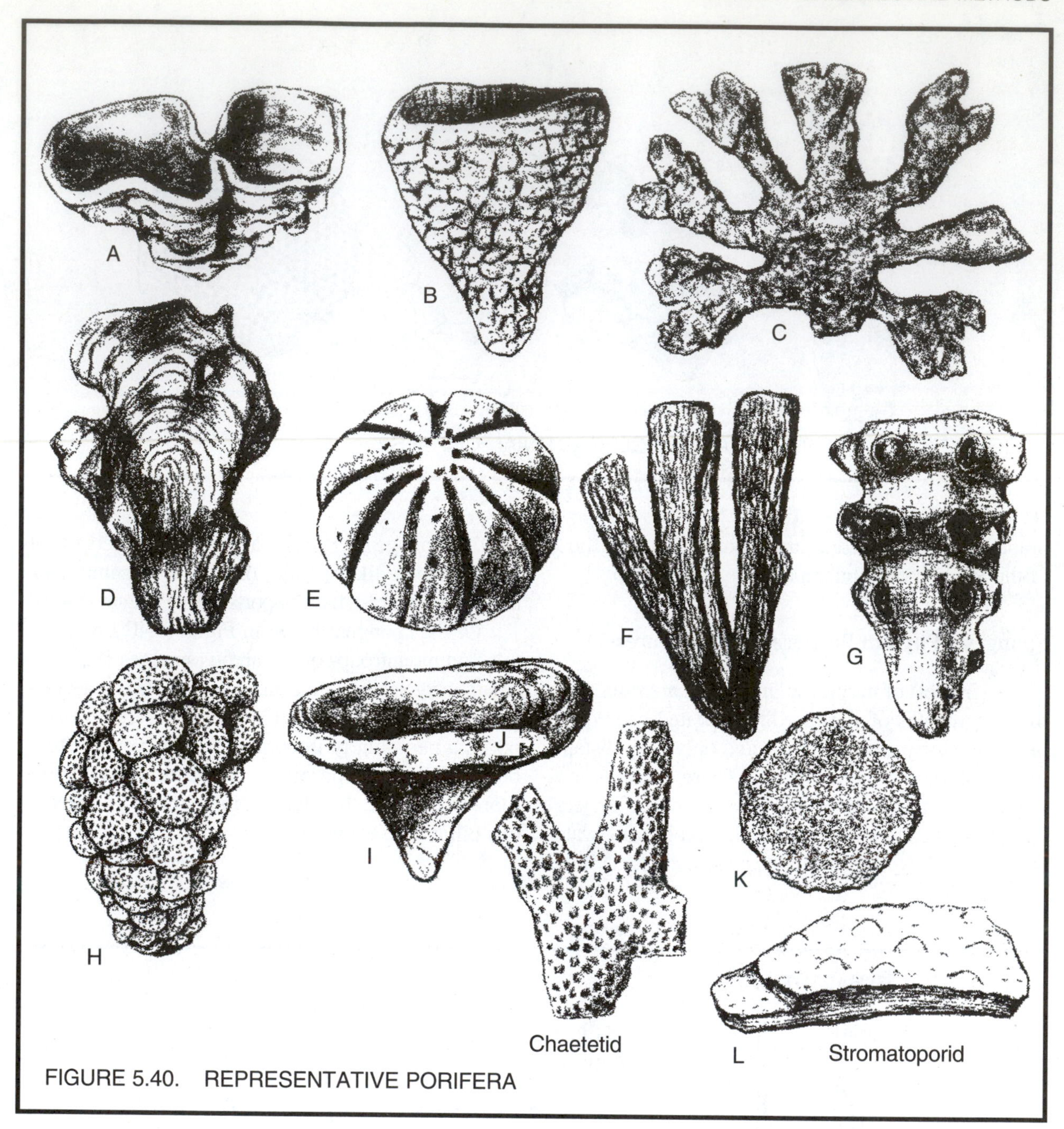

FIGURE 5.40. REPRESENTATIVE PORIFERA

VOCABULARY

Analogous feature, Carbonization, Cast, Convergent evolution, External mold, Genus, Homologous feature, Internal mold, Mold, Paleontology, Parallel evolution, Permineralization, Petrification, Recrystallization, Species, Trace fossil.

ANATOMICAL FEATURES

Ambulacra, Arm, Bilateral symmetry, Calyx, Cephalon, Colonial, Columnal, Corallite, Endoskeleton, Exoskeleton, Foot, Holdfast, Lophophore, Mantle, Muscle scars, Pedicle, Pedicle opening, Pentameral symmetry, Plate, Polyp, Pygidium, Radial symmetry, Ruga(e), Septum, Solitary, Spicule, Stem, Stipe, Stinging cell, Septa, Siphuncle, Suture, Thorax, Tabula(e), Theca, Tube Feet, Water Vascular System, Zooecium, Zooid.

ORGANIC GROUPS (Phyla in Boldface)

Subkingdom I

Arthropoda: Trilobita;
Mollusca: Cephalopoda—Ammonoidea, Nautiloidea, Coleoidea Belemnitida, Gastropoda, Bivalvia.

Subkingdom II

Brachiopoda: Articulata, Inarticulata; **Bryozoa**;
Echinodermata: Crinozoa, Blastozoa, Echinozoa;
Hemichordata: Graptolithina.
Cnidaria: Zoantharia—Anthozoa, Rugosa, Tabulata Scleractinia
Porifera: Stromatopora.

LABORATORY EXERCISES

5.1. Fossil Identification and Classification Self-Quiz—Species representative of various groups represented as fossils are listed to the left on worksheet 5.1, and a series of icons are shown to the right.

A) Circle those icon(s) in each row which match(es) the group named at the right.

B) Circle those icons in each row which have the skeletal type (endo/exo-skeleton) listed at the left.

C) Circle those icons in each row which have the symmetry type listed.

D) Circle those icons in each row which have the homologous features listed.

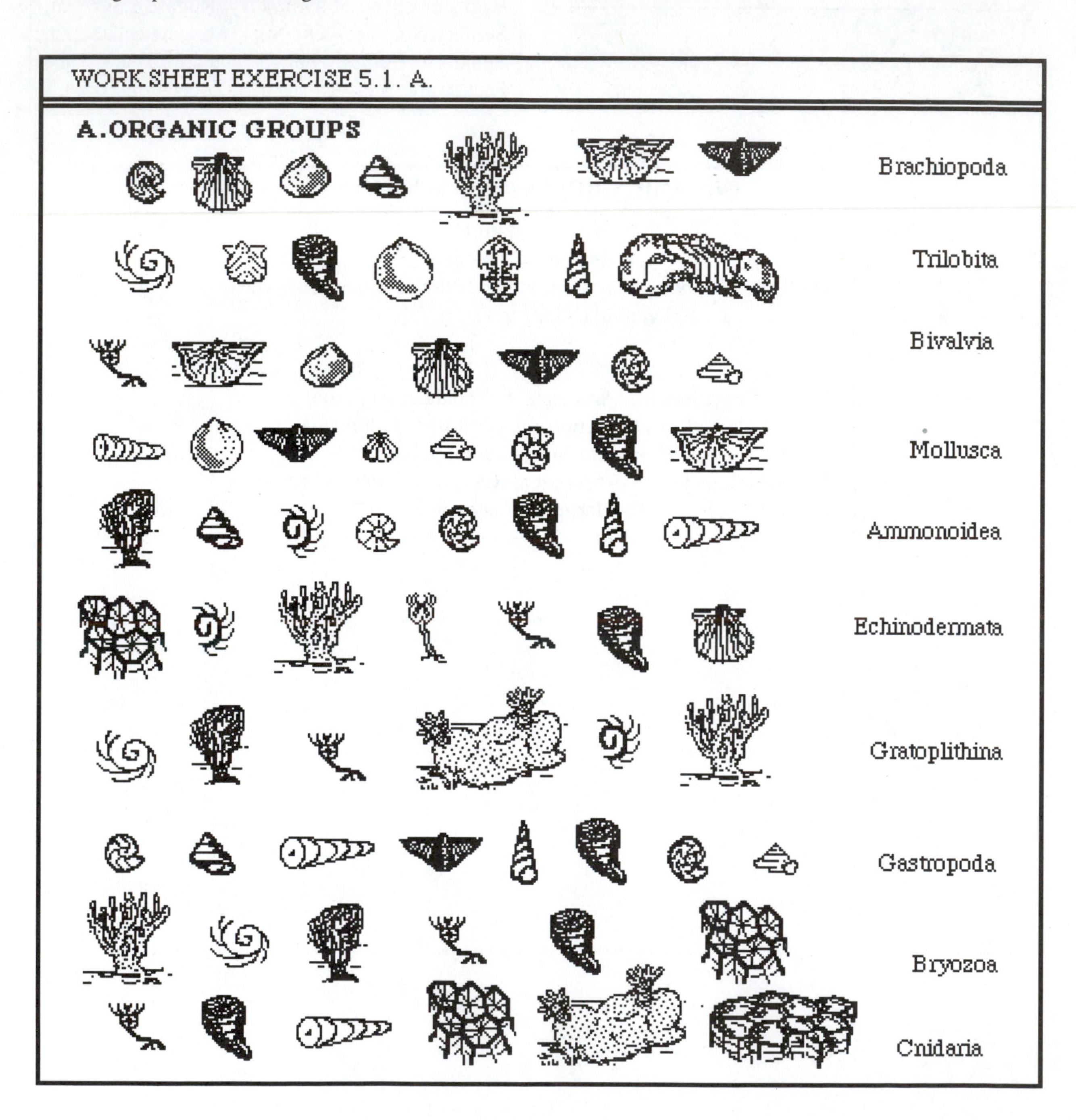

LABORATORY EXERCISES

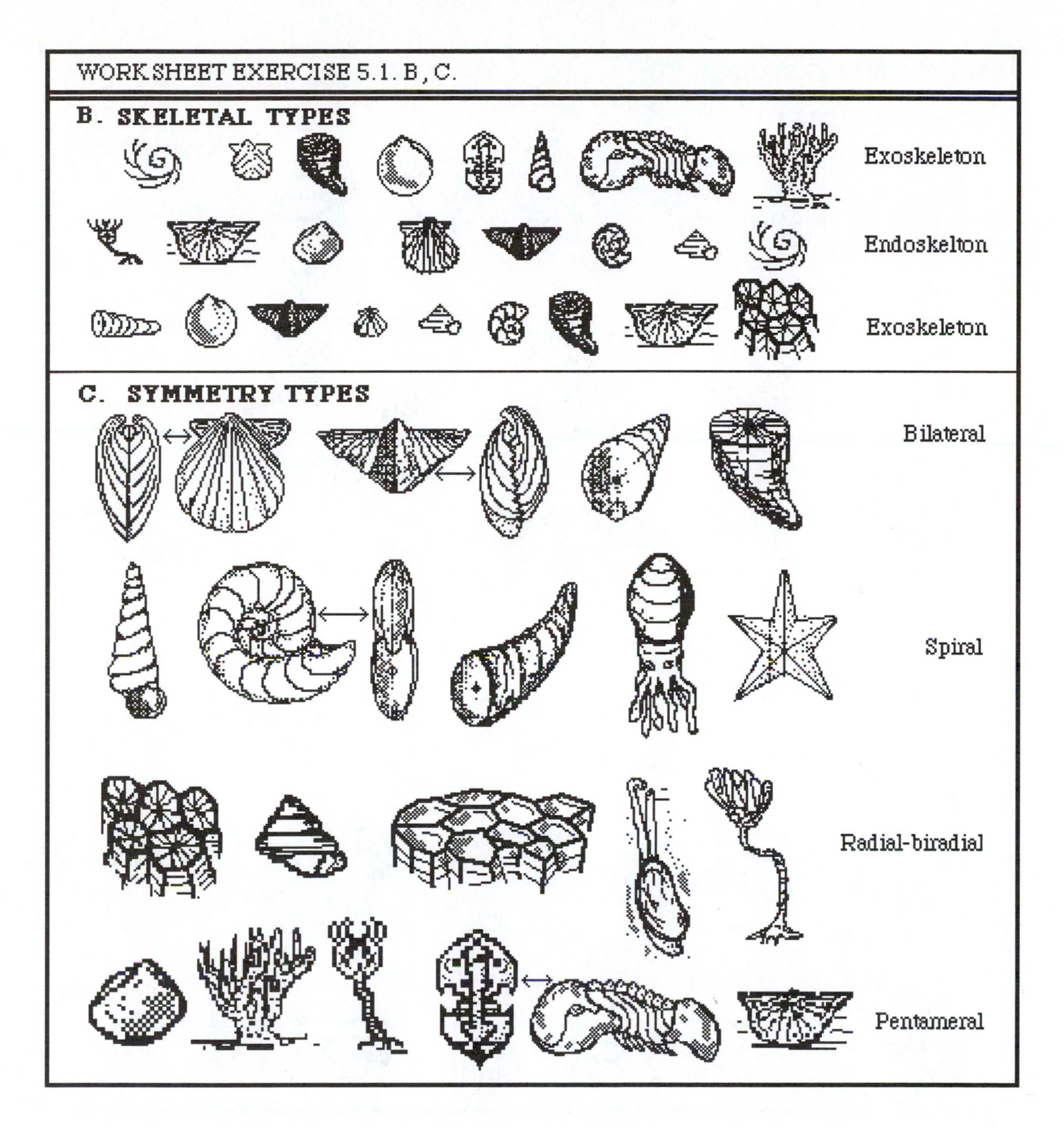

LABORATORY EXERCISES

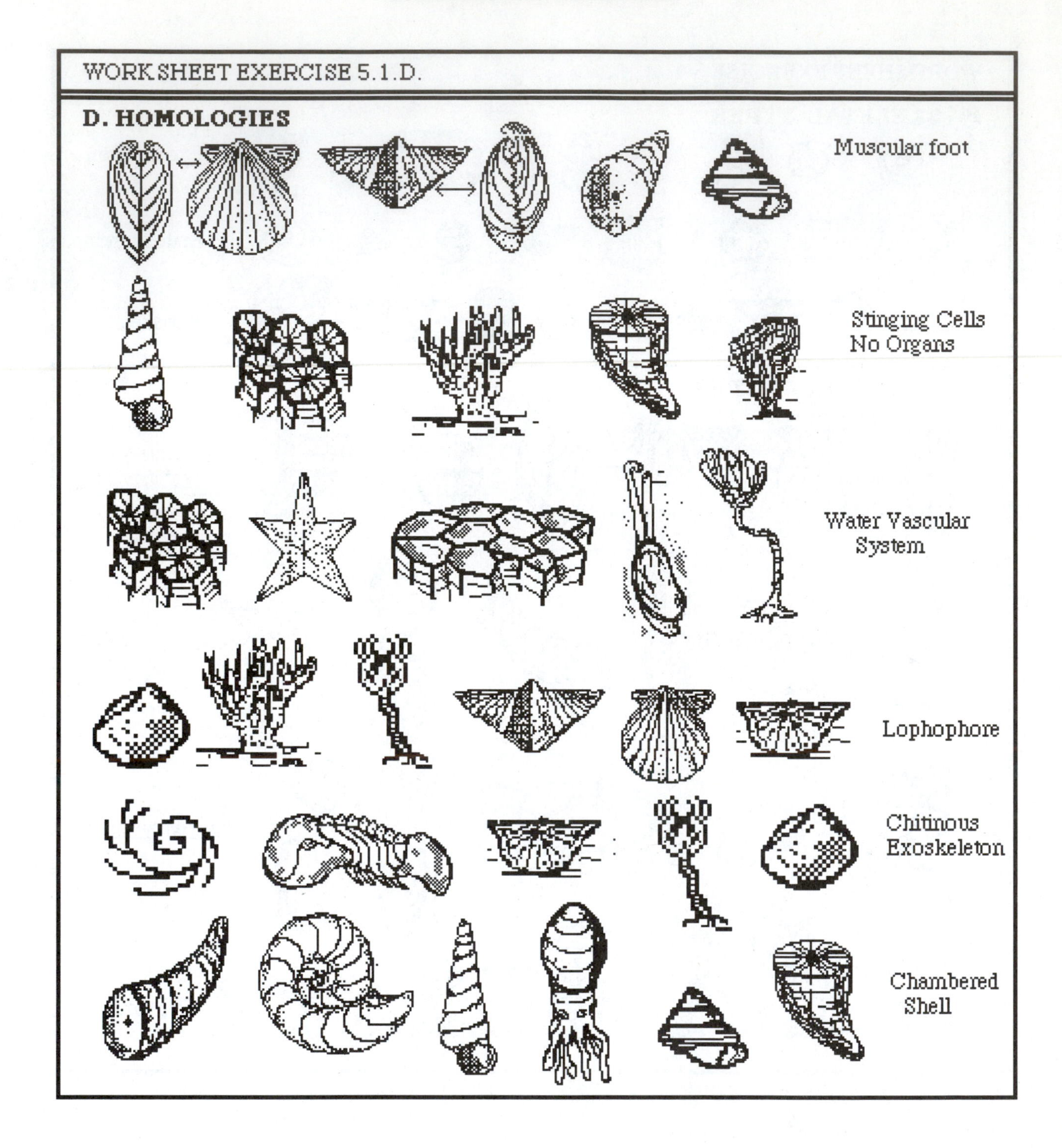

LABORATORY EXERCISES

5.2. Fossil Identification—Given specimens representing several of the commonly fossilized group, note its: 1) symmetry; 2) skeletal composition; 3) endo/exoskeleton; 4) particular distinguishing skeletal features; 5) phylum; 6) subpylum or class, as appropriate; 7) the principal bases for your identification; and, 8) sketch one specimen from each taxonomic group on worksheet 5.2. Label your sketches with the skeletal features which distinguish each from other groups and those which are shared by all members of each group. Your instructor may allow use of the dichotomous key in Appendix II, nevertheless, indicate those features which distinguish among groups and those which are shared homologies.

WORKSHEET EXERCISE 5.2. Fossil Description and Identification.

SYM-METRY	COMPO-SITION	End/Exo Skeleton	OTHER	PHYLUM	CLASS/ ORDER	Sketch
1.						
2.						
3.						
4.						
5.						
6.						

LABORATORY EXERCISES

5.3. Classification—Give a "complete" classification of two species selected by your instructor from the list of those identified in Exercise 5.2. Do so by listing all the higher groups to which each specimen belongs (as described above, from the most to the least inclusive level listed in the chapter discussion) according to the Linnaean hierarchical classification system.

A) SUBKINGDOM (if applicable)
PHYLUM
SUBPHYLUM (if applicable)
CLASS
SUBCLASS (if applicable)
ORDER (if applicable)
SUBORDER (if applicable)

B) SUBKINGDOM
PHYLUM
SUBPHYLUM
CLASS
SUBCLASS
ORDER
SUBORDER

5.4. Group Description and Identification—Your instructor will choose a specimen representing an organic group and pass it to the first student. Find and identify for the class, one anatomical feature, and pass it to the next student who will find and compare his/her assessment of the structure with that of the first. Discuss any disagreements or uncertainties with the aid of your instructor. Continue passing the specimen until your instructor indicates that the group has identified all possible major anatomical features. Only then should you identify the organic group to which the specimen belongs.

NOTES

LABORATORY EXERCISES

5.5. Specimen Comparison & Contrast—Your instructor will provide several pairs of specimens, a and b, numbered in sequence. Using worksheet 5.5: A) note by paired comparison and/or contrast, the distinguishing features of each; and, B) identify the group to which each of the pair belongs. If both belong to the same higher group (e.g., classes of Mollusca), note which features are homologies, distinguishing features, and subgroup membership. If each sample belongs to a different phylum, note which features (if any) are analogous, and which are distinguishing.

WORKSHEET EXERCISE 5.5. COMPARISON-CONTRAST.

Group	Symmetry	Composition	Endo/ Exoskeleton	Distinguishing Features	Homology or Analogy ?
1. A. Gastropod B. Cephalopod					
2. A. Brachiopod B. Bivalve					
3. A. Trilobite B. Ostracode					
4. A. Bryozoan B. Tabulate Coral					
5. A. Solitary Rugose Coral B. Porifera					
6. A. Colonial Rugose Coral B. Tabulate Coral					
7. A. Bryozoa B. Brachiopod					
8. A. Crinozoa B. Blastozoa					
9. A. Echinozoa B. Crinozoa					
10. A. Tabulate Coral. B. Stromatoporoid/chaetetid sponge					

LABORATORY EXERCISES

5.6. Specimens. Fossilization—Identify and note on worksheet 5.6 for each sample provided the: 1) organic group membership; 2) type of fossilization; 3) original composition of each skeleton, shell or test (see Appendix I), etc.; and, 4) presence of an exoskeleton or endoskeleton.

5.7. Specimens. Homology & Analogy—Your instructor will provide sets of similar looking fossils, containing members of one or several taxa. Identify the highest taxon to which each specimen belongs and state the homologous features on which your conclusions are based. If the specimens belong to two or more higher taxa, state the features anologous among groups.

EXERCISE WORKSHEET 5.6. Fossil Preservation and Composition.

Phylum, etc., (pre-identified)	Preservation Mode	Original Composition	Mineral Properties & Color	Fossil Mineral Composition	Endo/ Exo-skeleton
1.					
2.					
3.					
4.					
5.					
6.					
7.					
8.					
9.					

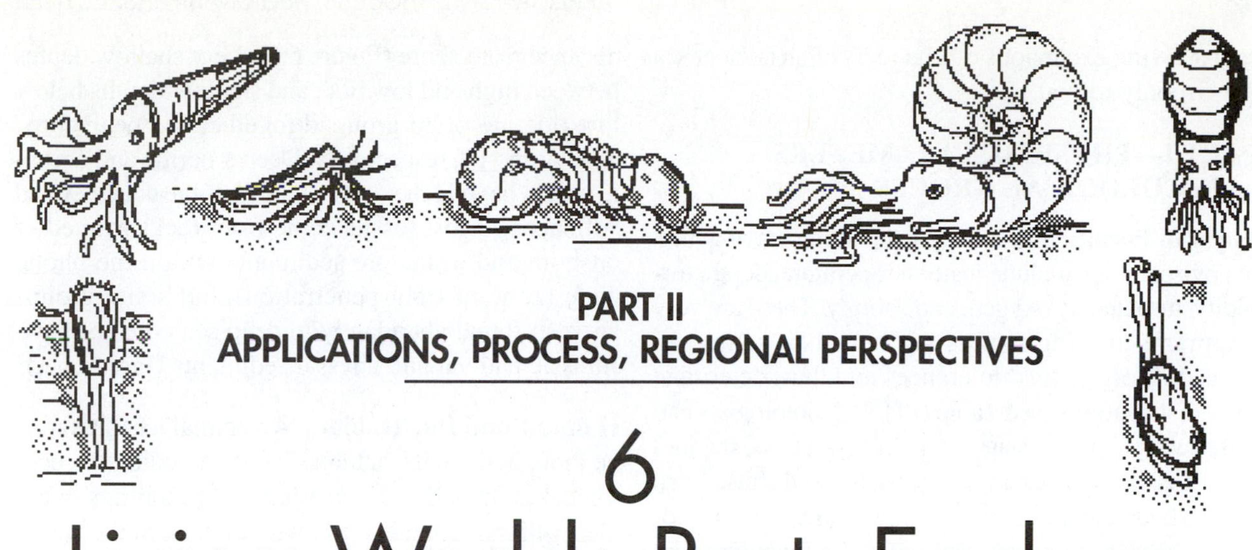

PART II
APPLICATIONS, PROCESS, REGIONAL PERSPECTIVES

6
Living Worlds Past: Ecology of Ancient Animals

Objectives
To demonstrate mastery of this material you should know how to: 1) discover the general life habits, feeding types and habitats typical of each fossil group; 2) infer general environmental settings/physical conditions from fossils; 3) determine whether an assemblage has, or has not, been transported or altered before burial; and, 4) determine the general tolerance limits of each group to physical parameters.

Using This Laboratory
As in Laboratory 5, an understanding of the ecology of fossils as discussed in your textbook requires an overview here of some unfamiliar topics; material which should be read prior to the laboratory period. Much of the content serves as a context within which your understanding of fundamentals should be enhanced and the ecology of fossils “brought to life”. Most of the essential material and terms are summarized and defined in tables. You are NOT expected to memorize the content or terms, but to use the reference material when completing the exercises. Several can be used as self-quizzes for pre-laboratory preparation. You will need to make frequent reference to Laboratory 5.

Ecology is the study of species diversity, abundance, and distribution, and the interrelationships among species and their environments. Paleoecology concerns that of ancient organisms. This laboratory is devoted to: 1) ecological concepts; 2) aspects of individual species’ ecologies (environments, life habits and habitats); and, 3) the interactions among species living together as communities and their relationships to the physical environment, and is thus, divided into three parts. We will learn how: 1) to reconstruct the physical conditions of environments from species’ ecological and physical attributes; 2) to determine if a fossil assemblage faithfully preserves a record of the animals that actually lived together when buried; and, 3) communities’ species compositions, numbers and abundances interact with, and respond to, one another and their physical environment. When studies of paleoecology and environments of sediment deposition (Laboratory 8) are combined, and a little imagination is added for flavor, the past, in all its dimensions, can be brought to life as if one were privileged to an H.G. Wells’ time-machine travel for scuba-diving excursions off the reefs of an ancient sea known only to you.

PART I—PHYSICAL PARAMETERS and ECOLOGICAL GROUPS

Physical Parameters—Physical parameters of aquatic environments include water temperature, depth, turbidity, turbulence, oxygen, and salinity. These are very important limiting factors because their states can exceed an organisms' tolerances and thus, determine their distributions (see Table 6.1). Paleontologists can infer such parameter states (e.g., fresh water vs. marine) from fossils and sedimentary rocks, and thus, learn much about animal life habits, environments and ecologies. These inferences are based on knowledge about the ecology of living animals related to fossil species, fossil associations, and features of the entombing sedimentary rocks (Laboratories 8-10).

Animals differ in tolerance of average physical conditions and of variation about these averages. Some can tolerate wide variation in a parameter(s)—e.g., temperature, salinity—and/or are found in a number of environments. Others are restricted to one environment and/or are intolerant of physical variations.

Ecological Zones—The marine realm is divided into a number of environments based on water depth and distance from shore (Figure 6.1). Very shallow depths between high and low tide, and shallow depths below low tide are often grouped together as "nearshore" and "onshore", respectively. Deeper neritic and oceanic waters beyond the continental shelf break are termed "offshore". Most fossils come from rocks formed of onshore and nearshore sediments within the photic zone (zone of light penetration), but some groups are also found abundantly in deep sea cores of post-Jurassic bathyal and abyssal sediments (Figure 6.1).

Habitats and Life Habits—An animal's habitat may be thought of as its "address": one including the ecological zone and the immediate surroundings occupied within. Animals may swim or float in the water, live on or burrow within sediment, or attach to the sea floor, other organisms (living or dead) or other hard objects (see Table 6.2A—C for terminology).

An animal's life habit is analogous to "what it does for a living"—that is, to obtain food (Table 6.2B) and to move about (or not) in the water, or on/within the sediment (Table 6.2C). The majority of invertebrates gather food by filtering suspended organic particles, plankton and dissolved nutrients from water, however, several other feeding methods are utilized by marine invertebrate animals (see Table 6.2B for terms and definitions).

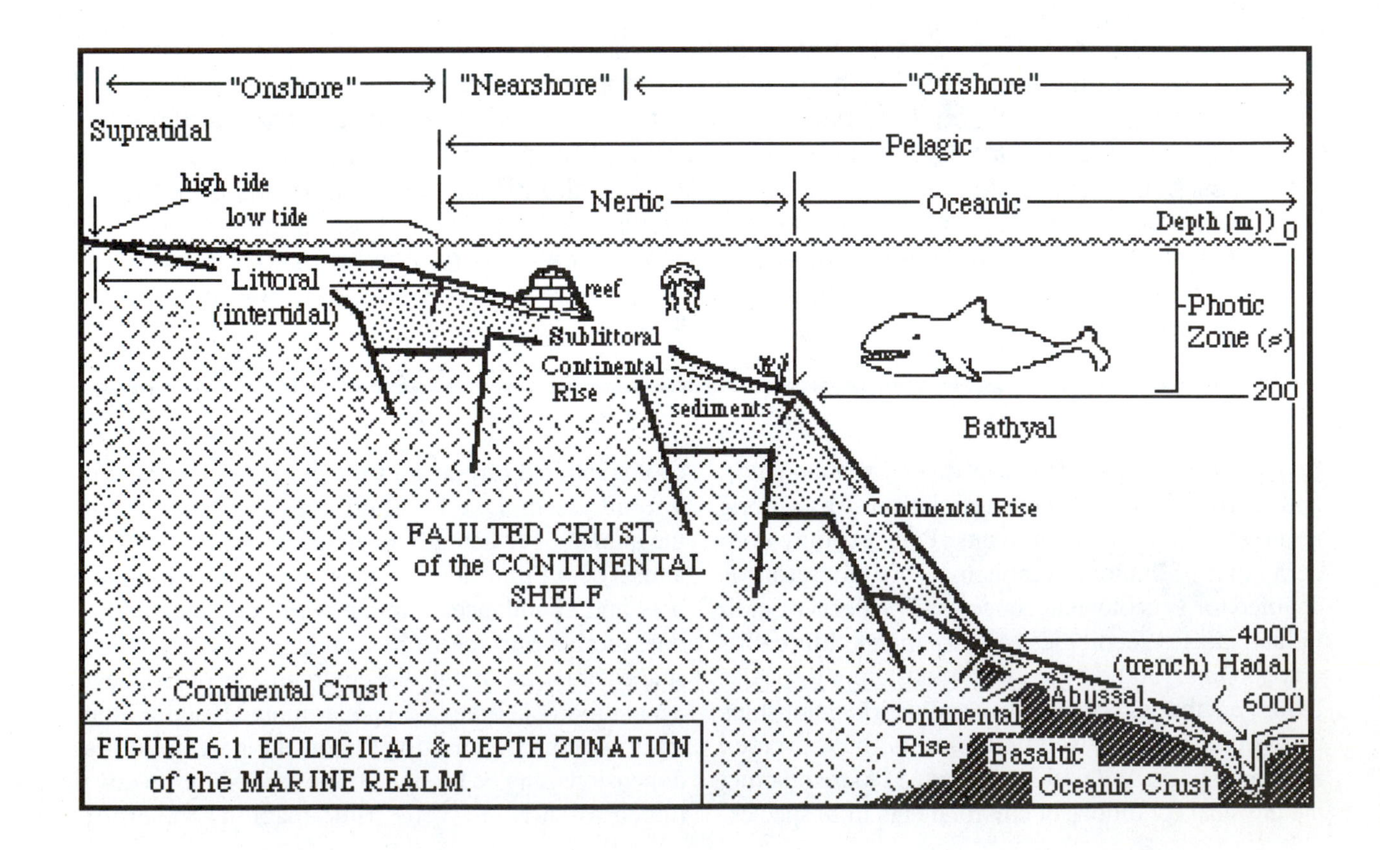

FIGURE 6.1. ECOLOGICAL & DEPTH ZONATION of the MARINE REALM.

TABLE 6.1. PHYSICAL PARAMETERS and EFFECTS on MARINE LIFE.

PARAMETER DEFINITION	SITES/ TRENDS	EFFECTS ±	GROUPS MOST EFFECTED/LIMITED
Turbidity Amount of Sediment Suspended	Greatest Near Detrital Sed. Sources	– Light Diffusion – Clogs Feeding & Respiratory Organs	–Algae, Phytoplankton*, Filter Feeders*. –Corals, Brachiopods, Bryozoa, many –Bivalves, Stalked Echinoderms, Sponges
Turbulence Degree of Water Agitation	Generally Highest Near Shore & Shallow	+ Enhance Oxygen, Food – Physical Damage – Prevent Food Settling	+ Attached Filter Feeders –Unattached, Fragile Species –Deposit Feeders*, Infauna*
Temperature	Latit. Gradient. ↓with Depth.	± Physiological Reaction Rate Control	± Many species limited to restricted temperature ranges, esp. Reef Corals.
Oxygen Dissolved in Water	↓with ↑Temp. ↑with Depth.	–↑ Oxygen, Reduced Respiration	– Majority Benthos Precluded by Low/ Lack Oxygen.
Depth	———	Linked to Temp., Sunlight & Oxygen	See Above.
Salinity Amount of Dissolved Salts	Fresh ≈ 0 ppth Brackish < 20 Marine 30–40 Hypersaline > 50 Marginal 40–50	Change osmotic pressure balance. Lethal if high or low.	– Echinoderms, Bryozoa, Corals, Reef Corals, Trilobites, most Brachiopods Cephalopods. * See Table 6.2

PART II—LIFE HABITS & HABITATS of INVERTEBRATE GROUPS

When studying the paleoecology of any species, inferences are drawn from several lines of evidence: by analogy with a living "relative"; by comparison with living species of similar shapes, structure and known functions; or, by comparison with a closely related fossil species whose ecology is understood. Sometimes such comparisons are not possible. Conclusions about one species must then be drawn from information about associated species and from analysis of sedimentary rocks (Laboratories 8–10).

The information below and that summarized in Table 6.3 includes generalizations and examples drawn from case and group studies of living and fossil species'

TABLE 6.2. LIFE HABITS and HABITATS of MARINE INVERTEBRATES.

A. HABITAT	TERM	ORGANISM	LIFE HABIT	ORGANISM
Open Ocean	Pelagic Zone	Phyto- & Zoo-plankton	Floaters Swimmers	Plankton(ic) Nekton(ic)
Sea/Ocean Floor	Benthonic Zone	———	Benthos	Benthos
On Sediment	Epifaunal	———	Epifauna	Epifauna
Within Sedimen'	Infaunal	———	Infauna	Infauna
On other Organisms	Eipbiontic	———	Epibiont	Epibiont

B. FOOD SOURCE	FEEDING HABIT	FOOD TYPE
Filtered water	Filter or Suspension Feeders	Dissolved Nutrients, Plankton
Sediment (on or in)	Deposit Feeders/ Scavengers	Fecal Pellets, Bacteria Dead Animals, Plants
Benthic & Pelagic	Herbivore	Algae, Phytoplankton
Benthic & Pelagic	Carnivore	Small Animals

C. MOBILITY	TERM
Immobile, Attached or otherwise secured	Sessile
Mobile	Vagrant
Swimming	Nektonic
Floating	Planktonic

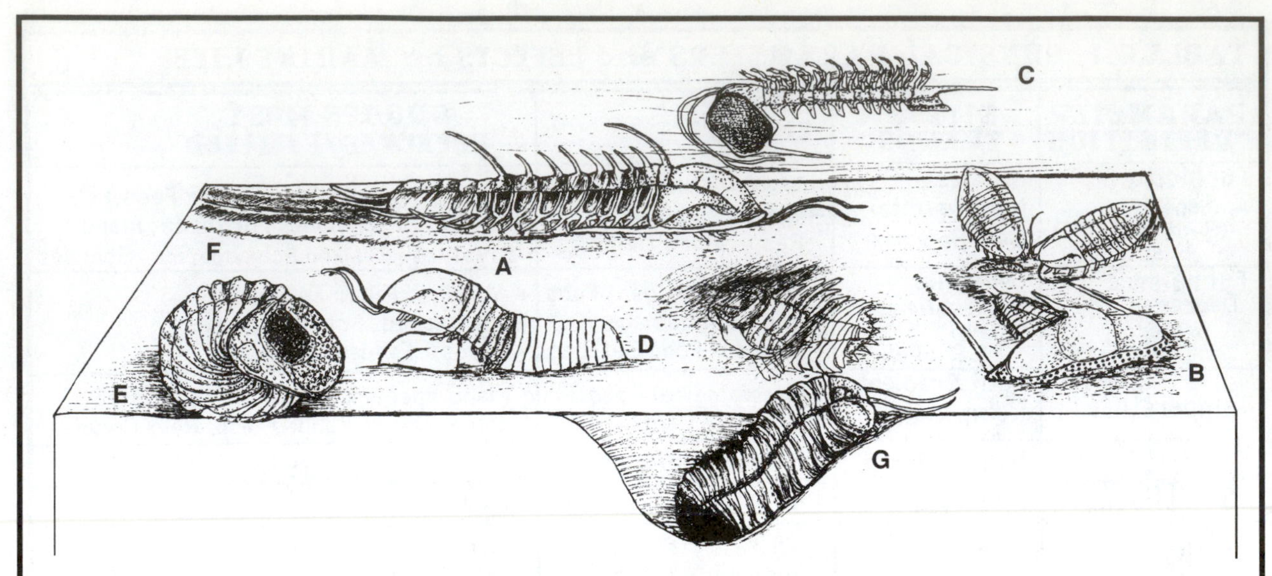

FIGURE 6.2 LIFE HABITS AND HABITATS OF BIVALVES. A) Benthic epifaunal, sessile & attached by byssus; B) epifaunal, sessile and gregarious; C) epifaunal, sessile & cemented to bryozoan colony; D) various infaunal burrowers; E) infaunal rock-borer; F) semi-infaunal (partially-buried); G) nektic (swimming); H) epibiont (*epi*=on; *bio*=life; i.e., attached) on alga; and, I) benthic epifaunal reef-formers. Carnivore (E), detritivores (some of D), filter feeders (all others).

ecology. Conclusions cannot always be drawn by inference from the general to the specific: one should be aware of the inferences used and base conclusions on as many lines of evidence as possible.

Bivalves—Bivalves inhabit all aquatic environments and many species exhibit tolerance of extremes in several physical parameters (Figure 6.2, Table 6.2). Most bivalves are either infaunal or attached epifaunal filter feeders, siphoning food from the water. Some are infaunal deposit feeders, siphoning food from the sediment-water interface (e.g., *Tellina*). Other deposit feeders ingest sediment, from which food is then stripped. A very few species are epifaunal carnivores, feeding on worms and arthropods.

Infaunal species usually burrow in sediment with their foot, but some use their shell or chemical secretions to bore into rock, wood or coral. Shallow burrowers have deeply convex shells, deep burrowers are less convex, and rapid burrowers are elongated (Figure 6.3). These molluscan bulldozers are also recog-

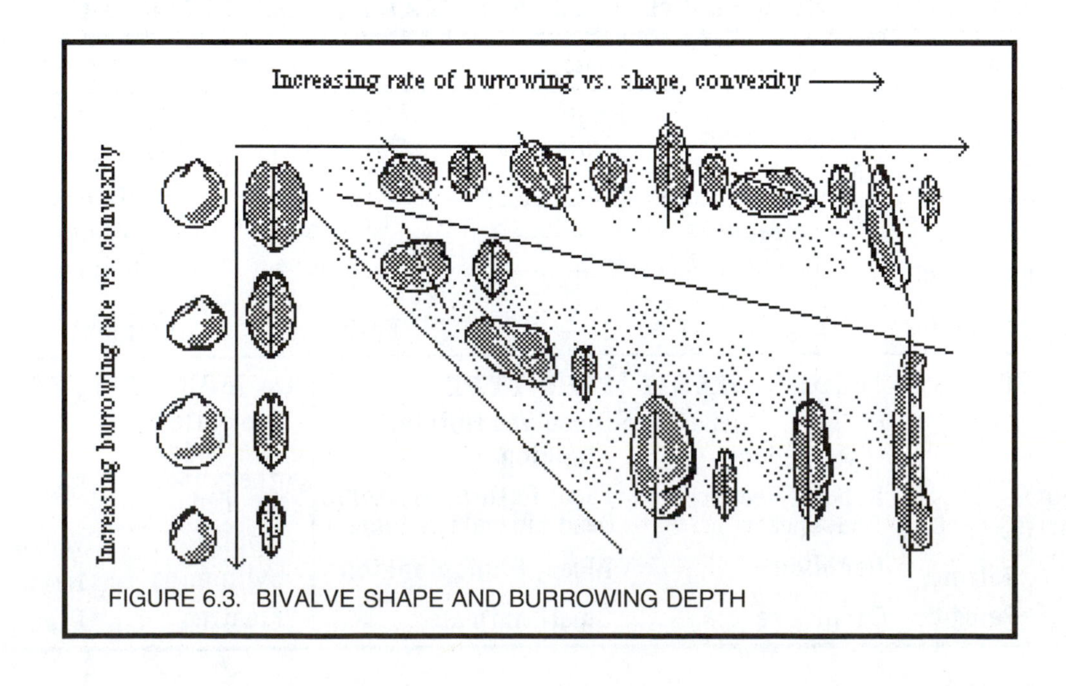

FIGURE 6.3. BIVALVE SHAPE AND BURROWING DEPTH

TABLE 6.3. GENERAL LIFE HABITS, HABITATS, ENVIRONMENTS					
GROUP	ENVIRONMENTS	FEEDING TYPES	MO-BILITY	HABITATS	TOLERANCES
Bivalves	Fresh, tidal flats & lagoons to abyssal.	Filter- & deposit, some carnivores.	Sessile and vagrant.	Benthic Epifaunal infaunal.	All salinities, Temp. Turbulence, turbidity & oxygen.
Cephalopods	Subtidal to Neritic.	Carnivorous.	Most nektic.	Pelagic, Benthic epifaunal.	Marine only, shallow water all O_2, Temps. & Turbidity & Turbulences.
Gastropods	Terrestrial, Fresh H_2O, Onshore to abyssal	Most herbiv., some deposit feeders & carnivores.	Most Vagrant, some sessile.	Pelagic, Benthic Epi- & infaunal	All salinities, Temp. Turbulence, turbidity & oxygen.
Brachiopods	Onshore to bathyal	All Filter-feeders.	Sessile.	Benthic Epifaunal & epibionts.	Generally marine, clear, oxygenated water. All temp. & depths.
Bryozoans	Some Fresh, most sublittoral.	All filter-feeders.	Sessile	Epifaunal & epibionts.	Marine, clear, O2, calm-turbulent.
Trilobites	Sublittoral.	Deposit-feeders.	Vagrant, Nektic.	Benthic Epifaunal, Pelagic.	Marine, clear - turbid H_2O, calm-turbulent. limited temp limited O_2.
Echinoderms	Sublittoral to abyssal.	Filter feeders. Echinoids - All types	Sessile. Echinoids vagrant.	Benthic Epifaunal, Infaunal.	Marine, clear, calm-turb., well oxygenated, all temp.
Corals	Open sublittoral marine. most in reefs.	Carnivorous, some symbiotic.	Sessile.	Benthic Epifaunal.	Marine, most narrow temps, depths, clear, turbulent, well oxygenated.

nized by the presence of an embayment of the mantle (Laboratory 5; Figure 5.12B), the depth of which is roughly proportional to the depth of burrowing. Many species are semi-infaunal, living partially buried and attached to some object within the sediment (Figure 6.2).

Sessile epifauna either attach to shell, coarse particles, wood or plants by a bundle of tough "threads", or shell cementation, or they are free-lying (Figure 6.2). Some of the former species (e.g., mussels) can detach and move about (e.g., scallops (pectens) may also "swim" short distances by clapping their valves together). Species attached by threads have a very small foot (and thus, attendant gape), but a gape marking the extrusion of attachment "threads", and markedly "asymmetrical" valves (Figure 6.2). These and cemented reclining species form shell beds or banks on tidal flats and below low tide in bays, lagoons, and estuaries. Such species may have highly irregular and variable shell shapes (e.g. Ostrea (an oyster), or in the case of rudists (Laboratory 5; Figure 5.13, Figure 6.2), resemble solitary corals.

Cephalopods—Much of what we infer about the habitats of extinct nautiloids and ammonoids is based on observations of squid and single extant nautiloid genus, *Nautilus* (Figure 6.4, Table 6.3). Like *Nautilus*, extinct cephalopods were carnivorous or scavengers, and possibly active a night. They captured prey with tentacles, and crushed it with a horny beak and (possibly) a set of mineralized "teeth".

Like cephalopods of today, all extinct species lived in marine waters. *Nautilus* is found in tropical southwestern Pacific at depths of 5 to 550 m (16 -1800 ft). By analogy with Nautilus, most of these "swimming" predators probably also crawled along and rested on the sea floor anchored by tentacles (Figure 6.4). The bizarre shapes of some, however, suggest vagrant and sessile bottom-dwelling.

Like *Nautilus*, related extinct shelled cephalopods had gas and liquid-filled chambers that provided buoyancy (Figure 6.4). In most species the body mass was balanced by the chambers when swimming, thus keeping the head and foot poised horizontally. However, some extinct species resembling diving bells apparently swam with the head directed toward the sea floor.

Gastropods—Gastropods are among the widest ranging invertebrates (Figure 6.5, Table 6.3). Most shelled gastropods are vagrant epifaunal or infaunal benthos. Those species with uncoiled or loosely coiled shells flattened on one side were (and are) sessile.

All feeding types are known: some are omnivorous, eating animals and plants, but the majority are

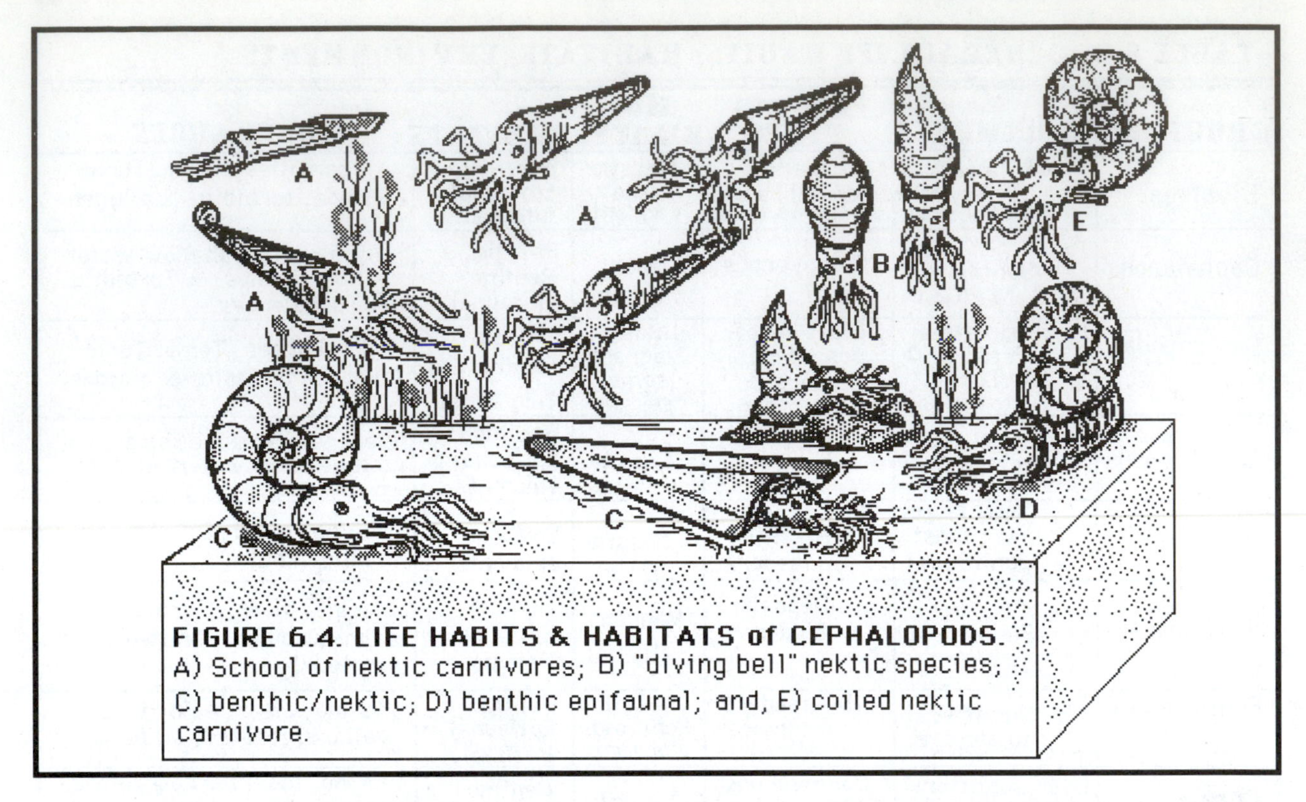

FIGURE 6.4. LIFE HABITS & HABITATS of CEPHALOPODS. A) School of nektic carnivores; B) "diving bell" nektic species; C) benthic/nektic; D) benthic epifaunal; and, E) coiled nektic carnivore.

carnivorous, scavenging, or herbivorous algae grazers (Figure 6.5). Carnivores of bivalves, barnacles, echinoids, and other gastropods either drill holes in the prey shells, pry them open, or casually "browse" on sponges and cnidarians. Some living marine species actually harpoon fish with a poison-tipped "tooth" on the end of the proboscis! A common Paleozoic genus *Platyceras* lived attached to crinoids, feeding on their feces.

Trilobites—Little is known of trilobite paleoecology because the group is extinct, but trace fossils provide some information about their behavior (Figure 6.6, Table 6.3). Trace fossil trails suggest that trilobites stirred up sediment and organic particles while crawling, then trapped and swept them with appendages to the mouth. Most species were probably benthic deposit feeders because they had small mouths and no "jaws". Trails which extend from pit-like structures on bedding surfaces show that trilobites dug into the sediment, possibly to make egg nests, to rest, or for protection or camouflage.

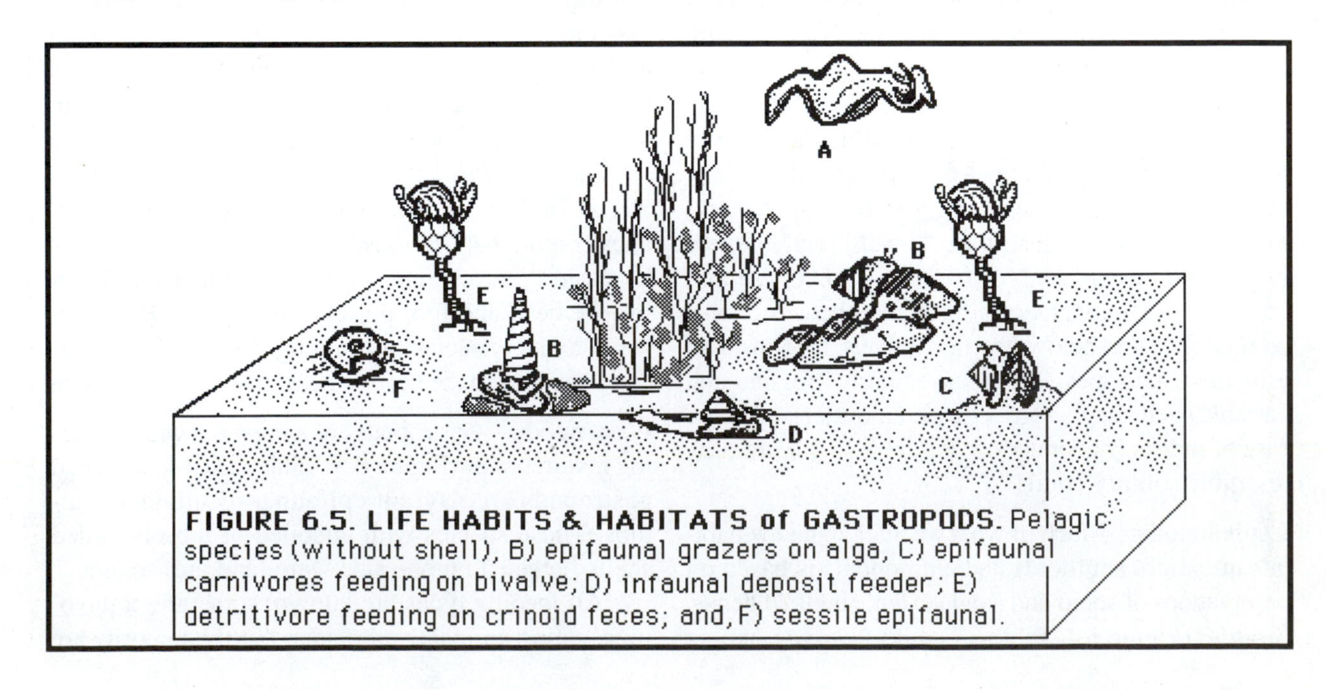

FIGURE 6.5. LIFE HABITS & HABITATS of GASTROPODS. Pelagic species (without shell), B) epifaunal grazers on alga, C) epifaunal carnivores feeding on bivalve; D) infaunal deposit feeder; E) detritivore feeding on crinoid feces; and, F) sessile epifaunal.

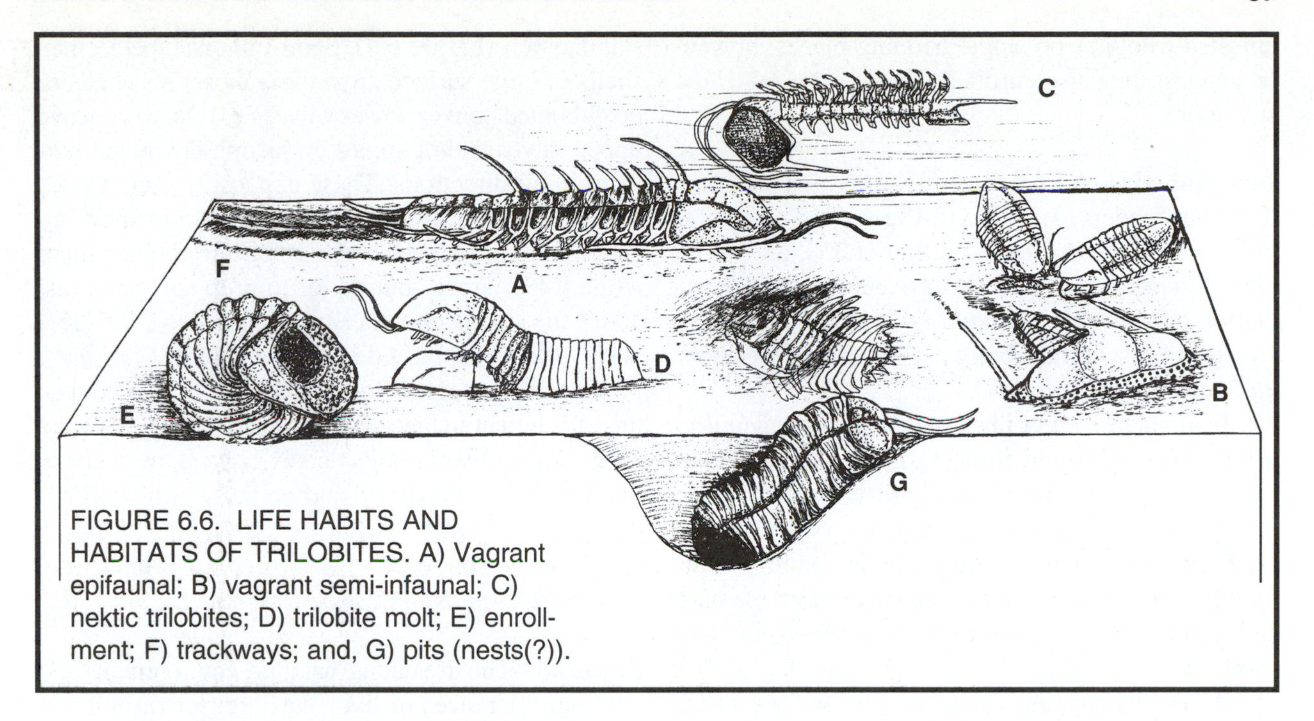

FIGURE 6.6. LIFE HABITS AND HABITATS OF TRILOBITES. A) Vagrant epifaunal; B) vagrant semi-infaunal; C) nektic trilobites; D) trilobite molt; E) enrollment; F) trackways; and, G) pits (nests(?)).

Trilobite appendages were atypical of arthropods; antennae were used for sensing, while the others functioned only for locomotion and respiration, lacking specialization for feeding and reproduction (Figure 6.6). Although most were vagrant epifauna, some swam far above the bottom, and others were (possibly) infaunal. Most were probably both epifaunal and nektonic, but their leg structures suggest little maneuverability or speed. Pelagic trilobites, like some living arthropods, may have swum upside-down or in turbid waters, or crevics. The small agnostid trilobites were blind and may have been pelagic nekton living below the photic zone.

Trilobite preservation also provides information about their behavior. Like other arthropods, they grew by molting, and thus, fossil trilobites are often fragmentary (Figure 6.6). They must have been especially vulnerable to predators when molting. Protection from predators may have prompted their curious but common "sow bug-style" enrollment. Numerous

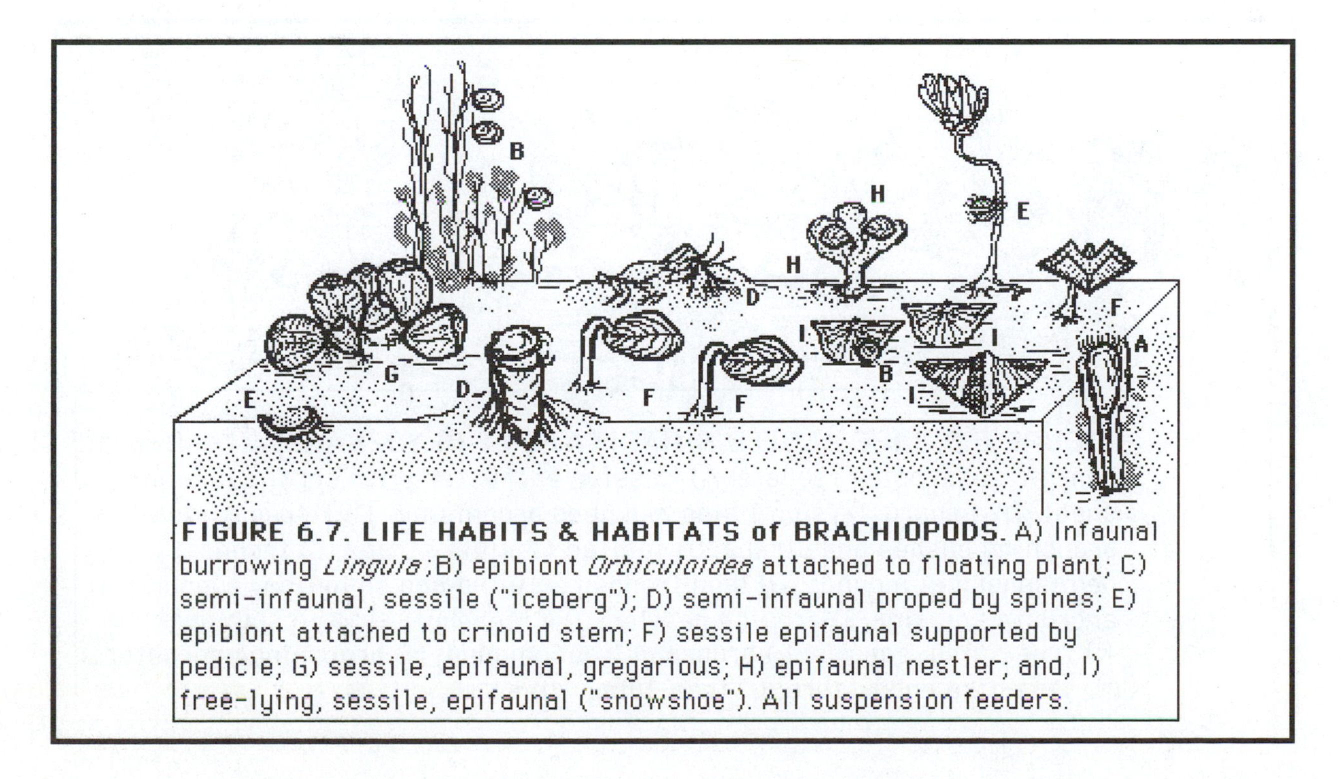

FIGURE 6.7. LIFE HABITS & HABITATS of BRACHIOPODS. A) Infaunal burrowing *Lingula*; B) epibiont *Orbiculoidea* attached to floating plant; C) semi-infaunal, sessile ("iceberg"); D) semi-infaunal proped by spines; E) epibiont attached to crinoid stem; F) sessile epifaunal supported by pedicle; G) sessile, epifaunal, gregarious; H) epifaunal nestler; and, I) free-lying, sessile, epifaunal ("snowshoe"). All suspension feeders.

enrolled trilobites on single bedding planes, however, suggest they also enrolled when storms raked the sea floor.

Brachiopods—All brachiopods were (and are) benthic filter feeders (Table 6.3). A few inarticulate brachiopods lived attached to other shells, and some *Orbiculoidea* lived attached to fixed or floating vegetation (Figure 6.7). *Lingula* is (and was) infaunal, living within a vertical burrow. All others were either epifaunal or semi-infaunal.

Inarticulate species like *Lingula* and *Orbiculoidea*, are commonly found in deposits formed in harsh on/nearshore environments and were tolerant of extremes in many physical parameters. Most species, however, tolerated only normal marine salinities, well-oxygenated and clear water (turbulent or quiet), because sediment easily obstructed filter-feeding and respiration.

All brachiopods are sessile and have various means of anchoring or stablizing on the substrate (Figure 6.7). Many live permanently attached by the pedicle, but some lose this attachment in later life and others always lay free. Those attached by pedicles commonly cluster, attaching to one another or larger animals. Others remain securely cemented to a hard object or partially buried within the sediment (semi-infaunal).

The spines of many extinct species probably aided in attachment to other animals, or for anchoring and stabilization (Figure 6.7). Semi-infaunal species had shells of large surface area: thin, expansive shells or deep-bodied convex lower valves, and flat to concave upper valves and/or spines projecting down and outward in all directions. These expansive shell shapes and spines could have acted as a "snowshoe" to keep the animal from sinking beneath the sediment while it lay largely buried, or to prop soft parts just above the sediment. Others may have lived like "icebergs", partially buried in the sediment. Many live permanently attached by the pedicle, but some lose this attachment in later life and others always lay free. Spines of another kept its deeply conical, horn coral-like lower valve propped and partially buried within the sediment, while its flat top valve flapped like a pop-open garbage can lid. Spines of others may have been used as light sensory sentinels.

Bryozoa—The life habits, habitats, environments and physical tolerances of byozoans are shown in Table 6.3. Because they are small filter-feeders and the lophophore (Chapter 5) is used partly for respiration, they are intolerant of frequent turbidity. All are epifaunal, and nearly all, sessile.

Bryozoans compete for horizontal encrusting space and for food by filtering from various levels above the bottom. Coexisting species subdivide food and space by tiering—growing to different heights above the bottom in the same way that forest trees and under-

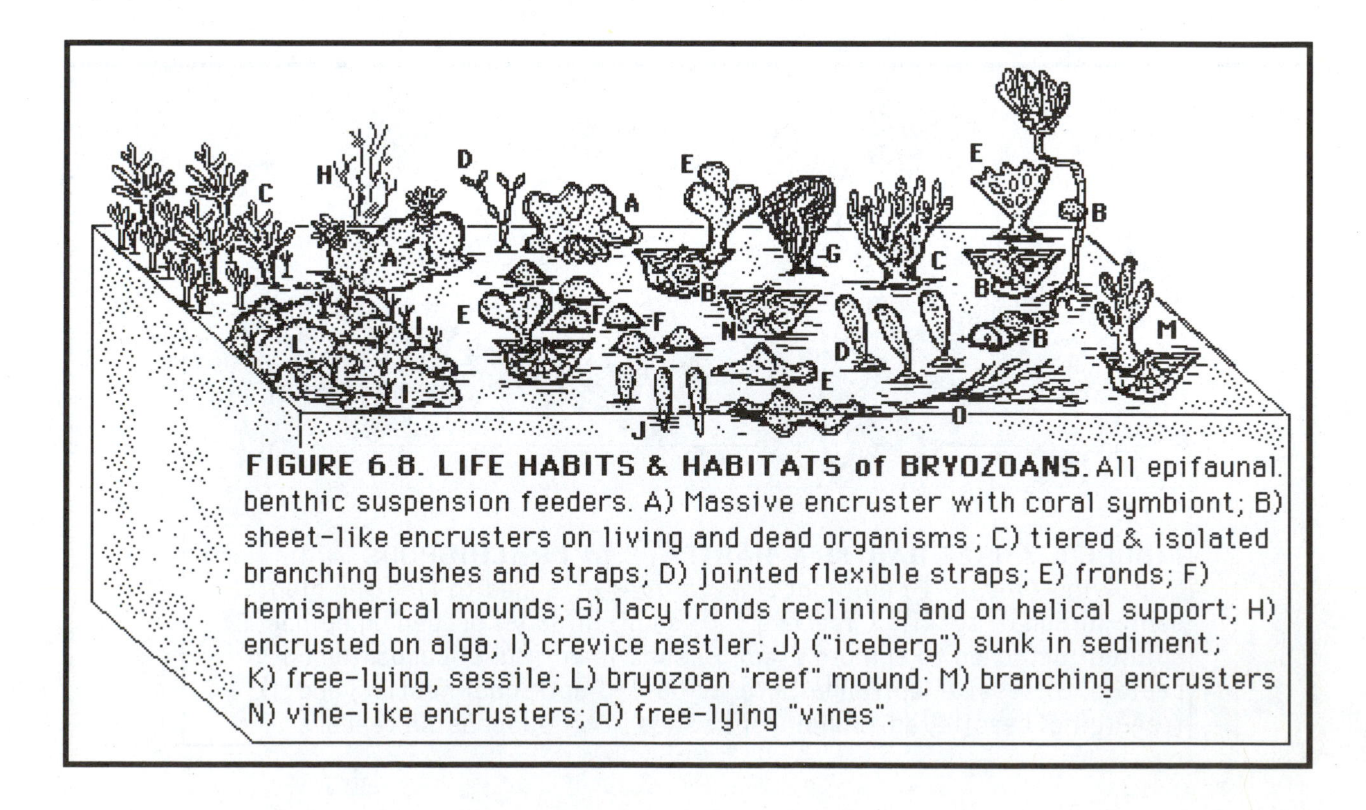

FIGURE 6.8. LIFE HABITS & HABITATS of BRYOZOANS. All epifaunal. benthic suspension feeders. A) Massive encruster with coral symbiont; B) sheet-like encrusters on living and dead organisms; C) tiered & isolated branching bushes and straps; D) jointed flexible straps; E) fronds; F) hemispherical mounds; G) lacy fronds reclining and on helical support; H) encrusted on alga; I) crevice nestler; J) ("iceberg") sunk in sediment; K) free-lying, sessile; L) bryozoan "reef" mound; M) branching encrusters N) vine-like encrusters; O) free-lying "vines".

growth subdivide vertical space and access to sunlight (Figure 6.8). Competition for space is most pronounced within reefs. Bryozoans are common in present reefs, but their small colonies comprise little of the reef itself. During the Middle Ordovician, bryozoans and stromatoporoid sponges built the first true (but small) reefs.

Shell fragments, pebbles, algae, live shells, or other bryozoans are needed for most larvae to settle and grow where sediment is muddy because colonies grow as either large low encrustations or erect branches from an encrusting base (Figure 6.8). Some however, are "stick-in-the-muds" anchored by partial burial of colony bases.

Bryozoans evolved many different colony forms (Figure 6.8). Some assumed different forms in response to the encrusted surface size and stability and, like corals, the strength of currents and waves. The more stable, sturdy encrusting, massive irregular, and hemispherically domed colonies tend to be more common than fragile erect colonies in shallow agitated waters.

Echinoderms—Blastoids, and most cystoids and crinoids were/are stalked sessile epifaunal filter feeders, and all echinoderms are, and were, exclusively marine (Figure 6.9, Table 6.3) Coexisting species apportioned vertical space and food resources by tiering. Crinoids and blastoids fed in the topmost tiers of sessile epifauna. Stems could be very long, but much of their length must have snaked along the bottom: one individual, for all but a meter or so of its estimated 17 m (50 ft) length!

Echinoids are mobile echinoderms and so have prominent bilateral, as well as pentameral, symmetry. Most echinoids are herbivorous grazers of algae, but a few are carnivorous or infaunal deposit feeders. Infaunal species, irregular echinoids (e.g., heart urchins, sand dollars), have bilateral symmetry, flattened, short spines, and petal-shaped regions of large pores for their tube feet.

Corals—Corals are sessile benthos usually presumed synonymous with reefs (Table 6.3). However, not all corals form reefs, and these actually comprise less of reef volume than assumed. Algae, tabulate corals, stromatoporoids, and bryozoans dominated Paleozoic reefs. Scleractinian corals, algae, sponges and bryozoa have been the principal reef builders since the Mesozoic.

Reef and nonreef corals occur in Paleozoic shallow water limestones. Solitary rugose and colonial corals inhabited shallow carbonate- bottom seas during the Paleozoic, but solitary species were not restricted to reefs and vicinity, and may have tolerated turbid environments.

All extant reef corals tolerate only marine water and are most productive in shallow (<90m (295 ft), photic zone), warm (25-29 °C), and well-oxygenated water. These limitations reflect requirements of the algal symbionts in their tissues. The algae increase skeleton formation rates, and analogous with many

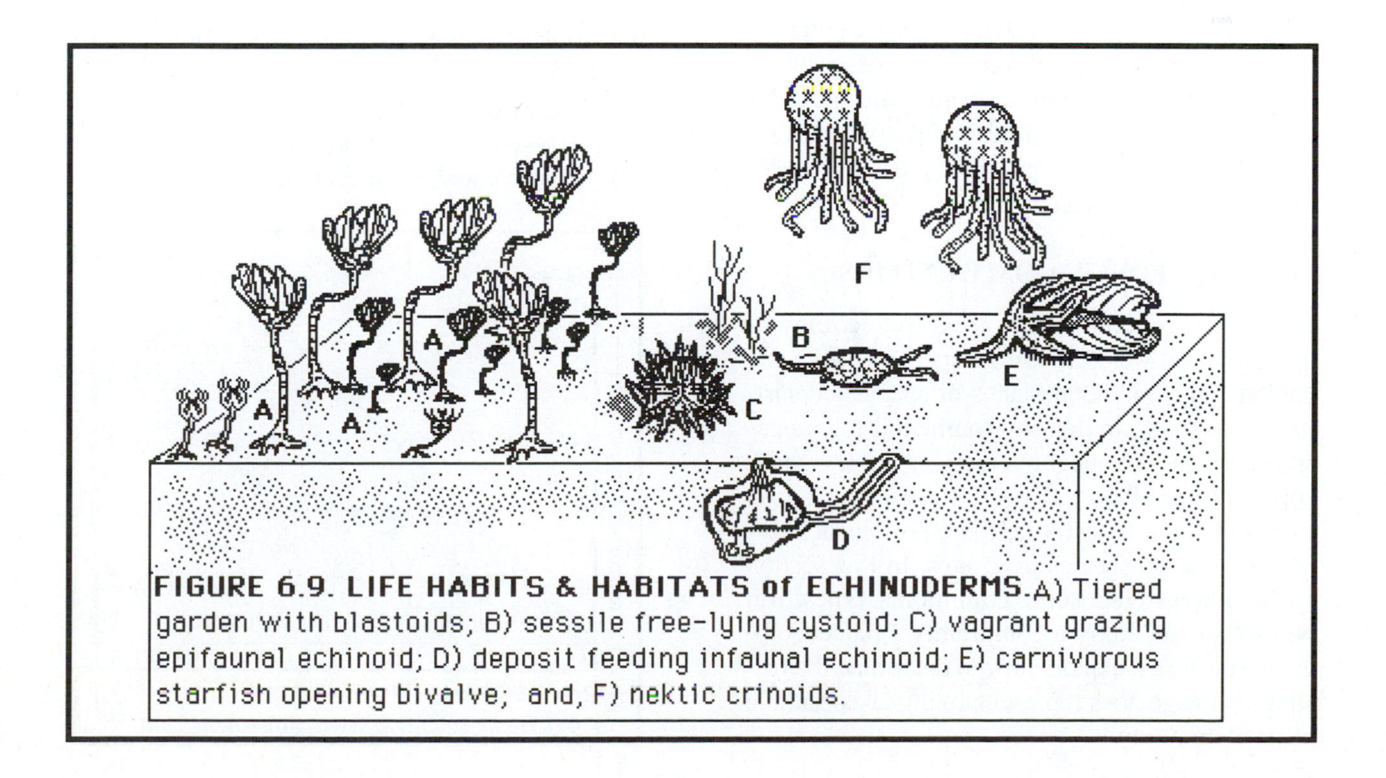

FIGURE 6.9. LIFE HABITS & HABITATS of ECHINODERMS. A) Tiered garden with blastoids; B) sessile free-lying cystoid; C) vagrant grazing epifaunal echinoid; D) deposit feeding infaunal echinoid; E) carnivorous starfish opening bivalve; and, F) nektic crinoids.

FIGURE 6.10. LIFE HABITS & HABITATS of CORALS. All are sessile, epifaunal suspension feedling carnivores. A) Tabulate and cololonial rugose corals and stromatoporoids; B) solitary rugose corals; and, C) scleractinian reef corals.

plants' metabolic dependence on nitrogen-fixing bacteria, are believed essential for transfering/recycling nutrients. Non-reef corals lack symbionts and some live in deep, cold waters. Several lines of evidence suggest Paleozoic reef-building corals also had symbiotic algae.

Corals display various colony forms (Figure 6.10), each a way of occupying and competing for space and food resources. Low mounds and massive corals occupy agitated waters like reef-fronts while delicate branching forms dominate deeper water and quiet back-reef lagoons. The flat colonies of reef species living in the lower photic zone provided greater surface area for algal photosynthesis.

PART III—FOSSIL COMMUNITIES

The above generalizations treat the ecology of groups as if it were the sum of species' habits and habitats. Natural assemblages of species coexisting because of similar environmental tolerances and required biological interactions are known as communities (Plates III—V; Appendix III). As such, communities carry far more ecological information than do species or groups. In this section we will investigate some community types. But before we can study a community's paleoecology, we must first explore how to determine if a fossil assemblage does represent an unbiased sample of its community.

Fossil Assemblage Formation

Many fossil deposits show signs of some disturbance (and thus, information loss) before burial. The definition of community requires that any study of fossil communities show whether co-occurring fossil species (assemblages) actually did live together, or whether exposure and/or current transportation removed some from (or introduced others to) the sample. Three extreme states of fossil assemblage formation are recognized (Figure 6.11);

1) instantaneous burial;
2) exposure without transportation; and,
3) exposure and transportation.

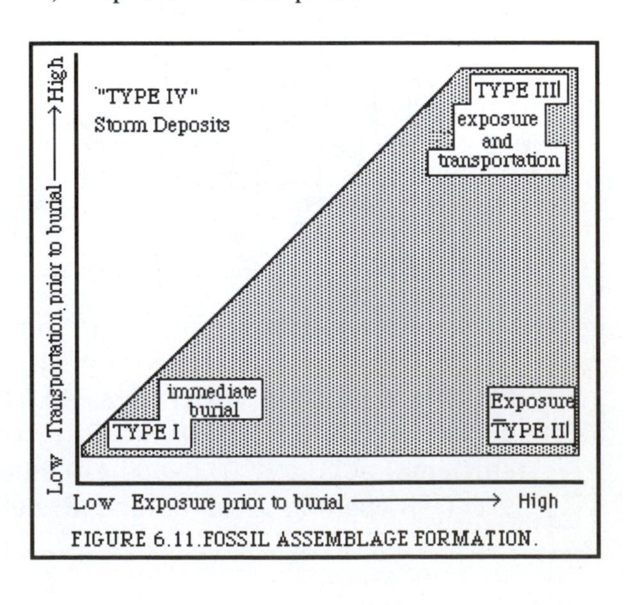

FIGURE 6.11.FOSSIL ASSEMBLAGE FORMATION.

Instantaneous burial is the ideal scenario, but if lacking, there will be inevitable exposure and possibly transportation of some or all species. Exposure on the sea floor leads to decomposition, breakage, predation and current sorting by size. Extended exposure leads to species' accumulations that may not reflect actual associations, numbers and abundances, because they may not have actually coexisted. In turbulent water, currents transport and sort shells by size and shape, and may even produce preferred orientations in burial.

The kinds of evidence indicative of each assemblage are listed in Table 6.4. Assemblages in regions of high sedimentation rates, and those that have been rapidly buried, leave relatively unbiased records of organisms with hard parts.

Assemblages long exposed show signs of disturbance and compositional bias (Table 6.4), but in some ways can resemble one that has been rapidly buried. Conclusions about such "communities" may be equivocal because some information has been lost, and they are probably time-accumulated mixtures of communities. The majority of fossil "communities" may be of this second type.

If a community were subjected to both exposure and transportation we could expect to see characteristics similar to those of an exposed assemblage and extreme compositional bias (Table 6.4). Assemblages like this do not provide information about the community(ies) in which the species lived, but they do add data for reconstructing the environment of deposition.

Species Diversity

Diversity (usually measured as species number) is an important attribute of communities. It is a measure of the number of species an environment can accommodate, and thus (among other factors (Laboratory 11)), the numbers of ways that organisms have evolved to coexist. It is also a reflection of the complexity of a community, proportional to the number of ways organisms can "make a living", habitats, and possibly, the number of organism interactions.

Generally, the more stable and predictable the physical environment, the greater the community's complexity and diversity. Few species tolerate extreme conditions (e.g., large, unpredictable variations in physical parameters), or the consequent fluctuations in other species' abundances. In diverse communities, most species are represented by few individuals. Whereas communities of low diversity have few species, the abundances of most of which fluctuate from few to very many. Like harsh mountain

TABLE 6.4. EVIDENCE of the MODES of FOSSIL ASSEMBLAGE FORMATION.

FEATURE	RAPID BURIAL	EXPOSURE	EXPOSURE & TRANSPORTATION
DIVERSITY	High or low in appropriate environments.	Some loss OR slight increase of diversity.	Diversity much decreased
FEEDING GROUPS	Many filter-feeders, fewer herbivores, deposit-feeders and carnivores.	Many filter-feeders, fewer herbivores, deposit-feeders and carnivores.	Domination of 1 group or unusual proportions of each.
FRAGILE STRUCTURES	Fragile, articulated structures preserved intact.	Some abrasion and breakage of fragile structures.	Most abraded & broken.
ARTICULATION	If shells disarticulated, left & right valves present in = no.	Much disarticulation, left & right valves present in ≠ no.	Most disarticulated. Sorting of left & right valves.
HABITAT GROUPS	Many epifaunal, fewer infaunal, planktic and nektic species. Perhaps abundant deep infauna.	Epifaunal, planktic & nektic sp. less abndt., infaunal sp. more abndt.	Most habitat groups not represented.
SHAPE & SIZE	Wide range of shapes & sizes.	Some size &/or shape sorting.	Much size and shape sorting.
FRAGMENTATION	Little or no breakage.	Breakage common.	Much breakage.
ORIENTATIONS	Many species in life orientations.	Most not in life orientations.	Preferred current-induced orienation or none.
ECOLOGICAL GROUPS	Species expected to coexist. e.g., no mixing of fresh & marine sp.	Species expected to coexist. e.g., no mixing of fresh & marine sp.	Unusual associations of species.
ABUNDANCES of SPECIES	Very uneven, few very abundant, many uncommon or rare	Abundances more even than in column 1.	Very even / much skewed by size/shape sorting.
SPATIAL DISTRIBUTIONS	Individuals clumped-nonrandom, heterogeneous distributions.	Homogeneous, random.	Very homogeneous.
LIFE HABITS	Species consistent with sediment type.	Species consistent with sediment type.	Inconsistent with sediment type.

environments, marginal marine habitats are unstable and unpredictable, and thus, support fewer species than shallow seas. Diversity in the oceans is generally lowest near shore, and greater in the shallow offshore.

More diverse communities tend to contain many species belonging to each of several food pyramid levels or links in the food chain. Phytoplankton (Table 6.2) form the food chain's base, and filter feeders and herbivores feed on phyto- and zooplankton. Carnivores feed on each other and on herbivores, deposit feeders and filter feeders, while deposit feeders and scavengers recycle waste and nutrients. The higher the diversity, the more species in each food pyramid level. Resources tend to be abundant but unpredictable in low diversity communities. Highly diverse communities are found in environments with limited, but predictable, resources (e.g., coral reefs; see below) which are finely apportioned among the many species. Limited resources result in specialized means of gathering food, food type and occupation of space.

Examples of Ancient Communities

Each of the environments mentioned in this chapter support distinct communities. Diverse level-bottom communities (Appendix III: Plate IV, Figure 1) usually occur offshore on detrital mud and/or carbonate substrates where physical conditions and resources tend to be stable and predictable.

Level-bottom communities contrast sharply with constructional communities (Appendix III: Plate III, Figure 1). In the latter, organisms build skeletons upward from the sea floor upon one another, thereby creating habitats at various levels and within the framework. Constructional communities include reefs and smaller organic buildups. As corals, stromatoporoids, bryozoans and algae grow from the bottom, their skeletons form a framework providing many habitats for other animals. The construction also allows vertical and lateral partitioning of space and food resources, thus, encouraging colonization by species with differing tolerances to turbulence and light intensity as well.

Reefs are the most diverse communities, not only because corals and other animals create additional habitats and their environments are stable and predictable, but most of all because their environment is resource (food) starved—regions of reefs are virtual "deserts " of the sea. Reefs are not so much gardens, as they are oases within resource-poor environments: situated in waters depleted of dissolved nutrients necessary for phytoplankton growth, and thus lacking the (usual) foundation of the food pyramid, the reef-coral—algae symbiosis becomes a proxy food chain base. Resource scarcity promotes new habitats through reef growth, and thus, ecological and species diversity.

Mesozoic and Cenozoic communities were similar to those of the Paleozoic, yet differed from them in several important respects (Appendix IIII: Plate V, Figures 3, 4). Brachiopods and crinoids have largely been replaced (not displaced) by bivalves and echinoids, bryozoans play a less important role, and trilobites have been replaced by crustaceans. The epifaunal habitat became further subdivided by tiering. The infaunal habitat became increasingly important and tiered (downward) with the evolution of infaunal bivalves whose bulldozing activities churned the substrate. The churning reduced sessile epifauna attachment sites and thus, their diversity. Predator-prey interactions also assumed greater importance, and the role of cephalopods was increasingly filled by predatory fish, crustaceans and gastropods.

VOCABULARY

Benthos, Brackish water, Biomass, Constructional community, Deposit feeder, Diversity, Ecology, Epifauna, Filter feeder, Food pyramid/web, Fresh water, Habitat, Infauna, Level bottom community, Life habit, Limiting factor, Marine water, Marginal marine, Nekton, Paleoecology, Pelagic zone, Photic zone, Phytoplankton, Plankton, Population, Sessile, Suspension feeder, Turbidity, Vagrant, Zooplankton.

LABORATORY EXERCISES

Note to Students: You may need the assistance of your instructor to identify some of the organic groups (& their life habits and habitats), present in the slabs for Exercise 6.5, 8 & 9 and/or in the Plates III-V (Exercise 6.5). If your instructor has assigned these/this exercise(s) as as pre-laboratory self-quiz, leave blank those questions which involve unidentifiable specimens/photographs, and seek your instructor's help during the laboratory session. Some questions may not have definitive answers or may not be answerable, depending on the slabs available.

LABORATORY EXERCISES

6.1. Habits and Habitats—Figures 6.12A & B show the outlines of the organisms in two Devonian communities illustrated in Appendix III, Plate III, Figure 1. Each organism outlined is labeled by letter (A-K). Determine and note on worksheet 6.1, the organic group (phylum, class, etc.), feeding type, & habitat & mobility categories to which each outlined, labeled organism in Figure 6.12A & B belongs. Do this by matching your answers to the above categories with the organismal group labels A-K on worksheet 6.1 under column labeled "Figure A" for Figure 6.12A, and column labeled "Figure B" for Plate III, Figure 1.

WORKSHEET EXERCISE 6.1A,B. PALEOECOLOGY–LIFE HABITS & HABITATS

GROUP A-K	TAXONOMIC GROUP		LIFE HABITAT (Ecological Zone)		LIFE HABIT (Mobility)		FEEDING TYPE	
	Fig. A	Fig. B	Fig. A	Fig. B	Fig. A	Fig. B	Fig. A	Fig. B
A								
B								
C								
D								
E								
F								
G								
H								
I								
J								
K								

NOTES

FIGURE 6.12A. DEVONIAN FAUNAL COMMUNITY

FIGURE 6.12B. DEVONIAN FAUNAL COMMUNITY

LABORATORY EXERCISES

6.2. Physical Tolerances of Organic Groups—Using the matrix charts of worksheet 6.2A-G, indicate which groups can be characterized as having many or all species belonging to a catagory, or that can/cannot tolerate, or are most/least tolerant of, the listed states of salinity, temperature, turbulence, water depth, oxygen content, turbidity, and sediment type(s). Do this by entering a number from 1 to 3 (where 1 = least and 3 = most tolerant/abundant) in the appropriate column. Leave blank the squares for species-parameter states that no species of a group can tolerate.

WORKSHEET EXERCISE 6.2A. SALINITY TOLERANCES

							6.2.B. TEMP.			6.2.C Turbidity			6.2.D. Depth				6.2.E. Turb.			6.2.F. Oxygen	6.2.G. Sediment	
	Fresh Water	Brackish Water	Marginal Marine	Marine	Marine Marginal	Hypersaline	Tropical	Temperate	Polar	Turbid	Periodic Turbidity	Clear	Intertidal	Sublittoral	Bathyal	Abyssal	Turbulent	Intermed.	Calm	Low Normal	Sand Silt Shale Carbonate	Most Physically Limited Taxa
Bivalves																						
Cephalopods																						
Gastropods																						
Trilobites																						
Brachiopods																						
Articulata																						
Inarticulata																						
Bryozoans																						
Echinoderms																						
Stalked																						
Echinoids																						
Corals																						
Tabulates																						
Rugosans																						
Scleractinia																						

WORKSHEET EXERCISE 6.3. LIFE HABITS & HABITATS

	FEEDING TYPES					HABITAT TYPES			MOBILITY TYPES				ENVIRONMENTS						
						Benthonic													
	Filter Feeders	Deposit Feeders	Scavengers	Carnivores	Symbionts	Epifaunal	Infaunal	Pelagic	Vagrant	Sessile	Nektonic	Planktonic	Tidal Flat	Lagoon	Delta	Shallow Marine	Clastic	Carbonate	Reef
Bivalves																			
Cephalopods																			
Gastropods																			
Trilobites																			
Brachiopods																			
Articulata																			
Inarticulata																			
Bryozoans																			
Echinoderms																			
Stalked																			
Echinoids																			
Corals																			
Tabulates																			
Rugosans																			
Scleractinia																			

6.3.Habits and Habitats—Summarize the feeding types and life habits of each of the groups listed in worksheet 6.3 by placing an "X" in the appropriate column. Indicate also the environments which each may have occupied.

LABORATORY EXERCISES

6.4. Community Types and Evolution—Plates III—V (Appendix IV) illustrate selected constructional and level bottom communities from the Paleozoic through the Cenozoic. The community types are labeled only in Plate III. You may need your instructor's help to identify some organic groups. Answer the following.

A) Determine the community types (level-bottom vs. constructional) shown in the 10 Figures of Plates III–V.

1. ______________________ 2. ______________________
3. ______________________ 4. ______________________
5. ______________________ 6. ______________________
7. ______________________ 8. ______________________
9. ______________________ 10. ______________________

B) Determine the diversity and identity of organic groups present in each scene.

1. ______________________ 2. ______________________
3. ______________________ 4. ______________________
5. ______________________ 6. ______________________
7. ______________________ 8. ______________________
9. ______________________ 10. ______________________

C) Determine the feeding types, life habits & habitats represented in each.

1. ______________________ 2. ______________________
3. ______________________ 4. ______________________
5. ______________________ 6. ______________________
7. ______________________ 8. ______________________
9. ______________________ 10. ______________________

D) Compare & contrast your above descriptions (including organic groups) of Paleozoic, Mesozoic and Cenozoic communities.

	Similarities	Differences
Paleozoic	______________________	______________________
Mesozoic	______________________	______________________
Cenozoic	______________________	______________________

E) Describe the changes observed in species diversity, composition, etc., among communities of these eras, & answer the following questions:

1) Which are constructional, & which are level bottom? Which seem to include both level-bottom & constructional types?

1. ______________________ 2. ______________________
3. ______________________ 4. ______________________
5. ______________________ 6. ______________________
7. ______________________ 8. ______________________
9. ______________________ 10. ______________________

2) Compare the organic group composition, life habits & habitats of the two constructional communities shown in Plates III-V. Indicate each community by giving the plate and figure numbers below.
a) (Plate & Figure #______) b) (Plate & Figure #______)

Group Composition ______________________ ______________________

Life Habits ______________________ ______________________

Habitats ______________________ ______________________

3) Which communities are tiered? ______________________

Which is tiered but not constructional? ______________________

Which animals are tiered in each? ______________________

4) Which communities (constructional/level-bottom and geologic period) have the highest, intermediate, & lowest diversities? ______________________

Categorize (relatively) the items below for constructional (C) vs. level-bottom (L) communities:

a) complexity; (C)__________________ (L)__________________

b) number of links between each feeding category (food web complexity);
(C)__________________ (L)__________________

c) relative numbers of species in each level of the food pyramid; &,
(C)__________________ (L)__________________

d) stability of members' population sizes & resource availabilities.
(C)__________________ (L)__________________

F) Compare diversities among:
1) Cambrian, Late Ordovician, Mississippian, Late Cretaceous, & Cenozoic level-bottom communities; &,
2) communities of Middle Silurian & Permian constructional communities.
1> # 2 or #2 > z# 1 (Circle answer)

Do either community types show trends (if so, which, and describe trend) over time in:

a) diversity; ______________________

b) sizes (# species) of each feeding group present; ______________________ or,

c) life habit/habitat groups present? Explain. ______________________

G) Relate the environments, & environmental stability-predictability of the two community types.
a) Constructional b) Level-bottom

Environments ______________________ ______________________

Stability-Predictability ______________________ ______________________

H) How do the diversities of sessile filter feeding, epifaunal benthos in Plates IV (Figures 1& 2) and V, (Figure 1) compare with that of Plate V, (Figures 3 & 4)?
Former > Later OR Later > Former. (Circle one).

1) What evolutionary events may explain the differing diversities? Explain. ______________________

____

2) Has there been any evident change in the organic group composition of any of the above life habit/habitat/feeding groups? ______________________

____

____

____

LABORATORY EXERCISES

6.5. Specimens—Fossil Assemblage Formation—Your instructor will provide three slabs containing fossil assemblages formed in different manners. As a class or in groups, first make & discuss observations of features pertaining to the way each fossil assemblage was formed. Then, each student will offer, in turn, an observation to the group, after which—& with the instructor's guidance—the class or group will determine which of the three types of assemblage formation (rapid burial, exposed, or exposed and transported) is represented in each slab. Use worksheet 6.5 to list the states of all criteria supporting your answers.

A) Which of the slabs represent"time-averaged" communities? ______________________________

____

B) In which communities are the original species compositions and population abundances most faithfully represented? ______________________________

FEATURES of ASSEMBLAGE FORMATION	SLAB 1	SLAB 2	SLAB 3
DIVERSITY			
FEEDING GROUPS			
FRAGILE STRUCTURES			
SHELL ARTICULATION			
HABITAT GROUPS			
SHAPE & SIZE			
FRAGMENTATION			
ORIENTATIONS			
ECOLOGICAL GROUPS			
ABUNDANCES			
SPATIAL DISTRIBCTION			
LIFE HABITS			

WORKSHEET EXERCISE 6.5. FOSSIL ASSEMBALGE FORMATION

6.6. Specimens-Ecology—Evaluate & list on worksheet 6.6 the: A) organic groups; B) feeding habits; C) habitats; &, D) physical parameter tolerances of the invertebrates in those two slabs which do represent samples of ancient communities (see exercise 6.5). Estimate the percentage of each feeding type present in each slab.

A) Do any of the slabs contain tiered assemblages? Which? ______________________________

B) Is there any evidence of infauna? ______________________________

C) By which feeding group were these communities dominated? ______________________________

6.7 Specimens–Environments—Characterize the type of environment represented by each slab in terms of salinity, turbidity, depth, oxygen, temperature and turbulence. Be sure to use both positive and negative evidence and to cite both. Record your conclusions on worksheet 6.6.

WORKSHEET EXERCISE 6.6.

SLAB	ORGANIC GROUPS	SALINITY	TURBIDITY	DEPTH	OXYGEN	TEMPERATURE	TURBULENCE
1.							
2.							
3.							

	FEEDING GROUPS	MOBILITY TYPES	HABITATS	DIVER-SITY	ENVIRON. PREDICT	ENVIRON. STABILITY	DISTANCE to SHORE	RESOURCE AVILABILITY
SLAB 1					RANK:	RANK:		RANK:
SLAB 2					RANK:	RANK:		RANK:
SLAB 33					RANK:	RANK:		RANK:

6.8. Specimens-Diversity—Determine each slab's diversity & rank it according to depth or distance from shore (onshore, "intermediate", offshore), environmental stability (High = 1; "Med." =2; Low = 3), & resource availability (High/Low), & environmental predictability (High = 1; "Med." = 2; Low = 3). Record the ranks on worksheet 6.6. Assume entire or sizeable fragments that look similar belong to the same species. Your instructor will provide help with fossil identification.

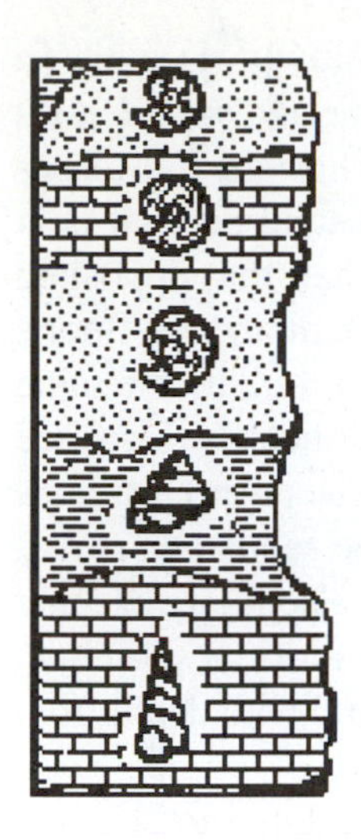

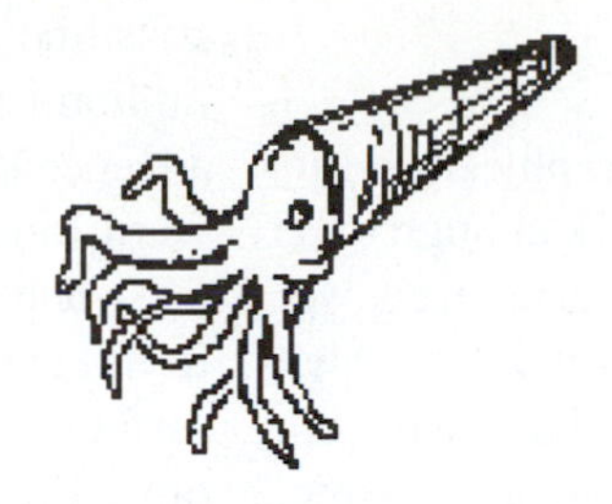

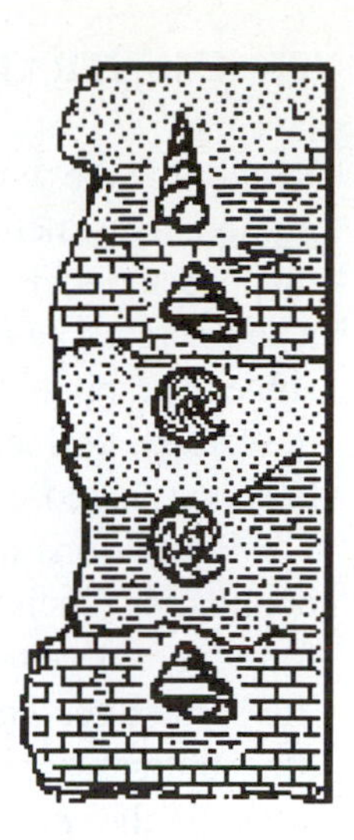

7
Fossils and Time

Objectives
To demonstrate mastery of the material you should understand: 1) how the geologic time scale was constructed; 2) why formations can be time transgressive; 3) the basic methods of biostratigraphy and the characteristics of guide fossil species; 4) which fossil groups are of most use in biostratigraphy, and why; and, 5) how to construct biostratigraphic zones.

Using This Laboratory
The material discussed through "Guide Fossils" is conceptual background. The last three sections concern the methods used to apply these concepts. Understand the concepts before continuing with the discussion of methodology. When completing the exercises, make frequent reference to the sections describing these methods, Table 7.1, and use the accompanying figures as examples. Use pencil to complete the correlation worksheet exercises.

In this laboratory we will see how fossils and the principle of biologic succession (Laboratory 3) are used to correlate strata and add the dimension of time to the rock record. In fact, it can be argued that were it not for fossils, geologists would not so readily have recognized that the Earth does have a history that preceded, and is independent of, human experience. In this chapter we will discover the principles of biostratigraphy and how they are applied to the fossil record to formulate a temporal sequence of strata and events in Earth history.

Recall that the principle of biologic succession reflects changing species compositions from one rock layer to the next; a phenomenon which results from, but does not require, organic evolution. The phenomenon stems from the fact that species compositions of natural fossil associations change with time, and are thus recorded in the successive rock layers. Changes of great magnitude, like mass extinction and radiation are commonly used to mark the boundaries of eras, and many periods and lesser divisions of Earth history. Sedimentary strata alone cannot provide a temporal sequence of events in most cases because the rock (re)cycle (with certain exceptions) merely produces more of the "same" few types over time. Cross-cutting relationships and unconformities (Laboratory 3) provide coarse, regional temporal frameworks, and radiometric dating is not always possible. Thus, it is the record of life that tells us each age of the Earth is unique and that there is a history to be excavated.

TIME in STRATIGRAPHY

Recall that the working unit of stratigraphy is the formation: a distinctive body of rock, thick enough to be mapped and widespread enough to be recognized over a large area. The definition's application is somewhat subjective, but formations can be seen, hammered on, described, defined, walked out and mapped. Time, however, is more elusive.

Formations are defined without regard to time boundaries. Their upper and lower contacts with other formations need not, and are not (usually), everywhere the same age. Thus, to relate formations (or other lithostratigraphic units; see Table 3.1) to time we must consider the relationships that can exist between the two. The commonest relationship is illustrated by Figure 7.1A,B. In part A, a thick deposit of nearshore sand extends many kilometers seaward and along the coastline at time t_1. As sea level slowly rises over a million years to time t_2, the shoreline and sand deposit advance landward (Figure 7.1B). Thus, the most landward (and upper) portion of this sand deposit is at least one million years younger than that deposited from the base to time t_1. At any instant, the sea floor's surface sediments were deposited at roughly the same time. But its slope changes as sediment is deposited and the sea advances (transgresses) landward. Millions of years later, when this distinctive sandstone is recognized as a formation, geologists also recognize that the strata are younger close to the old(est) shoreline, and increase in age away from it, and with depth from the strata's top to its base.

This concept can be understood in another way (Figure 7.2A). Imagine a volcanic eruption has blanketed the sea floor with ash at time t_1. This instantaneous event marks a planar surface that is everywhere

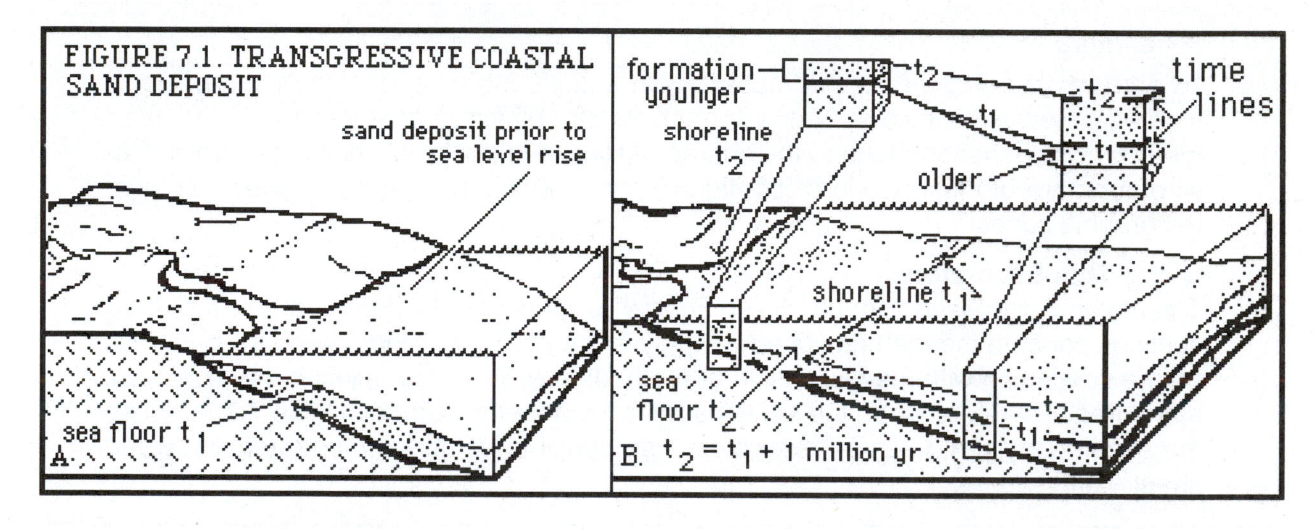

FIGURE 7.1. TRANSGRESSIVE COASTAL SAND DEPOSIT

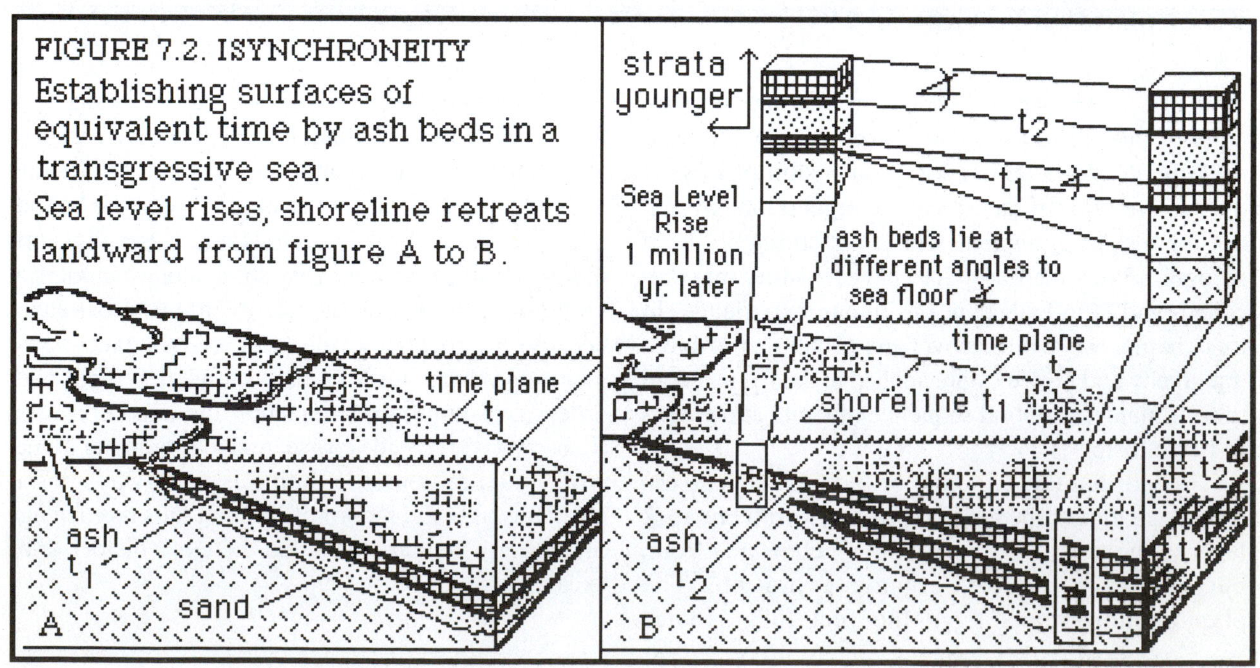

FIGURE 7.2. ISYNCHRONEITY
Establishing surfaces of equivalent time by ash beds in a transgressive sea.
Sea level rises, shoreline retreats landward from figure A to B.

contemporaneous within the formation. When the ash bed is traced over a large area, its surface parallels neither the upper nor lower formation contacts. Instead the bed lies at an oblique angle to the formation contacts. If a second ash fall occurred at time t_2, it would lie at an oblique angle to the first ash and the formation contacts (Figure 7.2B). Formations of this kind (and of Figure 7.1) are time-transgressive (or diachronous; *dia* = across, *chronos* = time) units because they are not everywhere of the same age. Each ash bed is the same age everywhere (or isochronous; *iso* = same), but the formation's cross-cutting by isochronous beds indicates its diachroneity. Alternatively, if the first and second ash beds had been (inappropriately) used to define the upper and lower contacts of this rock unit, the "formation's" boundaries would not be time transgressive, yet any traceable bed within it would.

Some instantaneously deposited strata (e.g., turbidites (Laboratory 2), ash beds, storm deposits, etc.) may have formation status, and be nondiachronous. As such, they are of great importance for they allow geologists to recognize isochronous surfaces, and with the use of fossils, to make temporal correlations among lithostratigraphic units in distant regions (see below).

GEOLOGY in FOUR DIMENSIONS

Although superposition (Laboratory 3) tells us that the uppermost rocks of a sequence were deposited after those below, their diachroneity precludes tracking sequential events through time over regions of a formation's outcrop, or among formations of other regions, by simply stacking rock units. Thus, it is necessary to define rock units encompassing strata deposited during the same intervals of time. Only then can time-equivalent rock units be recognized and regionally correlated. These are called chrono-stratigraphic units (see Table 3.1): divisions of strata that record a time interval, and defined as rocks subdivided on the basis of time. To fully understand chronostratigraphic units we must consider the relationship between sedimentation and time's passage.

Time is a directional linear continuum recognized by a sequence of events, but depositional events (and thus rocks), are not continuous through time: sediment deposition may cease, start again and vary in rate as time passes. Although rocks record depositional events (in time), sediments are not deposited continuously, or constantly in any region(s). Thus a rock sequence may span much more time than its thickness would appear to indicate had deposition been continuous (Figure 7.3). In other words, most sedimentary rock sequences contain innumerable short, and some long, depositional gaps, and if erosional unconformities are also present, much of a region's history may not be represented by rock at all. A chronostratigraphic unit then, records an interval of time, more or less of which is represented by rocks, within the unit's boundaries. Nevertheless, a more or

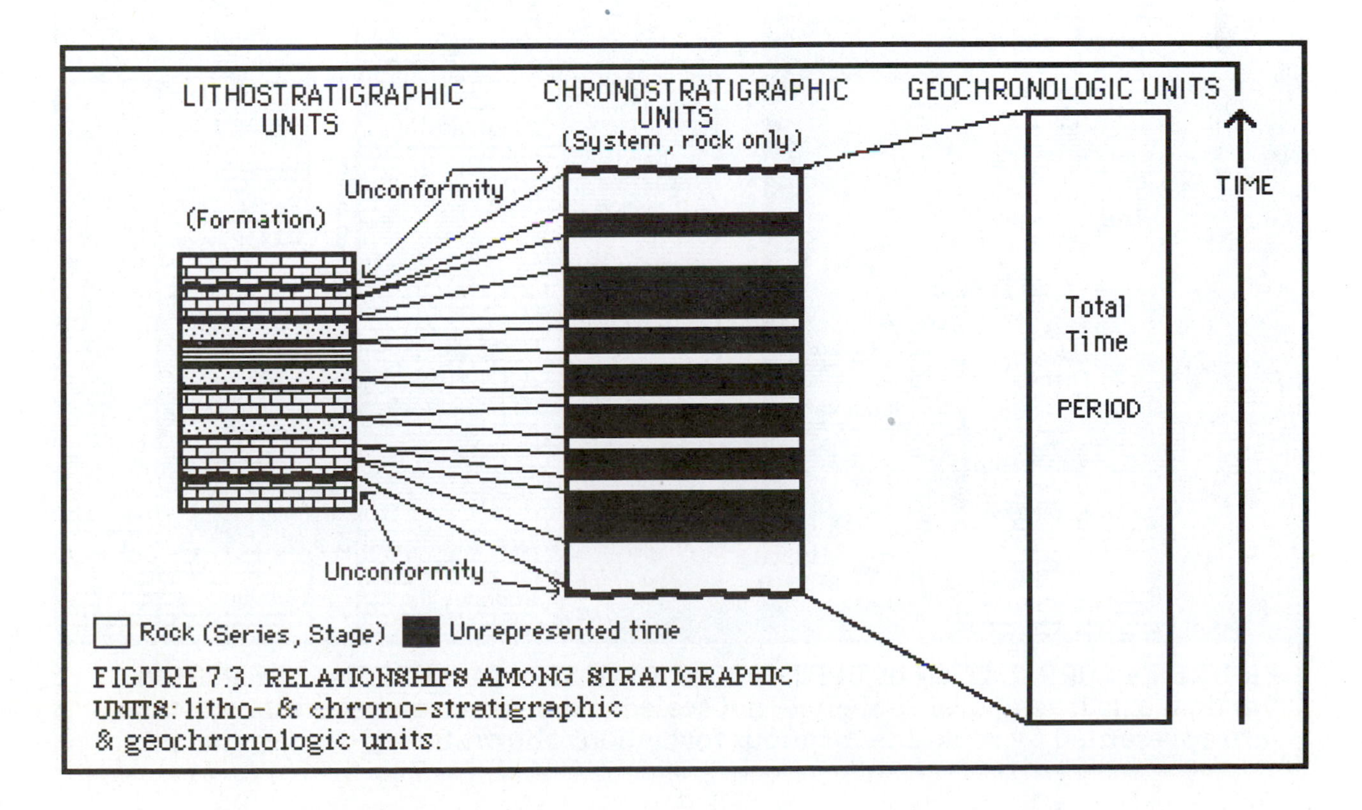

FIGURE 7.3. RELATIONSHIPS AMONG STRATIGRAPHIC UNITS: litho- & chrono-stratigraphic & geochronologic units.

less continuous record of any time interval was deposited somewhere, and thus, the total (continuous) stratigraphic record has been reconstructed by correlating sequences among regions.

To complete the logical chain from strata to time's passage, a link between chronostratigraphic units and geologic time is needed. Units of geologic time divide the temporal continuum. Each chronostratigraphic unit corresponds to a geochronologic unit or time interval in which it was deposited (Table 3.1). For example, the Ordovician System is that rock deposited during the Ordovician time interval (geologic period). Thus, geochronologic units are divisions of time distinguished and defined by the chronostratigraphic units within their boundaries. Just as periods are subdivided into temporal epochs and ages, systems are divided into corresponding series and stages. The hierarchy of lithostratigraphic units, however, is not equivalent to those of chronostratigraphic or geochronologic units.

Figure 7.4 illustrates the relationship between lithostratigraphic, chronostratigraphic and geochronologic units of the Ordovician in a region of the United States. Locally, long spans of time are not represented by rocks. A continuous regional sequence can, however, be reconstructed when formations are fitted into chronostratigraphic units by their fossil content, radiometric geochronology (Laboratory 4), marker beds (Laboratory 3), or methods described below. These methods bridge the gap between rock units and time.

FOSSILS in STRATIGRAPHY

Biostratigraphy is the correlation of stratigraphic units based on fossil content. The principle of biologic succession can, given certain qualifications, allow linkage of these units to geologic time. First the local and regional stratigraphic ranges of certain fossils are determined (Figure 7.5). Once sufficient exposures containing the distinctive species have been correlated, the geologic range of each can be plotted. A geologic range is that interval between a species' first and last appearance.

Once geologic ranges have been established, biostratigraphic zones are determined. These are bodies of rock defined by their fossil content. They can be

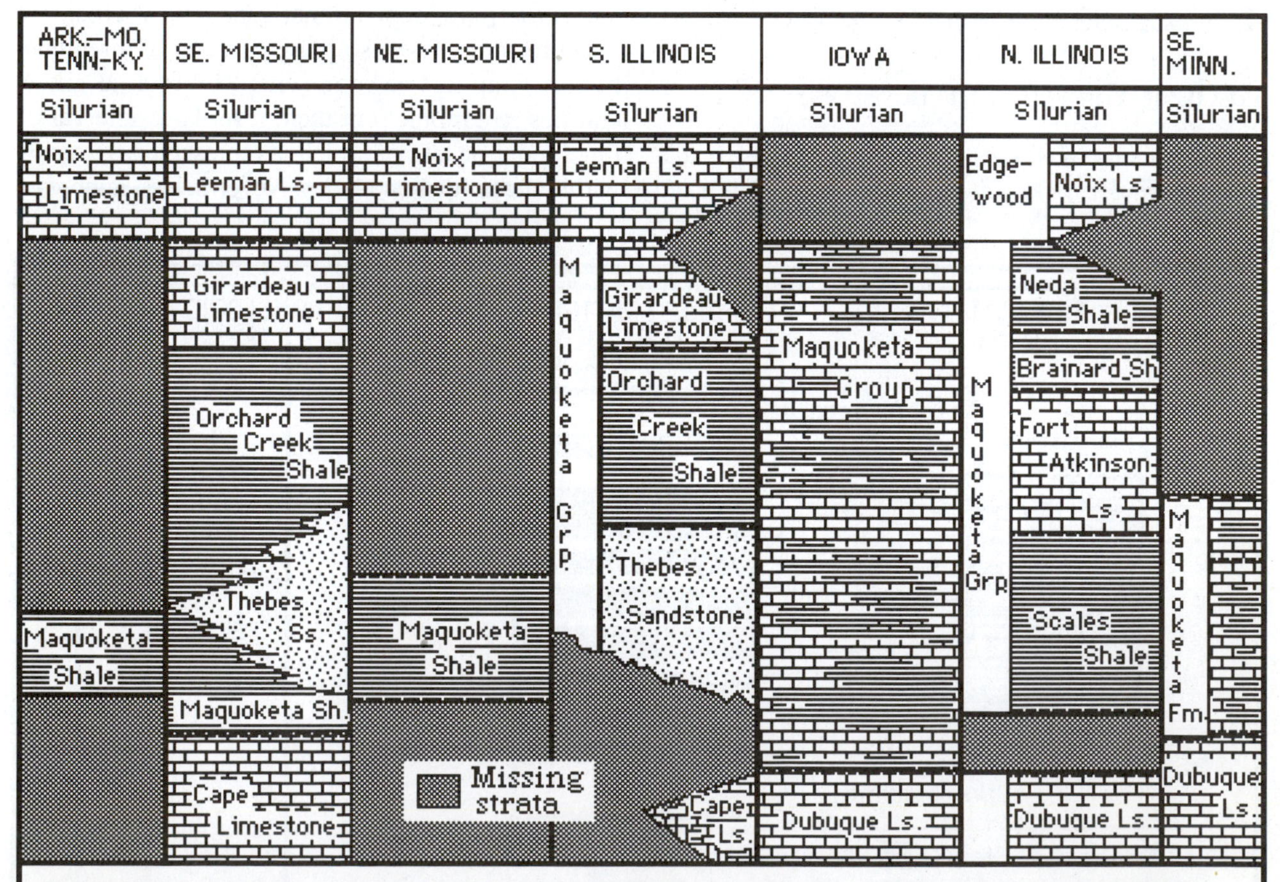

FIGURE 7.4. CORRELATION of UPPER ORDOVICIAN STRATA, MISS. VALLEY REGION. Vertical axis is temporal sequence, not scaled to time among regions. Much time unrepresented by rock. Diachronous formations shown by zig-zag/angled line.

linked to chronostratigraphic units (recall the principle of biologic succession), but differ fundamentally from lithostratigraphic units (Figure 7.6A). When tied to nondiachronous or radiometrically dated beds, the result is a chronozone recognizable outside the region by its fossil content (Figure 7.6B). Chronostratigraphic units may coincide with biostratigraphic units, and in this or other ways (below), chronostratigraphic units can be linked by biostratigraphic zones, and chronozones linked to geochronologic units (Figure 7.6C). To understand biostratigraphic zone construction we must know why certain fossils are useful stratigraphic tools.

Guide Fossils—Biostratigraphically useful species are known as guide fossils (or index fossils) because they can be used as guides for recognition of chronostratigraphic units. Guide fossils usually:

1) have widespread distributions;
2) short temporal durations (stratigraphic ranges) resulting from rapid speciation and extinction;
3) abundance throughout their geographic and geologic ranges; and,
4) are easily recognized.

There are several causes of widespread species distributions (see Laboratory 11). The larval stage of many invertebrates is long, allowing wide dispersal by currents. If a species having planktonic larvae is also tolerant of wide variations in physical parameters (Laboratory 6), it is more likely to be widespread. Most benthos' distributions are limited by substrate type or associated physical variables, and so are not the most useful correlation tools. In contrast, floating and swimming species tend to be widespread and very useful for correlation. Ironically, many widespread species also have very long durations because related species evolved slowly.

Correlations of strata and biostratigraphic zonations are usually based on multi-species methods because:

1) the biases of local species' ranges will be minimized;
2) the species used need not possess all (or any) guide fossil characteristics;
3) the interval of species' co-occurence can be recognized over a larger region; and most importantly; and,
4) temporal resolution is enhanced.

Biostratigraphic Methods—Range zones are intervals of rock identified by the range of a single species (or genus), and constructed by correlation among combined, less complete local range zones. In a hypothetical example (Figure 7.7), the geologic range of each species of the trilobite genus *Phacops is* determined by combining the local range zones of each species. The geologic range of the genus is the interval spanned by the combined geologic ranges of its member species.

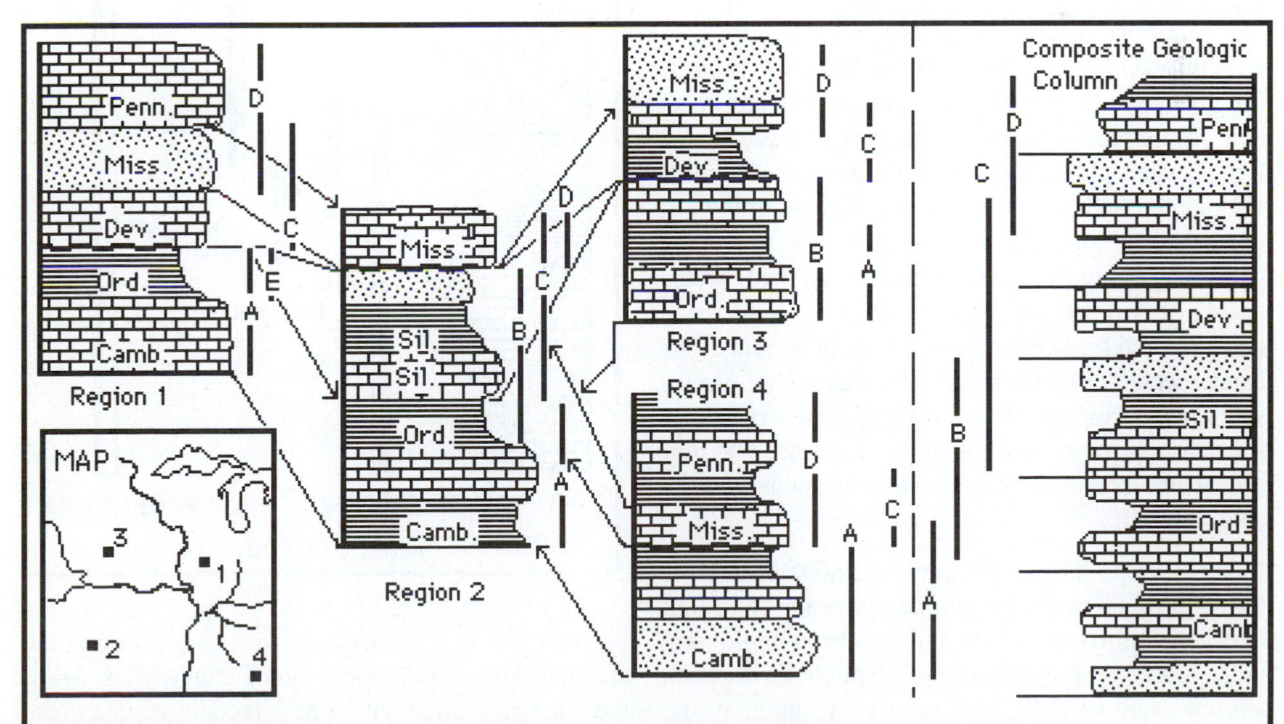

FIGURE 7.5. REGIONAL CORRELATION by GEOLOGIC RANGES

Ranges of several species discovered in each of several exposures in a region. Local ranges may be short. Geologic range= longest range (sp. A, region 4)/combined range (sp. B).

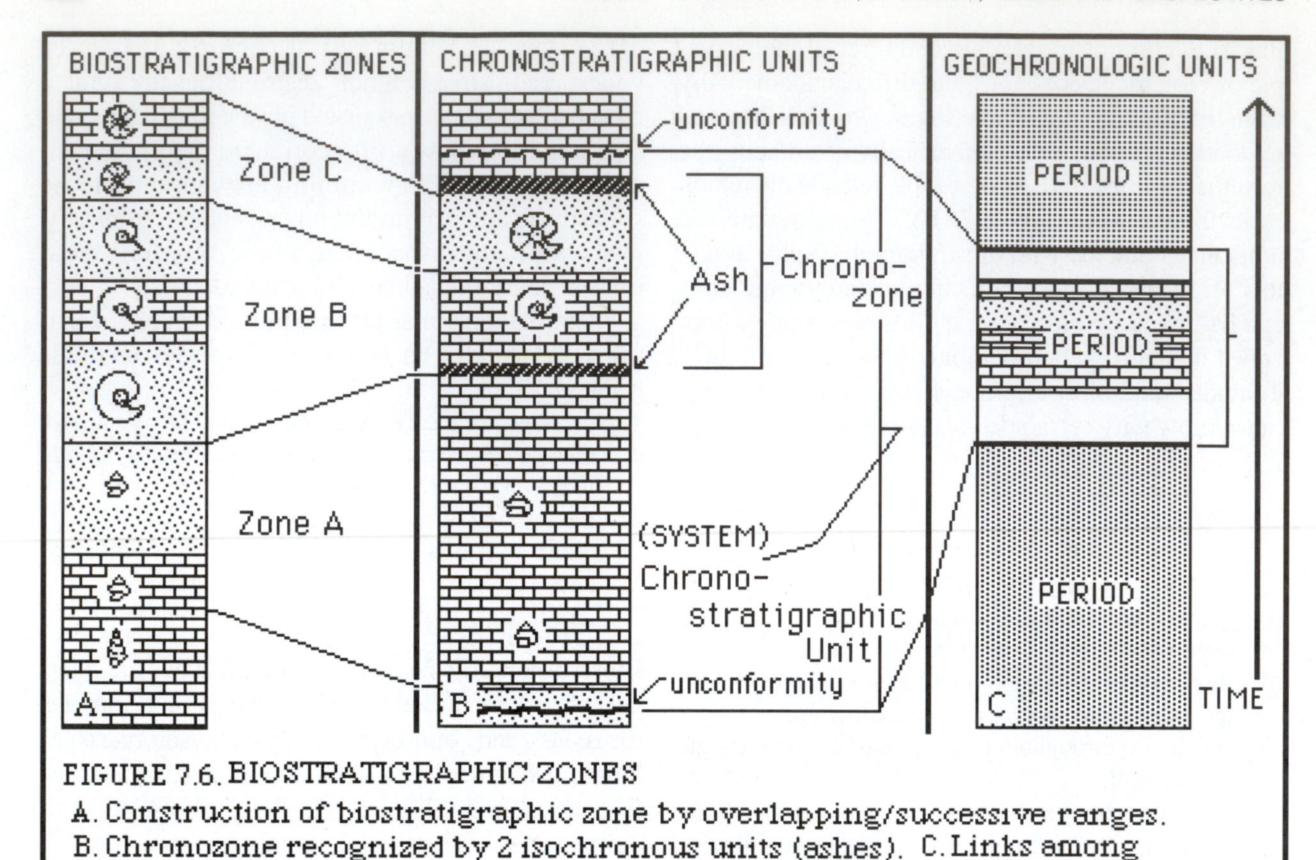

FIGURE 7.6. BIOSTRATIGRAPHIC ZONES
A. Construction of biostratigraphic zone by overlapping/successive ranges.
B. Chronozone recognized by 2 isochronous units (ashes). C. Links among chronostratigraphic units and geologic time by biostratigraphic & chronozones.

Because range zones are subject to error (above) and may encompass long time spans, interval zones, based on intervals created between the highest or lowest occurrence of a pair of species, provide improved resolution. Figure 7.8 shows four species' ranges. Single and combined geologic ranges span considerable amounts of time, but consecutive zones of overlap between pairs of species' ranges are considerably shorter and constrained by the highest, or lowest, occurrence of one. Another type of interval zone is defined as the interval between the first (or final) occurrences of two species (Figure 7.9). A body of strata characterized by a certain assemblage of several coexisting species, or by the ranges of three or more species is defined as an assemblage zone (Figure 7.10). Only the majority of an assemblage's species need be present at any locality for correlation with another region (Figure 7.10).

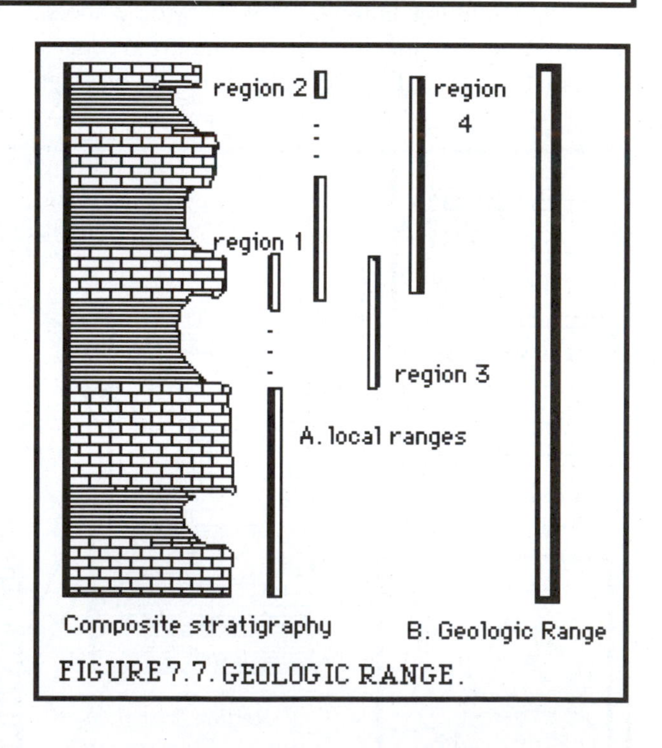

FIGURE 7.7. GEOLOGIC RANGE.

From Rock to Time—With these examples of biostratigraphic methods we can complete the final link between lithostratigraphic and geochronologic time units. If a biostratigraphic zone's boundaries are each shown to be of the same age everywhere by the physical tracing or radiometric dating of marker beds, then the zone delimits a time interval (= chronozone). Such a zone defines an interval of time that, like two volcanic ash beds, brackets a rock sequence in time and thus, links it to a geochronologic unit. Thus, where the same biostratigraphic zone is recognized elsewhere it can be linked to the recognized chronozone and, to a geochronologic unit. Figure 7.11 illustrates the results of various zonation methods applied to the stratigraphic

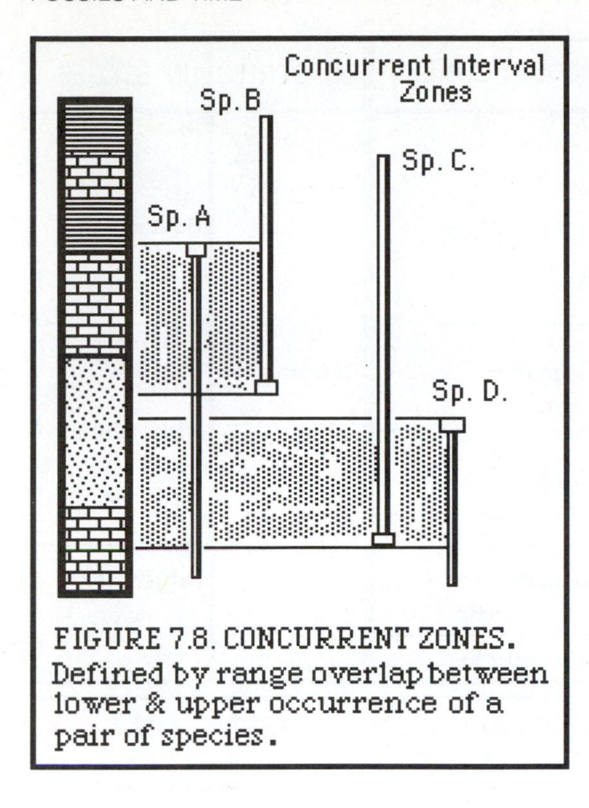

FIGURE 7.8. CONCURRENT ZONES. Defined by range overlap between lower & upper occurrence of a pair of species.

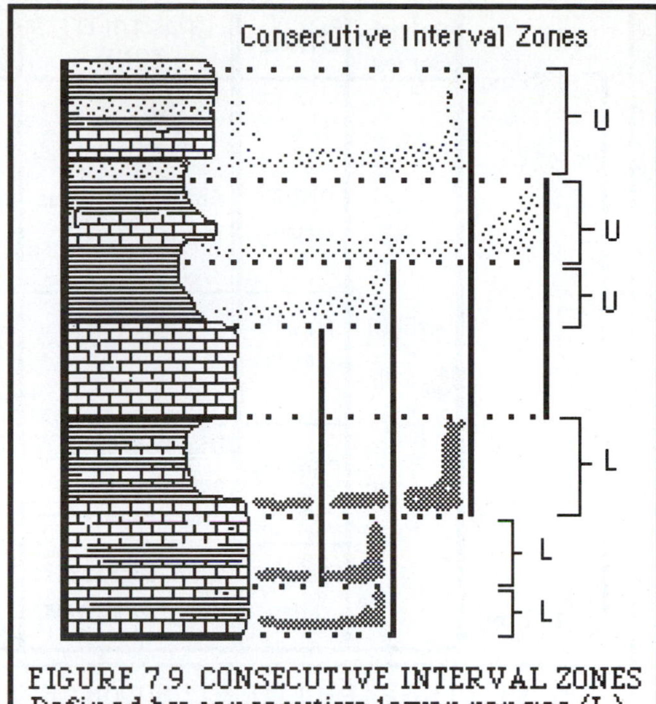

FIGURE 7.9. CONSECUTIVE INTERVAL ZONES Defined by consecutive lower ranges (L), upper ranges (U).

distributions of conodonts (Appendix I) and graptolites, and the linkage of these zones to the geologic time scale.

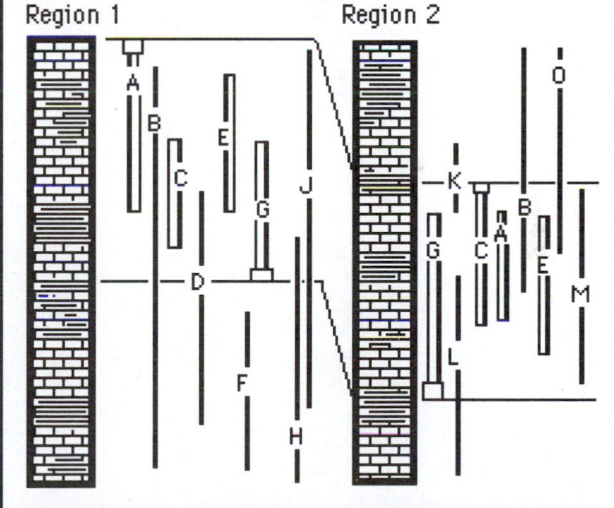

FIGURE 7.10. ASSEMBLAGE ZONES. Defined by co-occurrence of several species. Two regions correlated by occurrence of species A, B, C, D, E & G.

GEOLOGIC RANGES of ORGANISMAL GROUPS

The known geologic ranges of familiar groups are given in Table 7.1. Some are more useful for correlation than others (above). Often a group is most useful during the chronostratigraphic intervals in which it reaches high(est) diversity (e.g., trilobites zones in Cambrian and Lower Ordovician strata) because the many species were wide-ranging and of short duration. Other species are equally useful throughout their geologic range. You should now begin Exercise 7.1 by plotting the geologic ranges of species as you study Table 7.1.

VOCABULARY

Assemblage zone, Biostratigraphy, Biostratigraphic zone, Chronostratigraphic unit, Diachronous, Formation, Geologic Time, Geologic Period, Geologic Range, Geochronologic unit, Guide Fossil, Isochronous, Lithostratigraphic unit, Range zone, Stage, Lithostratigraphic System, Time transgressive formation, Zone.

AGE	British Series	STAGE	GRAPTOLITE ZONE	CONODONT ZONES	US. SERIES	US. STAGES
435 my		Hirnantian	Amplexograptus inviti	Gamachognathus		Gamachian
	Ashgill	Rawtheyian		Aphelognathus shatzeri	Cincinnatian	Richmondian
		Cautleyian	Digellograptus complanatus	Aphelognathgus divergens		
440 my			Climacograptus manitoulinensis	Aphelognathus grandis		Maysvillian
		Pusgillian				
442 my				Oulodus robustus		

FIGURE 7.11. GRAPTOLITE & CONODONT BIOSTRATIGRAPHIC ZONES - UPPER ORDOVICIAN U.S.A. & BRITAIN.

TABLE 7.1. GEOLOGIC RANGES of MAJOR FOSSIL GROUPS.

GROUP	GEOLOGIC RANGE	PERIOD(S) of HIGHEST DIVERSITY	COMMENTS
Nautiloids	U. Camb.- Recent	Ordovician	Nektonic
Ammonoids	L. Dev. - U. Cret.	Jur., Cretaceous	Nektonic $2 - 9 \times 10^6$ yr. interval zones
Bivalves	L. Cambrian-Recent	Cenozoic	many eurytopic, used to establish Cenozoic epochs.
Brachiopods	L. Camb.- Recent	Devonian	used for Devonian-Mesozoic
Trilobites	L. Camb.- U. Perm.	U. Cambrian -L. Ordovician	Cambrian Extinctions/ radiations used for zonation
Ostracodes	L. Camb.- Recent	throughout	microfossils, fresh & marine
Stromatoporoids	L. Ord.- U. Cret.	Sil., Dev., Jur.	————
Tabulate Corals	L. Ord.- U. Perm.	Silurian, Devonian	————
Rugose Corals	M. Ord.- U. Perm.	Silurian, Devonian	————
Scleractinian Corals	M. Trias.- Recent	Recent	————
Bryozoa	L. Ord.-Recent	Ord., Dev., Recent	————
Foraminifera	L. Camb.- Recent	Carboniferous -Cenozoic	many planktonic, micro-fossils, ocean sediments
Diatoms	Cret.- Recent	————	planktonic, microfossils, ocean sediments
Coccolithophores	Jur. - Recent	————	planktonic, microfossils, ocean sediments
Echinoderms	L. Camb. - Recent	crinoids and blastoids in Mississippian	————
Crinoids	M. Camb. - Recent		
Blastoids	M. Sil. - U. Perm.		
Echinoids	L. Camb.- Recent	Cenozoic	————
Graptolites	M. Camb.-L. Penn.	Ordovician - L. Devonian	many planktonic, black shales, deep water sed. zones 1 - 5 my. long.
Conodonts	U. Proter.- U. Trias.	Paleozoic	many nektonic, short zones, fission track dating

WORK SHEET EXERCISE 7.1.

Phanerozoic	Cenozoic	Quaternary		Recent	
				Pleistocene	
		Tertiary	Neogene	Pliocene	
				Miocene	
			Paleogene	Oligocene	
				Eocene	
				Paleocene	
	Mesozoic	Cretaceous			
		Jurassic			
		Triassic			
	Paleozoic	Permian			
		Pennsylvanian			
		Mississippian			
		Devonian			
		Silurian			
		Ordovician			
		Cambrian			

7.1. Geologic Ranges & the Geologic Time Scale—A) Plot the geological ranges of the taxa discussed in the last section of this chapter (Table 7.1) on worksheet 7.1. B) Construct zonations & indicate type(s) used.

7.2. Geologic Ranges & the Geological Time Scale—Plot the geological ranges over which the species discussed above are most useful on worksheet 7.2. Indicate the zone type(s)used.

WORKSHEET EXERCISE 7.2.

Phanerozoic	Cenozoic	Quaternary		Recent	
				Pleistocene	
		Tertiary	Neogene	Pliocene	
				Miocene	
			Paleogene	Oligocene	
				Eocene	
				Paleocene	
	Mesozoic	Cretaceous			
		Jurassic			
		Triassic			
	Paleozoic	Permian			
		Pennsylvanian			
		Mississippian			
		Devonian			
		Silurian			
		Ordovician			
		Cambrian			

WORKSHEET EXERCISE 7.3.A. RANGES of FOSSIL GROUPS & BIOSTRATIGRAPHIC AGE.			
SLAB	FOSSIL GROUPS REPRESENTED	RANGES of FOSSIL GROUPS	AGE
1			
2			
3			

7.3. Specimens—Your instructor will provide fossiliferous slabs or sets of fossils.

A) Identify the fossils in each (to the phylum, class or subclass level); and,

B) using the information in Table 7.1, determine the geologic period(s) during which each identified fossil group existed and plot these on worksheet 7.3B, below. Use the biostratigraphic methods discussed above to determine the period(s) represented by each slab and indicate these on worksheet 7.3A). [You may need your instructor's assistance for some identifications. After you have completed the exercise, your instructor may provide further information on the ranges of the organismal groups in each slab or set enabling you to check initial results and to further narrow the intervals of time represented by each.]

WORKSHEET EXERCISE 7.3B.					
EON	ERA	PERIOD		Epochs	PLOTS of GEOLOGIC RANGES
Phanerozoic	Cenozoic	Quaternary		Recent	
				Pleistocene	
		Tertiary	Neogene	Pliocene	
				Miocene	
			Paleogene	Oligocene	
				Eocene	
				Paleocene	
	Mesozoic	Cretaceous			
		Jurassic			
		Triassic			
	Paleozoic	Permian			
		Pennsylvanian			
		Mississippian			
		Devonian			
		Silurian			
		Ordovician			
		Cambrian			

LABORATORY EXERCISES

7.4. Biologic Succession and Stratigraphy—Determine the ages of the strata and/or stratigraphic relationships shown below and answer the question . Assume the stratigraphic level at which an icon of a fossil group appears is its highest occurrence unless otherwise indicated.

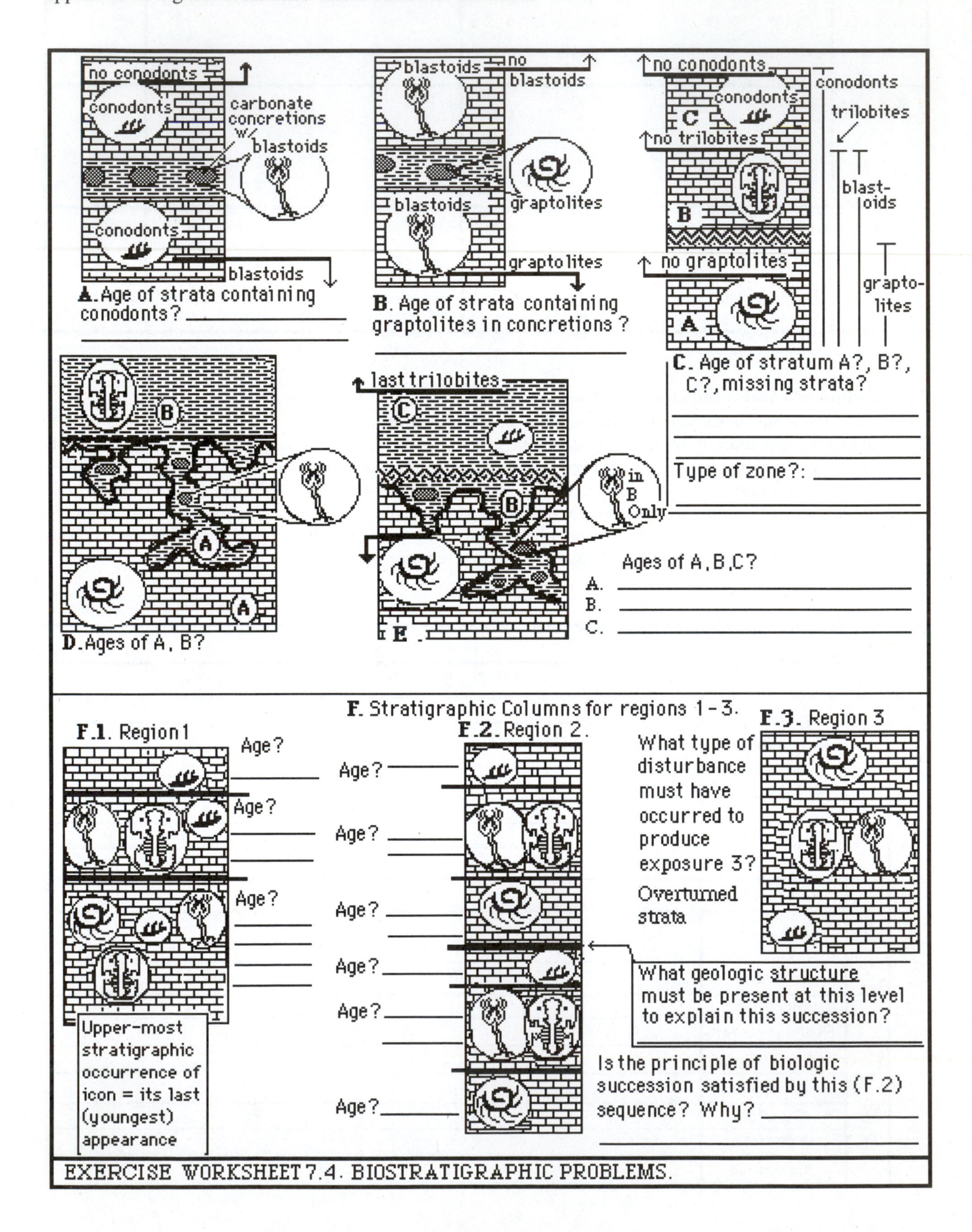

LABORATORY EXERCISES

7.5. Guide Fossils, Zonation, and Systems—Indicate the chronostratigraphic range and/or single most probable chronostratigraphic unit (if more than one period is possible) to which you could assign a lithostratigraphic unit containing (you may wish to use worksheets 7.1, 2 or 3):

A) Archaeocyathids (Appendix I), trilobites, gastropods, inarticulate brachiopods, and conodonts (Appendix I)?

____

B) Bryozoans, brachiopods, nautiloid cephalopods, bivalves, tabulate corals, crinoids, and graptolites? ______

____

C) Brachiopods, brachiopods, crinoids, nautiloid cephalopods, rugosan corals, conodonts, and blastoids? ____

____

D) Gastropods, ammonites, brachiopods, conodonts, and trilobites? ______________________

E) Ammonites, rudistid bivalves, and gastropods? ______________________

F) Coccolithophores (Appendix I), gastropods, bivalves, belemnites and ammonites? ______________

____

G) Fusulinids, (Appendix I) trilobites, crinoids, and brachiopods? ______________________

H) Diatoms (Appendix I), echinoids, bivalves and gastropods? ______________________

7.6. Biostratigraphic Zonation—Worksheet 7.6 illustrates four geologic columns and ranges of species in each. Construct lithostratigraphic correlations using dashed lines, and chronozones using dark solid lines among the four columns.

A) Construct:

1) concurrent zones (Figure 7.9) for species in the upper and middle portions of each column;

2) consecutive range zones for those in the middle portions of the columns (Figure 7.8); and,

3) assemblage zones (Figure 7.10) for those in the lowermost portions of the columns (3-4).

Correlate the lithostratigraphic units within the top, middle and bottom portions of the diagram, and determine the geometric relationship between the litho- and biostratigraphic units in A—D.

B) Which biostratigraphic zone (3-4) has a boundary paralleling an isochronous bed? Which is diachronous?

____

C) Which can be considered equivalent to a chronozone? ______________________

____

D) Infer the boundaries of the chronozone(s) in the stratigraphic column to the left using the interval zones at the far right. ______________________

____

E) Determine the stratigraphic level of the chronozone in column #4. ______________________

____

7.7. Guide Fossils and Zonation—Construct assemblage zones in worksheet 7.7 (next page) using only those species having life habits/habitats most conducive to widespread dispersal. Species restricted to certain rock types, and species tolerant of extreme environmental variation and not linked to lithology are both present.

A) Determine which are restricted (R) to a certain lithology and which are not (N).

R ______________________ N ______________________

______________________ ______________________

______________________ ______________________

WORKSHEET EXERCISE 7.6

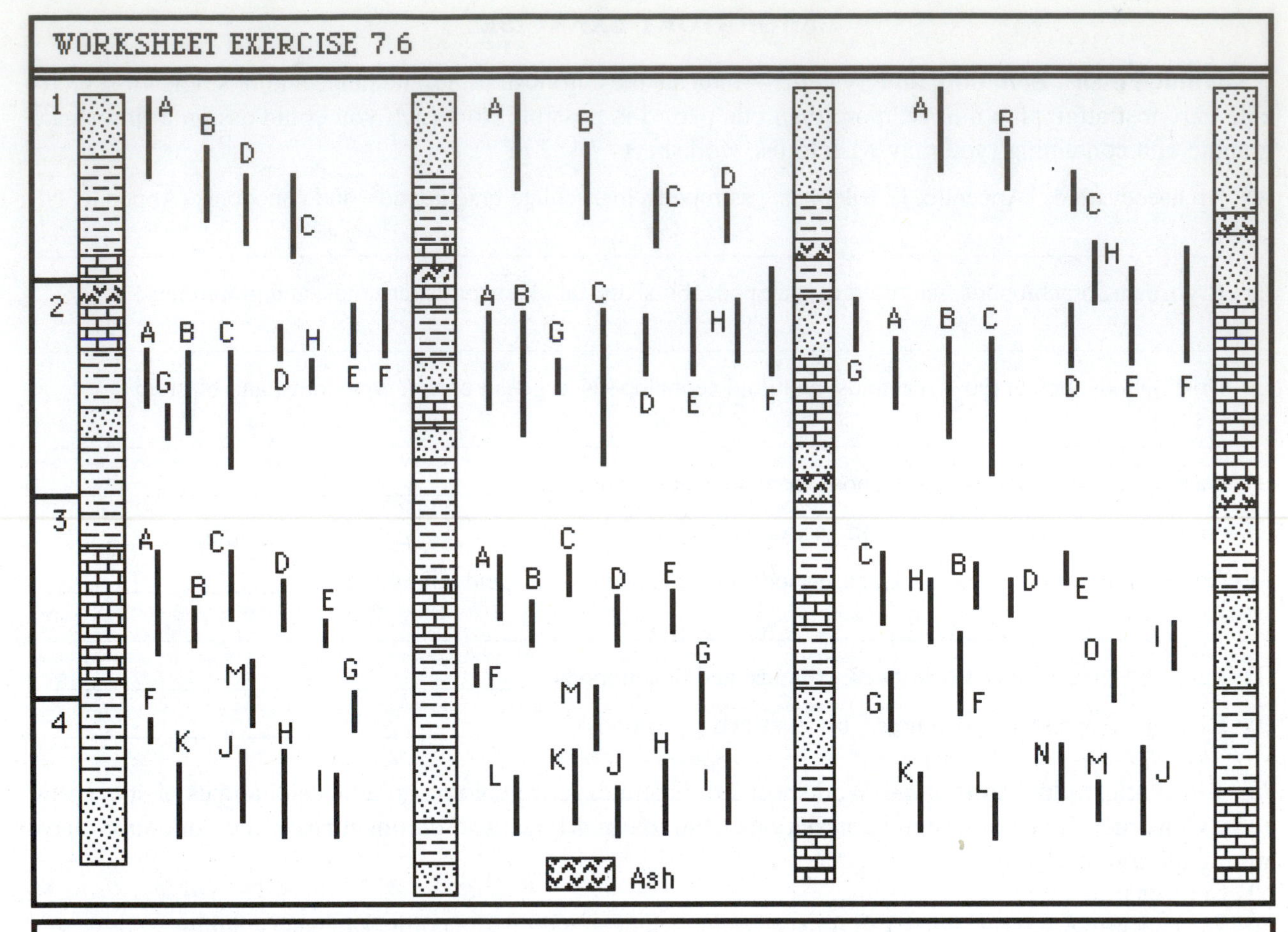

WORKSHEET EXERCISE 7.7.

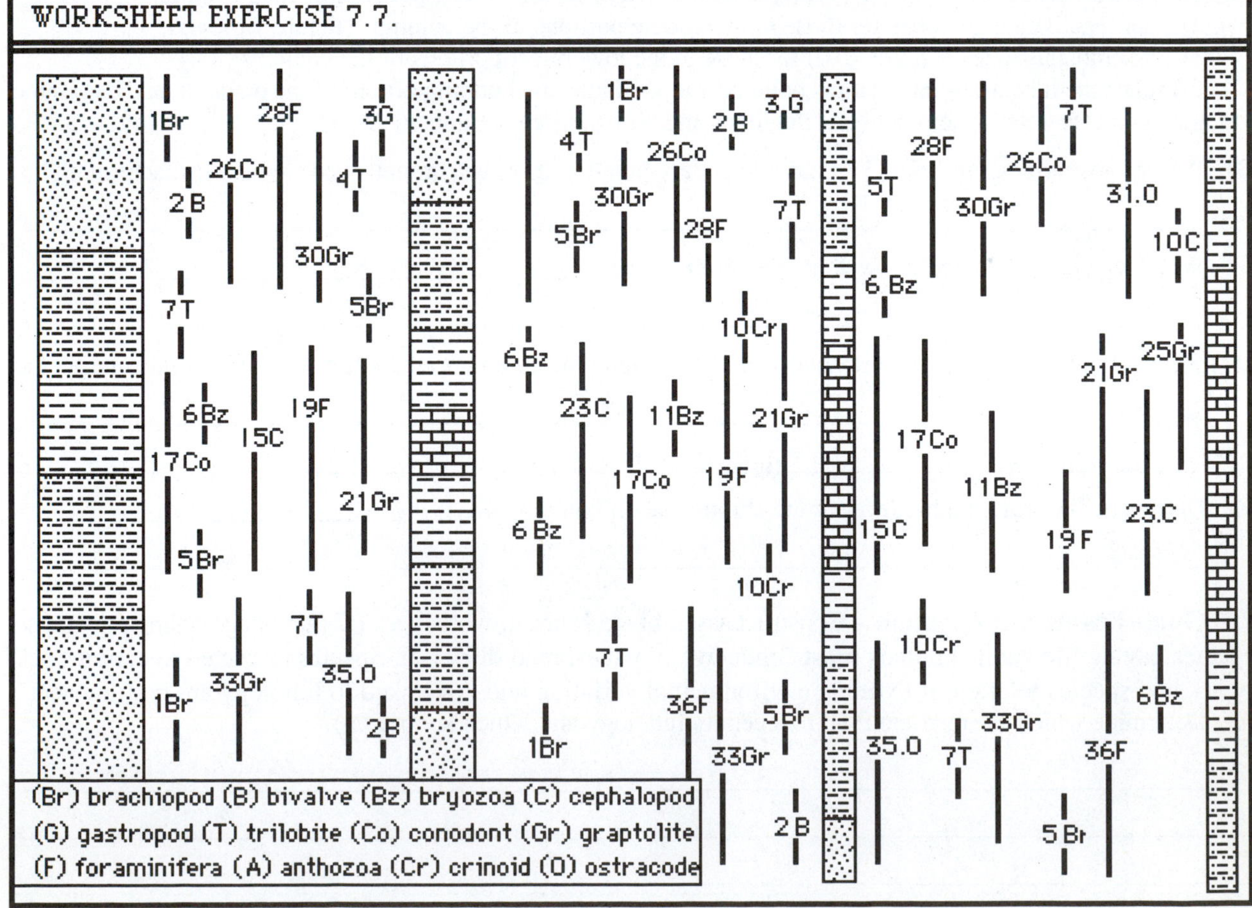

Exercise 7.7. Continued

B) Which range zone has better resolution and why? ______

C) Use the associations among stratigraphic and geologic ranges, lithologic associations, and groups of organisms to determine which species should be employed in constructing biostratigraphic zones. Those species which are associated with a lithology are located most closely on the figure to that rock type.

D) Which species are restricted to one rock type? ______

______

E) Which species are the best guide fossils? ______

______

F) Correlate and describe the relationship(s) between the lithostratigraphic and biostratigraphic units. ______

______

G) Describe the vertical stratigraphic distributions of the guide fossils and the species restricted to one lithology.

______

______

H) Which have the longest geologic ranges? ______

______

7.8. Biostratigraphy and Trends in Evolution: The Alien Time-Traveler's Thought Experiment—You are an alien biologist exploring the evolution of life on Earth by means of a time-machine. You visit the Earth during three nonsequential geologic periods, but your geochronometer has malfunctioned, and thus, you have no knowledge of the temporal order of evolution of the animals and plants you study in each visit. Neither do you have any knowledge of, or access to, the fossil record. You have only comparative anatomy, species lists, ecological data, and a tentative classification of life forms at your disposal.

A) One of your scientific goals is to place the observed faunas and floras in a linear sequence. Can this goal be reached, and if so, what methodology would you use?

B) Your second goal is to place these three faunas and floras in correct chronological order. Can a chronology be constructed using only the information given? Explain. [Note: neither answer is necessarily correct-defend that which you choose].

7.9. Biostratigraphic Zones, Lithostratigraphy & Chronozones. (See next page).
Questions below.

A) Describe the geometry/orientation of the lithostratigraphic units with respect to the ammonite zones. ______

______

B) Can a biostratigraphic zone contain more than one lithostratigraphic unit or rock type? ______

______

C) Can lithostratigraphic units cut across biostratigraphic units or vice versa? ______

______

D) What would the latter indicate about the temporal relationship(s) of the two types of units? ______

______

LABORATORY EXERCISES

7.9.A. Biostratigraphic Zones and Lithostratigraphic Units—Worksheet 7.9A illustrates 8 Jurassic ammonite zones in two widely separated regions (#1, and #2A—D). Use the geometric relationship between the ammonite zones and time in Region 1, and the distribution of ammonite species in columns 2A—D to determine the relationship between time and the lithostratigraphic (sandstone) unit in Region 2 A (marked by *). Correlate the sections of Region 2 using the ammonite zones. Correlate the sandstone marked "X" among sites A-D.

7.9B. Chronozones—Construct sequential chronozones across Region 2 by using the biostratigraphic zone membership of the sandstone unit to adjust the relative vertical (i.e., temporal) position of each column. Redraw the adjusted columns on worksheet 7.9B..

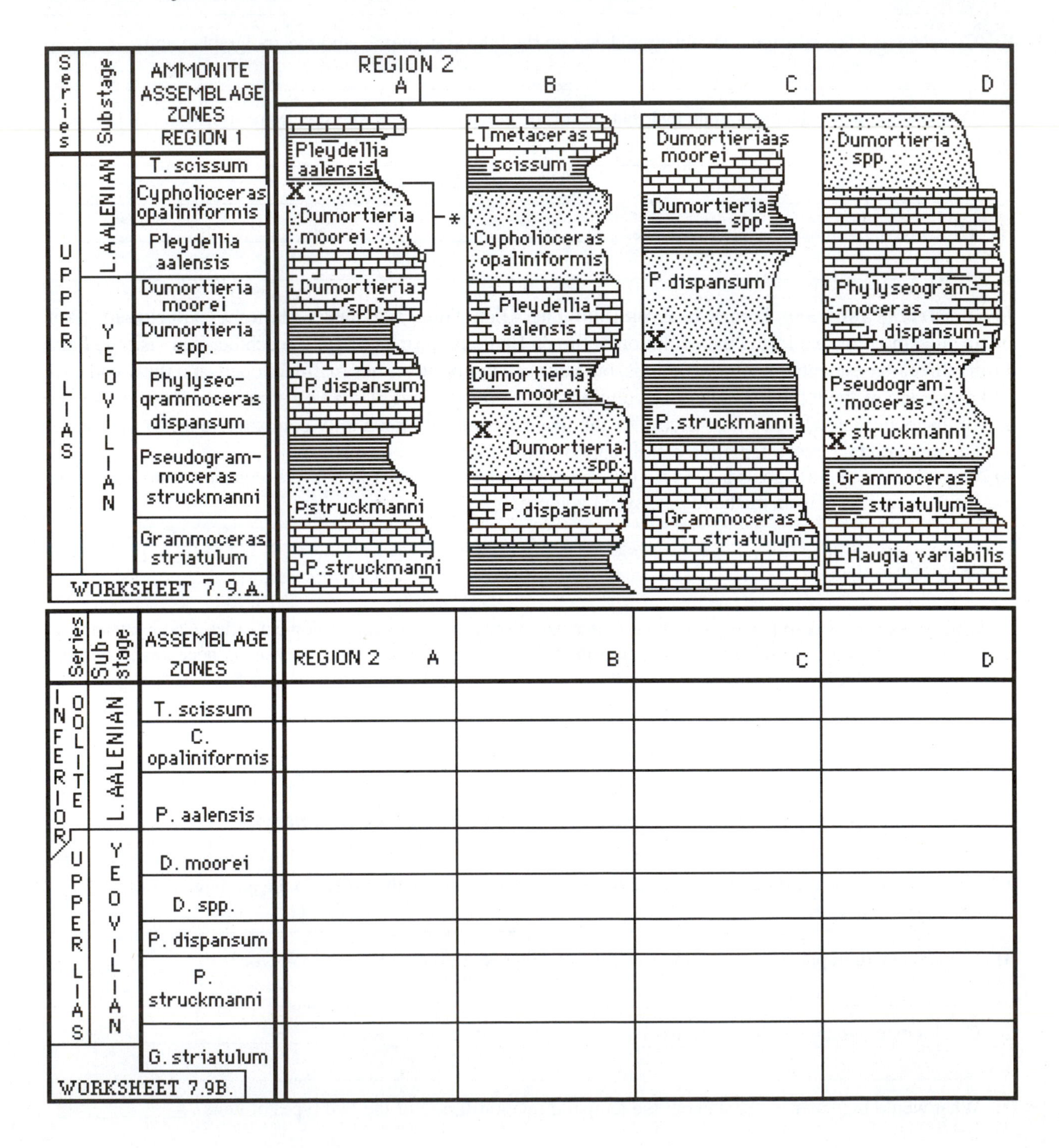

8 Ancient Environments

Objectives

To demonstrate mastery of this material you should: 1) understand the environmental conditions under which certain types of sedimentary rocks are deposited in association with one another; 2) understand and be able to apply Walther's Law of Facies; 3) be able to recognize transgressive and regressive facies; and 4) be able to broadly describe and recognize, with reference to the lithologic and facies descriptions and stratigraphic columns of this chapter, the sedimentary sequences typical of: a) alluvial fans; b) clastic coastlines; c) deltas; and, d) shallow-shelf carbonates.

Using This Laboratory

The discussions presented are not to be memorized for content, but understood as processes resulting in sequential strata that are records of environments past. In this chapter you will learn how to read the rocks for those clues that help geologists recreate ancient environments, and how the migrations of those environments can be traced through time and space along with the rise and fall of sea level. Use the figures as exemplary and summary descriptions of the kinds of environments of deposition to be recreated in the exercises. Again, much of the descriptive material on the latter is provided for contextual understanding, not memorization or expectation.

When geologists look at an exposure of rock, they see it in a manner wholly different from others. It is not that they see with a more technical or discerning eye, or with some professional particular in mind—all these may be so. But most importantly, an otherwise nondescript exposure of rock can, through geologic eyes, instantly transcend pedestrian anonymity to reveal a wondrously real glimpse of history. Within a geological instant, this search unearths clues enough to assemble a picture of the past; the geography, the life, the climate—the environment as it was at one place, at one time—the very place beneath one's feet. And each of these places with an identity and heritage is as centered in history as any in human time and experience. With such skills, you too may be so transported to slog through an ancient coal swamp; leave your footprints on a beach known only to you; or rest at the bottom of the sea, surveying the world as it was. You can take a trip through time.

ENVIRONMENTS of DEPOSITION

An environment of deposition is the sum of all physical, geographic, climatic, ecological, and chemical

ENVIRONMENT	DEFINITION	SEDIMENTS	CHARACTERISTICS
TERRESTRIAL	land	detrital, scant ls., evaporites.	lake & plant fossils
Alluvial Fan	ephemeral stream deposits, cone-shaped	poorly sorted conglomerate & breccia, arkose.	humid & arid regions of high relief.
Eolian Dunes	desert & coastal dunes	well srtd, rndd qtz sand	large cross-bedding* animal tracks.
Fluvial	rivers & floodplains	conglomerates, poorly srtd sands, clays & silts	channels w/in clays & silts; land, lake animals
Lacustrine	lakes	clays, silts, occas. ls., evaporites.	thin & well bdd, lamin., plants & lake animals
MARINE	brackish-hypersaline water, ocean & sea.	detrital and carbonate.	marine animals & plants
Coastal Plain	supra-shallow subtidal envs. gentle slope shore	clays, silts, sands, abundt. sed. source.	mud-cracks, ripples, cross-bedding *
Barrier Coast	barrier islands, lagoons, tidal flats, subtidal.	silts, clay, sand	mud-cracks, ripples, cross-bedding *
Deltaic	delta & assoc. envs.	clays, silts, subgraywacke, qtz. sand, coal, lenticular sand	marine & fresh, x-bddg.* mud-cracks, ripples,
Carbonate Shelf	supra- to shallow subtidal reefs, open shelf.	various ls., dolostone & evap. in supratidal, reef rock, assoc. evaporites.	marine fossils, mud-crks ripples, crossbedding*
Carbonate Shelf	open ocean/sea on shelf	ls., reef rock, shells of protistans	marine fossils

*sedimentary structure Lab 10

TABLE 8.1. MAJOR DEPOSITIONAL ENVIRONMENTS and SUBENVIRONMENTS.

parameters of a site that affect, or are reflected in, the characteristics of sediment and its accumulation. Carbonate, evaporite and coal-forming sediments originate where they are deposited, and as a result, their character is "entirely" the product of that environment. Other sediments are affected by the conditions of their source and transportation, as well their environment of deposition (Laboratory 2). The tectonic setting, whether an inactive trailing margin of a continent or an active zone of vulcanism and uplift, also affects general sediment types and the range of environments present (Laboratory 12). Geologists study sediments and sedimentation in present environments to find clues for use in recognizing similar ancient depositional environments. Among the clues used are sediment composition and texture (Laboratories 1 & 2), sequences, geometries, and spatial relationships of deposits (this and Chapter 9), primary sedimentary structures (Laboratory 9), and fossil content (Laboratories 5 and 6).

Your textbook discusses each of the broadest categories of depositional environments, thus our overview is limited to brief descriptions of the sediment types typical of each of those illustrated by Figure 8.1 and Table 8.1. After reviewing this material, study the discussions of large-scale dynamic processes and contexts in which such deposits are formed, and the spatial and temporal relationships produced by dynamic sedimentation. Our discussion then concludes with four examples of facies depositional settings that serve as examples for those investigated in the exercises.

FACIES, SEA LEVEL and CRATONIC SEQUENCES

Facies—Sedimentation leaves, as evidence of a depositional environment, a body of rock which has textures, structures, fossils or other features that together characterize that environment. A body of rock recognized by some or all such distinguishing features is known as a facies (sing. and pl.). A facies recognized by one or more lithologic characters is termed a lithofacies (Figure 8.2). For instance, a conglomerate deposit believed to be a rocky shoreline deposit lithofacies", while glauconitic sands of the shallow subtidal are called the "glauconitic sandstone lithofacies". If the depositional environment is recognized by an assemblage of fossils, it is called a biofacies (Figure 8.2). For example, a limestone containing bryozoans and stromatoporoids of small reefs, may be called the "bryozoan-stromatoporoid facies".

The facies concept is powerful because it is *dynamic*. Facies are not static, layer-cake strata, but the remains of adjacent environments whose ever-changing spatial relationships are recorded in three-dimensional geometries of rock bodies (Figure 8.3). They are representations of those materials and processes which produced the rocks we now wish to understand.

Sea Level—The most dramatic facies changes accompany the rise and fall of sea level. [Recall our discussion of formations and sea level transgression, Laboratory 3.]

To understand the spatial relationships created by such facies migrations, let's examine a simple scenario in which three facies migrate over a 10 million year interval. Imagine that you are standing on the sandy beach of a Cambrian shallow sea. Stretching for several kilometers offshore and hundreds of kilometers along it are quartz sands (Figure 8.4A). Some 40 km offshore the sands gradually mix with lime muds, and over the next kilometer they gradually become less sandy and more limey. Carbonate muds stretch for 100 km from here to near the shelf break where the sediments grade into a 20 km-wide band of clays. The facies lie in bands parallel to the shoreline, and although the widths of these three facies vary from place to place, so long as sea level does not rise or fall and there is a continual source of sediments, their respective widths and positions do not change substantially (Figure 8.4A).

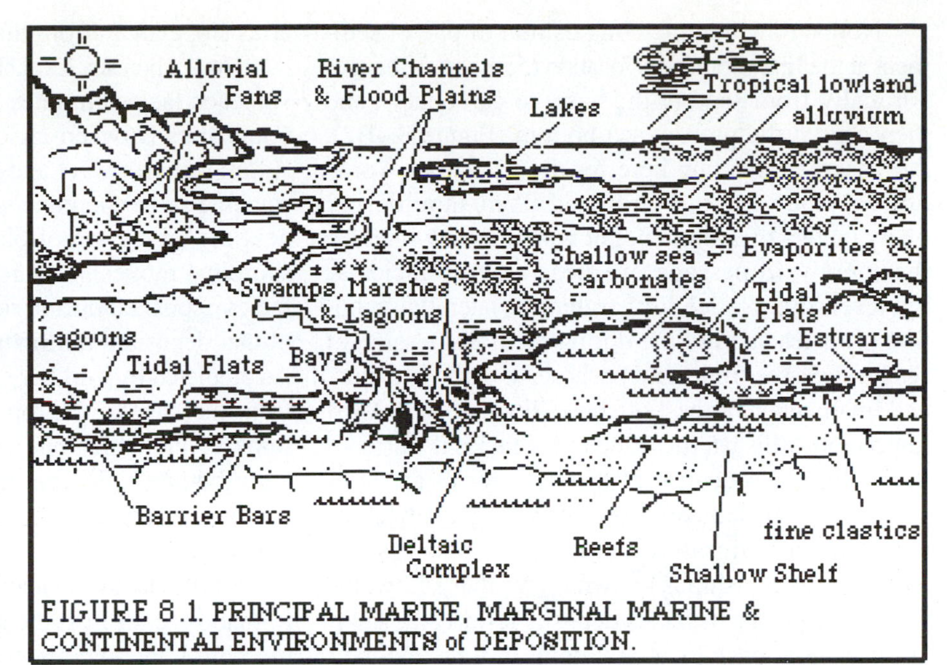

FIGURE 8.1. PRINCIPAL MARINE, MARGINAL MARINE & CONTINENTAL ENVIRONMENTS of DEPOSITION.

Now as sea level begins slowly to rise, the shoreline migrates inland and each facies also migrates toward the former shoreline. After two million years have passed, sea level has risen—100 m and the shoreline now lies far inland of its former location (Figure 8.4B1). A volcanic ash-fall serves as a marker of the spatial relationships among the facies as they existed (forming the sea floor) at one instant in time (Figures 8.4A, B1). Each facies has moved in the shoreward direction; movement known as transgression ("to move across"). Sea level continues to rise another 100 m over the next two million years, at which time a second ash-fall marks the farther shoreward advance of each facies (Figure 8.4B2). A third ash-fall, 1 million years later marks the farthest advance of the shoreline. Here sea level remains for sometime.

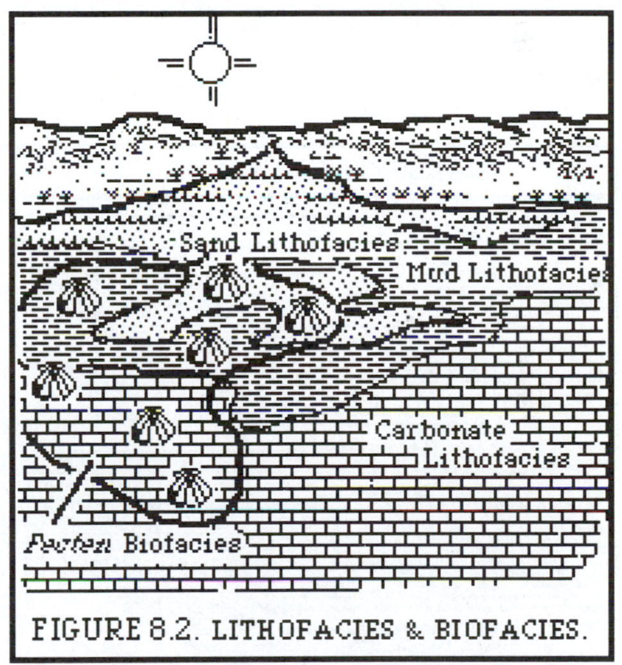

FIGURE 8.2. LITHOFACIES & BIOFACIES.

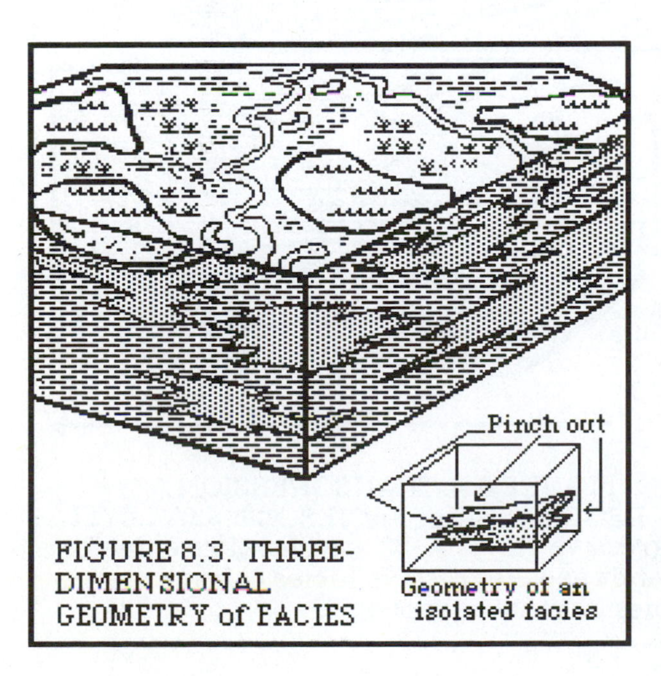

FIGURE 8.3. THREE-DIMENSIONAL GEOMETRY of FACIES

Notice now that the composition of pile of sediment at the first shoreline's location (5 my past) changes vertically from sand at the base, to carbonates, and then clay at the present sea bottom (Figure 8.4B2). An observer standing here for the past 5 my would not only be very old and tired from shouldering many meters of sediment, but would have witnessed the shoreward passage overhead of each facies as sea level rose. Notice further that the lateral distribution of facies at any time during the transgression of sea level passes from sand onshore, to carbonate mud, and finally clay, farthest offshore (Figure 8.4C). Most importantly, this lateral sequence of facies is mirrored in the vertical sequence of the same facies accumulated over our dedicated sedimentologist's head as each facies migrated landward. This spatial-temporal relationship is known as Walther's Law: the vertical sequence of facies deposits reflects the lateral facies relationships at any point in time and their migrations toward our hypothetical observer. The direction of facies migration is indicated by the slope of the boundaries between facies; when sea level rises, the facies transgress the old shoreline and their boundaries slope away from the shore (Figure 8.4A).

Now when sea level slowly falls over a 5 my interval, each facies migrates seaward (Figure 8.4C). The boundaries between facies then slope toward shore. Again, our stalwart and fossilized sedimentologist standing at the shoreline's original location would witness the regression of one and another facies as the shoreline moves seaward. The vertical sequence of facies deposited during regression is then the reverse of that deposited during transgression (Figure 8.4D). As predicted by Walther's Law, the sequence of facies from the base to the top of the entire section is the same as that which our geologist would encounter in a sea-bottom trek from the shore to the shelf break, and back again at any time during the 10 million year interval.

Recall that the contacts (boundaries) between most formations cut across the imaginary time planes defining simultaneous events. The boundaries between facies in our example also cut across the ash-falls that define singular instantaneous events in Figure 8.4. Regressive and transgressive facies are diachronous: forming at two times, neither their upper nor lower

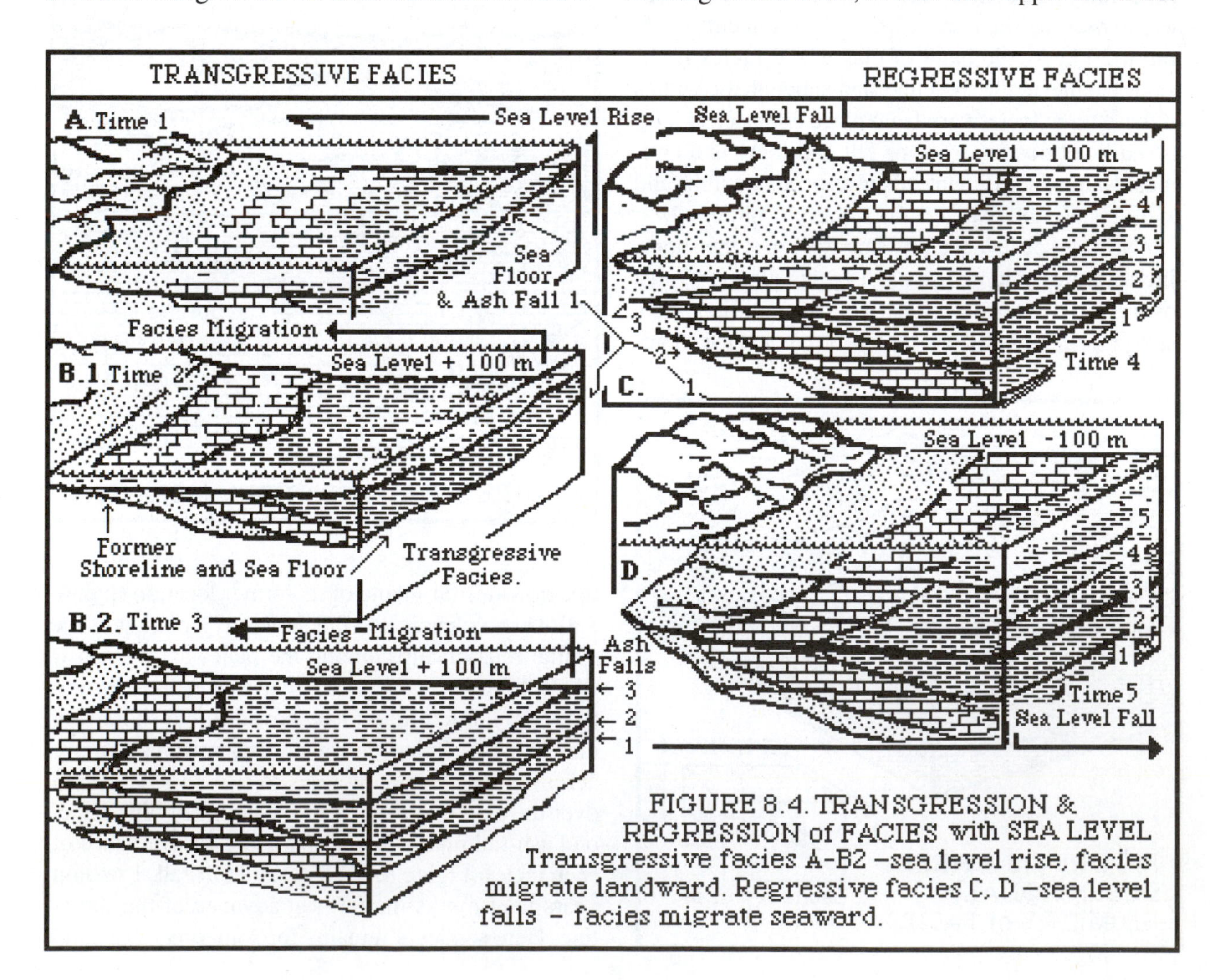

FIGURE 8.4. TRANSGRESSION & REGRESSION of FACIES with SEA LEVEL
Transgressive facies A-B2 –sea level rise, facies migrate landward. Regressive facies C, D –sea level falls – facies migrate seaward.

contacts are of the same age everywhere and each cuts across time "surfaces". During facies transgressions, the base of the sand at the shoreline is younger—in some cases much younger—than the base of the sand deposit formed farthest offshore. The reverse is true of the regressive facies; that portion of the sand facies nearest the most inland shoreline is much older than the base of the sand facies farthest to sea. Like formations, not all facies are diachronous (e.g., ash beds, storm deposits, and turbidites).

Local facies limited to one or a few sediment types and directly linked with one or several related environments show lateral migrations of their boundaries in response to two factors: 1) changes in the sources and quantities of sediment and thus, development and demise of local environments; and, 2) localized and limited changes in sea level. These facies movements, however, occur within the larger context of regional and global sea level changes over long periods of time. Before inspection of the characteristics of particular depositional environments, we need thus, to set them within their proper framework—the great and long march of shorelines marking the continent-wide oscillations of sea level known as <u>cratonic</u> <u>sequences</u>.

Cratonic Sequences—Sea level has risen and fallen many times during the Phanerozoic Eon. North America has seen six craton-wide trangression-regression couplets known as <u>cratonic</u> <u>sequences</u> (Figure 8.5). Many less dramatic changes in sea level have occurred within each of these sequences. Cratonic sequences are comprised of transgressive-regressive facies couplets of the highest order of magnitude, from maximal coverage of the craton to maximal withdrawal. It is readily apparent that they are time-transgressive. Associated with major advances and retreats in sea level are also cyclic changes in the volumes of different sedimentary rock types. Carbonate rocks tend to be more widespread during transgressive periods and times of maximal sea levels. Land-derived clastic sediments are more widespread during times of regressions (see below and Laboratory 12).

FIGURE 8.5. CRATONIC SEQUENCES & OROGENIES of NORTH AMERICA

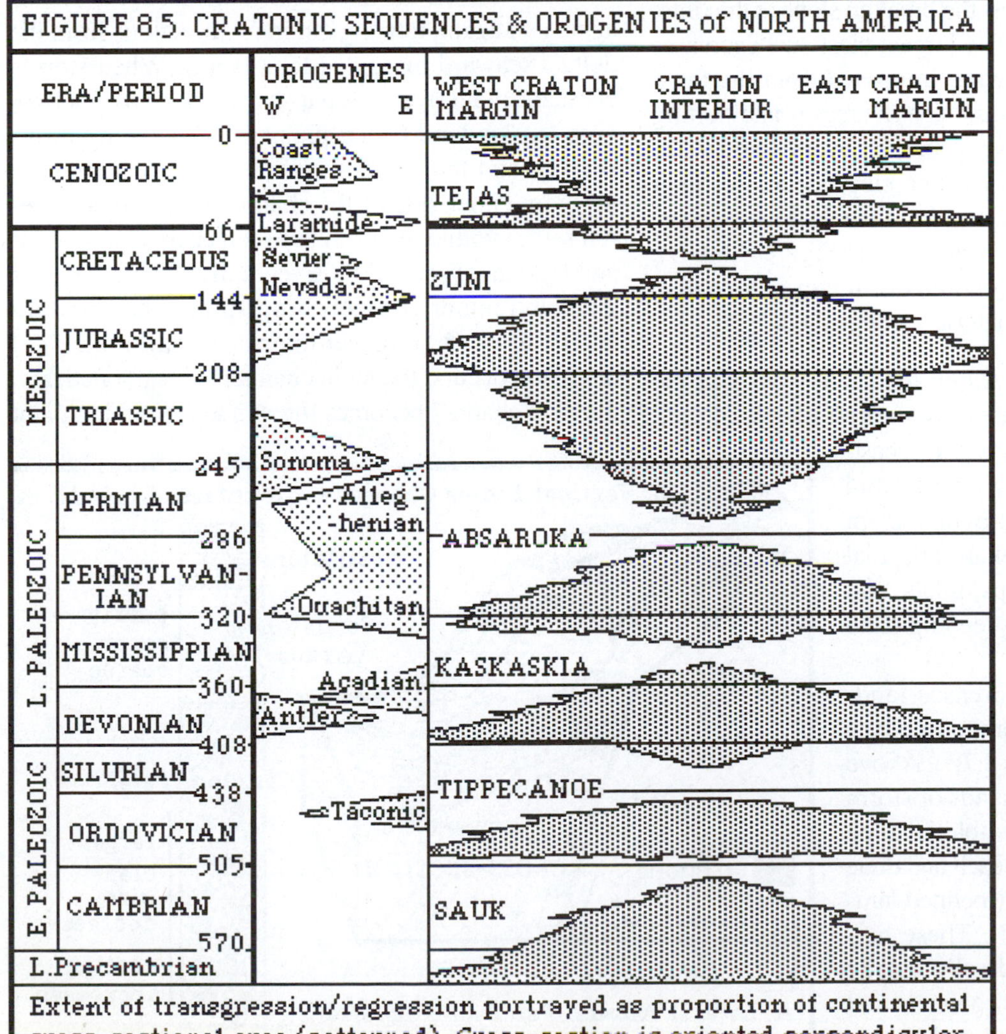

Extent of transgression/regression portrayed as proportion of continental cross-sectional area (patterned). Cross-section is oriented perpendicular to North - South (present directions) shorelines.

DETRITAL FACIES and ENVIRONMENTS of DEPOSITION

Within cratonic sequences are facies representing local sediment deposits in all variety of environments. To determine the nature of past depositional environments, geologists consider sediment type, sedimentary textures, structures (Laboratory 9), types of animals and plants and features of their ecology (Laboratory 6), and the sequences of facies in strata, and compare these and their associations with similar present-day conditions and well-studied geologic examples. Environments of deposition cannot be recognized by any single criterion: many features must be considered simultaneously. Four depositional models are discussed below.

Alluvial Fans—Alluvial fans are cone-shaped aprons of poorly sorted to unsorted sediments—deposited on land at the base of steep slopes, such as those that formed where a steep ephemeral mountain stream reaches a gently sloping valley floor (Figure 8.6A). When a stream gradient suddenly changes from steep to gentle, water velocity drops sharply, causing coarse and fine sediments fall out of suspension. The stream channel shifts from one side of the fan to another over time as each becomes clogged with sediment and leaves its banks to form a new channel. This sediment becomes a conglomerate (Laboratory 1).

A cross-section or slice through the interior of a fan shows some sorting (Figure 8.6B). Coarsest material is found nearest the break in slope at the (interior) base of the fan. The largest clast size decreases progressively, and sorting improves from the fans' head to its margin. Grain size decreases and sorting increases from the base to the top everywhere within the fan because its surface slope decreases as it builds out- and upward, and the stream's slope decreases as the source region is eroded (Figure 8.6A,B). As the surface slope decreases, material size also decreases, and finer material is deposited.

Clastic Coastlines—Along the sandy coasts of the Carolinas and Texas are long narrow islands of sand that form barriers protecting the coastline from open-ocean waves and currents (Figure 8.7A). Behind them lay lagoons, and landward are tidal flats broken by swamps, inlets, and the estuaries of rivers feeding sand and mud to the coast.

Sediment size increases landward from clay offshore to fine barrier beach sand (Figure 8.7B). Above the beach are coarser sands of storm deposits and finest wind-blown dune sands (Laboratory 2). Beach and dune sands are crossbedded (inclined laminations; Laboratory 9). These barrier islands are periodically cut by storm surges, forming channels and depositing shell, gravel and coarse sand.

Behind the barrier on the seaward face of the lagoon, is a wide apron of sand or mud forming a tidal flat. Within the lagoon are laminated silts and clays deposited in quiet water, and sands from tidal inlets (Figure 8.7B). Organic-rich black muds may accumulate in the lagoon and onshore swamps. Mud-cracked sediments and evaporites form here in arid climates. A vertical sequence of sediment types accumulated during a transgression of the sea is illustrated in Figure 8.7B. The regressive sequence is similar—but reversed—and thinner or incomplete because regressions are associated with erosion.

Deltas—As a river meets the sea its velocity is suddenly checked and much of its sediment load is dropped to form a delta (Figure 8.8). When the source river is large, a delta can extend hundreds of kilometers seaward and along shore, where subenvironments such as tidal flats, beaches, floodplains, bays, lagoons, lakes, swamps, barrier islands, and river channels (Figure 8.9) will develop within.

In all deltas, the main river channel splits in-two again and again, fanning out in all directions from the delta's seaward margins (delta front). When storms cause the main river to top its banks, floodwaters flow onto the lowland (flood)plain; water velocity drops sharply, and the coarsest sediments settle forming a levee on either side of the channel while finer material settles within the plain between channels, swamps and lagoons (Figure 8.8). Eventually the gradients of main and minor channels drop so low that they are easily choked with sediment: a major channel diversion occurs; the main channel is abandoned, and the new channel becomes the conduit of an incipient

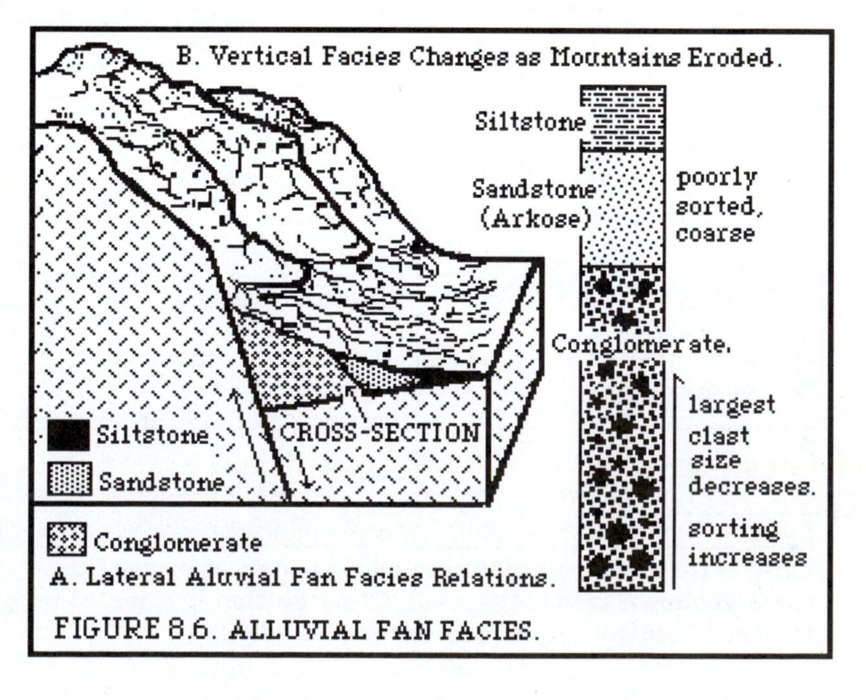

FIGURE 8.6. ALLUVIAL FAN FACIES.

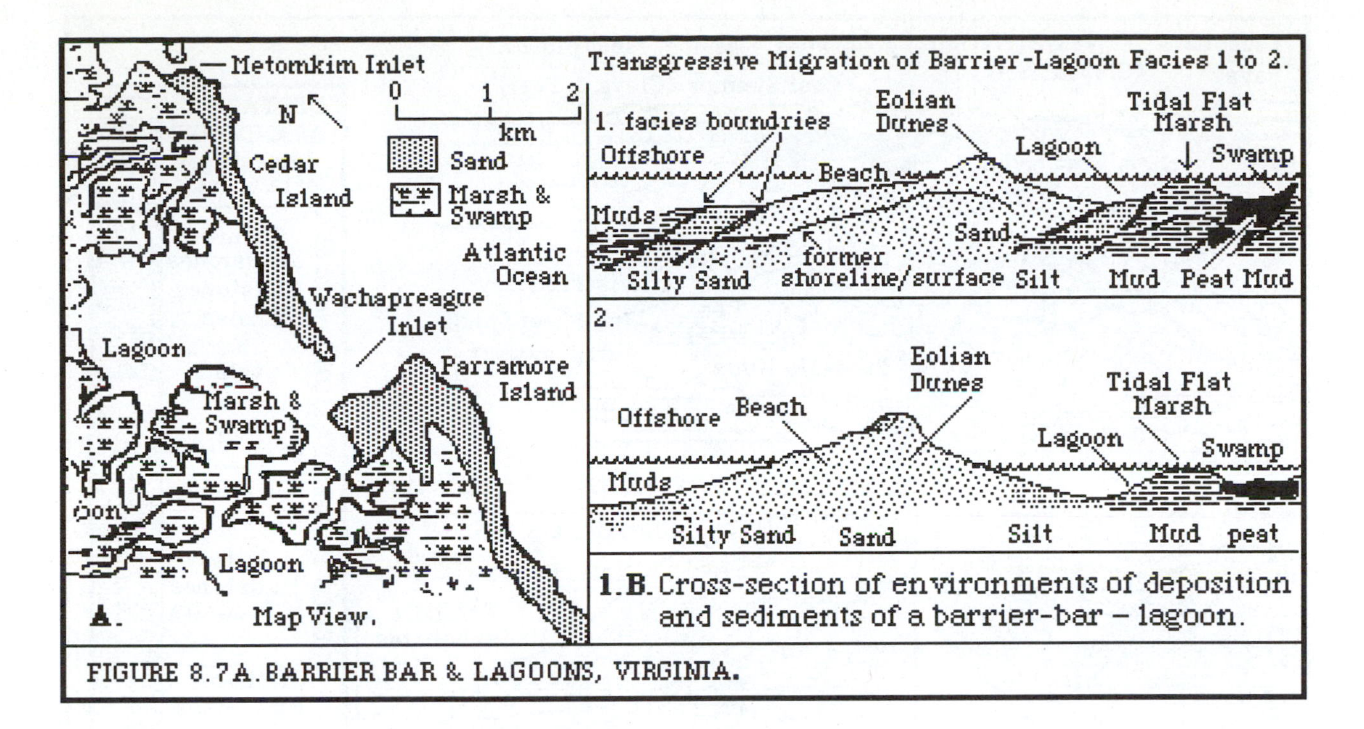

FIGURE 8.7A. BARRIER BAR & LAGOONS, VIRGINIA.

delta lobe. Sedimentation all but ceases in the old lobe, and currents erode portions of it. Active sedimentation resumes elsewhere and the new delta lobe builds outward.

Figures 8.9A,B show the distributions of sediment types within a three-dimensional block cross-section of a delta and representative stratigraphic column. As delta formation begins, sediments are deposited only below sea level in the prodelta. These beds of sediment, beginning with muds and grading upward to silts and sands of the channel, build outward (prograde or slope seaward) in a fan shape (compare with the alluvial fan above). These facies are followed by sediments found between the distributary channels, including clays and silts of floodplains and levees, organic-rich clays and silts of lagoons, lakes or bays, and thin soils and peats, occasionally cut by the sands of small interdistributary channels. A repeated sequence of clays and silts of interdistributary facies and sands of channel facies come to overlie and overlap one another several times as the interdistributary region is filled and the channel migrates (Figure 8.9A,B). Because so many environments within deltas contain similar sediments, evidence from fossils, sedimentary structures, and facies geometries often must be used to distinguish among brackish-water, fresh water and marine facies.

When a delta lobe is abandoned, rates of sedimentation decrease dramatically. The most seaward portion acquires a blanket of well sorted, fine-grained sand, shell debris, and marine muds (Figure 8.9). Limestones may be deposited seaward of the delta if conditions are suitable, and extensive peat beds may develop across the shoreward portion of the delta.

CARBONATES

Limestone Classification—Limestones may be classified by matrix or cement type, and the type,

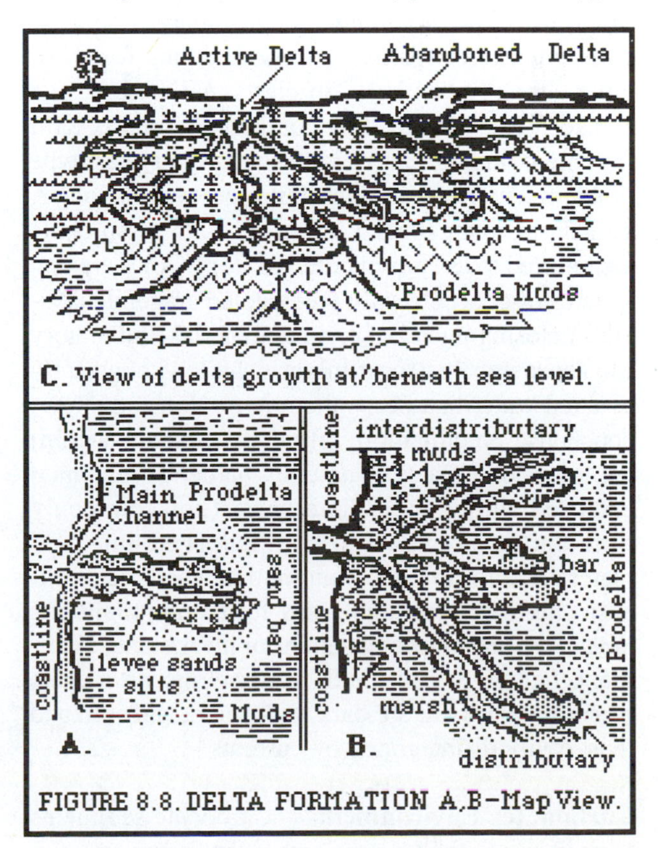

FIGURE 8.8. DELTA FORMATION A,B–Map View.

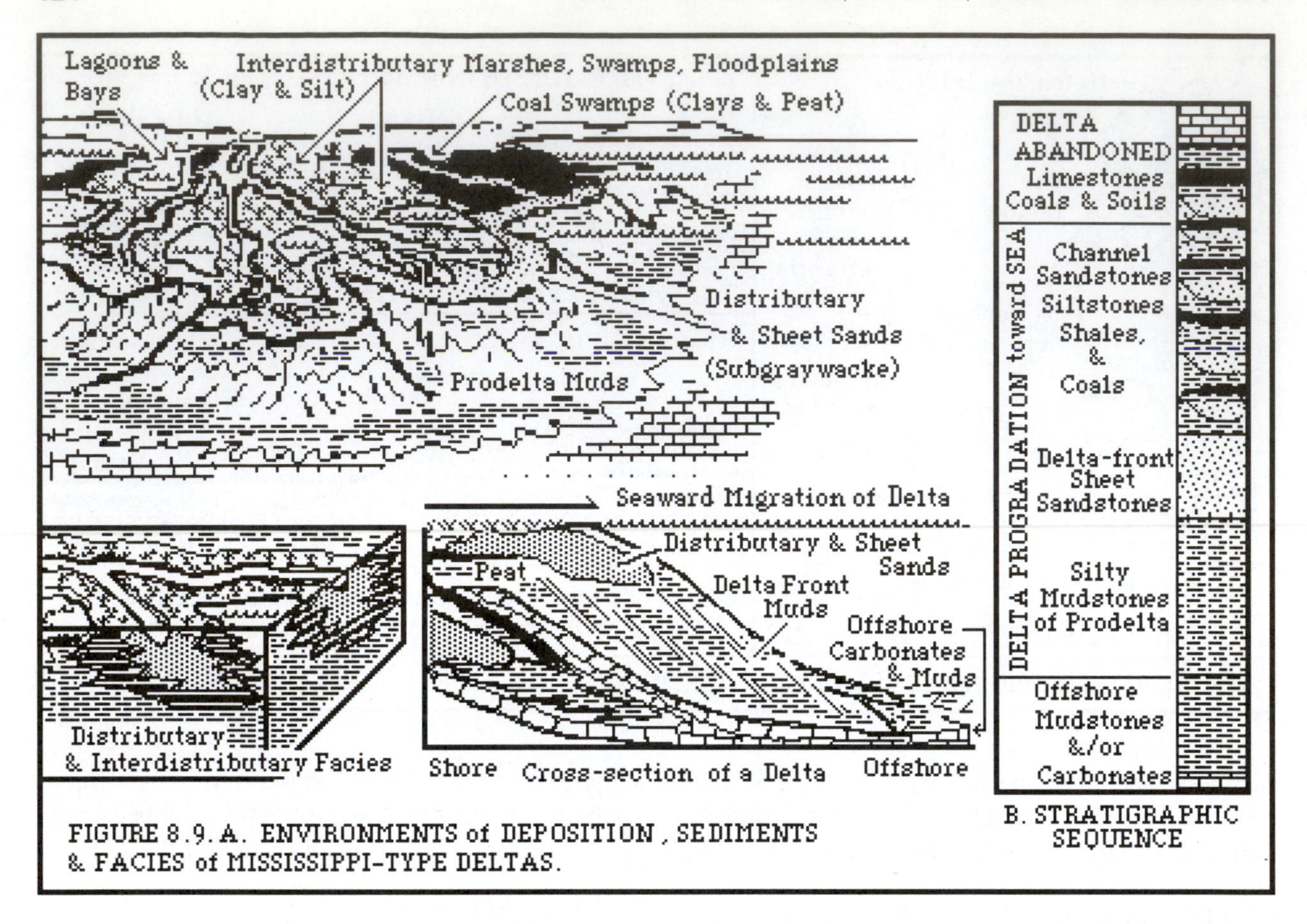

FIGURE 8.9. A. ENVIRONMENTS of DEPOSITION, SEDIMENTS & FACIES of MISSISSIPPI-TYPE DELTAS. B. STRATIGRAPHIC SEQUENCE

abundance, sorting and rounding of clasts (Table 8.2). They may consist of mud-sized crystals of calcite (micrite) or larger calcite crystals (sparite) only; of either micrite or sparite matrix containing fossils or other clasts; or "entirely" of clasts. A limestone can be succinctly described by combining a matrix term (micrite or sparite) and a prefix signifying clast type (Table 8.2). There are four clast types: fossils, oöids, intraclasts, and pellets. Limestones with fossil clasts are referred to as "bio(micrite/sparite)". Oöids are tiny spheres of concentrically precipitated aragonite crystals enclosing nuclei of organic material (Laboratory 2). These spheres form in very shallow, constantly agitated water that keeps them in motion and allows concentric precipitation. Intraclasts (*intra* = within) are lumps and plates of partially hardened lime mud enclosed within softer lime mud. They commonly form during storms when semi-hardened lime mud plates are torn from the bottom, that when redeposited within carbonate mud, may form an intramicrite. Pellets are tiny ellipsoidal grains of invertebrate excrement that, set within mud, form pelmicrite. Pellets resemble oöids but are dark, unlayered, and soft, and thus, easily disintegrated by currents.

Carbonates Environments—Carbonate sediments form in any suitably warm, marine environment (<3.5 to 5.5 km depth) if clastic sediments are absent and carbonate producing organisms are present. There are two general types of carbonate environments: 1) protected shelf lagoons like those behind the reefs of the Bahama Banks (Figure 8.10); and, 2) open shelves like that of the western coast of Florida (Figure 8.10). Shelf lagoons form in the quiet waters behind coral reefs, islands and shoals. Each of these shelf environments has (at least) five subenvironments: the supratidal zone, shore zone, marine platform, reef belt, and shelf slope.

In the supratidal zone are carbonate and pelletal muds (Laboratory 2) deposited during storms (Figure 8.11). Crusts of algal mats grow on, and bind, the sediment (see Laboratory 9). Extensive marshes and temporary pools develop on mud-cracked, evaporite-containing supratidal muds.

Tidal flats, channels, ponds and beaches form in the shore zone (Figure 8.11). Laminated pelletal muds and clast-free mud (Laboratory 2) are deposited by storm surges and bound by algal mats (Laboratory 9).

Muds with shells and intraclasts, and shell "hash" wash into tidal channels and ponds during storms. Beach ridges of laminated pelletal muds and shell fragments parallel the shore. Such descriptions are also applicable to noncarbonate environments of similar geographies.

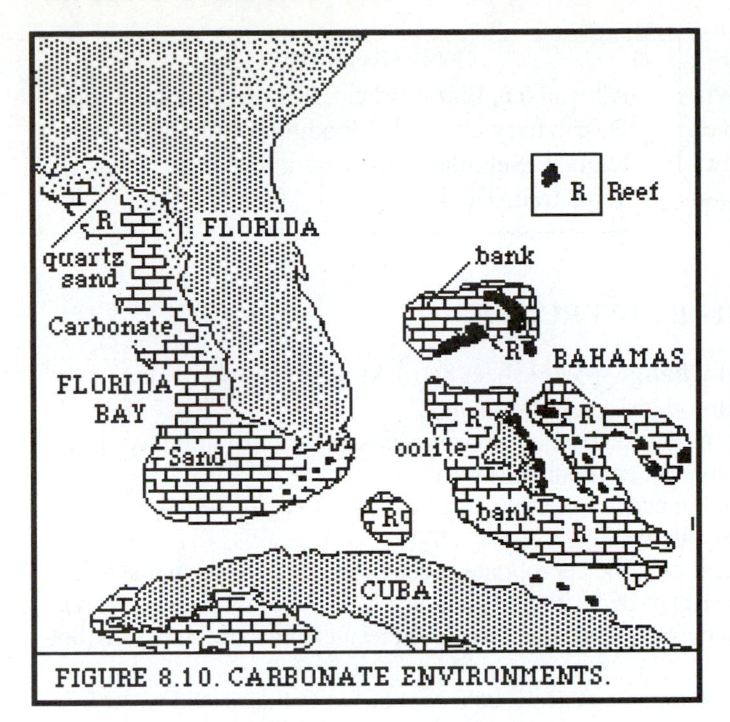

FIGURE 8.10. CARBONATE ENVIRONMENTS.

Marine platforms protected by fringing reefs or shoals, accumulate pelletal muds and shell-bearing muds (Figure 8.11). If unprotected by reefs, much carbonate sand and pelletal mud will be deposited in wide belts, along shoals, and the shelf edge. Oöids are widespread here and tend to form in very shallow water (3-5 m;10-16 ft) near tidal channel margins and along the shelf break.

The three major types of reefs are barrier reefs, fringing reefs and atolls. Sediments in reef facies are comprised of coral-algal framework rock, reef rubble, skeletal sands, and shelly muds.

Carbonates of the shelf slope are deep water sediments, often organic-rich and lacking shell fragments (Figure 8.11). In present shelf slope environments, foraminifera contribute much of the carbonate debris. Bedding may be contorted by mass flows of sediment and large blocks of limestone embedded in carbonate mud.

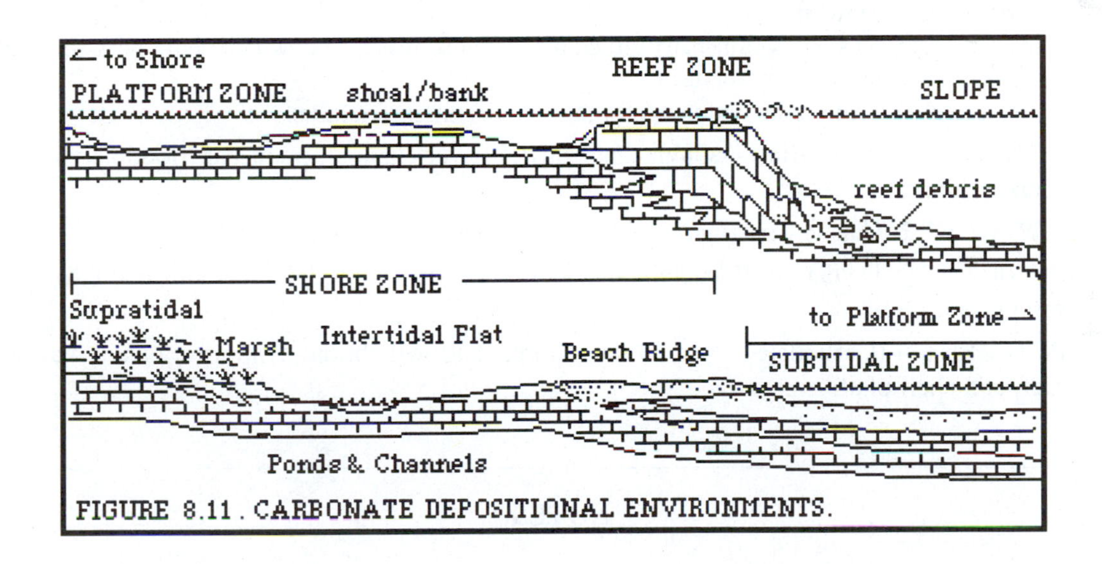

FIGURE 8.11. CARBONATE DEPOSITIONAL ENVIRONMENTS.

TABLE 8.2. Simplified version of Folk's Limestone classification scheme.

	CLAST TYPE	Fossil Fragments	Intraclasts	Oöids	Pellets
	PREFIX	"Bio"	"Intra"	"Oö"	"Pel"
MATRIX TYPE	Mud (Micrite) →	Biomicrite	Intramicrite	Oömicrite	Pelmicrite
	Crystalline (Spar)	Biosparite	Intrasparite	Oösparite	Pelsparite

VOCABULARY

Biofacies, Cratonic sequence, Depositional environment, Diachronous, Facies, Lithofacies, Offshore, Onshore, Regression, Transgression, Walther's Law.

ENVIRONMENTS

Alluvial fan, Barrier island, Clastic coastline, Delta, Distributary channel, Floodplain, Intertidal zone, Lagoon, Subtidal zone, Supratidal zone, Swamp, Terrestrial, Tidal flat,

LABORATORY EXERCISES

8.1. Depositional Environments and Temporal Change—Worksheet 8.1(next page) shows the major terrestrial and marine environments of deposition, and their subenvironments.

A) Locate on worksheet 8.1 the regions illustrating the four major environments of deposition: alluvial fans (#1), delta (#2), clastic coastlines (#3); and, carbonate environments (#4);

B) Study the figure and write in the space provided in each of the four vertical boxes (# 1, 2, 3 & 4), the name of each "Major Environment" illustrated directly above;

C) Write in the "lettered" blanks below each vertical box, the appropriate subenvironment (facies) name and/or sediment type(s) corresponding to the specific sites of deposition of each labeled on the figure (e.g., alluvial fan, conglomerate). List these in a probable sequence of deposition that would result from a transgression (bottom) through a regression (top);

D) Draw in each blank vertical box, from the base to top, the general sequences of facies that would result from a transgression (bottom half), and followed by a regression (top half)—Use standard lithologic symbols. In the cases of Major Environment 1 indicate only the sequence of facies that would result as the mountainous source region eroded to low hills; and,

E) Also indicate in the blanks any sedimentary structures, textures, fossils, etc. which could be associated with the facies of each depositional environment.

The choices of subenvironments within each major environment include:

ENVIRONMENT 1 (Terrestrial)—alluvial fan, fluvial (river channel, point bar, floodplain, lacustrine (lake), eolian dunes;

ENVIRONMENT 2 (Clastic Coastline) —tidal flat, barrier bar, lagoon, & shallow subtidal;

ENVIRONMENT 3 (Deltaic) —floodplain, channel distributaries, prodelta, lagoons and bays, marshes and swamps; and,

ENVIRONMENT 4 (Carbonate) —supratidal, shore, shallow subtidal evaporitic basin, marine platform, reef, and shelf platform.

NOTES

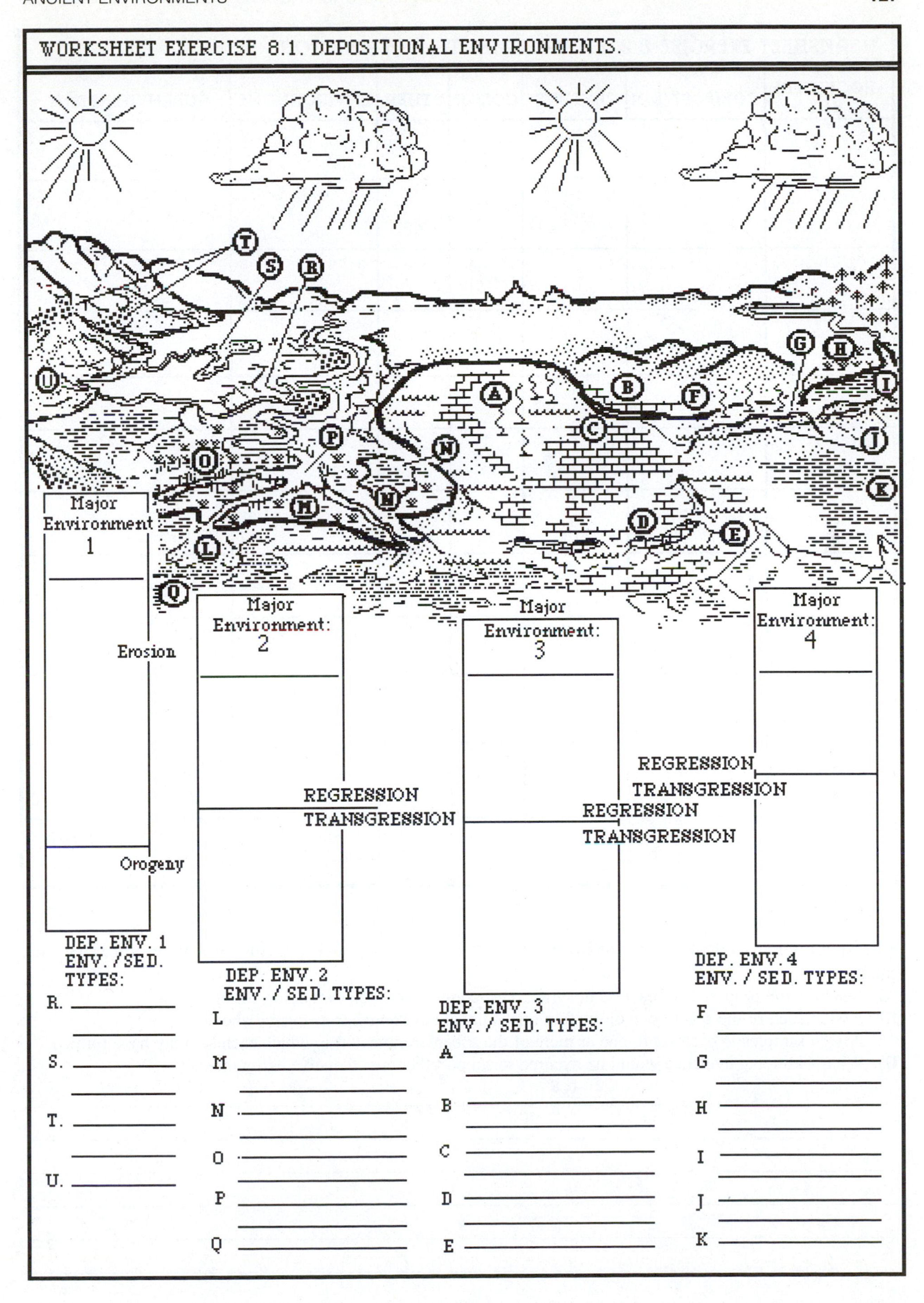
WORKSHEET EXERCISE 8.1. DEPOSITIONAL ENVIRONMENTS.
T
S
R
U
A
B
C
D
E
F
G
H
I
J
K
L
M
N
O
P
Q
Major Environment 1
Erosion
Orogeny
Major Environment: 2
Major Environment: 3
Major Environment: 4
REGRESSION
TRANSGRESSION
REGRESSION
TRANSGRESSION
REGRESSION
TRANSGRESSION
DEP. ENV. 1 ENV. / SED. TYPES:
R.
S.
T.
U.
DEP. ENV. 2 ENV. / SED. TYPES:
L
M
N
O
P
Q
DEP. ENV. 3 ENV. / SED. TYPES:
A
B
C
D
E
DEP. ENV. 4 ENV. / SED. TYPES:
F
G
H
I
J
K

WORKSHEET EXERCISE 8.2. ENVIRONMENTS of DEPOSITION.

ROCK TYPE & SET #	COMPOSITION	TEXTURE	COLOR	OTHER	ENVIRONMENT	SUBENVIRINMENT

8.2. Specimens. Depositional Environments—Your instructor will provide sets of sedimentary rock samples.

A) Describe the properties of each sample (textures, composition, relevant color, fossils, etc.).
B) Assign each of the sets to one of the four depositional environments discussed above.
C) Assign samples in each set to one or more of the subenvironment(s) in which each set may have formed.
D) What additional evidence would be required to support the more tentative assignments?

LABORATORY EXERCISES

8.3. Specimens. Facies Sequences—Your instructor will provide an ordered set of samples (#1 being that found at an exposure's base) taken from one of the depositional environments discussed above, and in Exercise 8.2.

A) What depositional environment(s) is/are represented by the sequence?

Cite your evidence and explain your reasoning. ______________________________

____

____

____

____

B) Does the sequence represent a transgressive facies or a regressive facies? Explain. ______________

____

____

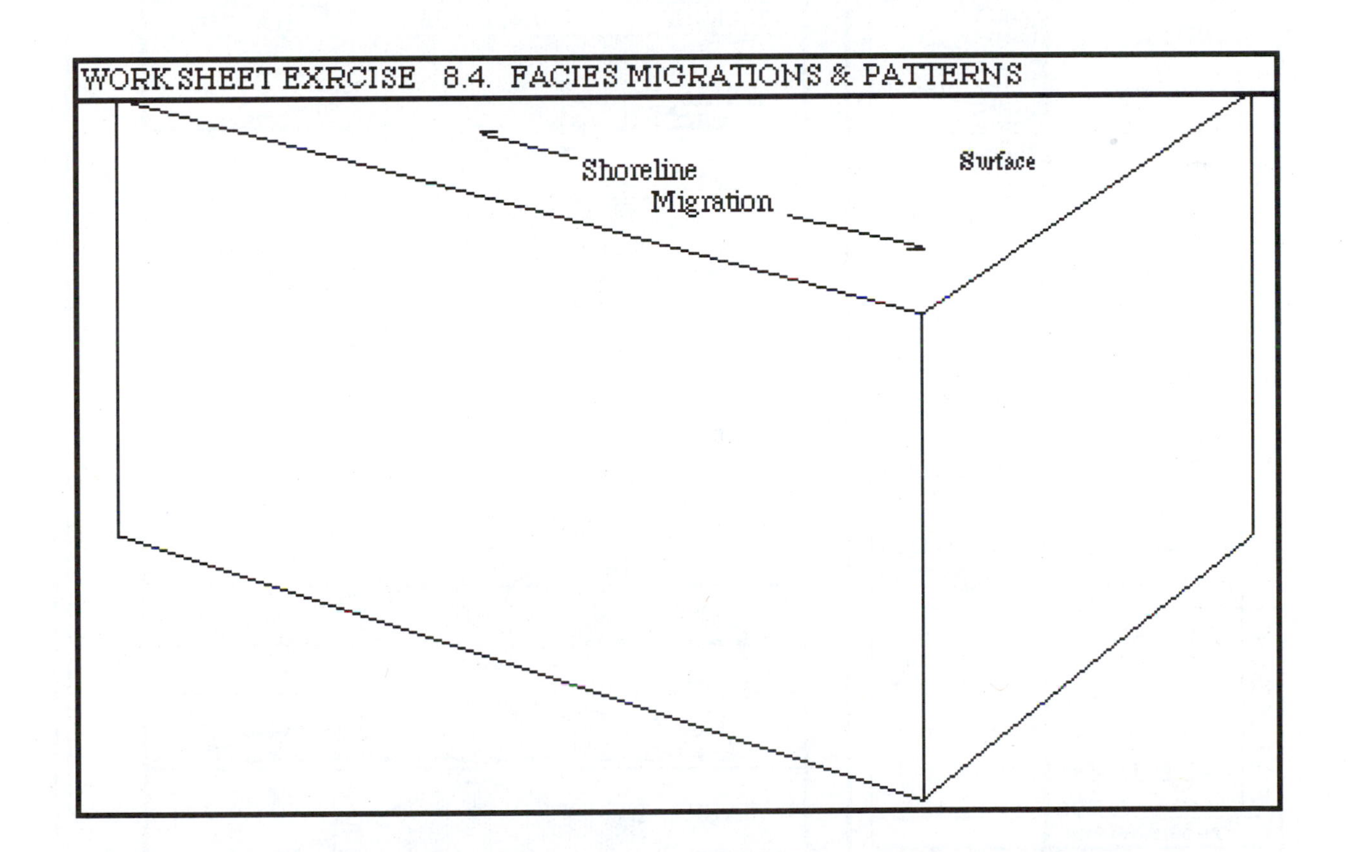

8.4. Facies Patterns in 3-Dimensions—Prepare on the outlined block diagram provided (above), a diagram of the transgression and regression of the facies along a clastic coastline environment (shallow subtidal, barrier bar, lagoon, and tidal - supratidal facies) as follows:

A) Use the distribution of facies illustrated in the cross-section in Figure 8.7 as a standard of comparison for the lateral (i.e. geographic distribution of facies perpendicular to the shoreline) at one moment in time, and draw this from left to right as a stratigraphic column lying on its side.

B) Imagine that sea level rises in three discrete steps, and then sea level falls in three steps. Illustrate these same facies as they would appear in a stratigraphic column, and in a cross-section of the sequence oriented:

1) perpendicular to the shoreline as sea level rises and falls; and,

2) parallel to the shoreline. Offset each of the next two "columns" (as in part A, above) in a stepwise manner to the left, and the final three to the right, as shown in the prepared block diagram.

C) Indicate the relationship between facies' geographic migration and temporal change, by drawing lines connecting the boundaries between each facies. Compare the sequence of facies in the horizontal and vertical columns.

D) Draw lines connecting strata of equivalent ages on the front face of the diagram (perpendicular to shore).

E) In which orientation, parallel or perpendicular to shore, is the temporal sequence of strata best shown at any single site? ______________________

In which orientation is the geographic distribution and sequence of facies best shown? ______________________

____

F) What law of geological process does the diagram illustrate? ______________________

____

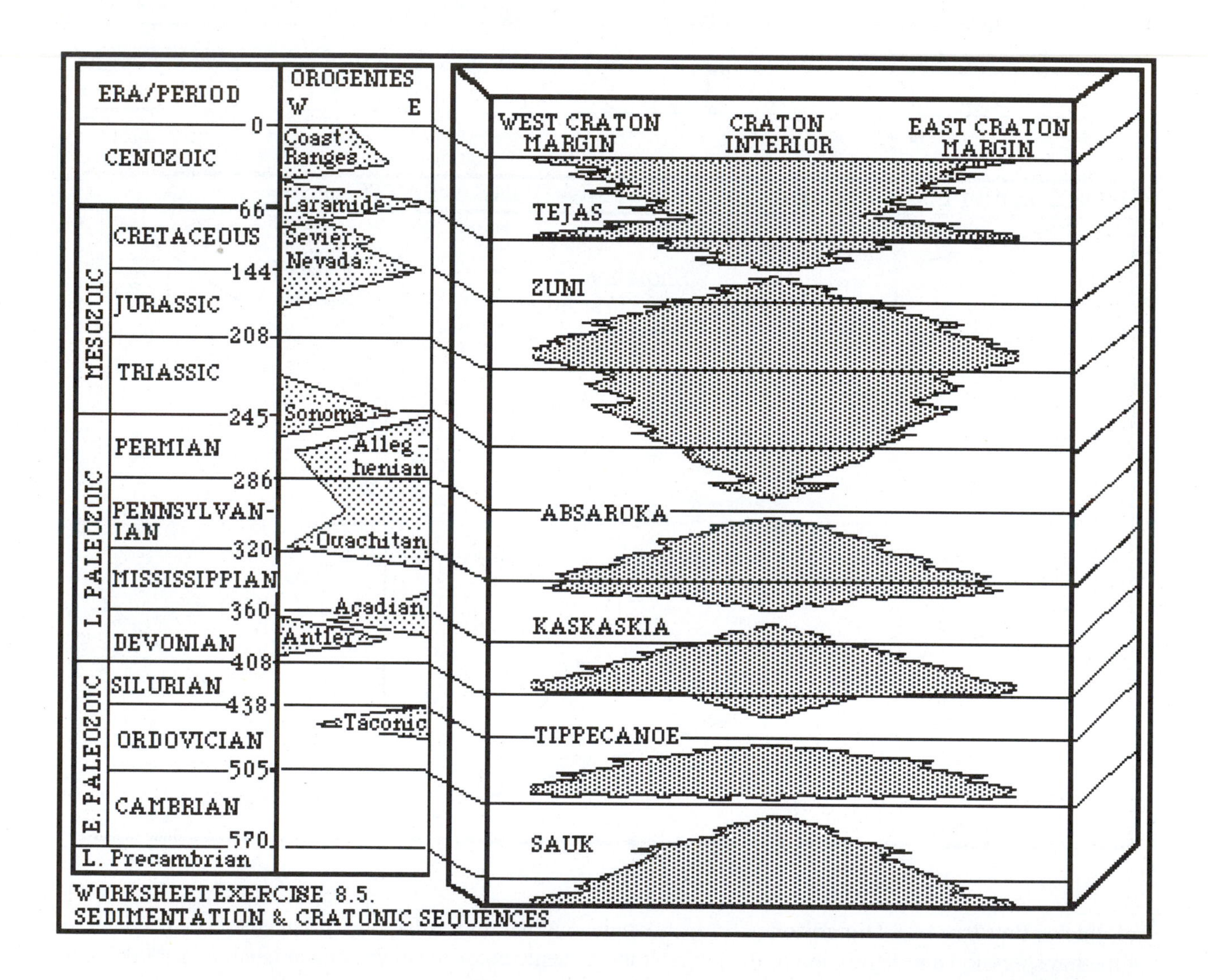

8.5. Walther's Law and Cratonic Sequences—Worksheet 8.5 shows the cratonic sequences of North America in latitudinal (E-W; front face) and longitudinal (N-S; side faces) cross-section. The central shaded portion represents the changing area of land. Complete the diagram as follows:

A) In the portions left blank, draw the transgressive and regressive facies patterns of the onshore to offshore sequence of rock types both parallel (side of block) and perpendicular (front face) to shore: sandstone, carbonates, and shale.

B) Use Walther's Law as a guide to drawing the facies relationships.

C) Vary the lateral extent (latitudinal) and proportional thickness of carbonate and clastic rock types according to their position in trangressive and regressive portions of each sequence.

D) Indicate levels where unconformities would be located. Use standard lithologic symbols for each facies (Laboratory 3).

E) The differing abundances of carbonates and clastics among the cratonic sequences resulted from what factors affecting terrigenous clastic sediment supply? ____________________

F) Compare and contrast the facies patterns in the latitudinal (perpendicular to shore) and longitudinal (parallel to shore) cross-sectional views. Which illustrates the sequence of strata that would be observed in one location throughout deposition of all six sequences? ____________________

G) Which view best shows the geographic (lateral) pattern of facies changes? ____________________

H) Which best illustrates the sequence of facies over time? ____________________

I) Where on the craton would you find the most complete ("continuous") stratigraphic record of the Phanerozoic? ____________________

J) Are unconformities diachronous or synchronous? ____________________

K) Which portion of each cratonic sequence can be characterized by a predominance of deposition/erosion? ____________________

L) Are transgressive and regressive deposits equally likely to be preserved ? Explain. ____________________

8.6. A Field Problem—You have measured and described the rock-stratigraphic units illustrated in Figure 8.12 (next page). Descriptions of rock types, thicknesses, compositions, textures, and fossil contents are written next to each unit.

A) Which of the four major environments is represented by this sequence of strata? Explain. ____________________

B) From the descriptions and stratigraphic sequence, determine which were deposited as part of a transgressive or regressive sequence. ____________________

C) Indicate which units should be recognized as facies. ____________________

D) Determine the most probable depositional environments represented by each unit (or combinations thereof). Use fossil content in your analysis. Explain your reasoning. ____________________

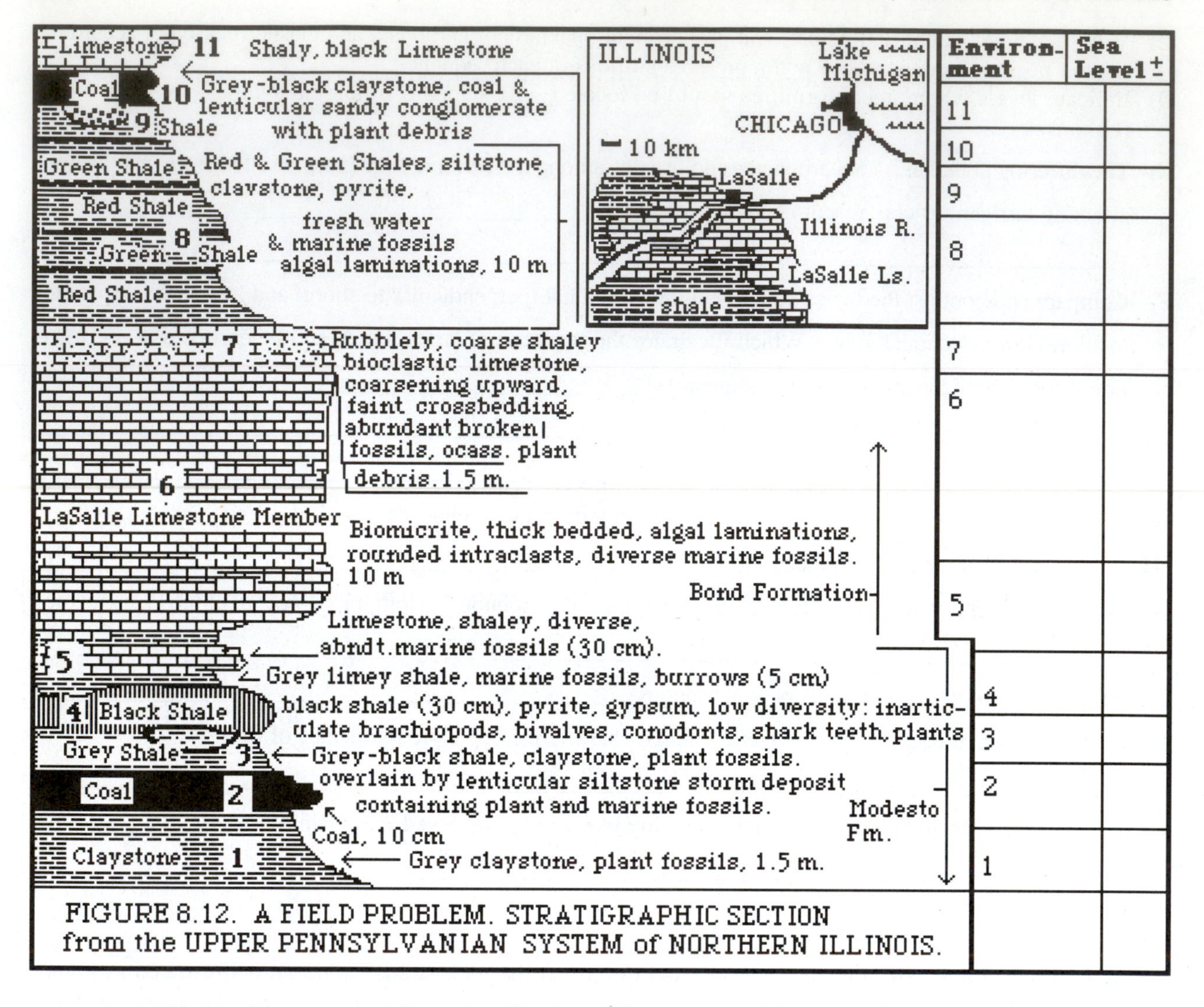

FIGURE 8.12. A FIELD PROBLEM. STRATIGRAPHIC SECTION from the UPPER PENNSYLVANIAN SYSTEM of NORTHERN ILLINOIS.

8.6. Carbonate Environments—Worksheet 8.6 (next page) illustrates the distribution of various carbonate sediments (A-I) in the region of Andros Island and the Great Bahama Banks. Determine the:

A) level of turbulence of each by matching the appropriate letter assigned to each sediment type (given in the key) with the ranks of relative water agitation given in the figure's table below;

B) type(s) of depositional environments in which each was deposited (match sediment types A-I with the five depositional environments listed in the table below); and,

C) the carbonate rock type formed by each sediment type. Do this by writing the appropriate rock type name next to the letter label for each sediment type on the table below in worksheet 8.6. Use carbonate rock classification of Table 8.2.

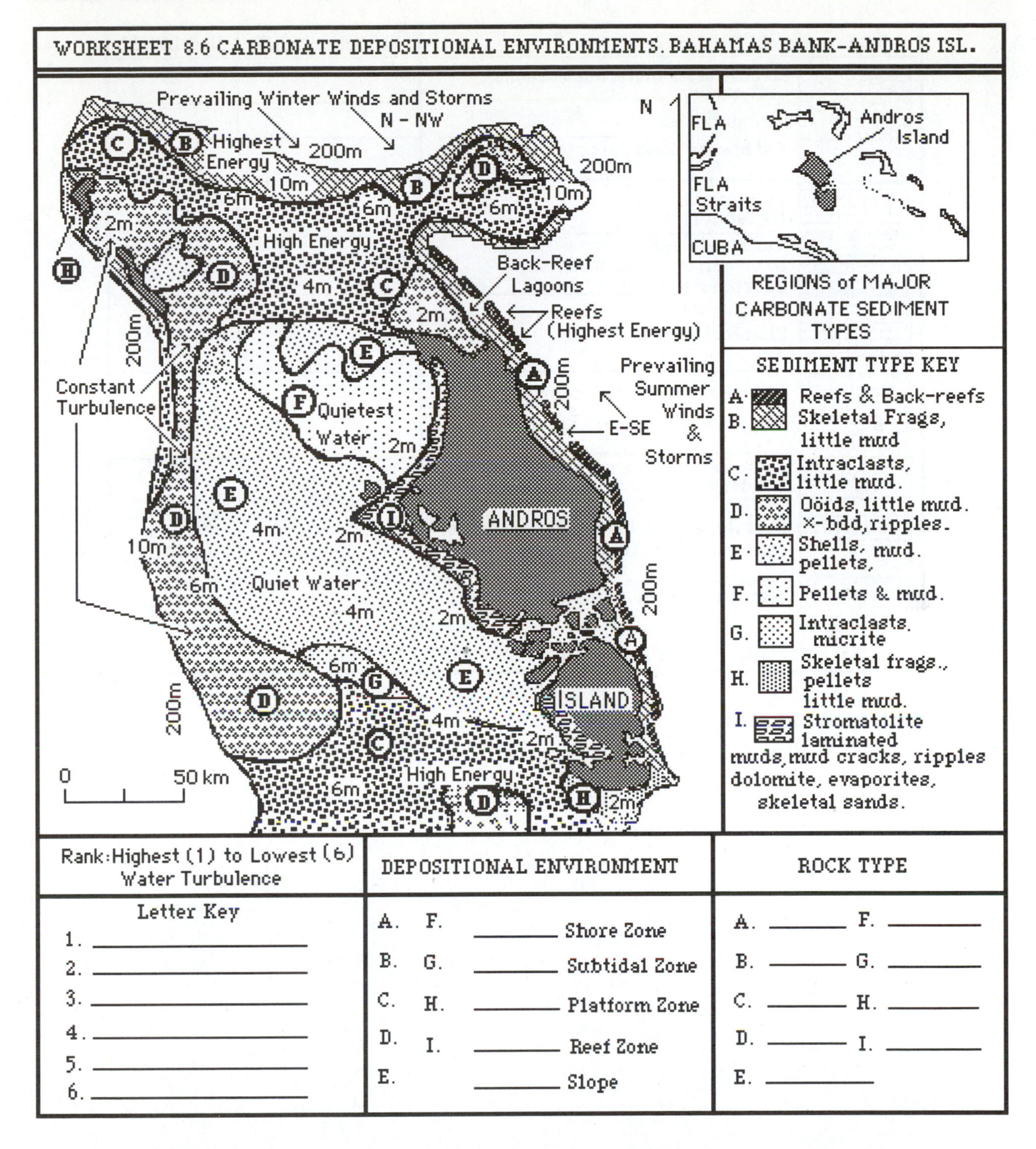

8.7. Carbonate Environments. A Field Problem—Figure 8.13 (next page) illustrates a stratigraphic sequence of carbonate rocks and descriptions of each lithology (including sedimentary structures and fossils, etc.).

A) Determine which of the 2 carbonate zones and 5 subenvironments discussed above are represented by each lithofacies (or combinations thereof), and label the figure accordingly.

B) Determine whether the sequence of facies (base to top) reflects a transgressive or regressive sequence, and indicate which facies are transgressive and which are regressive. Indicate your conclusion on the spaces provided.

LITHOLOGY	ENVIRONMENT	SEA LEVEL
Stromatolite mats, mud cracks salt casts		
Stromatolites		
Carbonate Sand , ripple marks		
Biomicrite Intramicrite		
Carbonate sand &/or oolite, Biomicrite, Biosparite		
Biomicrite		
Massive, nonbedded micrite large coral colonies, Biosparite		
carbonate sand & mud, high angle bedding		
Micrite, slump blocks of limestone, contorted bedding		Rise ↑ or ↓ Fall ?
Micrite, barren of fossils,		

FIGURE 8.13. CARBONATE ENVIRONMENTS – A FIELD PROBLEM

NOTES

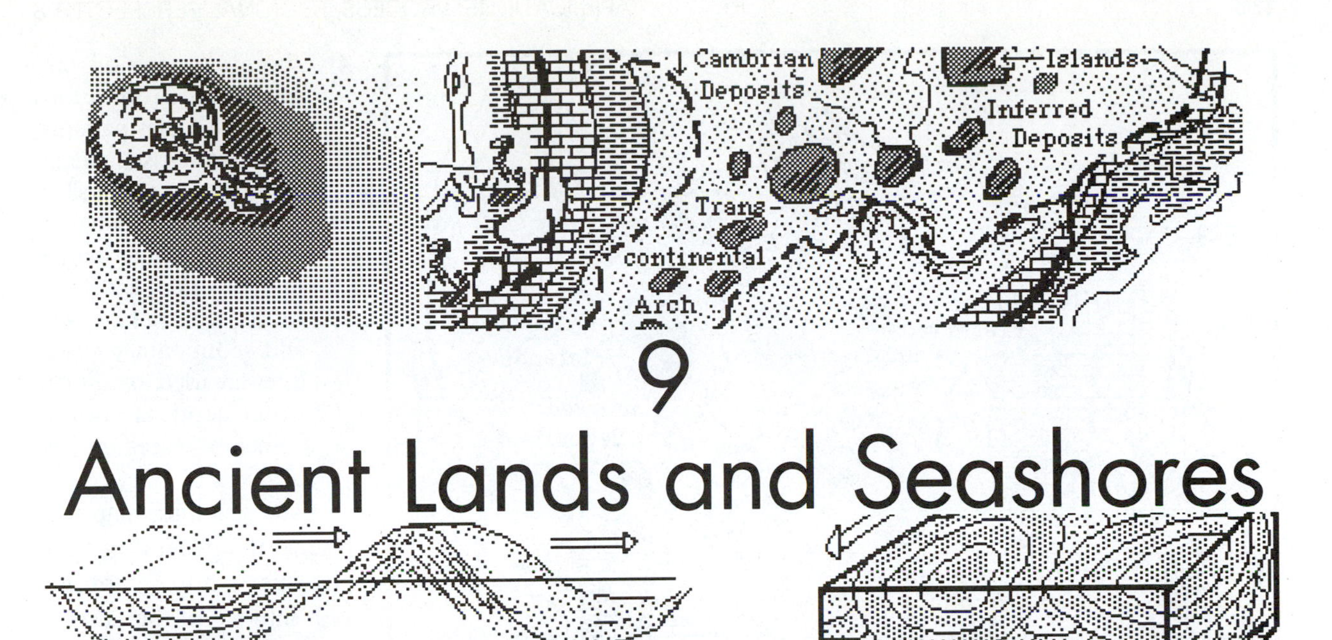

9 Ancient Lands and Seashores

Objectives

To demonstrate mastery of this material you should be able to: 1) construct paleogeographic and lithofacies maps; 2) apply criteria for mapping the locations of ancient shorelines and water depths; 3) infer general environmental conditions from primary sedimentary structure evidence; and, 4) make maps and graph the prevailing direction of, wind and current directions. Colored pencils, rose diagram and linear graph paper, and a compass are needed. The primary goal is to "experience" some of the ancient geographies of the North American plate during various geological periods by making maps from data.

Using This Laboratory

The first part of this chapter involves making and answering questions about paleogeographic and lithofacies maps. Secondly we will learn to use evidence from fossils, sediments and sedimentary structures to infer positions of ancient shorelines, relative depths of water in shallow seas, the directions of prevailing winds from data on the orientations of sand dunes and volcanic ash thickness, and deposits formed by the action of ancient hurricanes.

One of the ultimate research goals in historical geology is to synthesize information about periods of Earth history and display conclusions in the form of paleogeographic maps. These maps depict the locations and shapes of land masses and shallow seas, and are often combined with lithofacies maps to study different depositional environments, shoreline locations, and ancient climates (Laboratory 10). When geological information is so synthesized we gain, among other things, snapshots of the Earth much as if we had retrieved satellite images of the Earth from the distant past.

PALEOGEOGRAPHIC MAPS

Paleogeography is the reconstruction and mapping of past geographies. Traditionally, this has meant mapping of ancient seas and lands, and thus, the paleogeographer's major task was to locate the shorelines of these seas. Plate tectonic theory has radically changed this objective. Paleogeographers now chart the fragmentations, assemblies and travels of continents, and the births and deaths of oceans (Laboratories 10-12). Our exercises will focus on the locations of the ancient shorelines seas and the belt-like tracts of land created during past mountain building episodes.

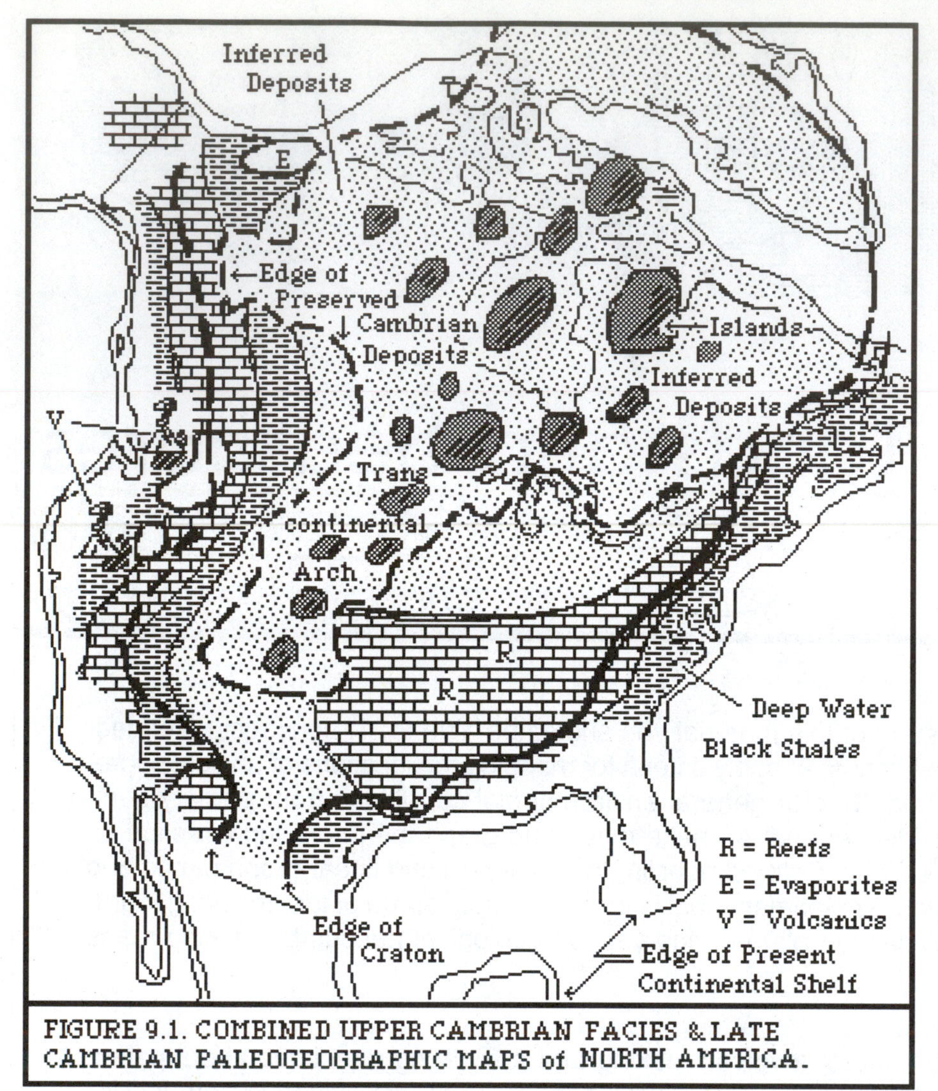

FIGURE 9.1. COMBINED UPPER CAMBRIAN FACIES & LATE CAMBRIAN PALEOGEOGRAPHIC MAPS of NORTH AMERICA.

The role of plate tectonics in creating these geographies (Exercises Figures 9.11A-C—9.13) and in similar geographically-based investigations of global scope is discussed in Laboratories 10-12.

Regional paleogeographic maps are made by plotting occurrences of rocks belonging to a chosen time-rock unit (Laboratory 7) of a certain age in a region; for example, the Croixan Series of Late Cambrian age, North America (Figure 9.1). The "rules" for constructing paleogeographic maps are listed below and illustrated in Figure 9.1:

1) areas lacking sedimentary rock are inferred to have been regions of nondeposition or extensive erosion;
2) the distribution and thickness patterns of facies that border the region of no data, and the history of uplifts, indicate whether strata are missing because the region was land, or because they were removed by subsequent erosion;
3) facies should become thinner toward a shoreline, should (generally) lie in bands paralleling shorelines, and the sequence of facies "should" grade logically toward a shoreline deposit;
4) shoreline positions are estimated from clues given by facies patterns, rock textures, sedimentary structures, and fossils;
5) fossils indicate whether certain rocks are marine or nonmarine; and,
6) fossils, sedimentary rocks and sedimentary structures are used to estimate water depth at various points in the region (see below).

The resulting map represents the general geography over a 5 to 20 million year interval of the past.

SHORELINES and WATER DEPTH

Methods of determining ancient shoreline locations and estimating water depth include inference from sedimentary structures, rock textures of sedimentary rocks, facies sequences, and fossils. In this chapter we will consider only depths of 200 m to 0 m (656 to 0 ft). It is usually necessary to combine lines of evidence to locate an ancient shoreline; any one approach may be inconclusive, or required data may not be available.

Sedimentary Structures—Primary sedimentary structures are evidence of processes or events which acted on sediment at depositional sites, and during or immediately after, deposition. We will discuss only those most common and which provide clues about water depth. Primary sedimentary structures include:

1) any type of markings on the tops or bottoms of beds that were made by the movement of various objects across the sea floor or by a current acting on the sediment; and,
2) a preferred orientation of laminations, bedding or grain sizes within a bed, that were created by water and sediment movements.

Mud cracks evince alternate wetting and drying that, in the marine realm, occurs very near shorelines. Most form as molds and casts of sediment (clay & lime mud) exposed and dried in the intertidal zone, and rapidly buried. When plates of dried mud are

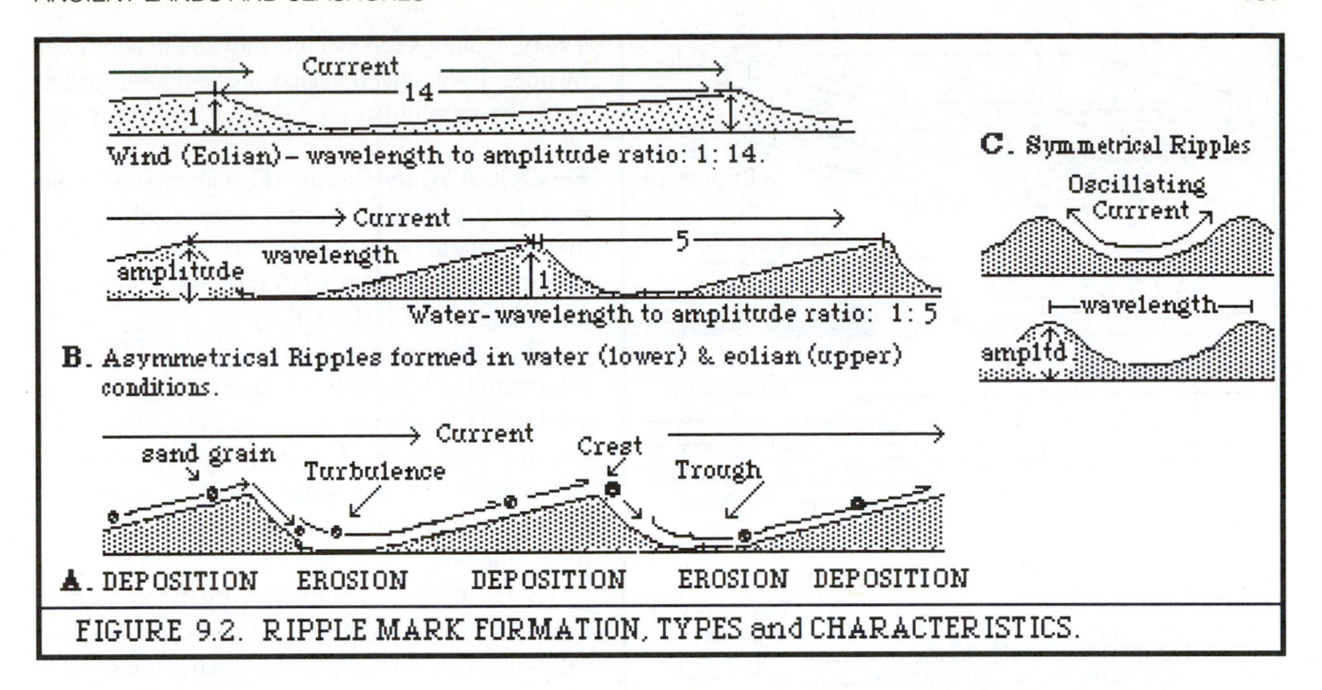

FIGURE 9.2. RIPPLE MARK FORMATION, TYPES and CHARACTERISTICS.

ripped from a tidal flat during storms they can be redeposited as intraclasts and form intramicrite (see Laboratories 2 & 8), or sorted from finer sediment and deposited as a flat-pebble conglomerate common intertidal deposits. Intraclasts are typical of supratidal deposits. Both, however, can form in deeper water environments.

Other structures formed in supratidal and intertidal deposits include trace fossils (footprints, animal tracks, trails), and raindrop prints. These may all form in dried lake beds, river flood plains and (except raindrop prints) subtidal marine deposits as well.

Ripple marks are parallel ridges of sand- or silt-sized sediment developed more or less perpendicular to the direction of wind or water flow (Figure 9.2A). Asymmetrical ripples form in unidirectional current flow (Figure 9.2B). The steep slope faces in the direction of current flow. Symmetrical ripples are formed by waves in the intertidal zone (Figure 9.2C), and can evidence former shorelines. However, waves can form symmetrical ripples in deep water (≥ 35 m = 115 ft), and currents can form ripples at extreme depths.

Ripples also form on the surface of sands deposited by wind. These are distinguished from water ripples by their longer wavelength (distance between ripple crests) and small amplitude (height measured from trough to crest; Figure 9.2B). Wind-blown ripples have a high wavelength to amplitude ratio (e.g., 14:1), while those of water formed ripples are much lower (e.g., 5:1).

A ripple is a small dune or moving sand ridge formed by fairly low velocity currents. As velocity increases, ridge size increases and the mound of moving sand becomes a dune. Structures called cross-bedding or cross-lamination are formed by laminated bedding within the dune that represent the sequential inclined "surfaces" of the moving dune. Of the several types of crossbedding, we discuss only two: trough and planar crossbedding.

Planar crossbedding forms on beaches, in wind-blown dunes, and other environments as a dune front moves across an inclined surface and planar "sheets" of sediment cascade down an inclined surface (Figure 9.3). The front face of planar crossbedding is linear, as opposed to the concave face of trough crossbedding (below; Figure 9.4), and is most steeply inclined in the direction of current flow. Crossbedding can form in water of various depths and in dunes above sea level (see below). Marine crossbedding usually indicates water depths of less than 100 m.

Trough crossbedding forms from the movement of barchan-like dunes, one of several dune types. The process is partly erosional, and only remnants of dune sediment are left in its wake (Figure 9.4A, C). Turbulence on the dune's downstream side continuously excavates (erodes) a trough or elongate depression ahead of it, and dune sand is deposited in the depression as inclined laminae as the dune moves downslope (Figure 9.4B). It is as though the dune were a bulldozer-dumptruck excavating sand and immediately filling the excavation as it travels (Figure 9.4B). Only inclined laminae of the trough—not the dune—are preserved. As dunes pass a point at different times, portions of old crossbeds are eroded and new laminae are deposited (Figure 9.4D), resulting in partial truncation of previous deposits.

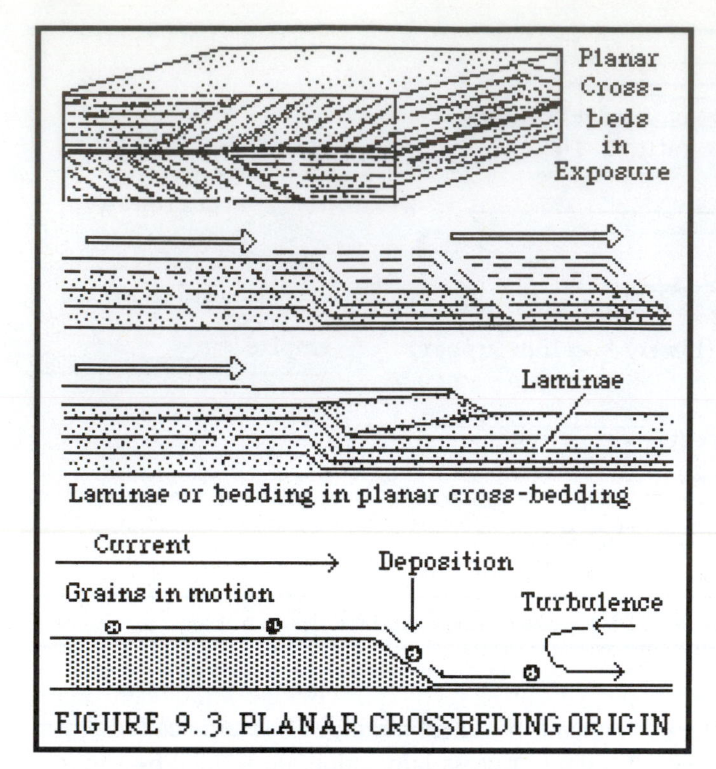

FIGURE 9.3. PLANAR CROSSBEDING ORIGIN

The orientations reflect dune movement controlled by tides. River channel deposit orientations reflect the largely unidirectional flow of river currents.

Fossils and Water Depth—Certain types of algae provide very good evidence of water depth because they are photosynthetic and thus, restricted to the photic zone ≈ 200 m (656 ft)), and commonly extending to ≤100 m (328 ft).

Different groups of algae require differing minimum light intensities. Calcareous green algae are largely limited to water ≤12 m (39 ft) deep (Figure 9.6A). The common Ordovician fossil *Receptaculites* (a green alga (?) see text, or archaeocyathid; Appendix I)) may indicate shallow water limestones.

Mats of blue-green algae (cyanophyte bacteria; see text) form mats and cabbage or mushroom-like laminated sedimentary structures called stromatolites (*stroma* = layered; *lithos* = stone: Figure 9.7). The thin, wavy laminations of limestone were deposited between successive mats as sticky filaments of "algae" (actually multispecies mats of bacteria) trapped and bound calcium carbonate precipitated from sea water (Figure 9.7A). They usually form at depths of ≤100 m, and are good depth guages because they are largely restricted to the supratidal, intertidal and shallow subtidal zones.

In a three-dimensional view, the laminae on three sides of a trough are inclined toward its axis, forming a concave structure with parabolic surface outline (top view; Figure 9.4D). This geometry allows determination current direction. Thus, when looking directly into the axis of the trough, the direction of current movement was either directly toward, or away from, the observer (Figure 9.4D). If the direction of laminae inclinations forming the sides of the trough are also determined, then the direction of current flow can be demonstrated exactly. Current flow paralleled the axis of the trough in the same direction as that of the laminae inclination on the trough's sides.

Trough crossbedding commonly forms in the tidal channels of barrier bars and in river channels. Current directions and trough crossbedding orientations recorded in tidal channels tend to cluster into two sets when plotted on a "circular bar graph" (rose diagram); each set is oriented at ≈ 180° to the other (Figure 9.5). One set of crossbedding has laminae inclined toward the shoreline, and the other set, inclined seaward.

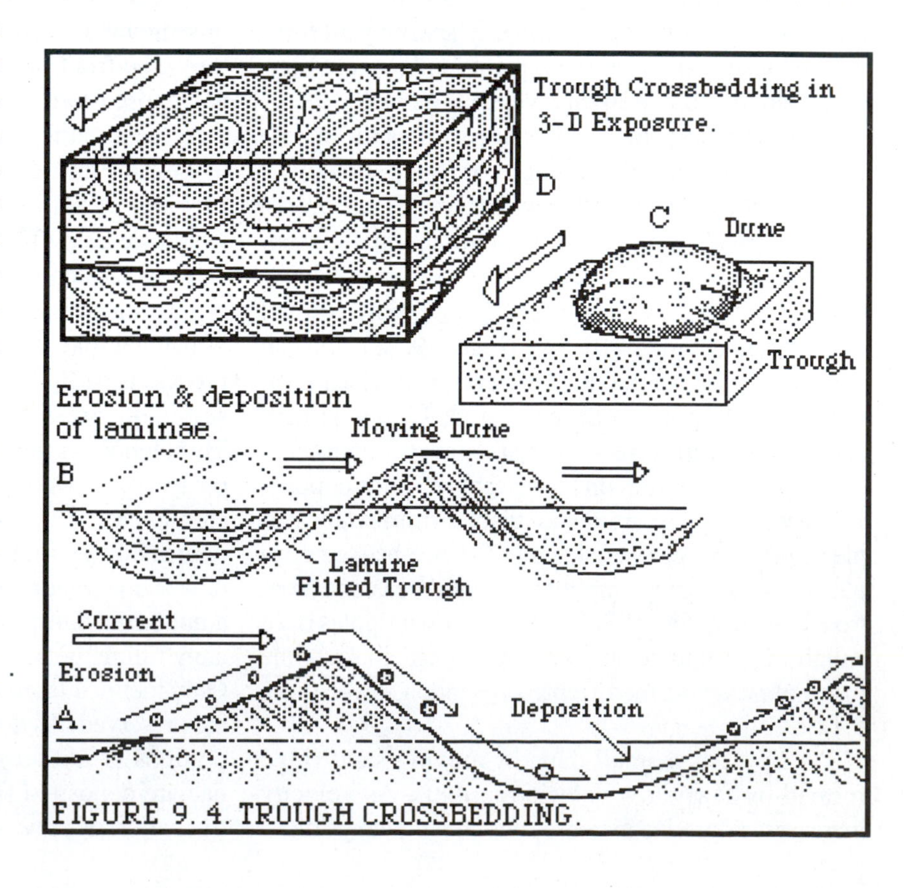

FIGURE 9..4. TROUGH CROSSBEDDING.

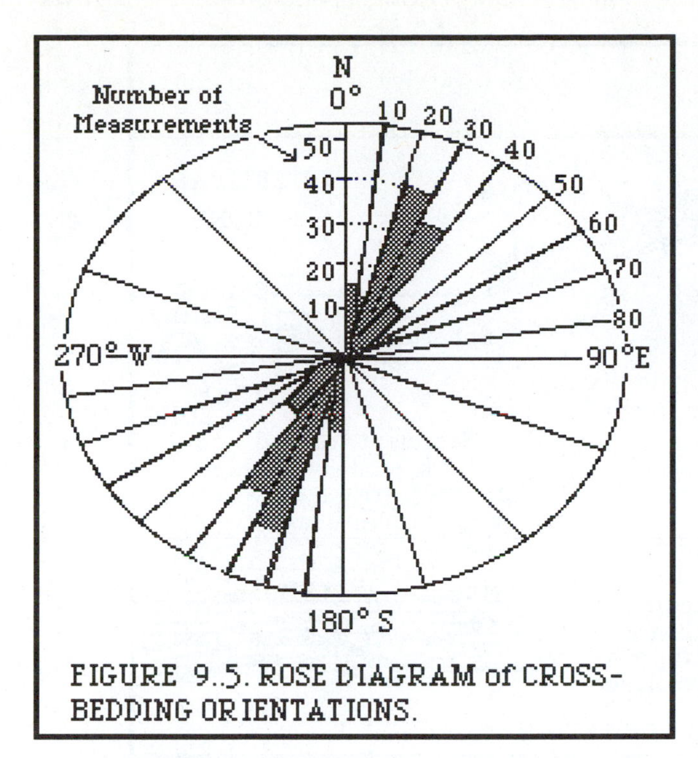

FIGURE 9.5. ROSE DIAGRAM of CROSS-BEDDING ORIENTATIONS.

Stromatolites are good indicators of water depth also because their shapes are readily molded by wave and current action (Figure 9.7B). Horizontal mats form in the supratidal zone where growth is undisturbed by wave action. Tidal currents in the intertidal zone scour and sculpt mats into club or mushroom-shaped pillars. and "cabbage heads". The pillars may be elongated perpendicular to the shoreline or aligned in parallel rows separated by scoured tidal channels. Waves in the shallow subtidal zone tumble algae into small spheres. Horizontal mats are often associated with mud cracks and thus, supratidal and tidal zones, but they can occur in the deeper water of the subtidal zone if grazing herbivores like gastropods (Laboratory 6) are absent.

Several other groups of organisms indicate limited depths; the best known and most diverse of which are reef-building corals supported by photosynthetic algae. These Mesozoic and Cenozoic corals (possibly Paleozoic corals as well) were limited to the photic zone (Laboratory 6).

Some animals show depth-related trends in shape and/or life habit. Colonies of reef corals grow as broad, flat plates in deep water, thereby increasing the surface area for photosynthesis (Figure 9.6B). Corals at shallower depths tend to have more erect colony forms. The ratio of erect to encrusting bryozoans (Laboratory 6) increases with increasing depth: encrusting forms are more common in shallow, agitated water; fragile, erect branching forms are more common in deeper, quieter water (Figure 9.6B). Corals may also show wave and current-energy related trends in shape. Deep, vertically, and rapidly burrowing bivalves are most common in the intertidal and shallow subtidal zones (Figure 9.6B), whereas shallow horizontal burrowers are more common in deeper water. Many benthic invertebrates (e.g., brachiopods) increase in diversity from the shore to deeper water.

If fossil taxa have living relatives, it may be possible to infer the depth limitations of extinct species. For example, species of foraminifera and ostracodes (Appendix I) are known to be restricted to particular depth ranges and are frequently used to infer water

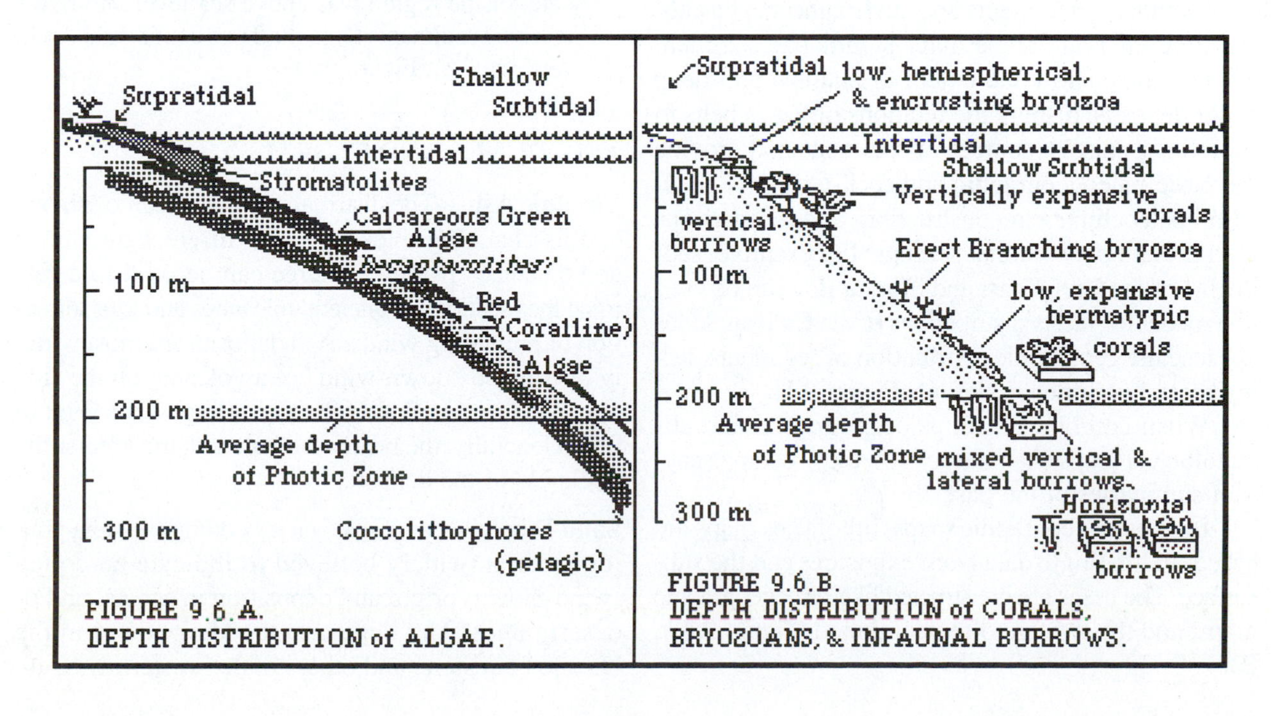

FIGURE 9.6.A. DEPTH DISTRIBUTION of ALGAE.

FIGURE 9.6.B. DEPTH DISTRIBUTION of CORALS, BRYOZOANS, & INFAUNAL BURROWS.

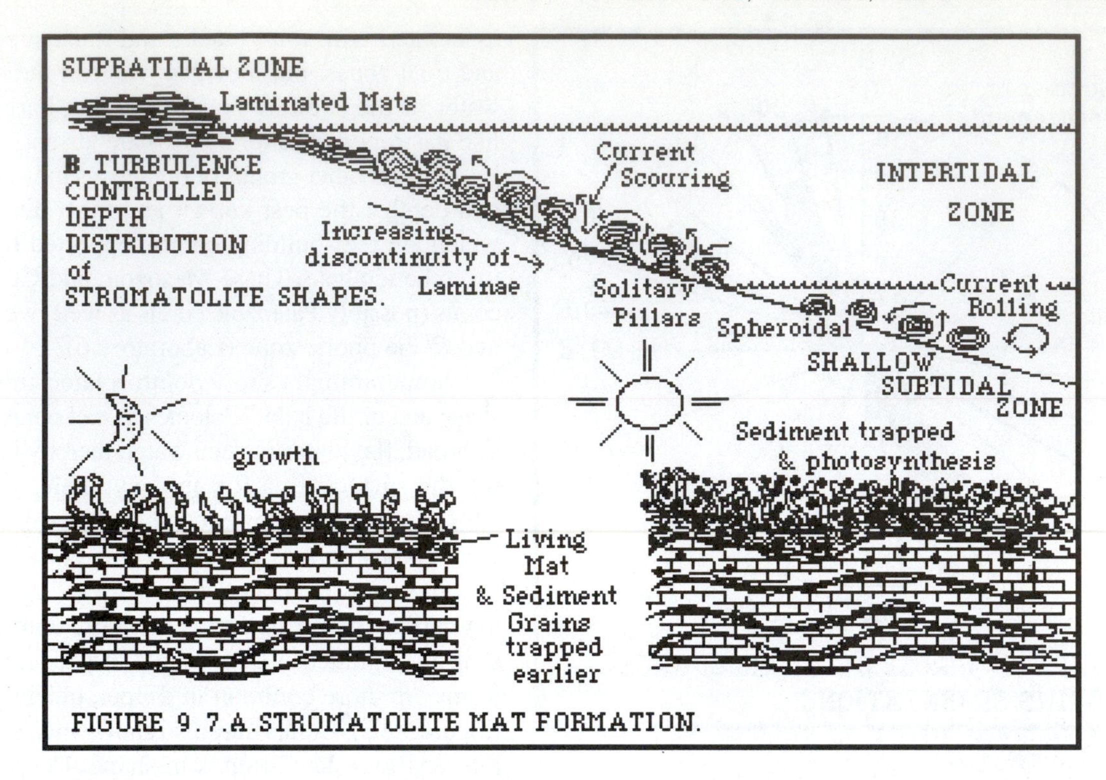

FIGURE 9.7.A.STROMATOLITE MAT FORMATION.

depths. If depth ranges of several species occurring in a single deposit are known, the depth of the deposit must have an upper limit of that set by the species with the narrowest depth tolerance.

LITHOFACIES MAPS

Maps of lithofacies (bodies of rock with characteristics of deposition in particular environments) are simple tools providing much information about geographic distributions of sediments and environments of a chosen time interval. These maps summarize sediment distributions, show the extent of shallow and deep water deposits, display the relations of facies belts to shorelines, suggest sources of sediment, and provide the basic paleogeographic and rock-type data needed for paleoclimatic reconstructions and comparisons with present depositional settings. For example, sediments of the Gulf Coast and Florida Bay can be used as a model for facies distributions of the Croixan Series because the geographic distribution of sediments during these two times is similar (Figure 9.1).

When combined with paleogeographic and climatological maps (Laboratory 10), they become snapshot summaries of the past.

Like paleogeographic maps, lithofacies maps are based on lithologic data from exposures and the subsurface. The maps are constructed like paleogeographic maps, and the two are often combined (Figure 9.8A, B):

1) determine which formations or groups belong to the chronostratigraphic unit (usually the series; Laboratory 7) of interest—a lithofacies map of the Croixan Series (Upper Cambrian) shows the general geographic distribution of facies over North America during the Late Cambrian;
2) plot the distribution of contiguous strata of the same rock type (e.g., sandstone, limestones); and,
3) determine if a region lacks deposits from the series of interest. If all deposits are absent in an area, the sediments were either removed by erosion at a later date, or the region was above sea level and never received sediments (e.g., the Transcontinental Arch, western U.S.; Figure 9.1).

ASH, DUNES AND STORMS

Volcanic Ash—The distribution of a volcanic ash bed and its changing thicknesses in different directions and distances from the source can be used to determine the locations of ancient volcanoes and thus, direction of prevailing winds. Ash deposits are most widely distributed "down-wind" of a volcano, on the side opposite the direction of the prevailing winds (Figure 9.9). Generally, the nearer the volcano, the greater the thickness of the deposit.

Sand Dunes—Large scale crossbedding (see text photographs) is widely believed to indicate an eolian (wind-blown) origin and deposition in coastal and/or desert dunes. The huge crossbedding found in the Navajo Sandstone and others in the western U.S. are

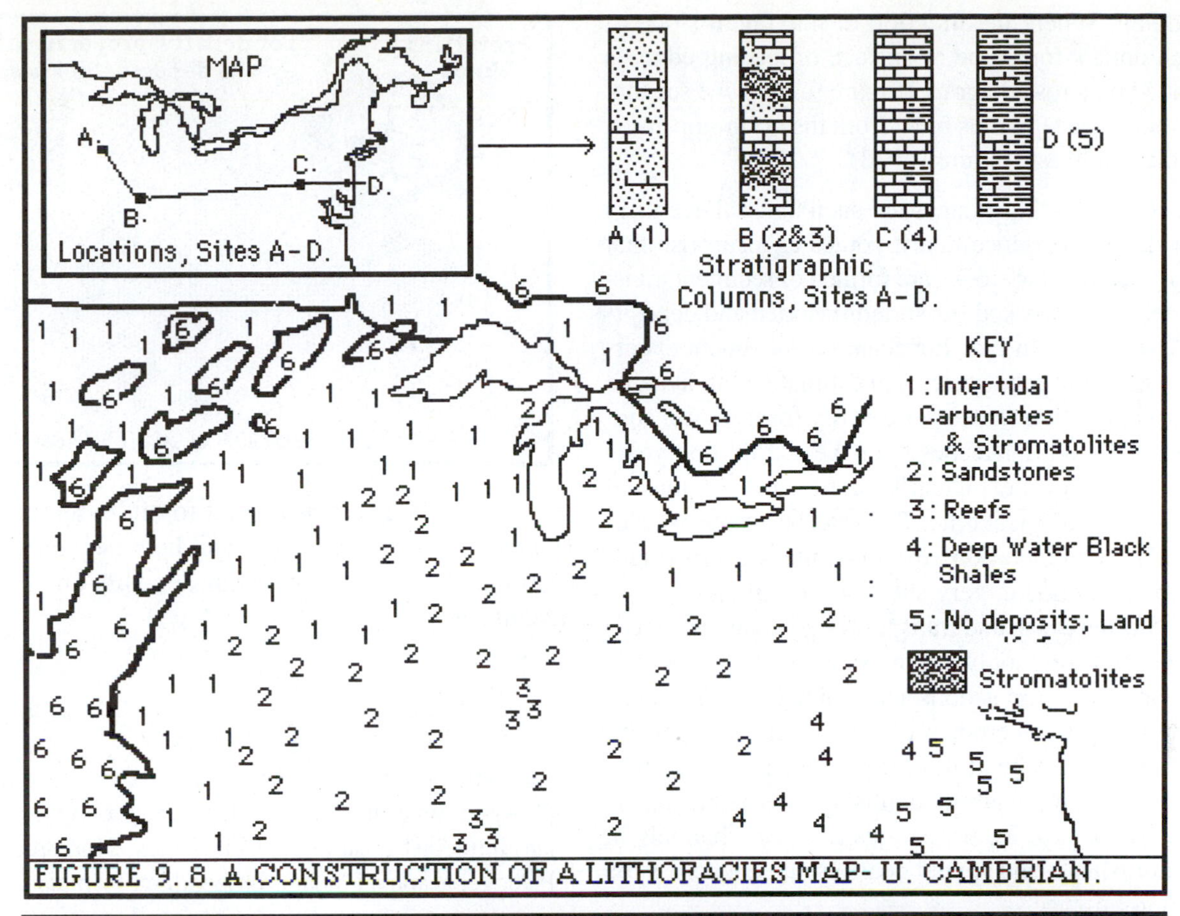

FIGURE 9.8. A. CONSTRUCTION OF A LITHOFACIES MAP- U. CAMBRIAN.

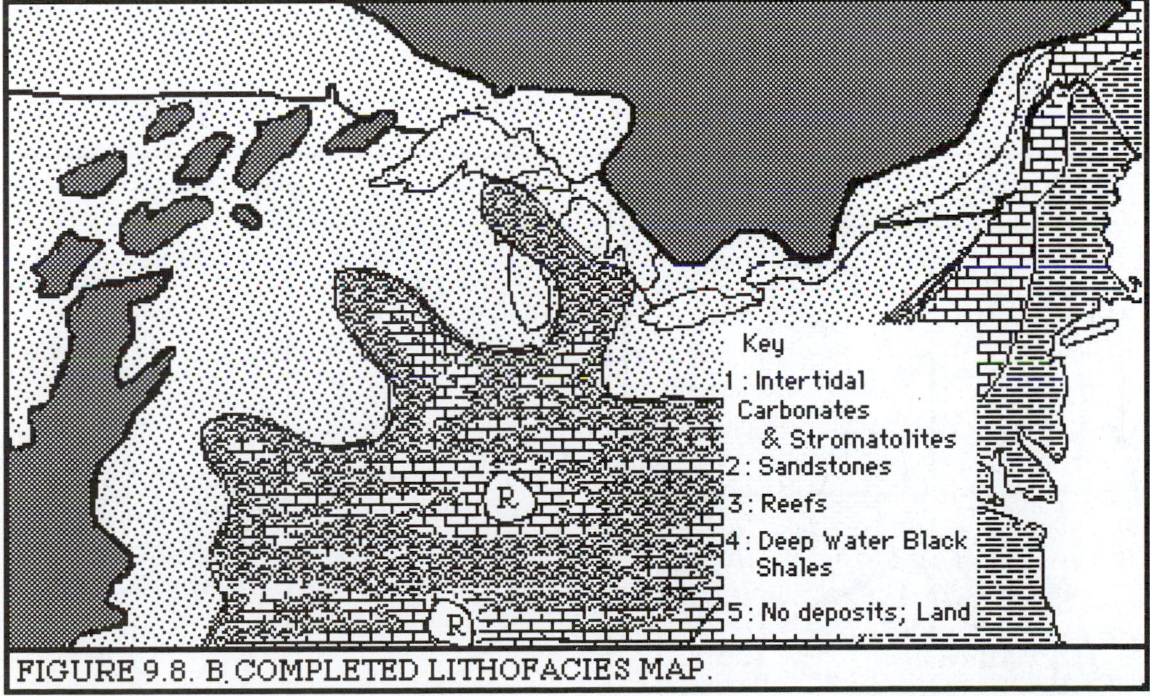

FIGURE 9.8. B. COMPLETED LITHOFACIES MAP.

composed of 90 percent quartz, and are very fine, very well rounded, frosted, and well sorted. Eolian dunes usually have steeply inclined laminae. Corroborating evidence is usually required to demonstrate origins of sediment sizes that could only be eroded and on land, however, because large similar crossbedding can also form in water.

The Upper Triassic Wingate Formation of the southwestern U.S. shows how wind directions may be determined from orientations of dune sand crossbedding. When the direction of maximum slope of each dune's foreslope or surface of leading edge, is plotted on a rose diagram (Figure 9.10A), we see that onshore coastal winds blew from the north-northwest to the southeast (Figure 9.10B).

Hurricanes—Hurricanes are such unusual events in our daily experience that it comes as a surprise that many sedimentary rocks are formed of sediment transported and reworked (re-sorted/unsorted and deposited) by storms. In fact, hurricanes are frequent events on a geologic time scale. [An estimate of an average of 16 major hurricanes per century for instance, suggests 320,000 hurricanes over the ≈ 2 million years of the Pleistocene!] Clearly, storms are not unusual events in geological time. They can be agents of voluminous sedimentation, and thus, many sedimentary deposits formed in very short periods of time.

Storms erode and transport large volumes of sediment, and commonly influence the final textures, fossil contents, compositions and sedimentary structures. Storm surges can erode sediment at a depth of 200 m, transport and spread it, along with shells and other debris, from the shallow subtidal to the supratidal zone (Laboratory 8). Return surges may cut channels in barrier islands and spread coarse sediment across the sublittoral zone.

Storm deposits typically result in deposition moved by strong currents within otherwise quiet water deposits. Consider at this point the various properties which storm deposited sediments and fossils assemblages (Laboratory 6) could exhibit and how they could be distinguished from non-storm deposits and fossil assemblages.

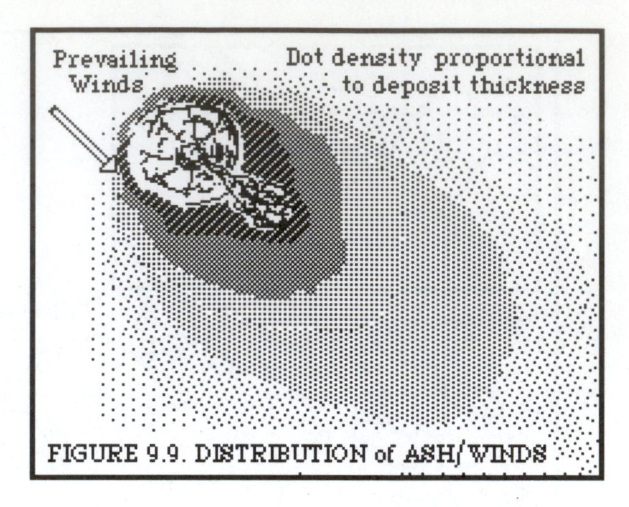

FIGURE 9.9. DISTRIBUTION of ASH/WINDS

VOCABULARY

Asymmetrical ripple marks, Bedding, Crossbedding, Cross-lamination, Facies, Flat-pebble conglomerate, Intraclasts, Laminae, Lithofacies, Mud cracks, Paleogeography, Planar crossbedding, Primary sedimentary structures, Ripple marks, Rose diagram, Stromatolites, Symmetrical ripples, Trough crossbedding.

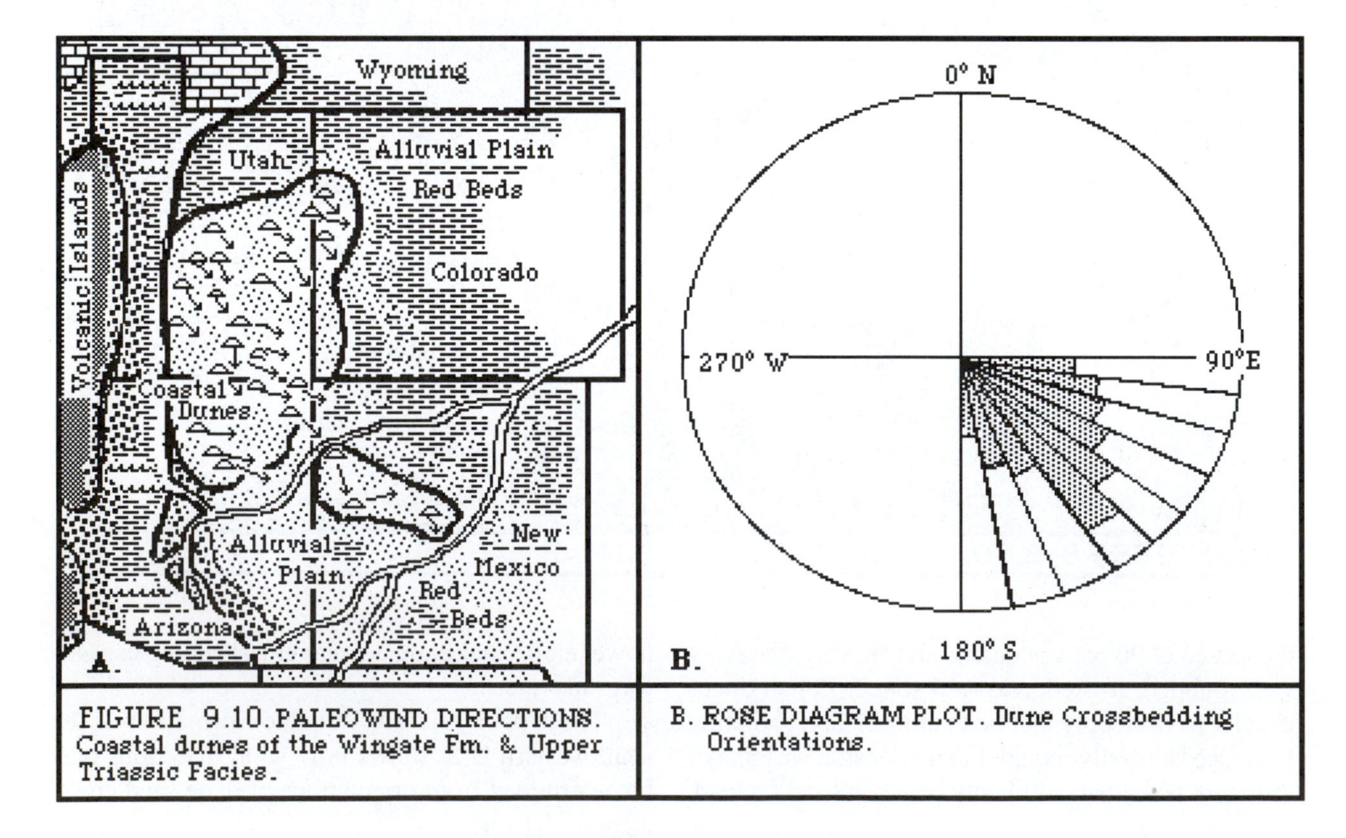

FIGURE 9.10. PALEOWIND DIRECTIONS. Coastal dunes of the Wingate Fm. & Upper Triassic Facies.

B. ROSE DIAGRAM PLOT. Dune Crossbedding Orientations.

LABORATORY EXERCISES

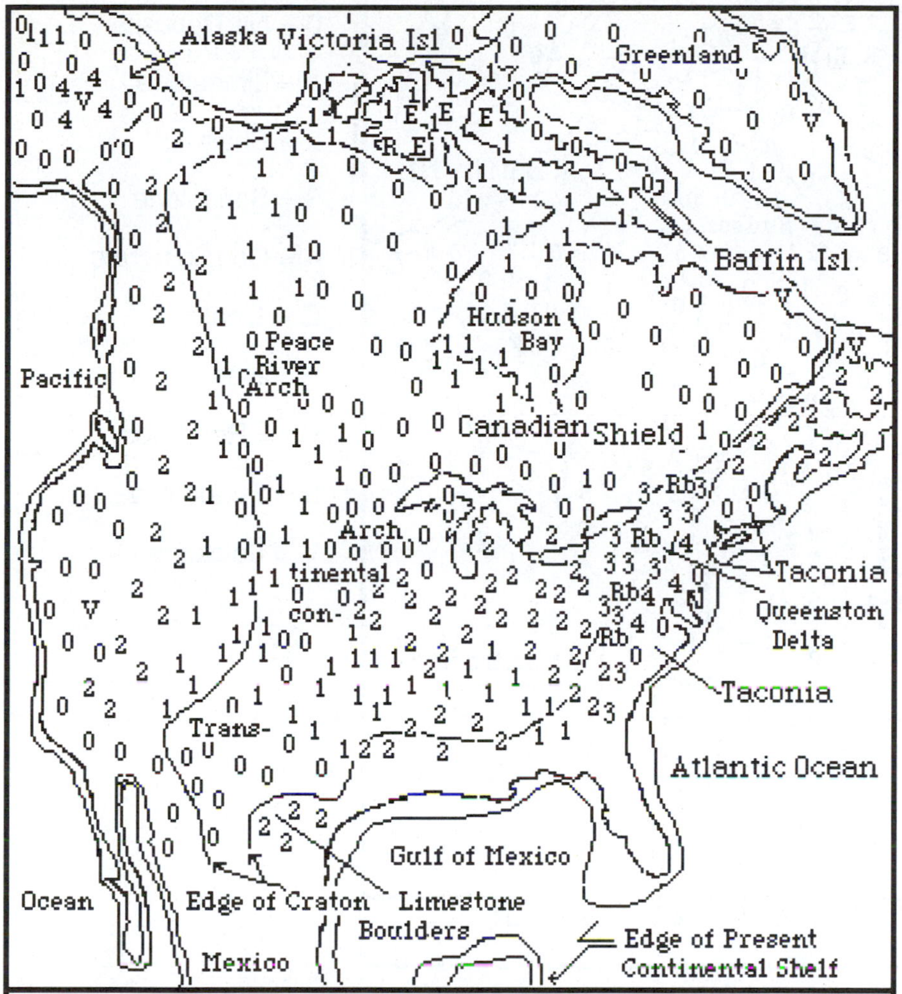

FIGURE 9.11A1. BASE MAP FOR UPPER ORDIVICIAN LITHOFACIES MAP.

KEY

#/Letter Code	Color
0 = No Deposits	
1 = Limestone	
2 = Shale	
3 = Sandstone	
4 = Conglomerate	
C = Coal	
R = Reefs	
Rb = Red Beds	
E = Evaporites	
V = Volcanics	

9.1. Lithofacies and Paleogeographic Mapping—Construct lithofacies maps of Figures 9.11A_1-C_1 and paleogeographic maps (Figures 9.11A_2–C_2) by encircling contiguous regions of a given rock type. In other words, draw boundaries between numbers representing different rock/sediment type. Indicate land areas and color-code the lithofacies. Indicate the types of primary sedimentary structures and fossils expect in each region (if not already indicated). Keys for maps are given beneath each, and "place-names" are labeled. You may need to refer to your textbook to answer some of the following questions:

A) Upper Ordovician—Refer to the maps of the Upper Ordovician, Figure 9.11A,B:

1) Were the Transcontinental Arch, Peace River Arch and Canadian Shield regions of land or shallow sea during the Late Ordovician? ____________________

Why or why not? ____________________

2) What kind(s) of additional data are needed to support your answer? ____________________

____

____

____

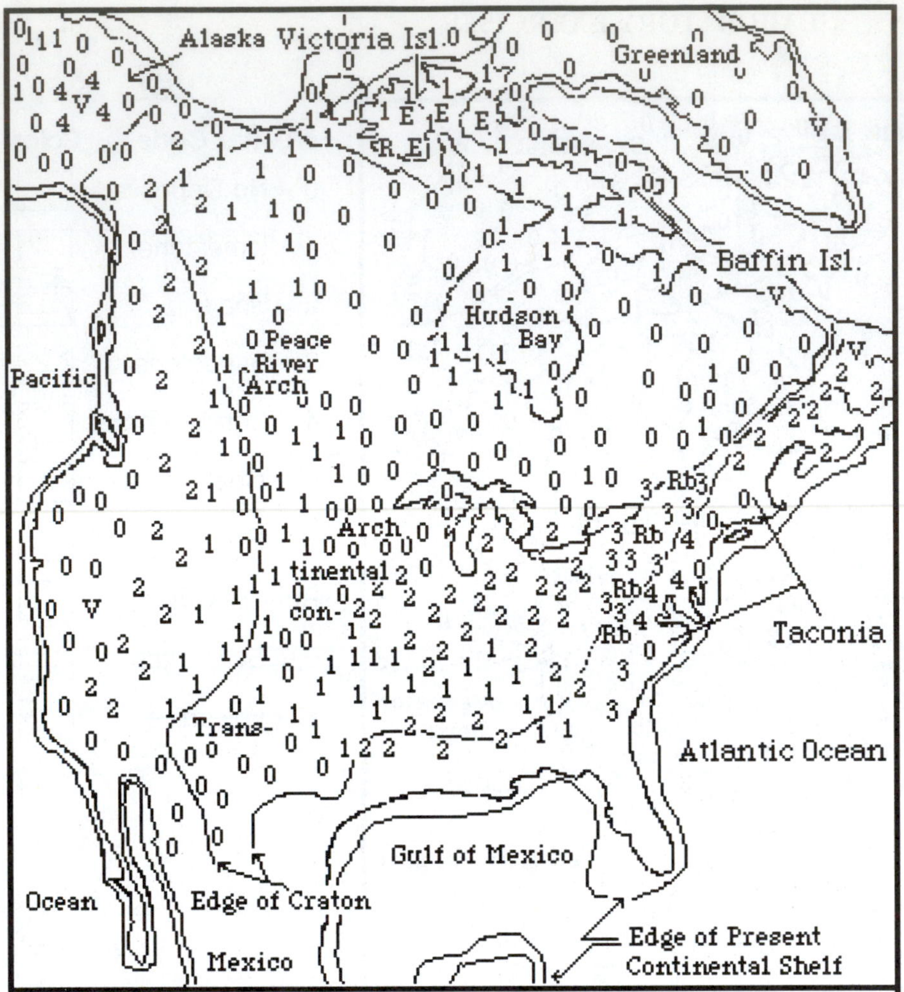

FIGURE 9.11A 2. BASE MAP FOR UPPER ORDOVICIAN PALEOGEOGRAPHIC MAP.

KEY

#/Letter Code	Color
0 = No Deposits	
1 = Limestone	
2 = Shale	
3 = Sandstone	
4 = Conglomerate	
C = Coal	
R = Reefs	
Rb = Red Beds	
E = Evaporites	
V = Volcanics	

3) Estimate the extent of the Late Ordovician sea by inferring the position of, and drawing shorelines.

4) Evaporite deposits occur between Baffin and Victoria Islands. What does this indicate about water depth and shoreline position in this region? Why? ______________________

____

5) What could the red sandstone & conglomerate facies from New York to Georgia indicate about the shoreline position? Why? ______________________

____

6) Locate the uplands of the Late Ordovician, Taconia, and the Queenston Delta deposits. ______________________

____

7) Where might volcanic islands have been located? ______________________

____

8) Indicate areas of the continent receiving deep water (continental slope) and shallow water deposits. ___

____

____

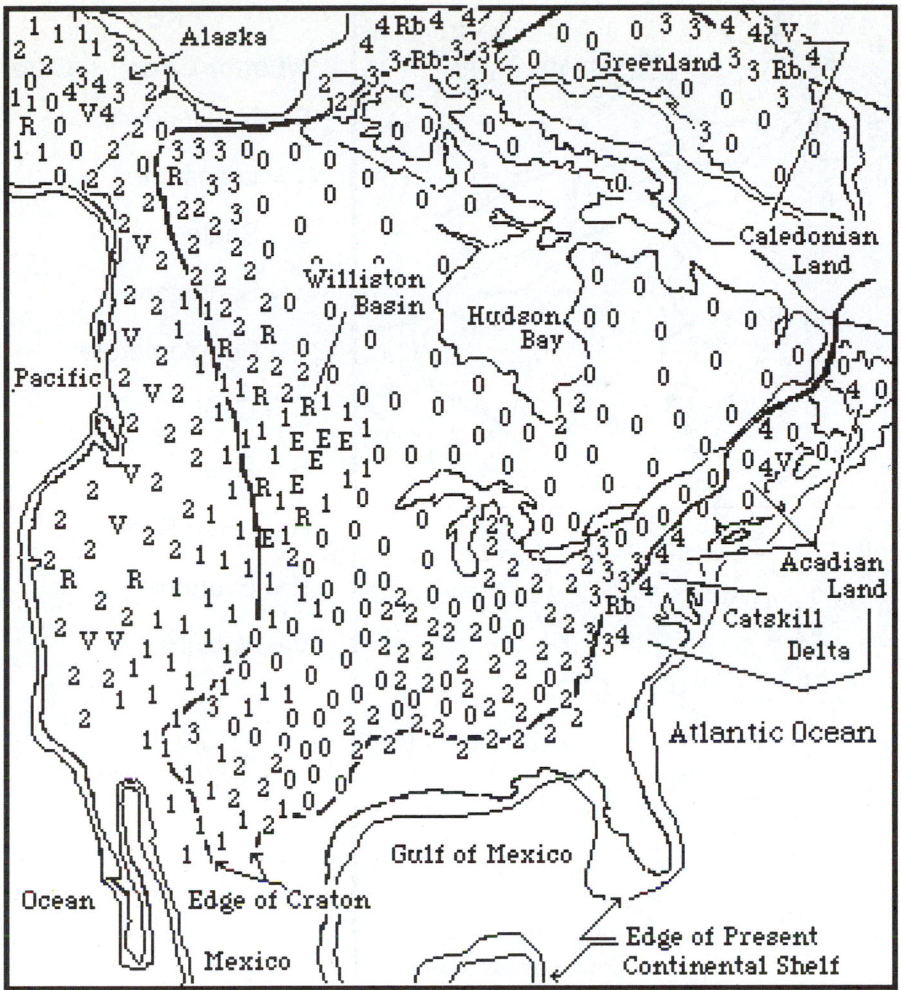

Alaska
Greenland
Caledonian Land
Williston Basin
Hudson Bay
Pacific
Acadian Land
Catskill Delta
Atlantic Ocean
Gulf of Mexico
Ocean
Edge of Craton
Mexico
Edge of Present Continental Shelf

FIGURE 9.11B$_1$. BASE MAP for UPPER DEVONIAN LITHOFACIES MAP

KEY

#/Letter Code	Color
0 = No Deposits	
1 = Limestone	
2 = Shale	
3 = Sandstone	
4 = Conglomerate	
C = Coal	
R = Reefs	
Rb = Red Beds	
E = Evaporites	
V = Volcanics	

B) Upper Devonian—Refer to the Upper Devonian lithofacies map, Figure 9.11B$_1$ and paleogeographic map (B$_2$):

1) What geographic feature lay to the east (present direction) of the Catskill red beds, along the NE coast of Greenland, & central Alaska? ______________________

2) Locate the source regions of alluvial plain & deltaic/coal swamp deposits.

3) Characterize the water depth & circulation in the Williston Basin. What kinds of sedimentary structures may be expected in this basin? ______________________

____

4) What tectonic feature is indicated by the volcanic deposits off the west coast of North America at this time? ______________________

5) Were these shale and volcanic facies deposited in shallow or deep water? ______________________

____

____

What sedimentary structures might you find in Devonian rocks of this region? ______________________

____

____

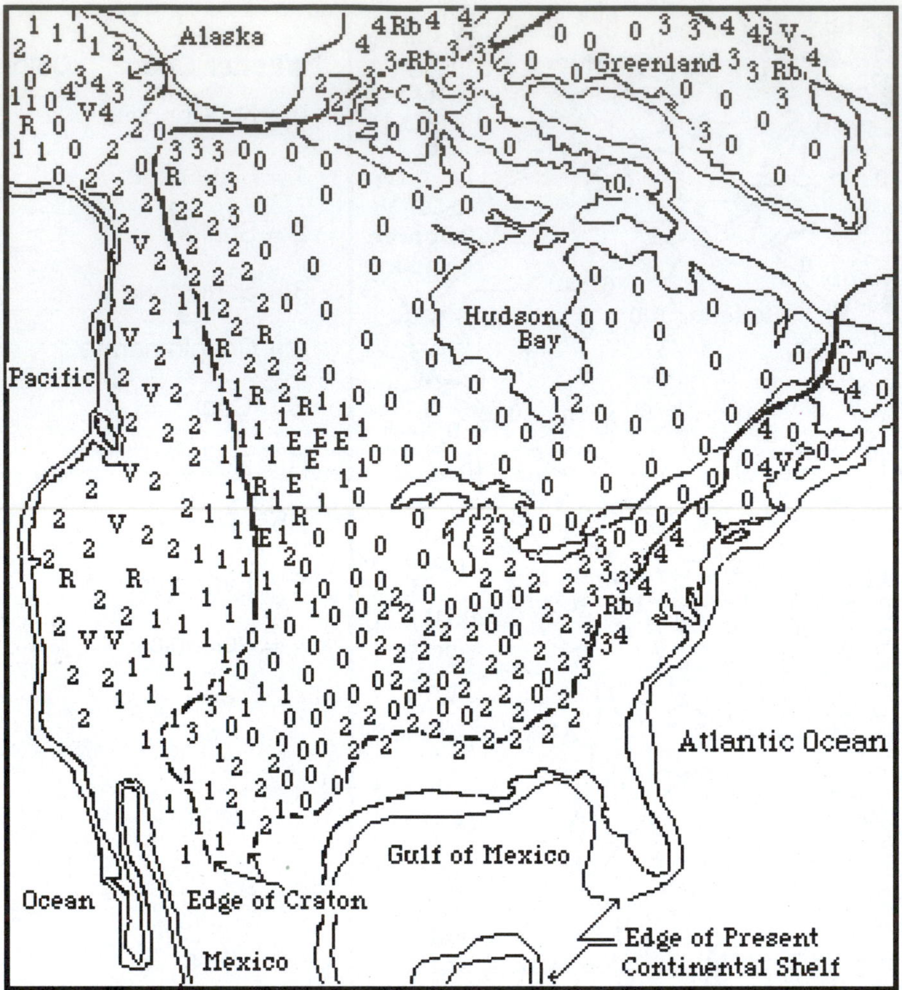

FIGURE 9.11B2. BASE MAP for UPPER DEVONIAN PALEOGEOGRAPHIC Map of North America.

KEY

#/Letter Code	Color
0 = No Deposits	
1 = Limestone	
2 = Shale	
3 = Sandstone	
4 = Conglomerate	
C = Coal	
R = Reefs	
Rb = Red Beds	
E = Evaporites	
V = Volcanics	

C) Middle Pennsylvanian—Refer to the Middle Pennsylvanian lithofacies/paleogeographic map, Figure 9.11C:

1) What were the sources of the red sands & conglomerates in the southern Rocky Mountain region and where were they located? ______________________________

2) Where were volcanic islands (island arcs) located? ______________________________

3) Locate Antlerland. What caused the Antler Orogeny? ______________________________

4) Locate the coal-bearing coastal plain & deltaic environments. Where might probable source areas of Illinois basin deposits have been located? ______________________________

5) If you discovered cross-stratification in the river channel sandstones of the deltas & coastal plains, in what general direction(s) would the crossbedding be oriented? ______________________________

____

6) Locate the regions uplifted during the Alleghanian Orogeny. What caused their uplift? ______________________________

____

7) Locate the Ouachita, Marathon and Colorado Mountains. What caused their uplift? ______________________________

____

LABORATORY EXERCISES

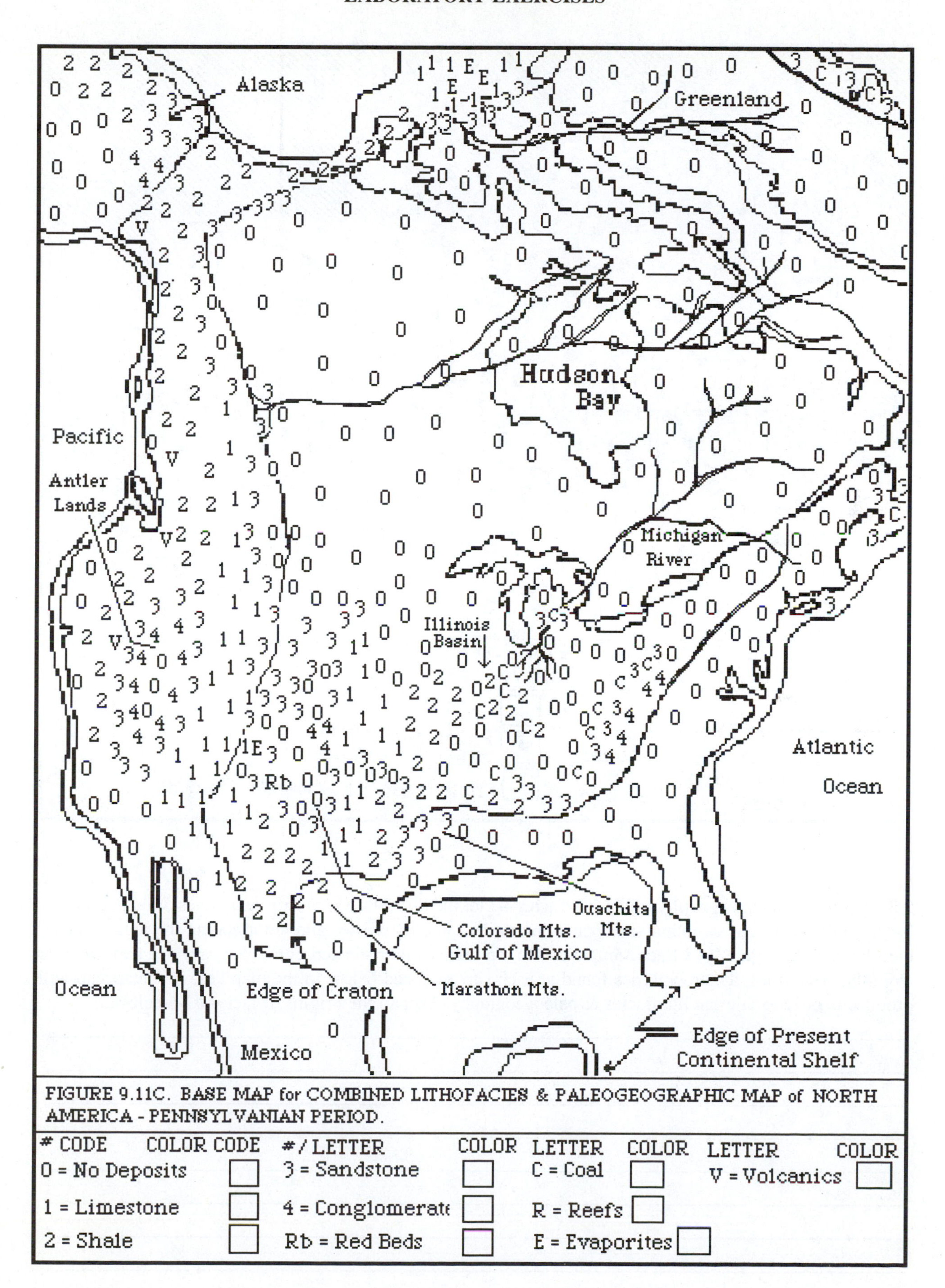

FIGURE 9.11C. BASE MAP for COMBINED LITHOFACIES & PALEOGEOGRAPHIC MAP of NORTH AMERICA - PENNSYLVANIAN PERIOD.

LABORATORY EXERCISES

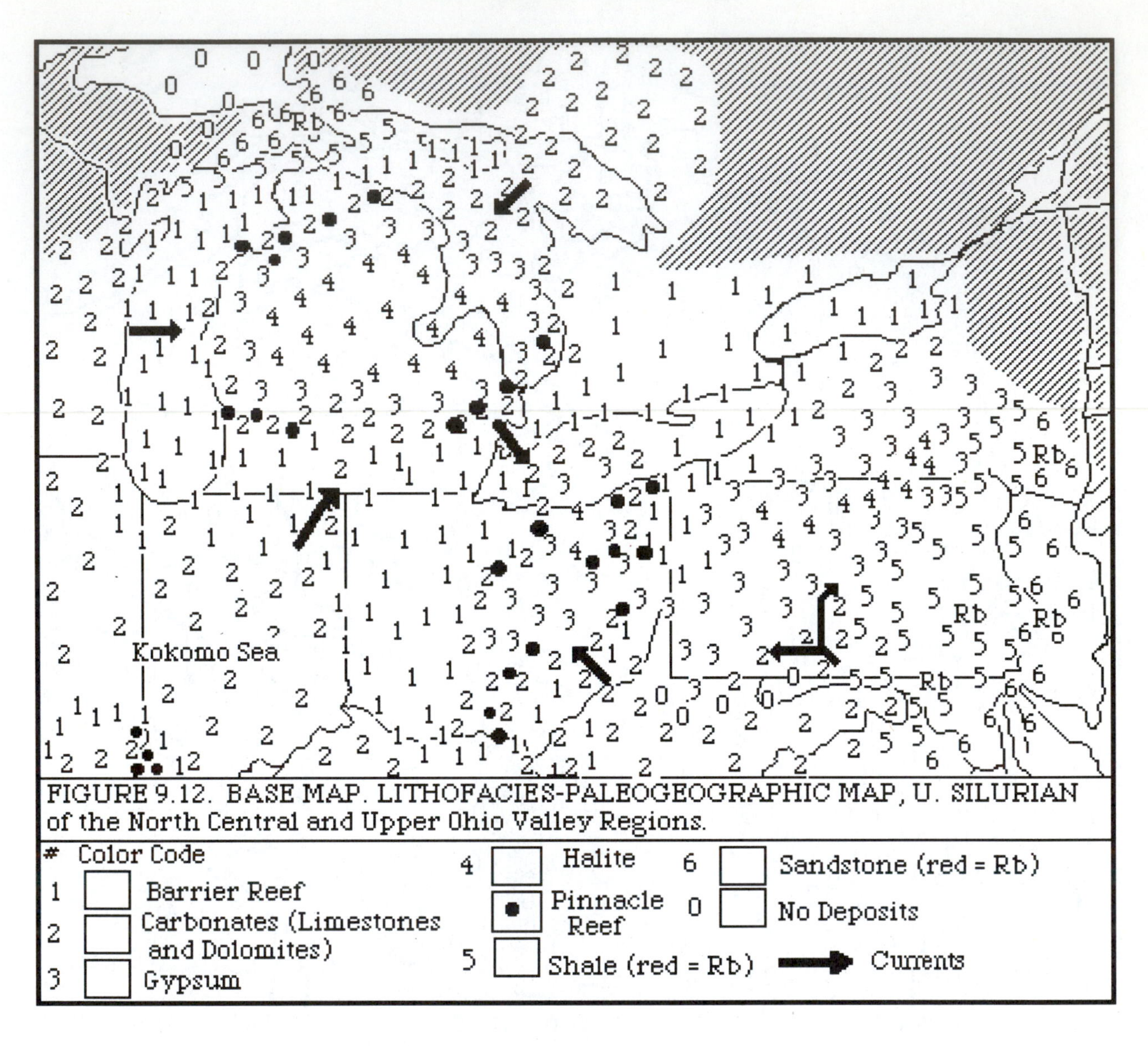

FIGURE 9.12. BASE MAP. LITHOFACIES-PALEOGEOGRAPHIC MAP, U. SILURIAN of the North Central and Upper Ohio Valley Regions.

9.2. Regional Paleogeography and Lithofacies—Plot the distributions of different facies & land/sea areas by encircling adjacent sampling sites belonging to the same lithofacies and color-coding the regions occupied by each in Figure 9.12 Each sample & site represents data collected either from surface exposures or from the subsurface. The rock types found at each site are listed below in the map key. The map you will produce is of Late Silurian lithofacies & paleogeography. Explain the origins of facies #3-6 below.

LABORATORY EXERCISES

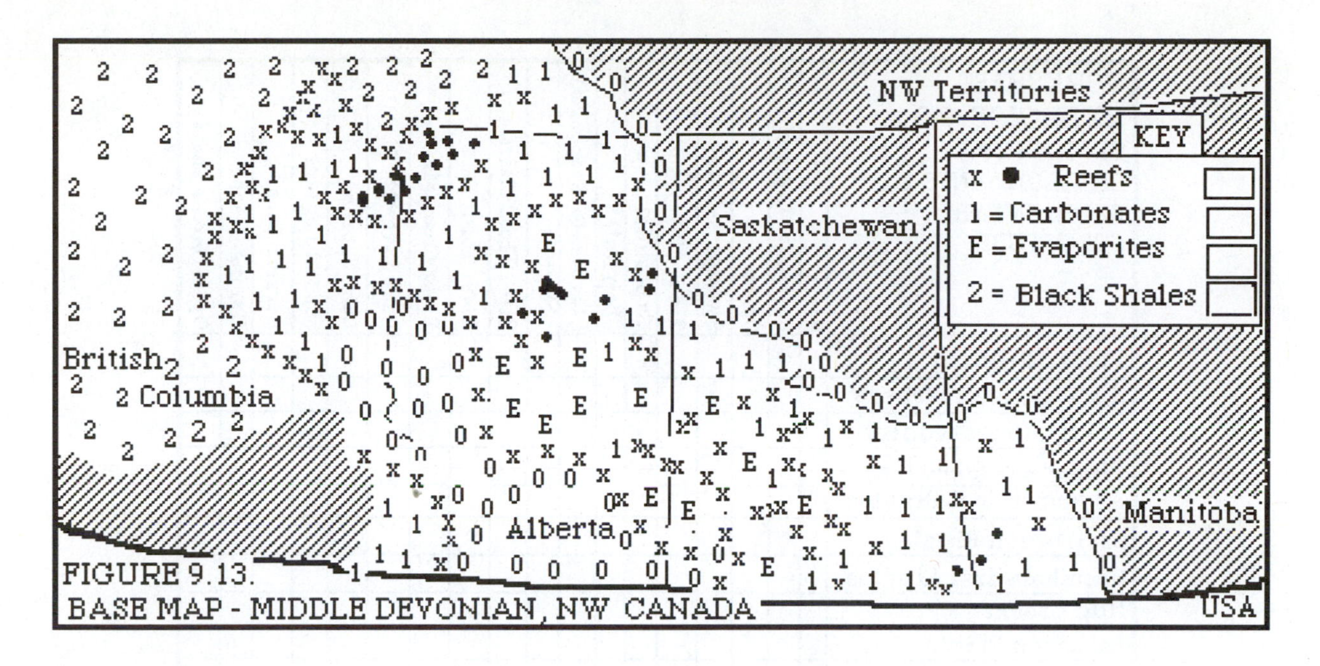

FIGURE 9.13.
BASE MAP - MIDDLE DEVONIAN, NW CANADA

9.3. Regional Paleogeography and Lithofacies—Plot the distributions of different rock types & land/sea areas on Figure 9.13 as above. The map represents Late Devonian lithofacies & paleogeography in the region of present Alberta, Canada. Rock types are shown in the figure key.

9.4. Fossils & Water Depth—You have discovered an Eocene deposit containing 8 non-reef coral genera, all of which have living relatives. Determine the probable depth of the water in which these corals lived using the depth ranges & temperature tolerance data of the living relatives of each genus Table 9.1. Plot the maximum-minimum depths & temperatures for (the living relatives of) each genus by drawing boxes enclosing regions of the graph (graph paper provided) bounded by temperature and depth limits. Determine the water depth of the deposit by finding the region of depth & temperature overlap of all genera.

TABLE 9.1. TEMPERATURE & DEPTH RANGES

DEPTH	(km)	MAX.	MIN.	TEMP. (°C)	MAX.	MIN
GENUS	A	1.4	0.00		15	4
	B	0.75	0.00		19	7.5
	C	0.6	0.1		26	6
	D	0.85	0.15		27	13
	E	1.2	0.00		2.65	7
	F	1.25	0.00		26	6.5
	G	1.4	0.5		25	5
	H	1.5	0.00		27.5	4.5

LABORATORY EXERCISES

WORKSHEET EXERCISE 9.5.A. SEDIMENTARY ENVIRONMENTS & SEDIMENTARY STRUCTURES.	Reefs	Back-reef Lagoons	Deltas	Swamps & Marshes	Tidal Flat	Supratidal	(Detrital)	(Carbonate-Evaporite)	Intertidal	River Floodplains	River Point Bars	River Channels	Lake Bottoms	Lake Shores	Coastal Lagoons	Barrier Bars	Marine Beaches	Open Marine Shelf
Planar Bedding																		
High-angle Crossbeds																		
Trough Crossbedding																		
Planar Crossbedding																		
Asymmetrical Ripples																		
Oscillation Ripples																		
Rhombohedral Ripples																		
Mud Cracks																		
Raindrop Prints																		
Flat Pebble Conglom.																		
Laminations																		
Salt Crystals & Casts																		

9.5. Sedimentary Structures—Complete worksheet 9.5A-C (above and following page) by placing an X in the appropriate box(es) to indicate (A) environment(s) & (B) rock types in which you would expect to find each of the primary sedimentary structures listed. Then determine the types of sediments which you could expect to find in association with the environments listed in part C. Blacken boxes to indicate a strong association.

A) Are any sedimentary structures uniquely associated with a single environment of deposition? ___________

____

____

____

B) What other information is needed to determine the nature of an ancient depositional environment? _______

____

____

____

C) Which sedimentary structures are strongly associated with shoreline environments? _______________

____

____

LABORATORY EXERCISES

WORKSHEET EXERCISE 9.5.B	Red Shale	Shale	Siltstone	Quartz Sandstone	Greywacke	Arkose	Bedded Chert	Conglomerate	Coal	Black Shale	Evaporites	Dolomite	Sparite(s)	Micrite(s)	Biomicrite
Planar Bedding															
High-angle Crossbeds															
Trough Crossbedding															
Planar Crossbedding															
Asymmetrical Ripples															
Oscillation Ripples															
Rhombohedral Ripples															
Mud Cracks															
Raindrop Prints															
Flat Pebble Conglom.															
Laminations															
Salt Crystals & Casts															

WORKSHEET EXERCISE 9.5 C. SEDIMENTARY ROCK TYPES & ENVIRONMENTS.	Reefs	Back-reef Lagoons	Deltas	Swamps & Marshes	Tidal Flat	Supratidal	(Detrital)	(Carbonate-Evaporite)	Intertidal	River Floodplains	River Point Bars	River Channels	Lake Bottoms	Lake Shores	Coastal Lagoons	Barrier Bars	Marine Beaches	Open Marine Shelf
Biomicrite																		
Micrite(s)																		
Sparite(s)																		
Dolomite																		
Evaporites																		
Black Shale																		
Coal																		
Conglomerate																		
Bedded Chert																		
Arkose																		
Greywacke																		
Quartz Sandstone																		
Siltstone																		
Shale																		
Red Shale																		

LABORATORY EXERCISES

9.6. Sedimentary Structures. A. Specimens—Identify each of the structures set out in the laboratory & determine the environment(s) of deposition in which each was formed. Observe and describe the composition(s) & textures of each specimen.

1) Which ripple mark specimen was made by a water/water currents/waves? Why? ____________

____

2) Identify the "direction" of current flow. ____________

3) What type(s) of crossbedding can you identify? ____________

Identify, if possible, the "direction" of current flow. ____________

Identify any erosional surfaces. ____________

4) Which structures could aid in reconstructing paleogeography? ____________

____

____

5) Which structures are commonly associated? ____________

____

____

6) In what rock types would you expect to find each of the structures? Why? ____________

____

____

____

____

7) Sedimentary structures & bedding are not often associated with infaunal animals (see Laboratory 6) Why?

____

____

____

____

LABORATORY EXERCISES

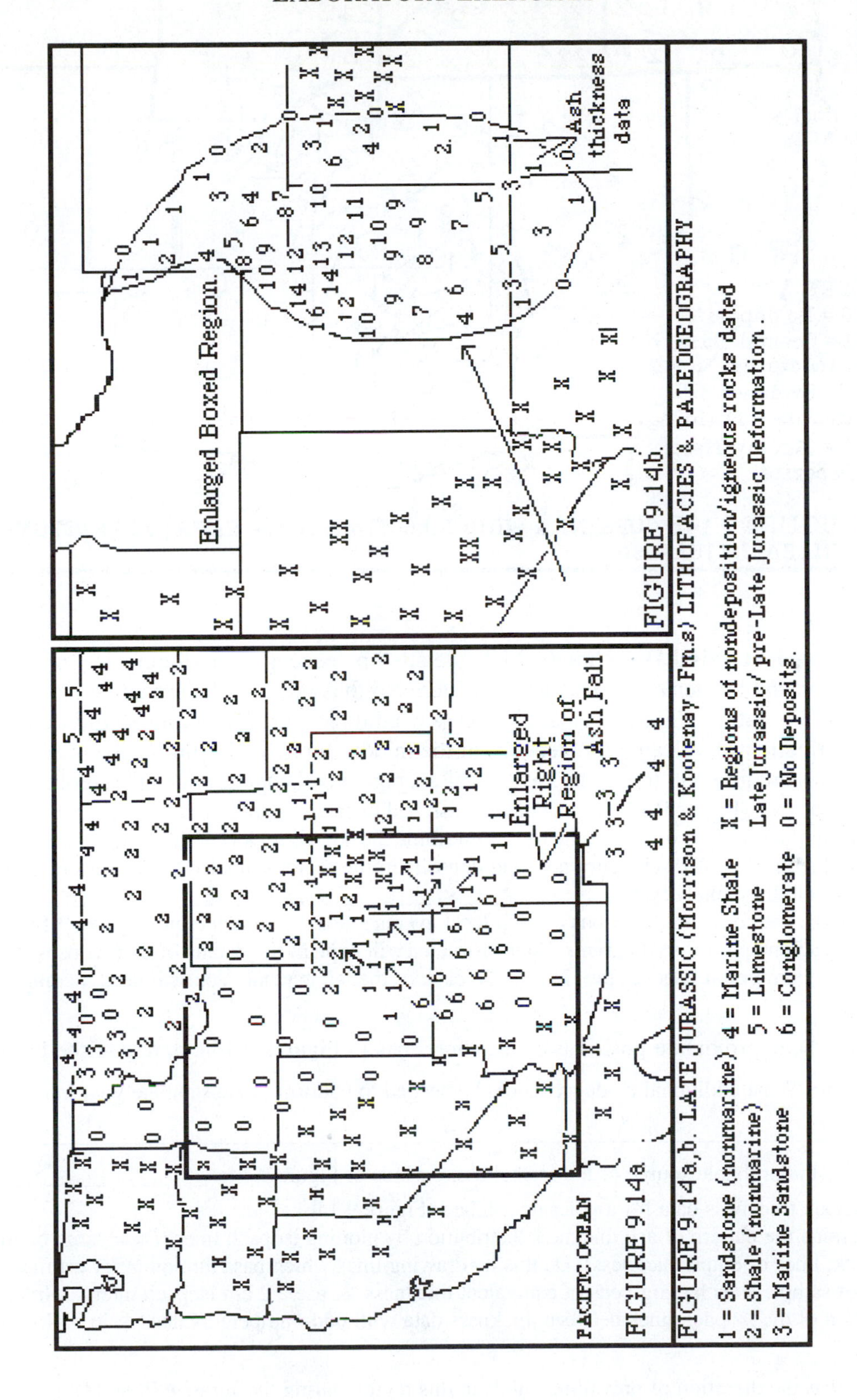

FIGURE 9.14a,b. LATE JURASSIC (Morrison & Kootenay - Fm.s) LITHOFACIES & PALEOGEOGRAPHY

1 = Sandstone (nonmarine) 4 = Marine Shale X = Regions of nondeposition/igneous rocks dated
2 = Shale (nonmarine) 5 = Limestone Late Jurassic/ pre-Late Jurassic Deformation.
3 = Marine Sandstone 6 = Conglomerate 0 = No Deposits.

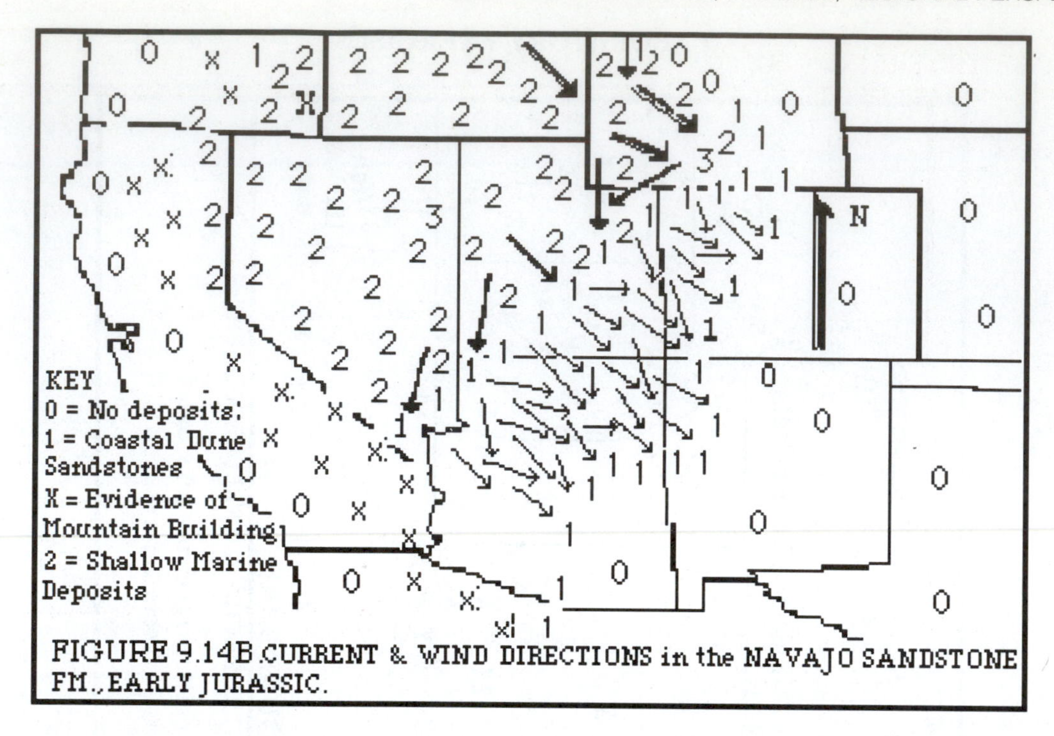

FIGURE 9.14B CURRENT & WIND DIRECTIONS in the NAVAJO SANDSTONE FM., EARLY JURASSIC.

9.7. Mapping Ancient Wind Directions—Make detailed paleogeographic-lithofacies maps of the Late Jurassic (Morrison and Kootenay Formations) of western North America—the formation famous for its fossils of sauropod dinosaurs—(Figure 9.14a, b) & map wind directions as determined from dune crossbedding & the distribution of volcanic ash deposits in the middle Jurassic Navajo Sandstone (Figure 9.14B; next page). Facies are given in the figure keys, local average crossbedding orientation (eolian and marine) are indicated by arrows (Figure 9.14a, c), & volcanic ash bed thickness data (cm) are given in Figure 9.14A.2).

A) Construct lithofacies maps of Figure 9.14a, c. Determine the source region of the conglomerates in the each, & suggest direction(s) of fluvial currents. Had deposits in the regions labeled "X" been preserved, of what rock type would they likely be? ____________________________

1) Determine the compass directions of each local average crossbed orientation (Figures 9.14a, c) & plot all data for each map on rose diagrams. Your instructor will explain the details of their construction. Rose diagram graph paper is provided at the back of the lab manual. You will need a compass to plot th data.

2) Locate the approximate positions of the shorelines in Figures 9.14a, c, & cite the bases for your decisions. What additional evidence would be needed to confirm or substantiate your inferences? ____

____

3) Indicate the probable source of the coastal dunes sands of Figure 9.14c. ____________________

B) Also given are thickness data for a volcanic ash bed (Figure 9.14b).

1) Determine the pattern of ash thickness distribution by plotting isopach lines (*iso* = same or equal; *pachy* = thick; lines of equal thickness). Do this by drawing lines which pass through &/or estimate the loca _ tion of sample sites having beds of equivalent thickness, & use a 2 cm isopach interval. Interpolate the position of the isopach lines between thickness data with odd-numbered values & in regions that lack _ data.

2) What was the direction of prevailing winds in this region during the Jurassic Period? ____________

3) What was the climate of the region like? Cite the evidence for your answer (see Laboratories 2 and 10).

____

4) Where were the sources of the volcanic ash located? Locate the zone of tectonic activity in the map region.

____

WORKSHEET EXERCISE 9.8.

SAMPLE SLAB	DESCRIPTION	STORM/NON-STORM DEPOSITION	EVIDENCE
1.			
2.			
3.			
4.			
5.			

5) What was the tectonic setting (i.e., active/passive margin?, rifting?, subduction?, etc.) ______________

6) Where within this region did mountain building occur? ______________

9.8. Specimens. Recognizing Storm Deposits—Your instructor will provide slabs which were formed under storm & nonstorm conditions.

A) Describe the states of preservation of the fossils, textures & sedimentary structures on worksheet 9.8, and determine which slabs represent deposition in storm/nonstorm conditions. [Refer to the discussion of fossil assemblage formation in Laboratory 6 for information on fossil preservation in storm deposits.] State your reasoning. ______________

B) What types & properties of sediment & sedimentary structures could be expected to form as storm deposits? ______________

C) What properties of fossil assemblage preservation & composition could be expected in storm deposits? ______________

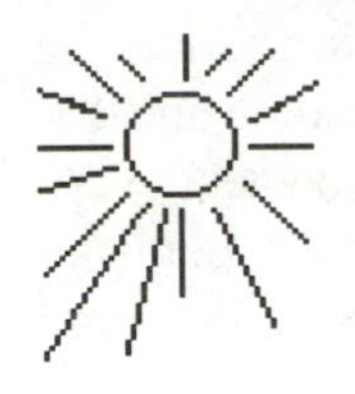

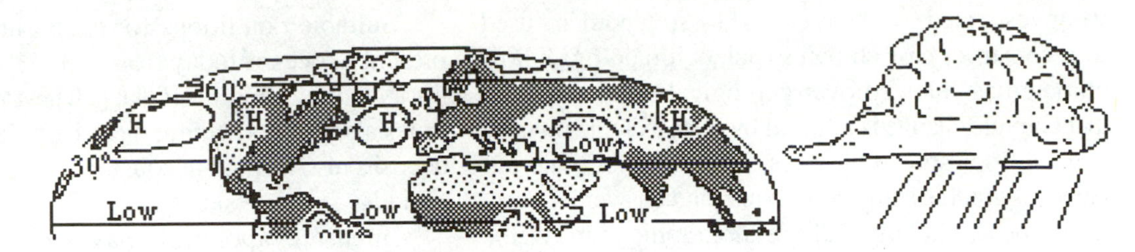

PART III
GLOBAL PERSPECTIVES: EARTH HISTORY AND PLATE TECTONICS

10
Ancient Climates and Oceans

Objectives
To demonstrate mastery of this material you should: 1) understand the principles of atmospheric and oceanic circulation; 2) be able to reconstruct in a very general way, hypothetical ocean circulation patterns from the guidelines provided; and, 3) be able to use rocks, fossils, and climatological principles as guidelines to reconstruct the broad outlines of past climates. Colored pencils are needed.

Using This Laboratory
The following material should be read for a general conceptual understanding. Neither memorization nor immediate grasp of the application of the principles set forth are expected. The fundamentals of climatology and oceanography are presented in a series of "rules" which serve as references and guidelines for completion of the mapping exercises. Only upon completion of the mapping exercises are you expected to have internalized these concepts. The exercises may be completed most simply and effectively using a step-wise approach: 1) a first approximation to reconstructing climatic zones should be most straightforward when based on the distributions of "climate sensitive" rock types; 2) refinements should then be added by reference to the summary tables; and finally (if your instructor chooses); 3) refinement by reference to the rules of paleoclimatic-oceanographic reconstruction, in order of discussion, from general to specific.

The primary goal of this laboratory is a sweeping view of climate change over geologic time through an imaginative application of fundamental principles: there are many possible "correct answers" and few "incorrect" within the guidelines set forth. For perspective, be aware that the many inferences, limited knowledge and sweeping scope of the questions relegate conclusions of technical studies to the realm of general hypotheses. The "take-home lesson" is foremost a greater familiarity with Earth history and plate tectonics through the medium of past climates. "Jump in head-first" with pencil and "practice" map copies, and attend only later to refinements of your maps.

This and remaining laboratories present material of global scope, each topic showing how various aspects of physical and biological evolution are linked to plate tectonics. Paleogeographic reconstructions (Laboratory 9) go far to complete our picture of the Earth's history, yet are not complete or realistic without knowledge of past climates as well. Here we investigate how the rock and fossil record are used to reconstruct past climates (paleoclimatology); how climate has changed over geological time; and, how that climatic change is affected by plate tectonic motions and seafloor spreading rates. Thus, we continue our paleogeographic study by making, and answering questions about, climatic and oceanographic maps of the past. The third section, explains some methods of estimating and measuring past temperatures from fossils.

ROCKS as ANCIENT WEATHER STATIONS

Several climatically "sensitive" rock types provide excellent clues about the climates in which they formed. Because these form only within limited climatic conditions, their geographic distributions enable us to sketch outlines of the past climates (see data summary, Table 10.1). For your reference, the rules used as guidelines in reconstructing paleoclimates are indicated by [#].

Evaporites—Evaporite rocks precipitate in sequence from anhydrite ($CaSO_4$), to gypsum ($CaSO_4$ $2H_2O$) and halite (NaCl) as evaporation increases and salinity reaches extreme values. Gypsum and anhydrite form in mean relative humidities from 93-76%, and halite (rock salt) from 76-67%. Low latitude coastal regions often have 70-80% humidities **[1A]**.

Suitable conditions for evaporite formation are present in oceans today from 5-35 °S latitude and 15-40 °N latitude (Figure 10.1), **[1B]**. Most ancient evaporites also formed within these latitude belts, but large deposits also formed in equatorial regions during some periods.

Intense evaporation may occur in arid regions with circulation restricted by coral reefs, narrow inlets (e.g., the Miocene Mediterranean), narrow basins, and bays enclosed by land with low humidity air masses **[2]**. High temperatures are often linked to, but

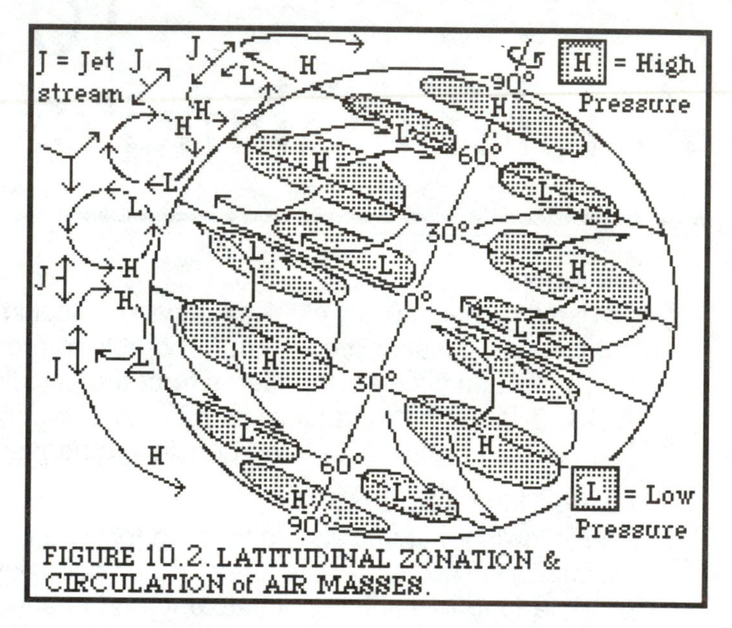

FIGURE 10.2. LATITUDINAL ZONATION & CIRCULATION of AIR MASSES.

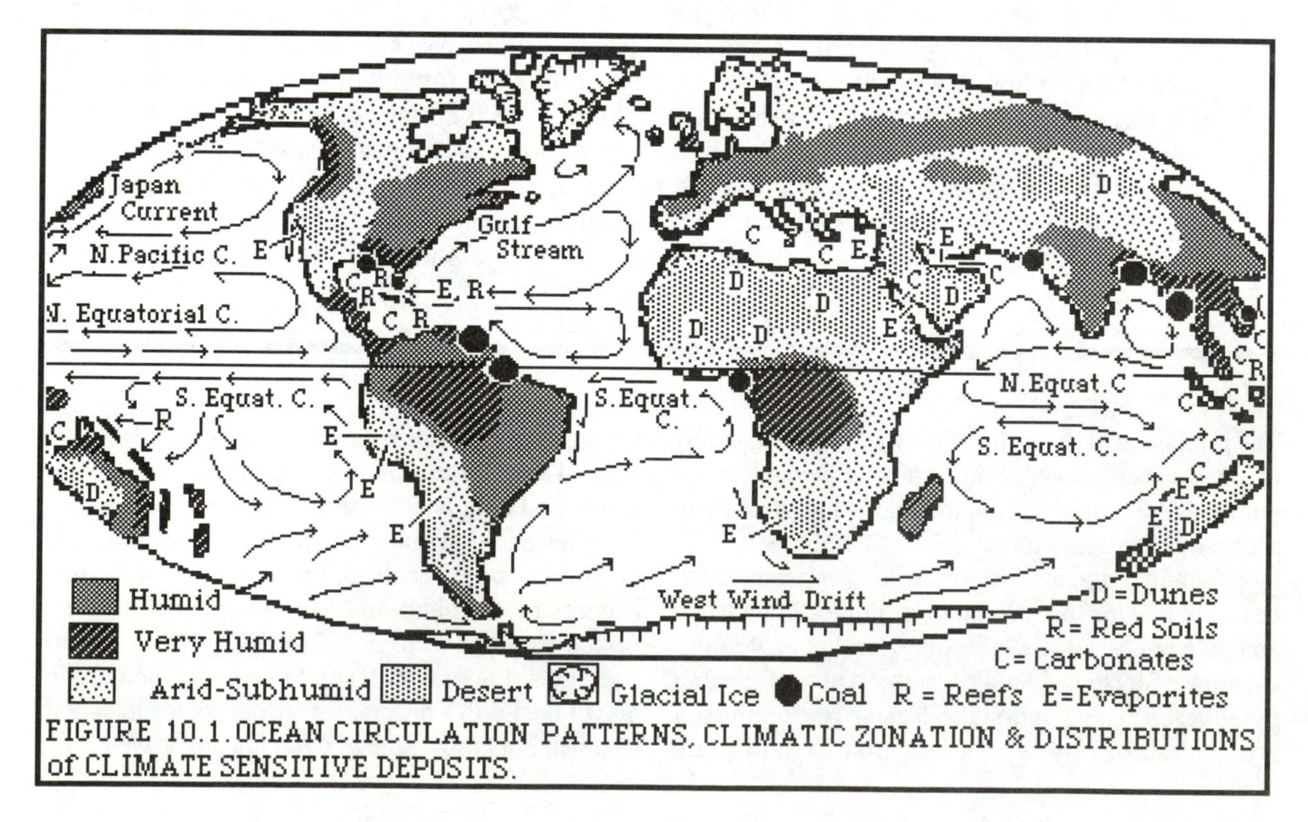

FIGURE 10.1. OCEAN CIRCULATION PATTERNS, CLIMATIC ZONATION & DISTRIBUTIONS of CLIMATE SENSITIVE DEPOSITS.

TABLE 10.1. CLIMATE SENSITIVE ROCKS.

ROCK TYPE	FEATURES	TEMP.	HUMIDITY	LATITUDE
Red Sandstones	Large High Angle x-bddng Very well rndd-Sorted Tracks of Land Animals	High	Arid	10/15-35/40° N & S.
Red Shales	Largely fossil-barren, plants -arid or wet features. vertebrates	High High	Wet -Humid Arid	0—10/15° N & S 10/15-35/40° N & S
Black Shales	Fossils Marine, Brackish but largely barren	Cool-High	Wet—Humid	Largely Lower Latit.
Coal	Land plants w/w-out annual growth bands.	High—Cool	Wet -Humind	0 to 70 or 85° N/S
Evaporites	Dolostn. w/ evaporites &/or Supratidal indicators.	High	Arid - Humid Halite - 67-76% Gypsum & Anhydrite -76-93%	5-35 °S & 15-40° N
Reefs	May be associated w/evap. & Dolostn.	20-22 °C or >	Arid or Humid	0 t0 30° N & S
Tillite	dropstn, striated & polished bedrock.	Cold-Cool	Arid to Humid	70—85° N & S

unnecessary for, evaporite formation. Evaporites will form as long as evaporation exceeds precipitation **[3]**.

Reefs—The distribution of reef-corals is limited by temperatures narrower than the range of warm shallow waters in which carbonate rocks are most readily formed (Laboratories 2 & 8). Reef limestones, chemical limestones, and dolostone commonly form(ed) along side evaporites, and in lowland &/or dry coastal regions receiving little or no clastic sediment.

Most coral reefs (Laboratory 6) now form in waters warmer than 20°C (68°F) and within tropical and subtropical latitudes 30° N/S of the equator (Figure 10.1) **[4]**. Reef corals are most abundant at temperatures of 25-29°C (76 to 82°F; average minimum of 20-22°C (68-70°F)), and are excluded below 16 or 17°C (60-62°F) or above 40°C (104°F).

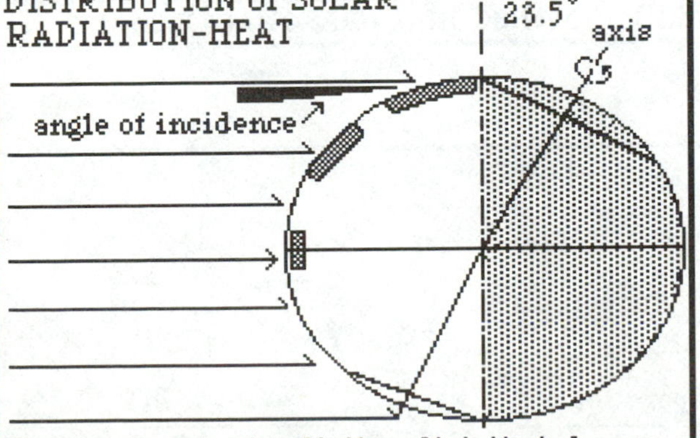

DISTRIBUTION of SOLAR RADIATION-HEAT

Equal amounts of radiation distributed over larger areas at higher latitudes, & thus, temperatures are lower.

FIGURE 10.3.

Red Beds—Red beds are often interpreted as having formed in warm, arid conditions (Laboratory 2), but Fe-rich soils are common in humid low latitudes (≤30°N/S). High temperatures are needed to oxidize iron minerals to the hematite of red sediments. Thus, red beds generally indicate warm climates in low latitudes (Figure 10.1) **[5]**.

Additional evidence of warm climates may come from fossils and other climate-sensitive rocks (see text). For example, red beds are commonly associated with evaporites and carbonates. When such associations occur, they are indicative of warm, arid conditions. If found associated with coals, red beds, origin the unusual altered soil form, oölitic iron (hematite spheres), are indicative of a warm, humid climate.

Coal—Coal is formed by compaction of abundant plant material, and thus indicates humid climates and conditions inhibiting decomposition (Laboratory 2) **[6]**. Coals form most readily in warm, humid climates (e.g., coastal Thailand; Figure 10.1). Large deposits can also form in cool, humid conditions (e.g., peat bogs of Maine).

Plant growth and coal formation is slow in temperate latitudes. Growth rings in fossils from high latitudes indicate seasonal changes of temperature and humidity levels. Coals associated with tillites (see below) can indicate the temperate conditions, while those occurring with red shales indicate warm temperates and a general absence seasonality.

Black Shales—Extensive black shales deposits formed at times of epicontinental seas and oceanward to the outer shelf (e.g., Devonian—Carboniferous and Jurassic—Cretaceous). Such shales indicate low levels of oxygen (Laboratory 2), and their ubiquity indicates a general warming and the ocean's loss of temperature stratification. Increased temperatures may be inferred because warm water holds less oxygen, oxygen that would have furthered organic decay **[7]**. These deposits are also associated with times of increased atmospheric carbon dioxide and coals of humid environments.

Tillites—Tillites formed from unsorted or poorly sorted glacial debris known as till, (rounded and faceted rock fragments of all sizes from boulders to mud). Tills are found in various glacial landforms, drowned glacial river valleys, and ice-rafted marine till deposits. Gouged, striated bedding planes indicate ice-embedded boulders dragged along bedrock by glaciers, and so can be used to determine directions of advance. Such evidence of glaciation indicates desert-like conditions and glacial cold (≤ 0°C (32°F)), but not necessarily frigid, climates **[8]**.

ORIGINS of CLIMATE: The ATMOSPHERE and OCEANS

The Earth's present climate is exceedingly complex and only partly understood. Still, some basic principles certainly apply to present and past climates, and these do provide glimpses of actual conditions among latitudinal (≈ E-W/W-E) belts and occasionally, on regional scales. Central to reconstructions of past climates are the latitudinal patterns of atmospheric and oceanic circulation. These form the basis of a climatic system which is modified, in turn, by the locations and sizes of the continents and oceans.

Climate originates from, and is driven by, the Earth's axial tilt and rotation which join in controlling the distribution of atmospheric and oceanic heat; by the sizes,

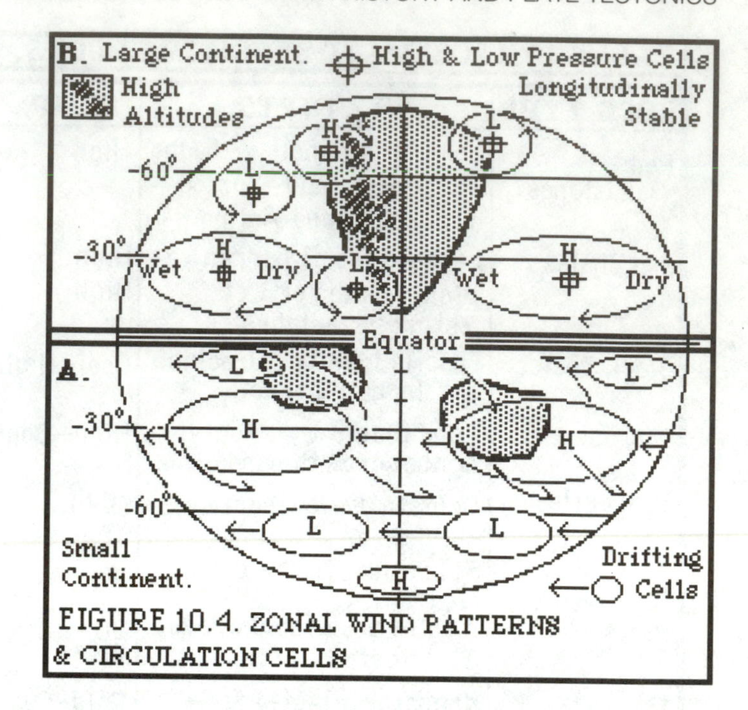

FIGURE 10.4. ZONAL WIND PATTERNS & CIRCULATION CELLS

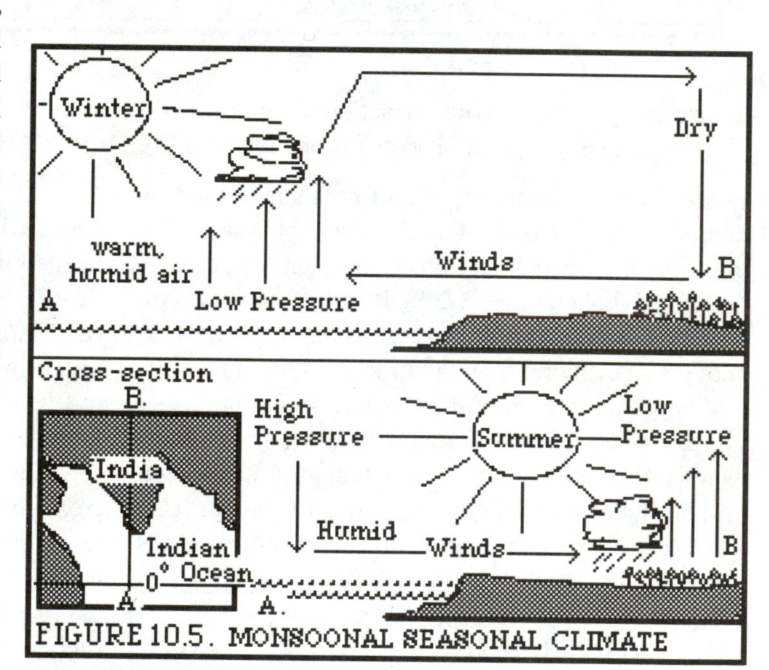

FIGURE 10.5. MONSOONAL SEASONAL CLIMATE

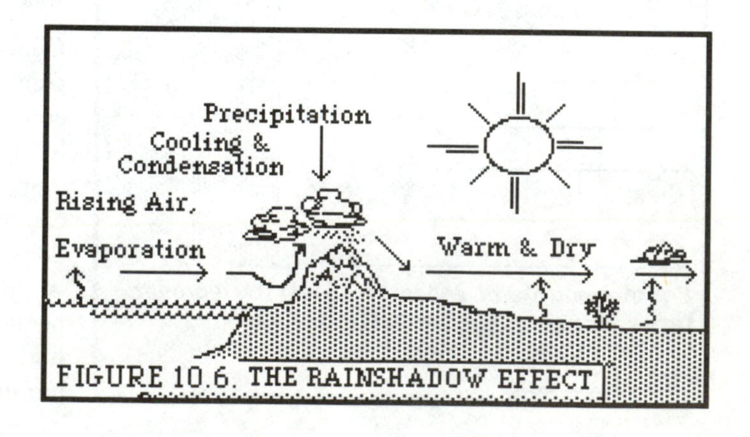

FIGURE 10.6. THE RAINSHADOW EFFECT

TABLE 10.2. CLIMATIC FEATURES of the ATMOSPHERE & OCEAN

A. Hemisphere

South	North	Air Pressure Belt	Wind Systems
65 – 90 °	65/75–90 °	Polar High	Polar East Winds
25/30 – 65°	30/40–65/75°	Subpolar Low Pressure Trough	Polar East Winds; Extratropical Westerly Winds
10–25/30 °	10–30/40°	Subtropical High Pressure Trough	Extratropical Westerly Winds; Tropical Easterly Winds
0–10°	0–10°	Equatorial Low Pressure Trough	Tropical Easterly Winds

B. Continent Size/Relief/Location

Continent Size/Relief/Location	Climatic Features
Small, Low, Subtropical to Subpolar, within Latitudinal Belts.	Moderate Temperatures & Rainfall —Dominated by Ocean Circulation Effects.
Large, Elevated, Pole to Pole Crossing Latitudes	Continental, greater seasonal variations, accentuated aridity/humidity —Dominated by Pressure Cell Effects Western Coasts Arid where Rainshadow.
Low, small, equatorial	Monsoon Seasonality.
10/15 –35/40° N and S —Subtrop. High Pressure	Arid - Desert.
Polar Landmasses 65-90 ° N & S	Cool to Glacial.
5 °S to 15 °N	Tropical Wet to Monsoon.
High Sea Level	Mod. to Warmer—Lower Global Temp. Gradient.
Low Sea Level	Continental—Steeper Global Temp. Gradient.

C. Oceanic Circulation

Latitude	Direction
0 – 20° N & S	East to West
≈ 40° N & S	West to East
≈ 60° to 90° N & S	Circumpolar —Enhanced Cooling
Large, Meridional Continents	Meridional, Strengthened on western shores Eastern & High Latitude Shores Moderate & humid to wet.

number, and distributions of continents and oceans, and the level of the latter; and lastly, by astronomical parameters discussed in your text. Atmospheric and oceanic circulation patterns are closely linked and similar because both are products of the same phenomena.

In this section we discover how to reconstruct the broad patterns of oceanic and atmospheric circulation and how these patterns have influenced past climates. The basic "rules" of climate and ocean reconstruction discussed below are identified by [#], and illustrated in Figures 10.2–8 and Table 10.2.

ATMOSPHERIC CIRCULATION

The Coriolis Effect and Centrifugal Force—The direction of airmass flow (and ocean currents) at the surface includes a deflection to the right or left due to the Earth's rotation and gravity's action on the oceans and atmosphere moving over a rotating surface. These deflections are due to the Coriolis Effect and centrifugal "force" (masses moving somewhat freely over and thus, experiencing reduced friction with, a large spinning body will differ in velocity at various distances from the rotation axis; see your text for further explanation). Sufficient for our purposes is the first rule of atmospheric and oceanic circulation (Figure 10.2): large air and water masses at lower latitudes are deflected to the right in the northern hemisphere, and to the left in the southern hemisphere **[1]**.

Temperature Gradients and Latitudinal Winds—Because the Earth's axis is tilted, heat is unevenly distributed across latitudes (Figure 10.3). The angle of sunlight incidence on the surface is lower in high latitudes, and thus, equal amounts of radiation are distributed over larger areas in high latitudes than in low latitudes. Thus, temperatures are lower in high latitudes.

There has always been a temperature gradient to a greater or lesser degree—from an equatorial maximum to polar minimums—and this gradient establishes the latitudinal pattern of atmospheric circulation [**2; see Table 10.2**].

Hot equatorial air is less dense than cold and, thus, expands and rises. As it does, surface air from higher latitudes moves toward the equator, replacing the rising mass and creating a low-pressure trough within about 10 degrees of the equator (Figure 10.2) [**3**]. Due to the Coriolis Effect, cooler low latitude winds (tropical easterly Trade Winds) are deflected from the northeast and southeast above and below the equator respectively to the equatorial low [**4**]. Rising equatorial air masses cool aloft, become denser and descend at about 30 °N/S [**5**]. Descending air masses create subtropical high-pressure ridges from 25-30 °S, and 30-40 °N [**6**]. From here masses travel either poleward or equatorward.

Equatorward air deflects again to the southwest and northwest. Poleward air deflected to the right forms the westerlies from 30-60 °S and 40-60 °N. As these masses heat and rise, subpolar low-pressure troughs form (≈ 65 °S and 65-75 °N) [**7**]. Subsequently, they travel equatorward, sinking at ≈ 30 °N/S, and pole ward, sinking to form polar high-pressure caps [**8**]. The polar air rotates clockwise (N.Hemisphere) or counterclockwise (S. Hemisphere; Figure 10.2). In summary, a pattern of tropical and mid-latitude winds like that of today (Figures 10.1, 2, & 7), the Coriolis Effect, and some kind of latitudinal zonation apply to any period of Earth history, and with the principles discussed below, are guidelines for reconstructing past climates (Figures 10.1, 2, 4, 7, 8; Table 10.2).

CONTINENTS and SEA LEVELS

Land masses affect the sizes and locations of latitudinal high- and low-pressure circulation cells (troughs and ridges, above) and precipitation levels. Their locations would not differ greatly from the pattern established above in a world without continents during those times when continents were small, dispersed and low (Figure 10.4A). However, when zonal wind patterns stretch across large, enmassed, and elevated continents, the latitudinal patterns described above are fragmented and pressure cells "lock" in position (Figure 10.4, 10.8) [**9**]. In such cases, especially when one or more continents stretch ≈ N-S and span several latitudinal zones as today (Figure 10.1,3 & 7) and in the Permo-Triassic time of Pangea, smaller pressure cells align longitudinally (≈N-S/S-N), thus producing a temperate zone characterized by climatic extremes and temperature seasonality.

When equatorial land surfaces receive more heat than the surrounding ocean, latitudinal wind patterns over the oceans are disrupted (Figure 10.4, 5) and seasonal differences in the heating of land and water arise [**10**]. Monsoon seasonality occurs when land warms more rapidly than water, low-pressure air over tropical lands heats, rises, and is replaced by cooler air from over the adjacent ocean (e.g., Indian subcontinent). As the humid air rises, it cools and moisture condenses causing high precipitation . As Autumn approaches the Indian Ocean receives more direct radiation than the subcontinent. The ocean air is heated more than the land, it rises and is replaced by dry seaward winds from the high-pressure cell over the continent descending from the Himalayan Mountains and Tibetan Plateau. Extensive mountain chains and plateaus such as these and the Rocky Mountains & Colorado Plateau accentuate temperate climates and general cooling by fostering arid conditions within and around high altitudes during summer low-pressure trough, and winter high-pressure ridge formation.

When climatic patterns are temperate, western coasts will generally be dry, and eastern coasts,wet (Figure 10.1) [**11**]. North-south trending mountain ranges accentuate this pattern when they lie across latitudes because air masses must rise to move across mountain ranges and, as they do, moisture condenses as precipitation (Figure 10. 6 & 7). Most moisture is lost on the windward side of the mountains (e.g., Sierra Nevada Mountains) and thus, the arid downwind side is is enveloped in a "rainshadow" (e.g., Basin and Range Province 10.1 & 6)).

Today's major deserts (regions receiving ≤ 25 cm (10 in.) precipitation and < evaporative loss; e.g., Sahara) lie within 10 or 15-35 °N/S of the equator and the subtropical high-pressure ridge (Figure 10.1) [**12**]. Ancient deserts were also common within these latitudes. Dry air descending in this high-pressure belt draws moisture from the land as it warms. Such regions persist, grow larger and more arid as evaporation increases and vegetation cover decreases with temperature. Polar regions also lie in high-pressure zones [**13**], and are literal deserts despite glaciation.

The differing potentials of land and sea for retaining heat have important effects on the total heat budget. Sea water absorbs heat more slowly than land, but retains it far longer and distributes it over a larger area. In contrast, much of the heat received by the land is re-radiated to the atmosphere. As the area of ocean increases during transgressions, so too does the total heat that the oceans retain. Thus, the equator-pole temperature gradient decreases and the mean global temperature rises [**14A**]. When the continental area increases as sea level falls, there is a net heat loss and, thus, the temperature gradient is steepened and temperate zones develop—as today **[14B]**. This effect is heightened if the poles lay near or within continents; more heat is lost from polar continental than from polar seas, the temperature gradient steepens further, and the mean global temperature decreases. Heat loss is also typical of deserts. Lacking plant and cloud cover, there is a net heat loss by nightly re-radiation. For these reasons, and the mid- to high-latitude positions of many, deserts may also also have low mean winter temperatures. Conditions favoring all of these latter states may combine to clothe the Earth in extremes of temperature and rainfall.

SEAS, "SEASONS" & CIRCUMPOLAR CIRCULATION

Ocean current patterns are important for several reasons: 1) they absorb, retain, and redistribute heat and moisture, thus affecting all aspects of climate; 2) they are the medium of transport and repository for sediment, dissolved gases (e.g., CO_2 and O_2), and critical nutrient elements (C, P and N); 3) currents transport larvae and may bar their migration (Laboratory 11); and, 4) circulation patterns control phytoplankton productivity—the food chain's foundation and the major source of oxygen.

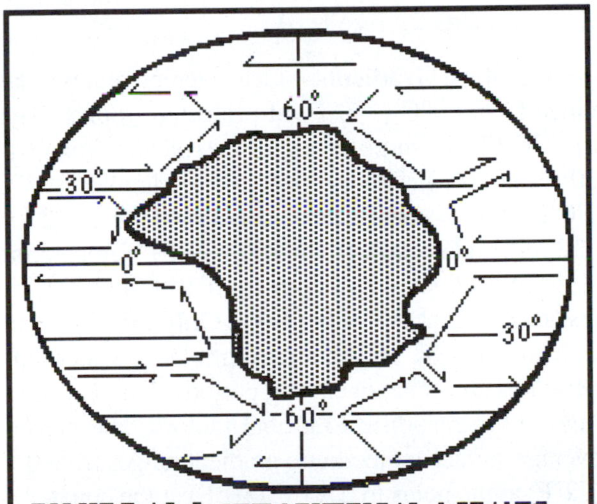

FIGURE 10.8. HYPOTHETICAL OCEANIC CIRCULATION PATTERN

Large scale currents (i.e., circulation), originate from the Coriolis Effect and density differences due to temperature and/or salinity that result in water mass stratification. Because research is only now revealing the climate effects of deep current patterns, our focus is limited to shallower patterns arising from the Coriolis Effect and land mass positions.

In general, past circulation was governed by the same rules as those of today (Figures 10.1 & 8), but modified by land mass positions and sizes. The "rules" of reconstructing circulatory patterns discussed below are identified by **[#]**.

The Coriolis Effect causes latitudinal circulation patterns broadly similar to those of the atmosphere (Figures 10.2 & 8). Currents are generally east-west within 20° N/S of the equator, and west-east at about 40 °N/S. From 60 °S to the Antarctic shores flows a circumpolar current **[1]**. The Arctic Ocean is also nearly isolated by continents and has a circumpolar circulation.

In general, circulation is presently more dominated by large "N-S" arranged continents (Figure 10.4B), than when they were small and scattered "E-W" (e.g., Early Paleozoic; Figure 10.4A), but less so than during the Permo-Triassic when Pangaea and the world-ocean Panthalassa stretched nearly pole to pole, (Figure 10.4B, 8) **[2]**.

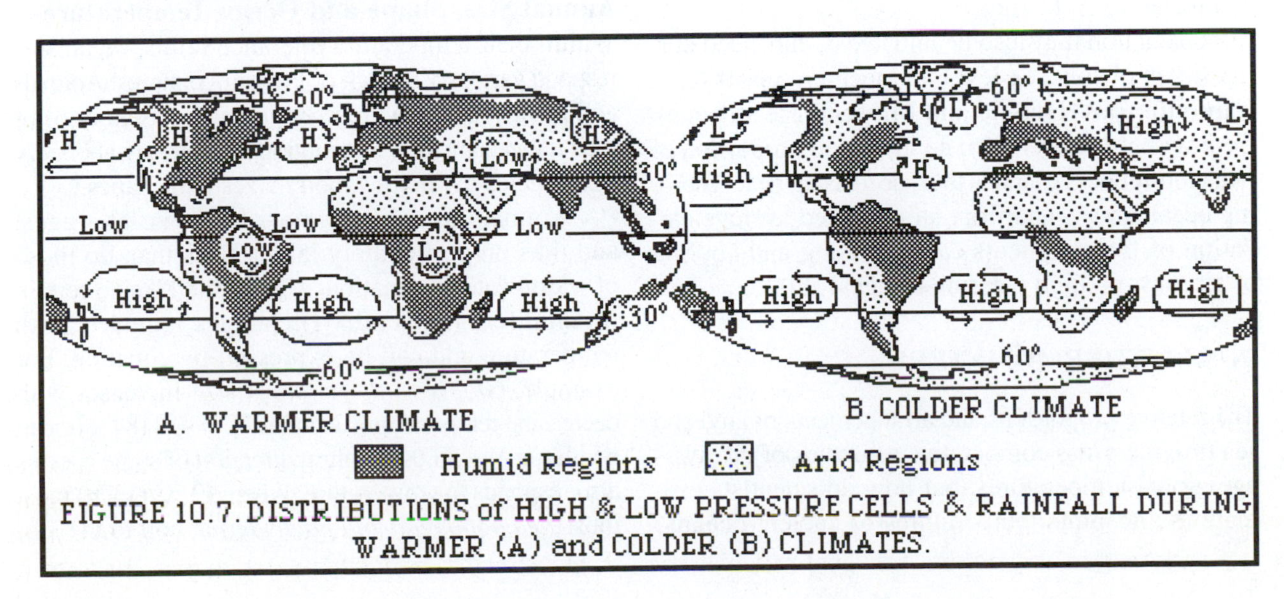

FIGURE 10.7. DISTRIBUTIONS of HIGH & LOW PRESSURE CELLS & RAINFALL DURING WARMER (A) and COLDER (B) CLIMATES.

From N-S currents on the western sides of oceans arise important west-east currents that have a dramatic influence on heat distribution and the climate. The Gulf Stream, the Japan Current, and others are warm currents formed of hurricane-like circulation cells which are narrowed and strengthened on the western sides of continents (Figure 10.1) **[3]**. These currents fan-out on their return and bring much warmer climates to higher latitudes on opposite eastern coasts.

Physical barriers accentuate temperature and salinity differences and, generally, restrict circulation to isolated basins (Figure 10.1), **[4]**. The spectacular effects of highly restricted circulation can be seen in some ancient ocean basins such as the thick Miocene evaporite deposits in the Mediterranean Sea (see text). Such circulation was also present during the break-up of Pangaea when in the narrow arms of the incipient North and South Atlantic Oceans.

Because water retains heat longer than air and the volume of the oceans is enormous, global climate is surely and strongly linked to ocean temperature and circulation. Ocean-atmosphere heat exchange is not clearly understood, but major climatic change does occur in response to plate movements that modify circulation. For example, when landmasses block circumpolar circulation, potentially cold high latitude currents are warmed by mixing with warmer currents from lower latitudes **[5]**. If no landmass blocks a circumpolar current (e.g. as the Australian Plate moved northward), that continent will cool further, becoming insulated and isolated (Figure 10.1). Isolation and refrigeration by this current then causes cooling of the high pressure cap over the landmass, which perpetuates and accentuates cooling of the current. The Pleistocene glaciation is partly due to the cold circumpolar current blocking warm water from the north and isolating Antarctica.

Glaciation may also be initiated or enhanced during times of low sea level and tectonic uplift (e.g., Carboniferous-Triassic and Miocene-Pleistocene). Cooling here results from an increased proportion of radiation-reflective land—thus, decreasing heat absorbing ocean area—and from an increased average elevation of the continents causing rising and cooling of air masses (high pressure troughs) **[6]**.

FOSSIL THERMOMETERS

The relative proportions and arrangements of land and sea provide gross comparative estimates of the average global temperatures, but how do scientists estimate the absolute temperatures of ancient oceans?

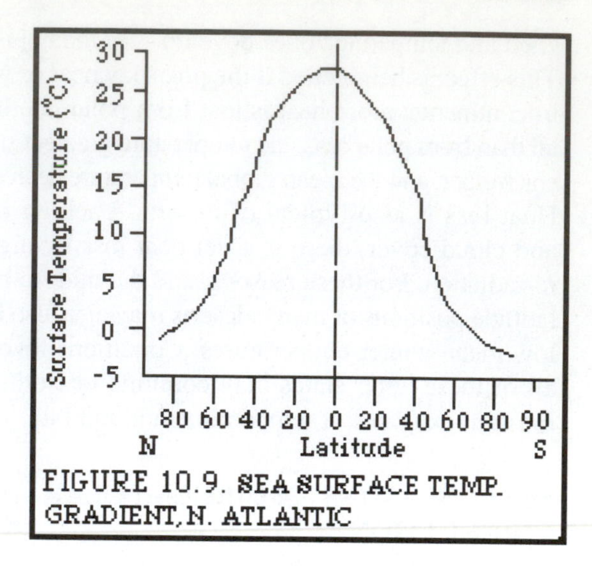

FIGURE 10.9. SEA SURFACE TEMP. GRADIENT, N. ATLANTIC

These are estimated by: 1) inference from temperature tolerances of closely related living species; 2) shape and size changes correlative with temperature; and, 3) oxygen isotope ratios, a method analogous to radiochronometry (see text).

Temperature Gradients—Ocean temperatures today range from 60°C (140°F) in protected tropical bays to -2°C (28°F) in polar regions; typically from 2 to 30°C (35-86°F), with an upper average limit of ≈ 33°C (90°F). Most ocean water lies below depths of 2 km (1.2 mi), and is cooler than 4 °C (39°F). Temperature steadily declines with depth from 100 m (328 ft) to less than 4°C at the ocean floor. Only the upper 100 m is affected by seasonal cooling. Deep water temperatures are stable over long periods of time, changing only in response to global factors. The present average surface temperature gradient ranges from 0°C (32°F) at the poles to 27.5°C (80°F) at the equator or 3.1°C (5.3°F) per 10° latitude (Figure 10.9).

Animal Size, Shape and Ocean Temperature—Within their temperature tolerance limits, organisms respond variously to high or low temperatures. Animals and plants reflect the temperatures (Laboratory 6) of their habitats by their distributional patterns (Laboratory 11), species diversities, body sizes and shapes.

Animals of colder waters breed at later ages, and thus proportionately larger sizes than do those of warm water populations. Test ("shell") size certain foraminifera (Appendix I) changes regularly with temperature and can be expressed by equation. For example, *Globigerina bulloides* size increases with decreasing temperature such that °C = -0.1184 x (mean width (μm)) + 45.06. Coiling direction of some species also responds to temperature: when ≤9 °C (13°F) most most *Neogloboquadrina pachyderma* coil to the left, from 10-15°C (50-59°F) both forms exist, and > 15°C, most will coil to the right.

Temperatures may be inferred by analogy with ecological properties of communities and tolerances of related species. Species diversities (e.g., corals, bryozoans, bivalves, foraminifera), increase from poles to equator (Figure 10.10) and diversity gradients can be used to infer relative water temperatures. Organisms such as reef corals have narrow temperature tolerances. If an extinct coral species was a reef-former (Laboratory 6), we may assume its temperature tolerance was within the range of living species. In some cases, living species with known tolerances also occur in the fossil record. Ancient temperatures can then be inferred by assuming that temperature tolerance has not changed over time (e.g., Cenozoic foraminifera).

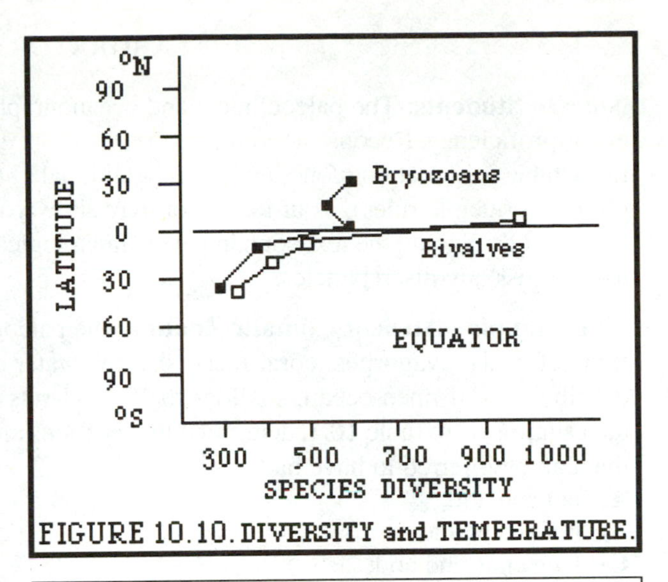

FIGURE 10.10. DIVERSITY and TEMPERATURE.

VOCABULARY

Climatically "sensitive" rocks, Coriolis effect, Desert, High pressure ridge, Low pressure trough, Monsoon, Paleoclimatology, Pangaea, Rainshadow effect, Red beds, Temperate latitude/zone, Tethys Seaway, Till, Tillite.

LABORATORY EXERCISES

Note To Students: The paleoclimate and oceanographic mapping exercises may be approached at two levels of proficiency. Reconstructions may be based on: 1) "climate sensitive" rock type distributions and summary table information alone; and/or, 2) additional climatological-oceanographic principles (Table 10.2 and climatic principle rules). Your instructor may allow you to work in groups, each concentrating on one of the four periods. Should the text contain paleoclimatological reconstructions, compare them with your own results and discuss any discrepancies.

10.1. Mapping Ancient Climatic Zones—The paleogeographic maps in Figures 10.11A-D show the locations of coals, evaporites, coral reefs, shallow water carbonates, tillites, and red beds. Also shown are the distributions of open ocean, shallow shelf, lowlands and uplands. Using these distributional data and the data summary of Table 10.1, determine the locations and general areal distributions of the continental regions that can be inferred to have had:

A) hot and arid;
B) hot and humid;
C) temperate and arid;
D) temperate and humid;
E) cool and arid;
F) cool and humid; and,
G) polar (glacial and non-glacial) climates.

Make initial maps using pencil and on xerox copies or tracings of the worksheet maps. Turn in finished color-coded maps (see map key) from the worksheets later.

Uplands **Lowlands** **Shallow Seas**

R = Reefs, Rb = Red beds, C = Carbonate rocks, E = Evaporites, ● = Coal

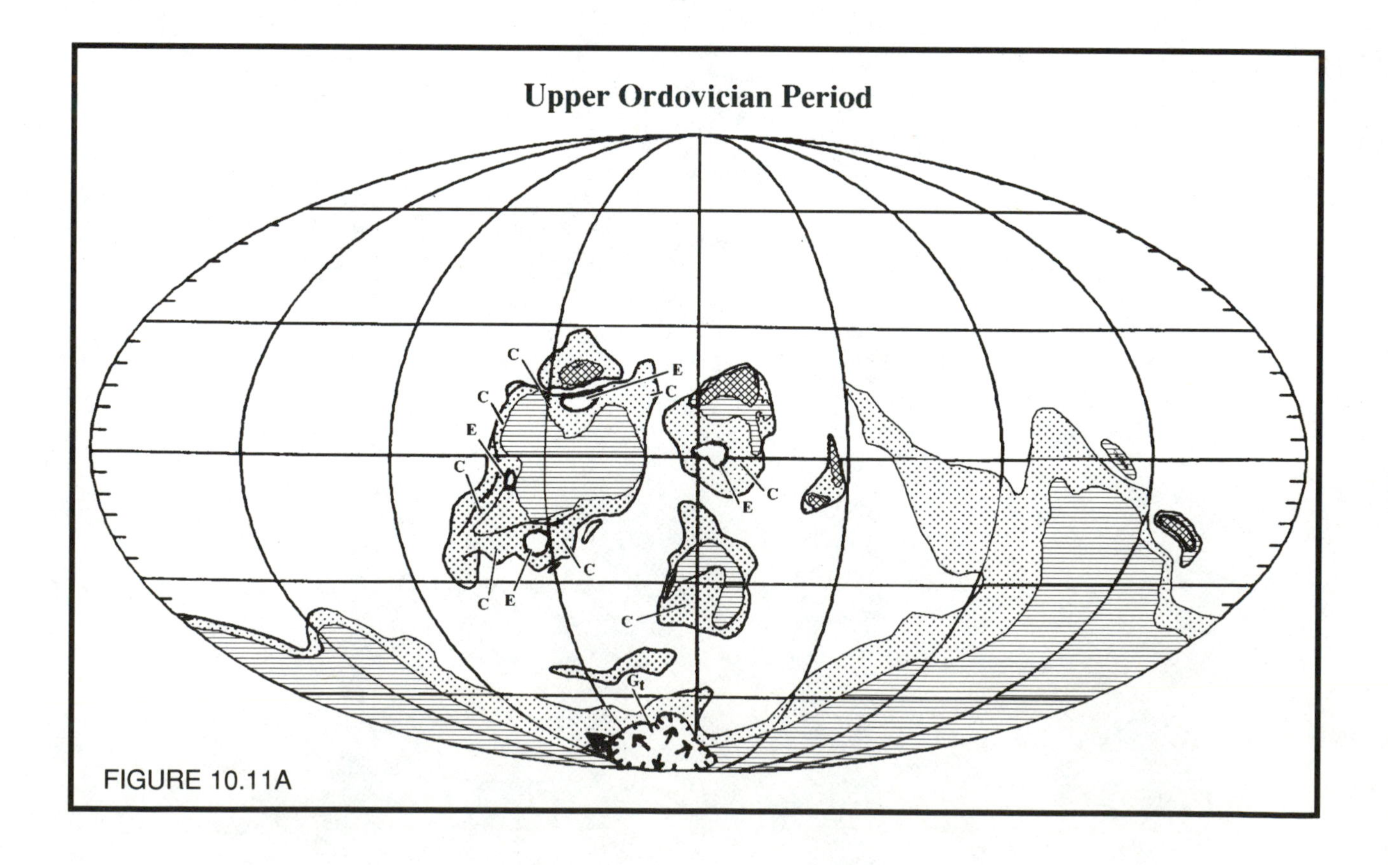

FIGURE 10.11A

Uplands **Lowlands** **Shallow Seas**

R = Reefs, Rb = Red beds, C = Carbonate rocks, E = Evaporites, ● = Coal

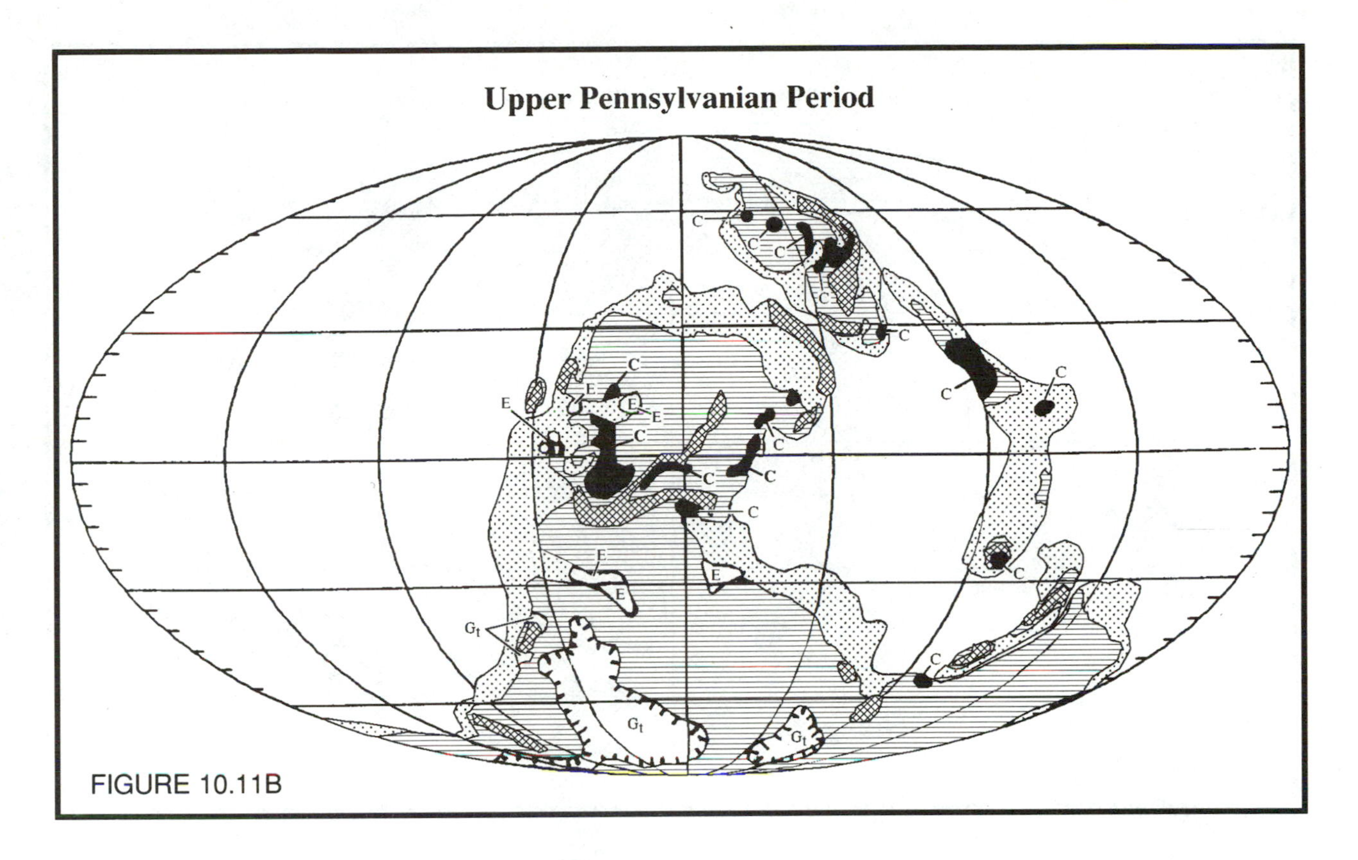

FIGURE 10.11B

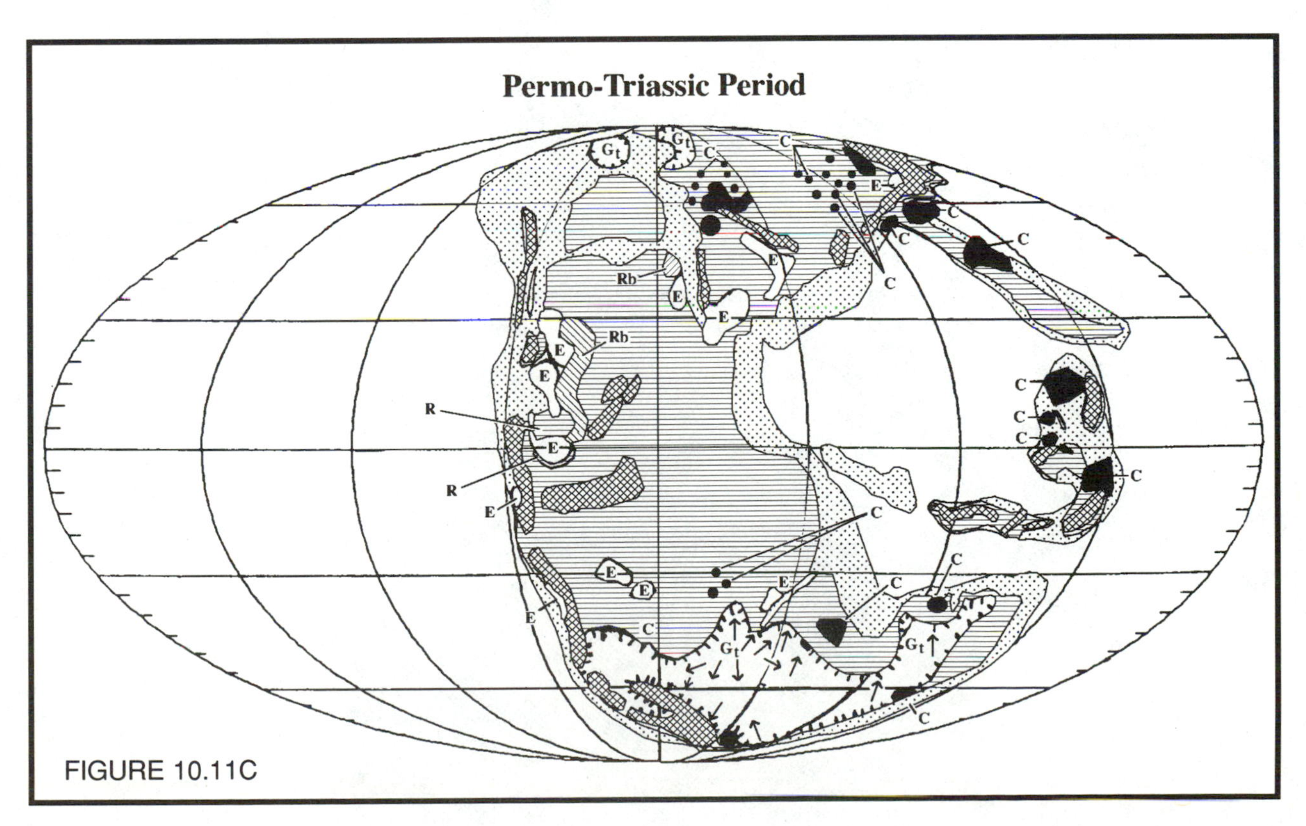

FIGURE 10.11C

Uplands Lowlands Shallow Seas

R = Reefs, Rb = Red beds, C = Carbonate rocks, E = Evaporites, ● = Coal

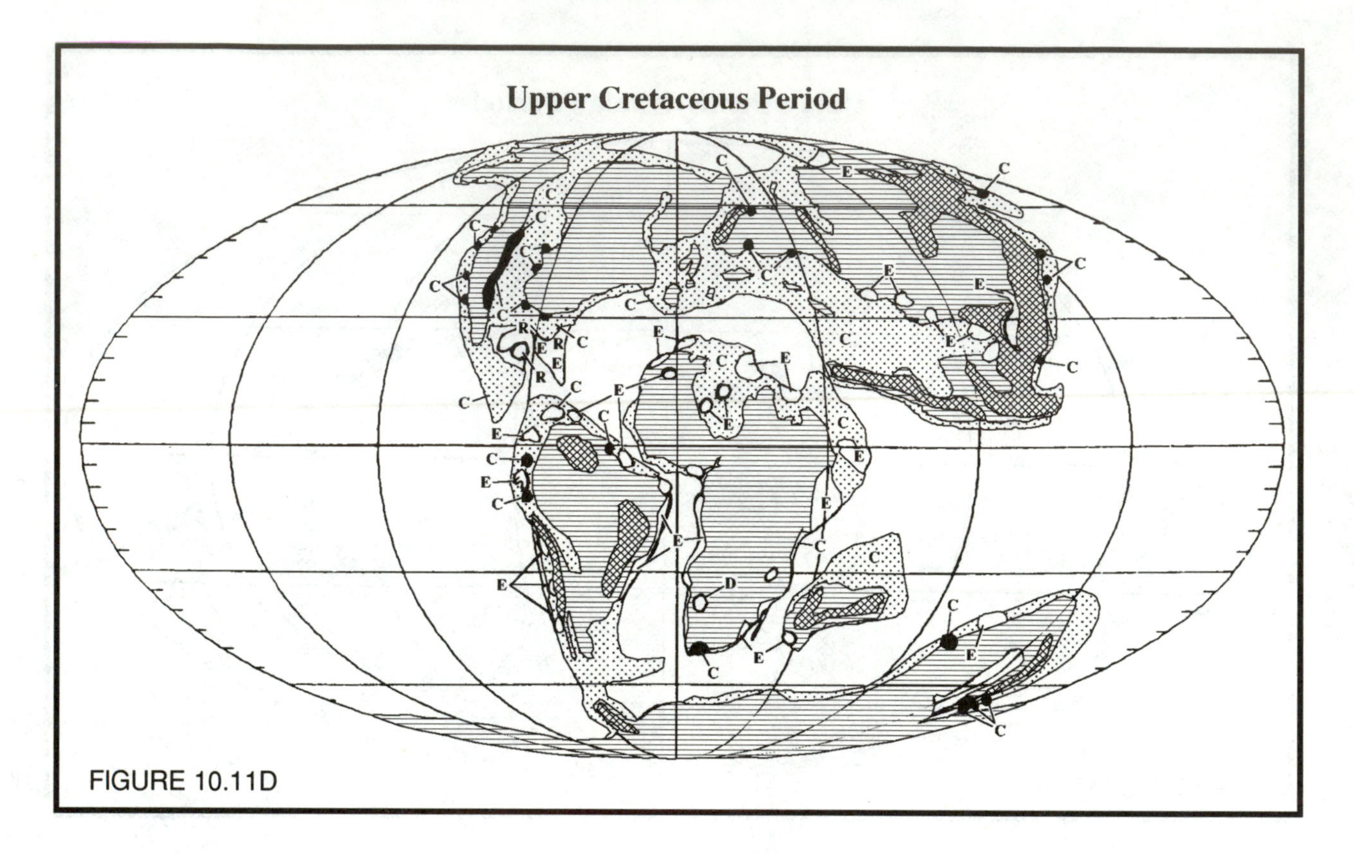

FIGURE 10.11D

LABORATORY EXERCISES

10.2. Specimens: Climate Sensitive Rocks—Your instructor will provide samples of climate sensitive rock types. Describe and identify each, and indicate the evidence which links each to one (or more) climate types on worksheet 10.2.

WORKSHEET EXERCISE 10.2. CLIMATE SENSITIVE ROCK TYPES		
SAMPLE	CLIMATIC INDICATOR PROPERTIES	CLIMATE
1		
2		
3		
4		
5		
6		
7		
8		
9		
10		

10.3. Mapping Ocean Circulation—Reconstruct the general ocean circulation patterns for the globe during the periods for which climatic maps (Figure 10.11A-D) have been made. Refer to the "rules" of oceanic circulation, and to Figures 10.1, 10.8, and Table 10.2. Your instructor may allow you to work in groups.

LABORATORY EXERCISES

10. 4. Mapping Ancient Climatic Zones—Follow the instructions for Exercise 10.1, but add to your data base the effects of atmospheric circulation patterns, continental factors by following the guidelines in Figures 10.1, 7 & 8, the "rules" of application ([#]).

You may be instructed to add the following to the reconstructions described in Exercise 10.1: monsoonal regions, deserts, and rainshadows of the Earth during each period.

A) Compare and contrast the prevailing climates in low to mid latitudes and high latitudes during the the four periods.

Ordovician

Low-mid latitudes ______________________________

High Latitudes ______________________________

Pennsylvanian

Low-mid latitudes ______________________________

High Latitudes ______________________________

Permo-Triassic

Low-mid latitudes ______________________________

High Latitudes ______________________________

Cretaceous

Low-mid latitudes ______________________________

High Latitudes ______________________________

B) During which period(s) was/were the temperature gradient(s) steepest, and during which was it most gradual? ______________________________

____

C) Does there appear to have been a dramatic climate change from the Cretaceous to the Recent? Describe briefly.

____

____

____

What bearing could your observations have on the causes of the Cretaceous-Tertiary extinctions (see text) of dinosaurs and other Cretaceous animals and plants? ______________________________

____

____

D) What was/were the prevailing climate(s) of North America during each period?

Ordovician ______________________________

____

Pennsylvanian ______________________________

____

Permo-Triassic ______________________________

____

Cretaceous ______________________________

____

Exercise 10.4. (Cont.).

D) During what period(s) were coals most extensive? tropical coals? ______________________

temperate climate coals? ______________________

How could vast coal deposits have formed in high latitudes close to glaciated regions during the Permo-Triassic Periods? ______________________

____

E) Why were limestone deposits so widespread in North America during the Ordovician Period? __________

____

F) What does the widespread occurrence of black shales in the Devonian, Carboniferous, Jurassic, and Cretaceous Periods indicate about ocean temperatures and thermal stratification at these times? ______________________

____

____

G) During which periods/epochs were evaporites most abundant? ______________________

____

What climatic and geologic factors combined to create conditions favoring evaporite formation? ________

____

____

H) During which and how many periods/epochs did extensive continental glaciers exist? ______________________

____

____

10.5. Species Diversity and Climate.

During which of the periods investigated should species diversity have been highest, and for which should it have been lowest according to the relationship between diversity and global temperature gradient?

____

____

____

____

TABLE 10.3. TEMPORAL & LATITUDINAL VARIATION in FORAMINIFERA COILING DIRECTION.

AGE (MY BP)	STRATIGRAPHIC SECTION A (LAT. 47° N)		STRATIGRAPHIC SECTION B (LAT. 33° N)		STRATIGRAPHIC SECTION C (LAT. 8° N)	
	%L	%R	%L	%R	%L	%R
0 Holocene	90	10	70	30	10	90
1	90	10	90	10	10	90
2 Pleistocene	90	10	90	10	10	90
3	90	10	90	10	2	98
4	90	10	30	70	0	100
5 Pliocene	90	10	10	90	0	100
6 Miocene	10	90	30	70	10	90
7	90	10	90	10	10	90
8	10	90	30	70	2	98
9	10	90	10	90	0	100
10	10	90	10	90	0	100
11	5	95	30	70	0	100
12	0	100	10	90	10	90
13	0	100	10	90	10	90
14	0	100	30	70	2	98
15	0	100	10	90	0	100

10.6. Foraminifera Coiling Direction—Table 10.3 lists data on changes in percentages of left- and right-coiled test of *Globigerina pachyderma* from three deep sea cores (Upper Miocene—Recent). Use the relationship between coiling preference (% of population) and water temperature to:

A) recognize/define cold, temperate and tropical biofacies (indicate on plot below); and,

B) plot and correlate these biofacies among the three sections as you would lithofacies (Laboratories 2 and 7).

C) How many periods of cooling and warming are indicated? ____________________

____

____

____

D) When did each period of cooling occur? ____________________

____

____

____

E) Were the transitions between warm and cold temperatures gradual or rapid? ____________________

____

____

____

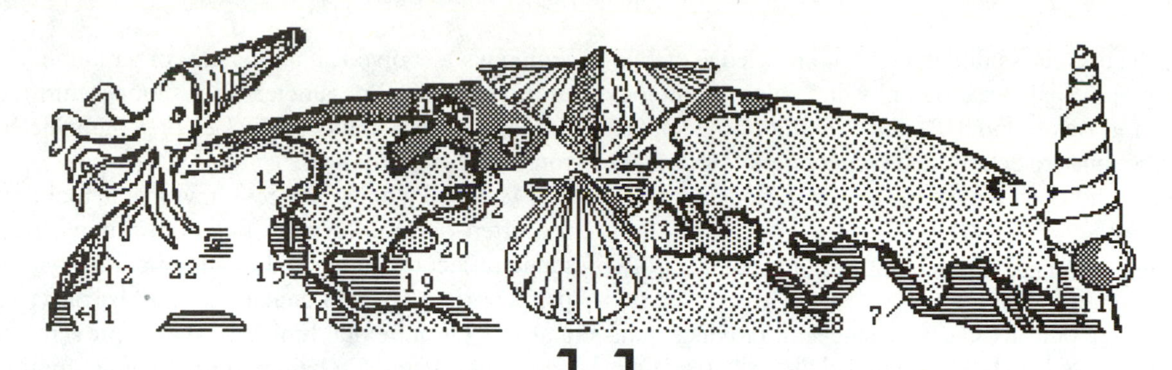

11
Tectonics and Life's Diversity

Objectives

To demonstrate mastery of this material you should know how to: 1) estimate the number and locations of ancient biotic provinces from paleogeographic maps; 2) relate global diversity to biotic province number; and if your instructor so chooses, 3) relate species diversity to habitat area; and, 4) calculate continental shelf area and, thereby predicted species diversity. A calculator is needed for the latter exercises.

Using This Laboratory

The recommended approach to this material and laboratory exercises is similar to that discussed in Laboratory 10. Precise guidelines for mapping exercises are given for your reference and confident conceptual understanding will follow only after doing these. The exercises should be treated foremost as imaginative excursions into some of the factors which have controlled global diversity, particularly those linked to plate tectonics. Some exercises are pointedly designed and aimed to address current concerns of habitat loss and extinctions. These are compared and contrasted with past extinction events to demonstrate relevance.

Paleobiogeography is the study of the past geographic distributions of extinct organisms. Animal and plant distributions are used in paleontological and evolutionary studies and in reconstructing past geographies. For example, Permo-Carboniferous distributions of the *Glossopteris* flora and the lake-dwelling reptile *Mesosaurus* provided convincing evidence (to Wegener and Suess) that India and all southern continents were once part of a supercontinent—Gondwana (see text). More recently, distributions of marine invertebrates have been used as evidence that 25% of western North America originated as Panthalassa Ocean islands and microcontinents (displaced or exotic terranes) which collided with the western U.S. coast (see text). And, species' distributions are central to studies of their origins and evolution (see text). For example, species' distributions in South America, the Galapagos Islands and the East Indies were among the observations that led Darwin and Wallace, respectively, to formulate the theory of organic evolution by natural selection.

In this chapter we first consider the factors which help and hinder species' migrations. Second, we see how these factors lead to the development of biotic provinces, each large regions of marginal and epeiric seas with characteristic species compositions. Finally, we see how a habitat's geographic area, provinciality and plate movements regulate global diversity.

The exercises build upon your knowledge of paleogeography, paleoceanography and paleoclimatology (Laboratories 9 and 10), and illustrate how species distributions are determined by geography (tectonics), climate and ocean currents. In the various exercises you may:

1) map estimates of the number and locations of past marine provinces;
2) use tectonically caused changes in province number to predict changes in global diversity over time;
3) predict relative diversities in a hypothetical continent-ocean geography based on the influence of plate tectonic features;
4) calculate the effect of changes in habitat area on extinction rates of shallow sea invertebrates and terrestrial species of the Brazilian rainforest; and,
5) use geological evidence of habitat change to infer evolutionary relationships among species

DISPERSAL and GEOGRAPHIC RANGES

Marine invertebrate species' geographic ranges are determined by their tolerance of physically extreme conditions, abilities to disperse widely, and their availability of migration routes and habitats. Species are categorized by the extents of their geographic ranges and ecological tolerances or habitat requirements. Many species are limited to one geographic region, while lesser numbers range over vast areas. Species attain such wide ranges by evolution of either tolerance of a wide range of conditions, or by widespread dispersal mechanisms. Species limited to one region have evolved an intolerance of variation in one or more physical parameters (e.g., temperature), or a limited dispersal ability (or by lack of geographic/environmental opportunity).

Maps of species' ranges show that actual ranges are often far smaller than potential ranges. Actual range reflects both the ability and opportunity to disperse. Various types of faunal (or floral) barriers (physical or environmental limitations) may prevent occupation of all suitable habitat. For example, mountain ranges physically bar dispersal of many lowland species, and intolerance of high temperatures in western U.S. intermontaine valleys may also limit species' ranges and dispersal among cooler habitats of mountain ranges (Laboratories 6 & 7).

Physical barriers may be as formidable as a deep ocean, or as slight as a short stretch of forbidding coastline. Whether a physical feature prevents dispersal depends on an organism's ability to traverse it and the ecological obstacles (e.g., lack of food) to be overcome. The chief physical barriers are of tectonic origin: the arrangement, elevations/depths of, and links among, continents and ocean basins. Currents, also determined by plate positions (Laboratory 10), may limit marine organisms' ranges to single water masses, and thus, can be important physical barriers. Environmental barriers include temperature, depth, and salinity. Influential biological factors include larval period length, and type, and adult locomotion type (Laboratory 7). We will consider only depth, temperature, and tectonics.

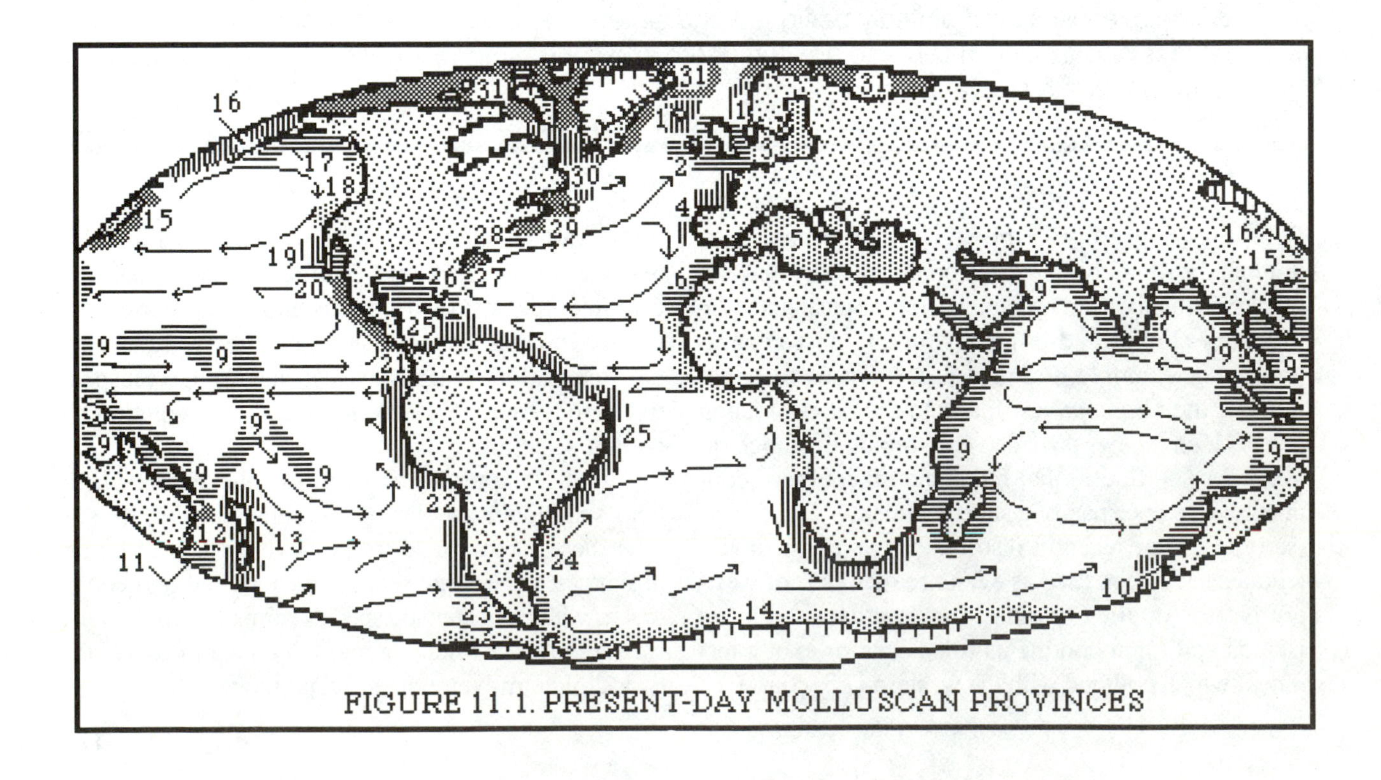

FIGURE 11.1. PRESENT-DAY MOLLUSCAN PROVINCES

All marine organisms are limited by depth in some way. For example, many groups must live within the photic zone (≤200 m (656 ft)) by direct or indirect dependence on photosynthesis. Shallow water benthos are limited to continental shelves; most cannot cross the deep ocean. Land masses (e.g., Isthmus of Panama) are impenetrable barriers to dispersal between ocean basins.

Temperature is a barrier to nearly all animals and thus, many species' ranges lie parallel to latitudinal zones of atmospheric and oceanic circulation (Figure 11.1). The most effective temperature barriers are between warm tropical waters and those of cooler mid-latitudes (≈ 30-45° N/S) where seasonal variation (especially from 15-18°C (58-64°F)) is a most effective restraint, and is linked to lower diversities. Temperate species' ranges invariably parallel this barrier. Cold water species are restricted by interposing warmer waters.

Barrier effectiveness differs among species, and so too, their dispersal in response to certain barrier types. There are three basic types of "routes" across barriers:

1) a corridor;
2) a filter bridge; and,
3) a sweepstakes route.

These routes differ in the ease and, thus, the frequency of use. A barrier with an open easy passage for many species is termed a corridor (e.g., the Isthmus of Panama, uplifted in the Late Pliocene, became a corridor for North and South American mammals (Figure 11.2)). Only select species can cross filter bridges (e.g., birds reach islands while large mammals can't). A sweepstakes route is a low-probability passage between widely separated habitats (e.g., a chain of islands linking continents across an ocean). Corridors or filter bridges for one species may be a sweepstakes route of others.

LIFE CYCLE of SPECIES' RANGES

Distributions can only be understood within geologic time and history's influence on dispersal barriers and pathways. Early in species' existence, are typically small, geographically isolated populations. Population size and geographic range expand and decline alternately, fluctuating over the species' lifetime, and a final decline often precedes extinction. Barriers to dispersal also arise, change location, and disappear over long intervals of time. If one or more barriers have broken down in the recent past, a species' current realized range may be less than its potential (e.g., the northward march of "the killer bees"). Other species may have highly fragmented ranges with populations isolated one from another by regions of unsuitable habitat. These "refuge" populations are isolated because environmental or geographic factors which changed after originally cohesive, extensive ranges were established. Thus, species' present ranges are the sums of their past ranges, opportunities for dispersal, and changes in environment and geography.

Geographic distributions affect the evolutionary histories of species. Related species having similar anatomies or ecological requirements usually occupy adjacent areas separated by a barrier. When a new barrier emerges, a single large population separates into two, one or both of which may later evolve a new species (Figure 11.2; see text). Prior to the Pliocene uplift of the Isthmus of Panama 1.65 to 5 mya, Caribbean and East Pacific marine fauna moved freely between basins. These related, but now separate, faunas are distinctly different today. Should the barrier

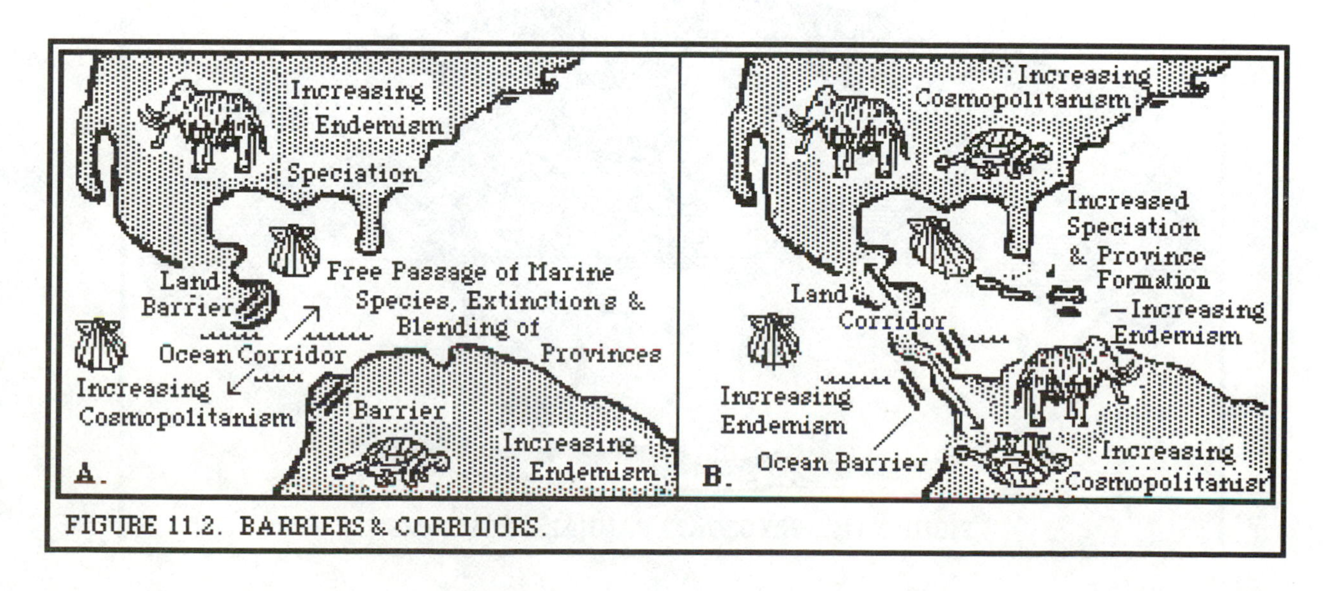

FIGURE 11.2. BARRIERS & CORRIDORS.

falter, competition may lead to evolution or extinction of similar/related species (Figure 11.2).

GEOGRAPHIC COMMUNITY COLLECTIVES: BIOTIC PROVINCES

A biotic province is the basic unit of biogeographic range: a region in which communities maintain characteristic species compositions. They are commonly delineated by the geographic boundaries of each species' range or as the only area in which ≥50% of species are endemic. Province boundaries are zones along which many species reach their geographic limits and community compositions change rapidly. Such definition is difficult to apply and boundaries are typically based on the distributions of one organismal group (e.g., bivalves, bryozoa; Figures 11.1, 11.3).

Provinces often strongly reflect temperature barriers. Temperate and tropical provinces may differ by 80% of their species. If coastal temperature changes gradually, only a few provinces are likely to be present. If temperature changes rapidly, then provinces should be sharply defined.

Figure 11.3 illustrates the general features of the present major marine provinces that serve as guidelines for province delineation (and here for exercise reference **[#]**). Equatorial provinces tend to be twice as extensive as those of midlatitudes **[1].** Polar provinces are extensive because temperature variation is slight **[2]**. East-west provinces may stretch over great distances **[3]**. East and west sides of oceans may support different numbers of provinces if temperature gradients and circulation patterns differ **[4]**.

PLATE TECTONICS' EFFECTS on PROVINCIALITY

Land and sea distributions and, thus, physical barriers and dispersal routes, are of tectonic origin. Thus, plate movements have greatly influenced changing provinciality and diversity. The locations and orientations of subduction zones, island arcs, spreading ridges, and intraplate volcanoes (see text) can act as barriers to, or pathways of, migration among oceans (Figure 11.4). Of all factors, continent number and position have the greatest effect on provinciality. Using the principles discussed above, past province numbers can be estimated by continent numbers, locations, and ocean circulation patterns.

A method for predicting past biotic province number is illustrated in Figures 11.1, 5, and 6. The "rules" for constructing those provinces are indicated by: **[#]**. Provinciality will be high when continents are many and small, and spread across zones of latitude. When assembled in supercontinents such as Pangaea, barriers would be few and, thus, province number low. Ocean currents have typical water temperatures, salinities, and other physical factors, and they disperse larvae. Thus, we can expect a province to be largely restricted to single current systems **[6]** .

The method of estimating province number is illustrated by a hypothetical single continent and a simple current system (Figure 11.6). The continent stretches to 50° N and S, and the current systems include: 1) a westward equatorial current; 2) a current with intensified west-east return flow reaching from 30-40° N/S; 3) a high-latitude westward return flow; and, 4) a clockwise /counter-clockwise polar current from 60-90° N/S.

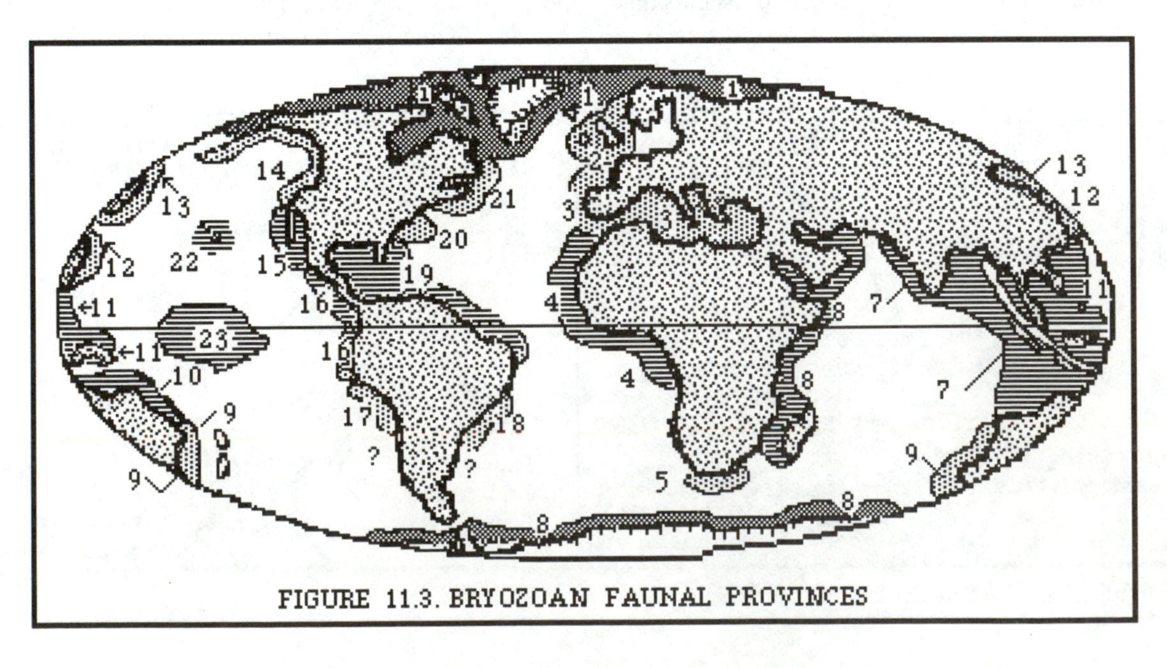

FIGURE 11.3. BRYOZOAN FAUNAL PROVINCES

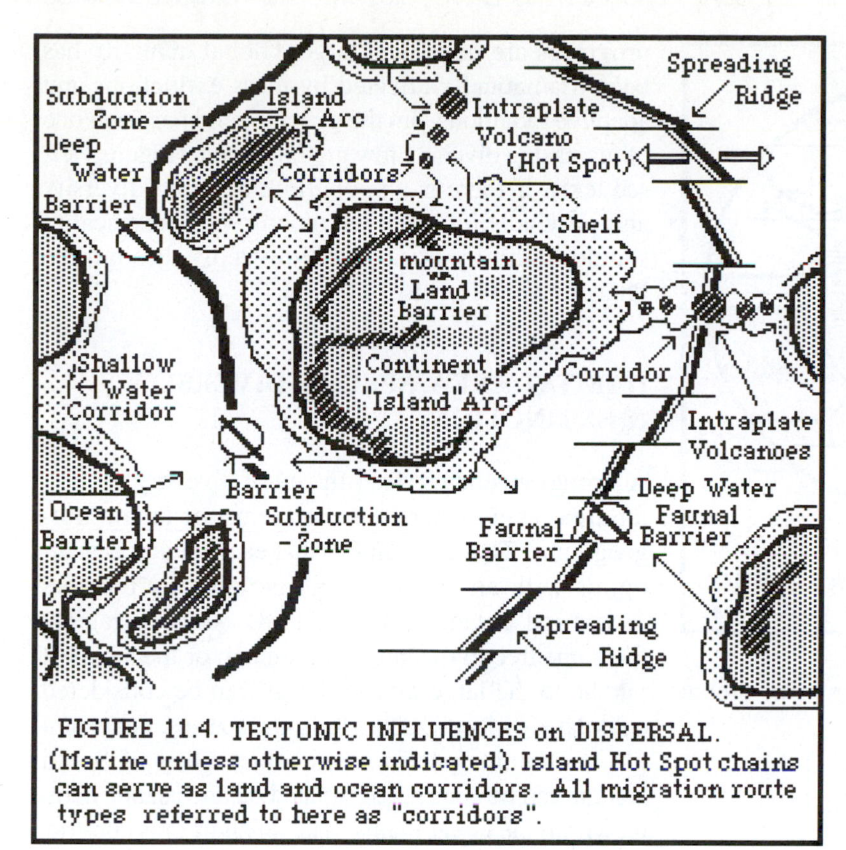

FIGURE 11.4. TECTONIC INFLUENCES on DISPERSAL. (Marine unless otherwise indicated). Island Hot Spot chains can serve as land and ocean corridors. All migration route types referred to here as "corridors".

This basic pattern will be modified by coastal geography and additional continents. Given this large hypothetical continent, there are 8 large provinces: two polar; eastern and western tropical; and, four temperate midlatitude. In other words, there should be one province for each climatic zone or current system on each coast parallel to meridians, and one province for each coast paralleling latitude lines [7].

Realistic estimates can be made by analogy with present provinces and by adjusting for scale and distance. For example, the Mediterranean province is separate from the North Atlantic (Figure 11.1, 11.3). By analogy, we infer that semi-isolated basins as large, or larger than, the Mediterranean, held their own provinces **[8]** . Province size is also limited by distance, even without other barriers. For example, the tropical region from the Indian Ocean to the Western Pacific spans 15,000 km and is divided into three provinces. Long-distance larval dispersal in an open ocean thus appears to break down beyond distances of about 4,000 km **[9]**. Consequently the vast Panthalassa Ocean probably supported more than one province within any latitudinal zone. Recall also that today's tropical provinces tend to be twice as wide as temperate provinces **[10A]** and, polar provinces may be extensive **[10B]**.

During periods of high sea level (e.g., Cretaceous, North America) barriers among oceans were fewer and less effective because shallow seaways provided corridors and,thus, provinces would be few **[11]**.

Province number and size also vary with average global temperature. The temperature gradient from equator to pole was steeper during times of glaciation than today, and much more gradual during times of global warming (e.g., Pleistocene vs. Mesozoic, Laboratory 10). Thus, a steeper temperature gradient increases province number and decreases province

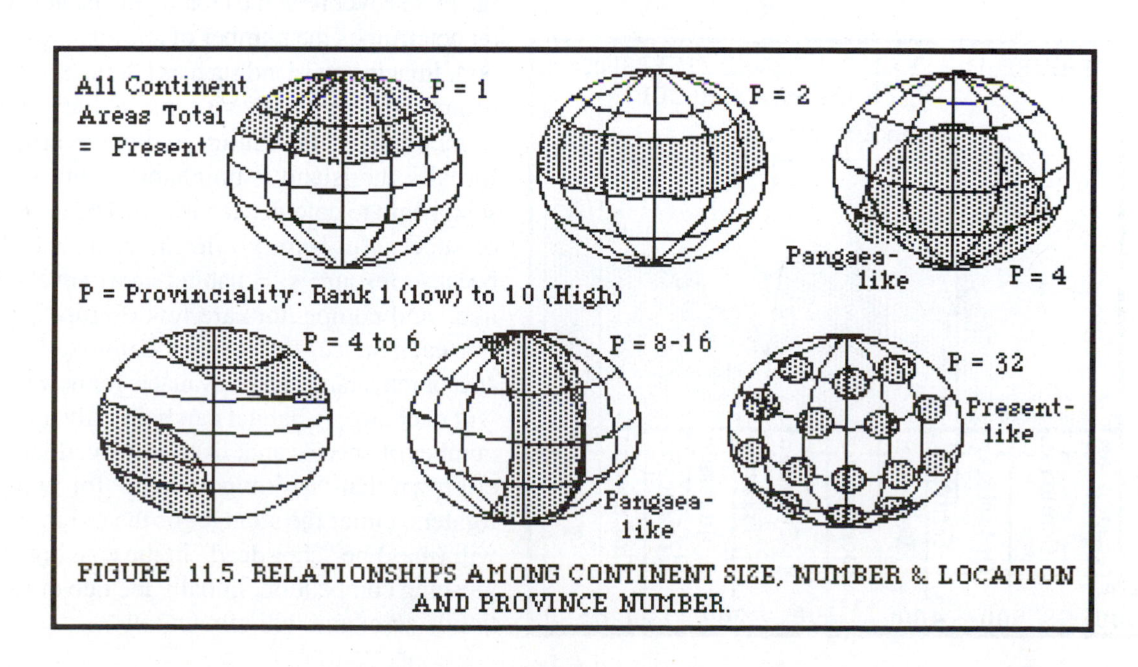

FIGURE 11.5. RELATIONSHIPS AMONG CONTINENT SIZE, NUMBER & LOCATION AND PROVINCE NUMBER.

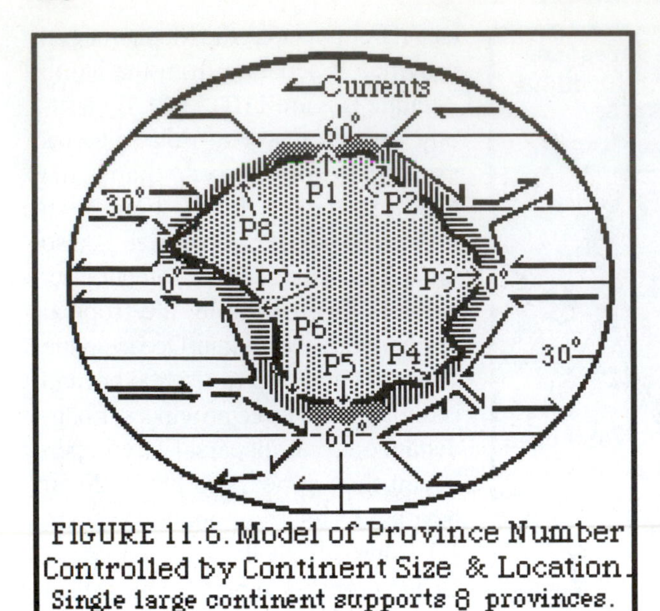

FIGURE 11.6. Model of Province Number Controlled by Continent Size & Location. Single large continent supports 8 provinces.

sizes, while the converse is true in times of global warming **[12]**.

CHANGING PROVINCE NUMBER & GLOBAL DIVERSITY

The number, size, and location of provinces reflect many aspects of life's history. Most importantly, provinciality is closely linked to global diversity (Figure 11.7). Provinciality and diversity are linked because diversity is determined by available habitat area (see below), and by climate and barriers. The larger is the a province number, and (to a point) the greater the area of each, the greater the global diversity. Habitable area and diversity are maximized (to a point) when provinces are many and large. Global diversity has been dramatically affected by mass extinctions and adaptive radiations, but the general trend has been one of increasing diversity toward the present (Figure 11.7; see text). If there is a strong link between diversity and provinciality, plate tectonic-driven provincality should faithfully track global diversity through the Phanerozoic (Exercises 11.1, 2).

HABITAT AREA, SPECIES DIVERSITY & EXTINCTION

A strong, secondary determinant of diversity is habitat area, as described by the theory of island biogeography. The term implies that each habitat is analogous with an "island", and isolated by barriers (water and distance), (Figure 11.8). Any size region, from one tree to the largest continent, or the smallest puddle to the largest ocean basin, can be considered an isolated "island" habitat for some species. For example, mountain species in the Black Hills of South Dakota can be considered isolated on an "island" habitat within the Great Plains "sea" of short grass prairie. The number of species an "island" can support depends on:

1) its area;
2) the number of species able to reach the "island" (immigrations); and,
3) the number of extinctions due to competition for space and resources. In other words: diversity = ((#species already present + # new immigrants) - # extinctions).

Now let's work through a hypothetical example to discover how the area of an "island" habitat determines the number of species it can support. Imagine an island such as ("little") Krakatau recently (re)formed by volcanic activity after its tremendous 19th century explosive eruption. Initially the island is uninhabited, but species soon begin to colonize the island. The initial rate of successful immigration is relatively high because resources (= habitat area) are "unlimited" and competitors are few (Figure 11.8A). For each successful species colonization, the area of remaining habitat available for new species will decrease. A habitat can hold only a limited number of species and, like the proverbial western town that "isn't big enough" for two gunfighters, either the stranger or the old town gun will (may) be "shot-dead" in the ensuing extinguishing competition. Initially the rate of extinction is zero, and will remain low.

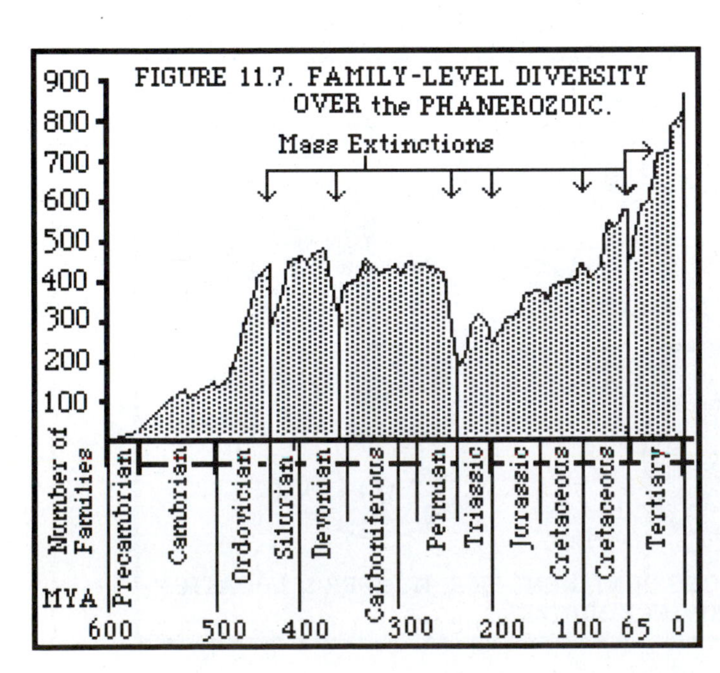

FIGURE 11.7. FAMILY-LEVEL DIVERSITY OVER the PHANEROZOIC.

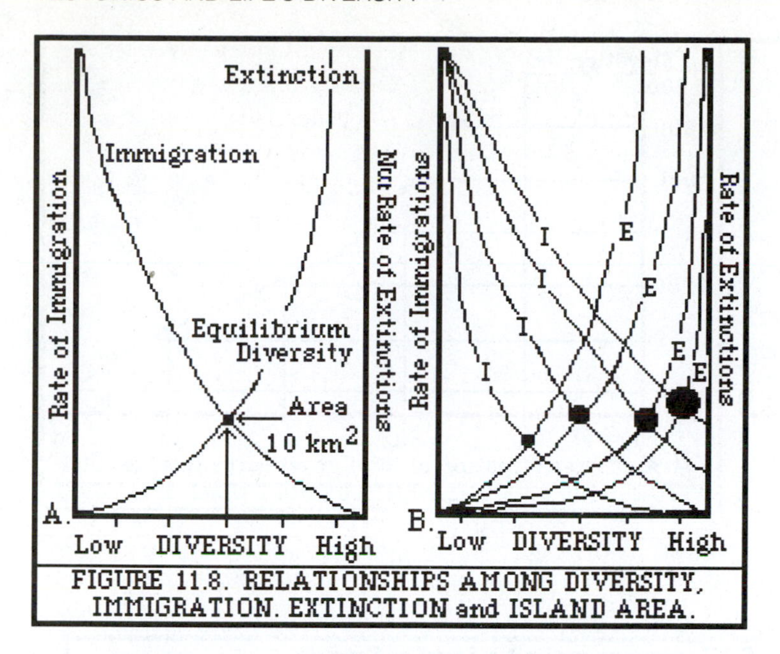

FIGURE 11.8. RELATIONSHIPS AMONG DIVERSITY, IMMIGRATION, EXTINCTION and ISLAND AREA.

By analogy to a modified game of musical chairs in which the number of *participants* changes rather than the number of chairs, while participant number steadily increases and participants are initially able to claim more than one chair each, the number of players will climb while the personage changes. Eventually participants will outnumber chairs, and some tenacious individuals will retain seats while entrance and elimination of others continues. The species composition will continue to change, but the number of species supported by an island of a certain area fluctuates about a constant value that is "just right" for an island of that area.

Although there are additional factors in the theory, we will consider diversity to be directly proportional to island size, and described by: $S=k A^z$; where S is species diversity; A is the habitat area; k is a constant (the y-axis intercept); and, z is a constant for the group of organisms in question (= the line's slope). Because the relationship is exponential, data are plotted on a log-log scale (Figure 11.9), and may also be expressed as: $\log S = \log (k) + z \log (A)$.

Of what consequence is this relationship for our study of the Earth's history ? As sea level has risen and fallen over the Phanerozoic, the habitable shallow sea area has varied accordingly. while diversity is low. But competition and lack of suitable habitat space for immigrants cause the extinction rate to increase as diversity increases. (Figure 11.9B). Eventually an equilibrium number of species is reached—a balance between successful immigrations and extinctions for an island of a certain area.

Thus, marine species diversity should have changed in proportion to rising and falling sea level as well as with changing province number: maximum diversities should occur during times of maximum inundations, and high province number, while widespread extinctions should occur during maximum regressions.

[Note: although loss of habitat area is one important mechanism of mass extinction, there are other viable and known explanations, including climatic change, vulcanism and asteroid impacts, among others (see text).]

SOUNDING PAST SEA LEVELS

To determine how diversity may have changed with changing shelf area (see text for causes of sea level change), we need a method of estimating past sea levels. When the percentage of land at different depths-elevations (Figure 11.10A) is plotted, there are two maxima:

1) between sea level and an elevation of ≈1 km= average land elevation; and,
2) at ≈ 4 km below sea level, the average sea depth (Figure 11.10A). These levels reflect concentrations of crustal rocks differing in density; less dense silica-rich continental crust (≈ 2.8 gm/ccm), and denser Fe, Mg-rich oceanic crust (3.0 - 3.3 gm/ccm). Differing densities causes them to "float" at different levels on the asthenosphere. If the percentage of a continent's crust below sea level during a given time is calculated by noting the area of marine deposits (Table 11.1), the hypsometric curve can then be used to determine the absolute sea level (Figure 11.10B). Given the shelf area covered at any time (x-axis), the sea level can be read from the y-axis.

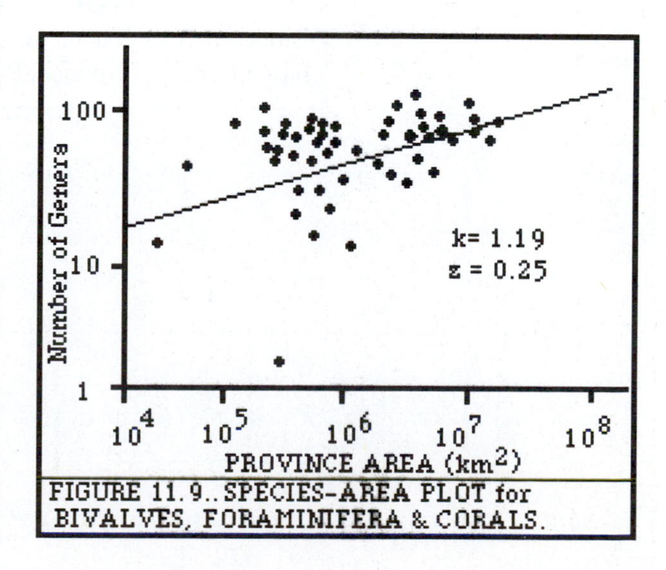

FIGURE 11.9.. SPECIES-AREA PLOT for BIVALVES, FORAMINIFERA & CORALS.

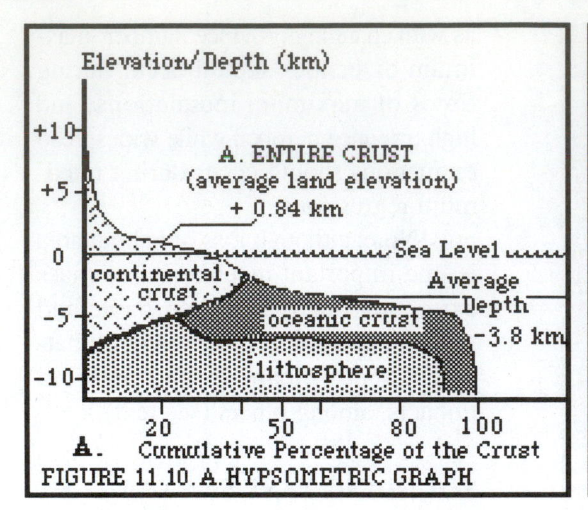

FIGURE 11.10. A. HYPSOMETRIC GRAPH

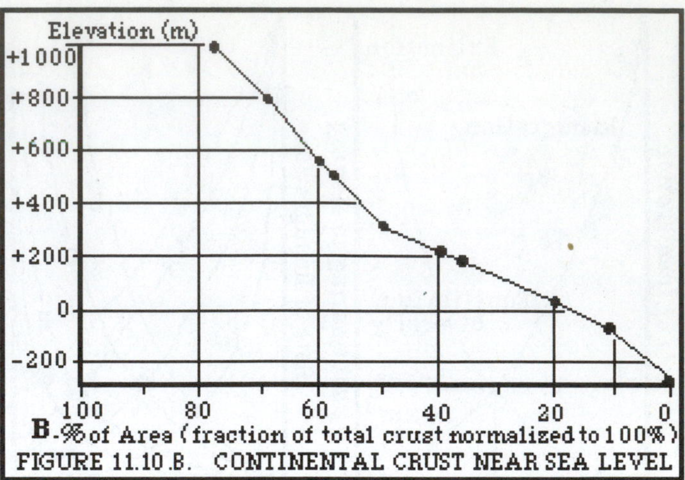

FIGURE 11.10.B. CONTINENTAL CRUST NEAR SEA LEVEL

TABLE 11.1. Estimated Areas of Phanerozoic Epieric Seas.

Stratigraphic Interval	Estimated Area ($x\ 10^6$ sq km)	% Total Area	Stratigraphic Interval	Estimated Area ($x\ 10^6$ sq km)	% Total Area
Quaternary	28	15.9	Pennsylvanian	59	33.5
Neogene	38	21.5	Mississippian	70	39.8
Paleogene	48	27.3	U. Devonian	61	34.7
U. Cretaceous	63	35.8	M. Devonian	70	39.8
			L. Silurian	80	45.5
U. Jurassic	55	31.3	U. Ordovician	62	35.2
U. Triassic	34	19.3	M. Ordovician	83	47.2
U. Permian	49	27.8	U. Cambrian	72	40.9
L. Permian	66	37.5	L. Cambrian	73	41.5

Total Area presently = 176×10^6 sq km. Present Sea Level covers 28.3×10^6 sq km.

VOCABULARY

Biotic province, Faunal barrier, Habitat area, Island biogeography, Province, Paleobiogeography.

LABORATORY EXERCISES

11.1. Tectonic Controls of Provinciality: a Hypothetical World—Figure 11.11 illustrates a hypothetical arrangement of continents, shallow shelf, deep oceans, spreading ridges, mountains and island arcs, and their distribution among climatic belts.

A) Identify, encircle and label each province, faunal barrier, and route of migration.

B) Rank the continents in order of decreasing shallow marine faunal diversity, and decreasing land animal diversity.

PROVINCIAL DIVERSITY RANKS:

	Marine	Land		Marine	Land
Wegenerland	______	______	Morgonia	______	______
Wilsonia	______	______	Hessland	______	______
Matthewinia	______	______	Hoffmanland	______	______
Vinleland	______	______			

11.2. Province Number & Global Diversity—A) Determine the approximate number of provinces from paleogeographic maps of each of the four geological periods shown in Figure 11.12A-E. Begin this exercise by reviewing the principles of climate reconstruction in Laboratory 10 &/or its exercise maps.

[If you have not completed the exercise in Laboratory 10 you may base your results on continent number and position, and general information on the global temperature gradients during the periods in question (see text and laboratory 10).]

Refer to the rules for delineating provinces and determining their number (Figures 11.1 & 11.3-6). Give number labels to, and indicate the geographic extent of, each province by encircling each numbered province. Begin with pencil and use copies or tracings before making final maps. [You (or on your instructor's direction) may also want to transfer to these maps the major ocean current patterns (if) plotted in Laboratory 10 exercises.]

Important: Realize that there are no "correct answers" for such questions, but only broad generalizations and educated inferences.

B) Identify and mark geographic features that may have been barriers, corridors, or isolated basins.

C) During which periods would species having widespread distributions be common, and during which would those with restricted ranges have been high? Explain. ______________________________

____

____

____

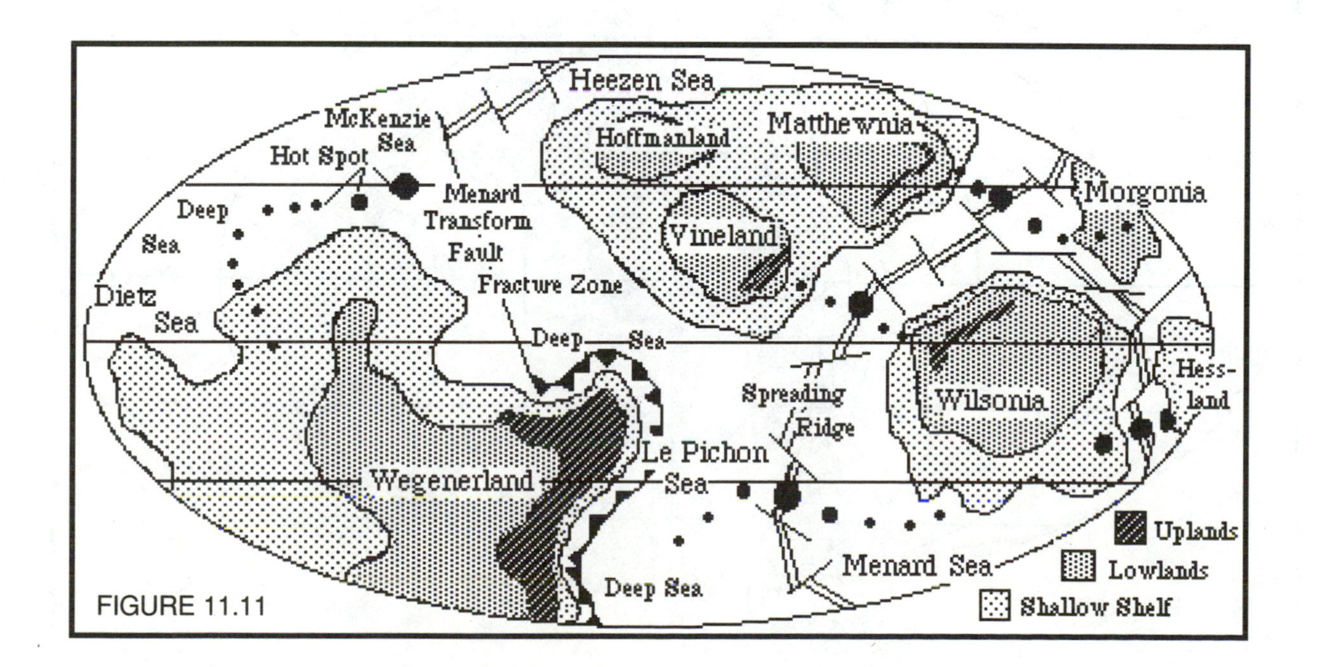

FIGURE 11.11

LABORATORY EXERCISES

Figure 11.12A (Ordovician Period) and B (Pennsylvanian Period)

Uplands **Lowlands** **Shallow Seas**

R = Reefs, Rb = Red beds, C = Carbonate rocks, E = Evaporites, ● = Coal

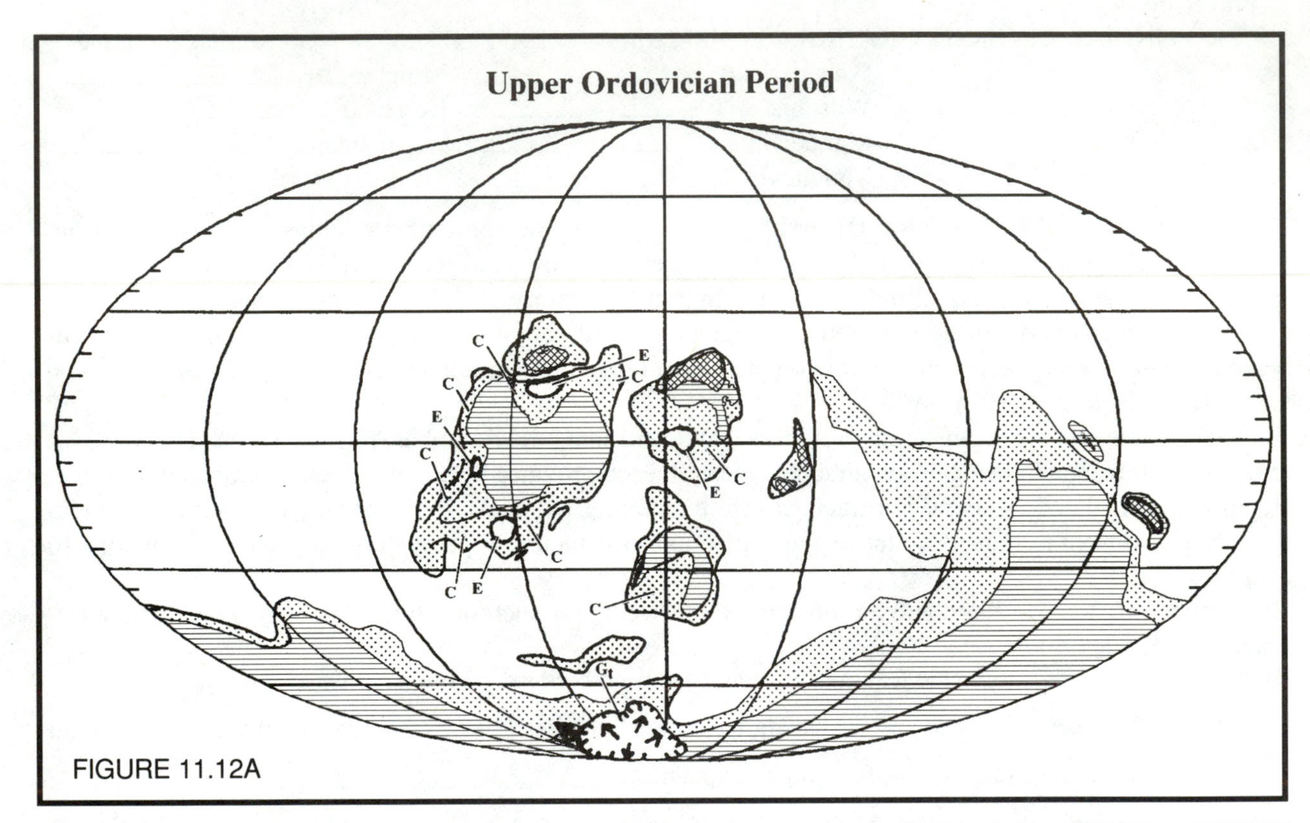

FIGURE 11.12A

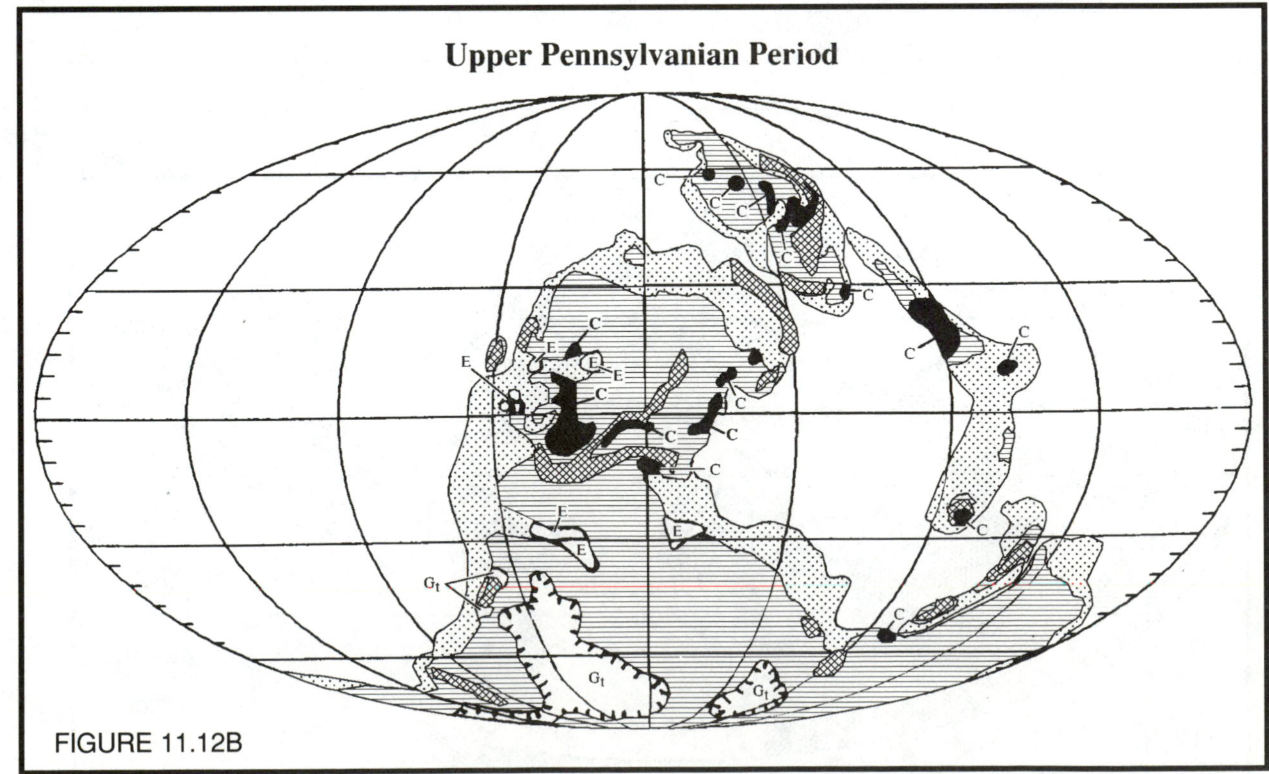

FIGURE 11.12B

LABORATORY EXERCISES

Figure 11.12C (Permo-Triassic Periods) and D (Cretaceous Period)—Land-elevation key as in Laboratory 10.

Uplands **Lowlands** **Shallow Seas**

R = Reefs, Rb = Red beds, C = Carbonate rocks, E = Evaporites, ● = Coal

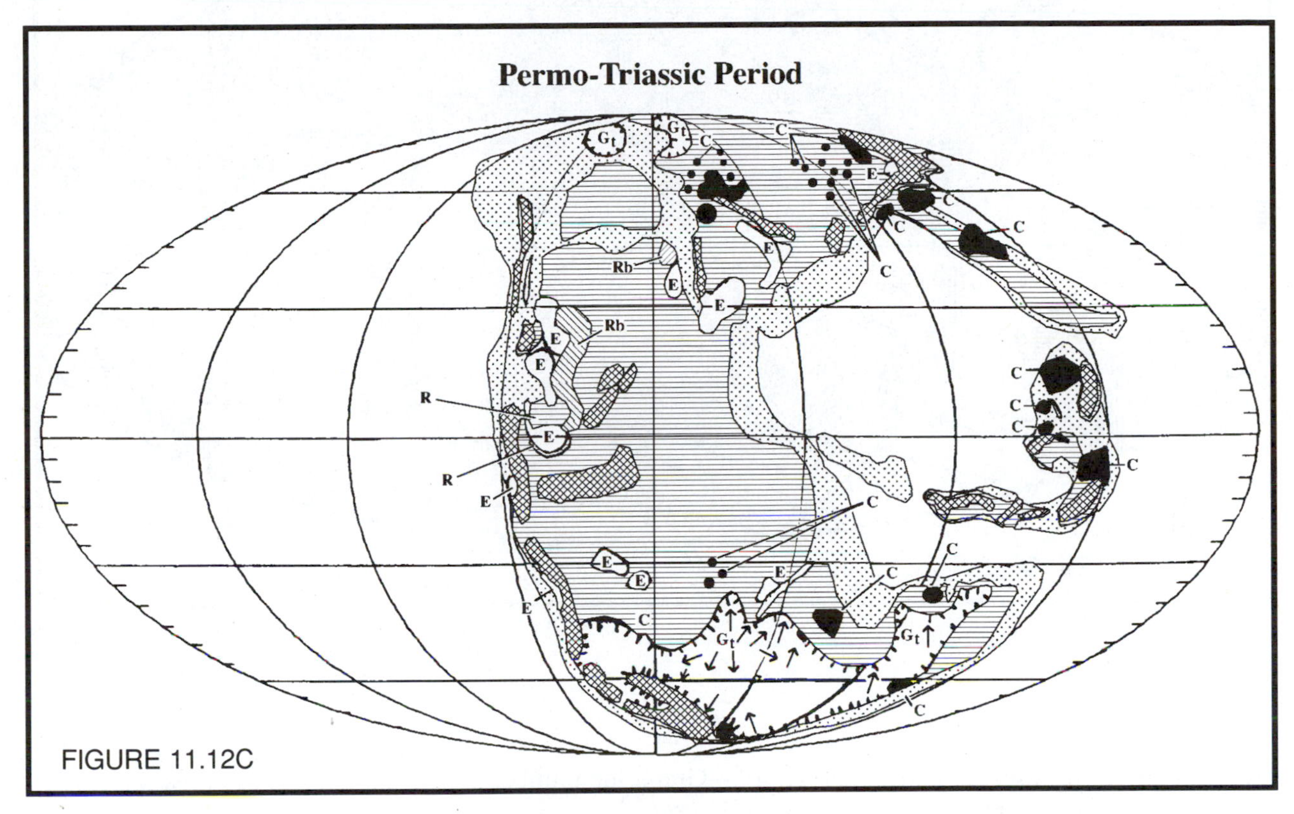

FIGURE 11.12C

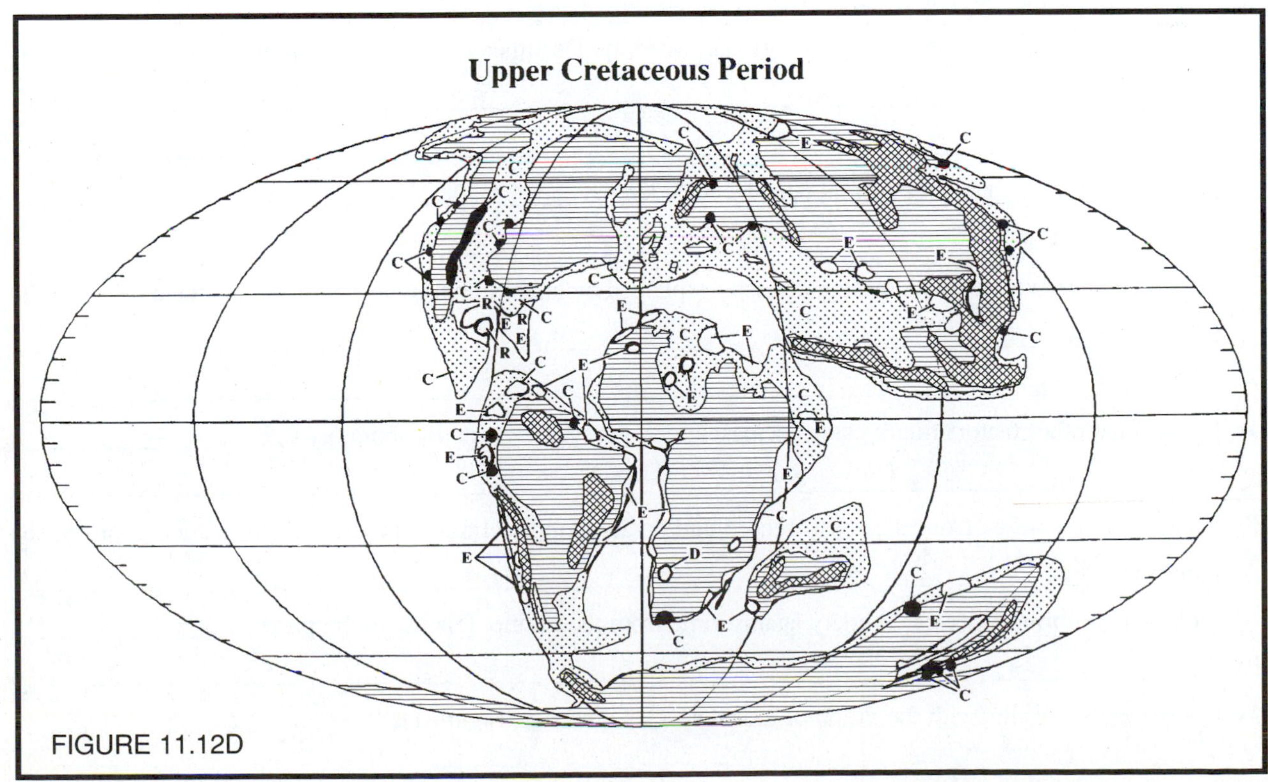

FIGURE 11.12D

LABORATORY EXERCISES

Figure 11.12E (Quaternary Period)—Land-elevation key as in Laboratory 10.

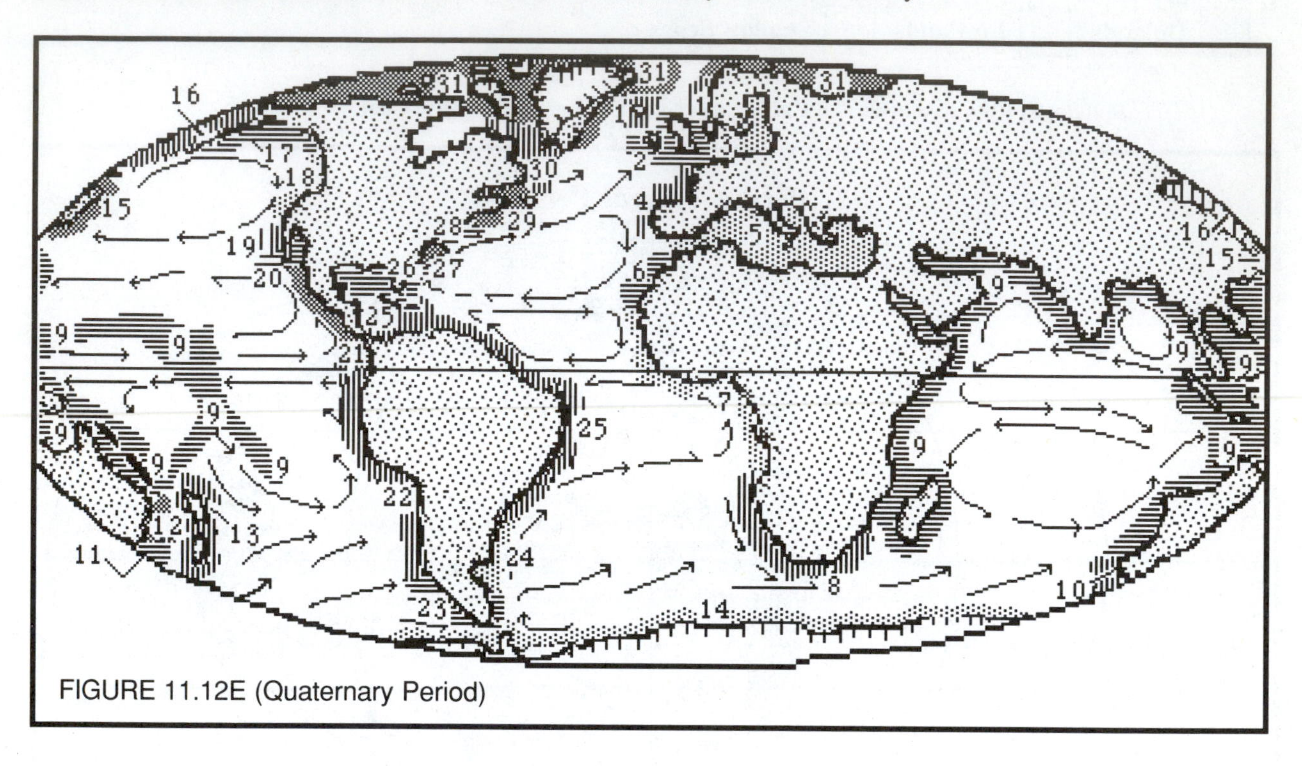

FIGURE 11.12E (Quaternary Period)

11.3. Province Number & Global Diversity—Graph the number of provinces from Exercise 11.12A-E against time using the same scale as Figure 11.7.

A) Compare them to the graph of known global diversity. Do times of high/low provinciality coincide with high/low diversity? ________________________________

B) If (or where) not, what other factors need be considered? ________________________________

C) Is the known trend toward increasing diversity explained by increasing province number, habitat area, or both? Explain. ________________________________

D) If not, what other factors might cause increasing diversity (see text and Laboratory 6)? ________________________________

E) What types of biases from the fossil record and/or geographic/climatic models might influence province number? ________________________________

F) Plot the equilibrium species diversity against time from the Permo-Triassic to the present. ________________________________

G) Compare your results with the graph of global species diversity (Figure 11.7). ________________________________

LABORATORY EXERCISES

H) Do your results indicate that mass extinctions (see text) always coincide with major sea-level regressions?

____

I) When were sea levels high during a mass extinction or low without a mass extinction? ____________

____

J) Suggest causes for each mass extinction not associated with major regressions. ____________

____

K) Which, if any, were associated with global cooling or warming (see Laboratory 10 or textbook)? ______

____

____

____

11.4. Habitat Area, Extinction & Rainforests—There are at least 5 million (mostly undescribed) species —some estimates are as great as 50 million—on Earth, most of which inhabit the world's rain forests (Figure 11.13). Rainforests produce oxygen, absorb CO_2, & profoundly affect the global distribution of heat and rainfall (see Laboratory 10). Ecologists have estimated that in 1989 deforestation caused ≈ 5,000 - 17,000 extinctions each year (≈ 14 - 47 per day, or 50,000 - 170,000 per decade!), & the number of species endangered ranges from 375,000 to 1,250,000. A century ago, rainforests covered 6.2 million square miles (16 million sq km). Presently 3.6 million square miles (9.32 million sq km) remain. An area the size of Ireland - 23,000 sq mi (59,570 sq km) is destroyed each year. Another 1.5 million square miles (3.88 million sq km) could be destroyed by the year 2000.

The Brazilian rainforest covered 8.55 million sq km before ≈ 1970, and in 1989 covered 5.44 million sq km (a 17 % loss/decade). Perhaps only 335,500 sq km will remain by the year 2000.

A) Use the species-area equation with a constant z = 0.32, and the present diversity and area to solve for k (y-axis intercept) & derive the species-area line for the rainforest ecosystem. Plot your work on log-log graph paper, your instructor will demonstrate use of log-log graphs).

S = ____________________

z = ____________________

k = ____________________

B) Using the species-area equation, determine the equilibrium species diversity & the number of extinctions that:

1) have taken place between 1500 and 1970 & from 1970 to 1990;S = __________

2) will take place from 1990 to 2000; and, S = ____________________

3) from 2000 to 2050, given the present deforestation rate. S = ______________

FIGURE 11.13. RAINFORESTS of SOUTH AMERICA

Exercise 11.4., Cont.

C) What happens to the rate of extinction as habitat area loss increases? ______________________

____

____

D) What percentage of the Earth's species is projected to suffer extinction from deforestation by 2000? By 2050? ______________________

E) What is the rate of extinction per one million years? (#/10 yrs / 10^6 yr.) ______________________

F) Compare your results with the rates of extinction during the Late Permian (230,400 of 240,000 species over approximately 10 million years), and Cretaceous-Tertiary (1,310 of 2,776 marine genera (assume 200 species per genus) over 8 million years). Before calculating extinction rates for these events, multiply the number of known species from each period by three to compensate for the soft-bodied species which are not preserved (studies suggest this figure is ≈ 70%). During which period was extinction rate highest? ____________

Note: If, however, much of the Cretaceous-Tertiary extinctions occurred instantaneously as the result of one or more asteroid impacts (see text), the rates are not comparable.

G) In what important ways (at least two) does the present extinction differ from past mass extinctions ? _____

____

____

____

____

F) It has taken global diversity an average 10 million years to increase and equilibrate from mass extinction events in the past. Could we expect global diversity to rebound faster, or more slowly, after the current mass extinction has ceased? ______________________

CALCULATIONS

LABORATORY EXERCISES

13.5. Marine Habitat Area and Sea Level—

A) Using the % continental shelf area coverage data in Table 11.1 and the hypsometric curve (Figure 11.10B), determine the absolute sea level in North America for each time interval by reading the sea level values correlated with percentage coverage.

B) Plot your data against the geologic time scale.

C) During which periods were sea levels highest and lowest? Refer to your text and discuss the possible cause for these fluctuations in sea level. [Note that the continents have stood both higher and lower within the asthenosphere in the past &that the North American continent has changed size through accretion & collision. Nevertheless, assume the results apply equally to all time periods.]

D) Given the constants of the species-area equation for marine invertebrates, k =1.19, and z= 0.25, the data on the area of shallow shelf which was submerged during each geological period (Table 11.1) and the total area of shallow shelf (derived by reading Figure 11.10B), determine the species diversity for each period. [Plot the line that describes the species-area relationship using areas of 103 —106 sq km as x-axis divisions, and y-axis divisions of 1, 10, 10^2, and 10^3 species on log-log graph paper.]

11.6. Extinction and Regional Habitat Loss—A mass extinction occurred at the close of the Ordovician Period. It is often attributed to glacial cooling of Gondwanaland. Yet, with glaciation also comes reduced sea levels & potential habitat loss. Changing tectonic conditions may also create/destroy habitat. The number of habitats present in a large region (or "islands within islands") is also an important determinant of diversity. Figure 11.14A-B show paleogeographic/facies maps (Laboratory 9) for the Ordovician and Silurian of North America.

A) Compare the facies of the two periods. Which had more habitats? Which facies were lost at the close of the Ordovician?

B) To what can the loss be attributed?

C) Use the map scale to estimate the area of habitat loss/gain for each of the major facies.

D) Calculate the increase/decrease in equilibrium species diversity in each facies (k =1.19, z = 0.25) between the Late Ordovician and Early Silurian using Table 11.1 and Figure 11.10B.

E) Is diversity affected more by the loss/gain of a portion of one large habitat, or elimination of smaller habitats?

F) What happens to extinction rates as habitat area grows smaller?

CALCULATIONS

LABORATORY EXERCISES

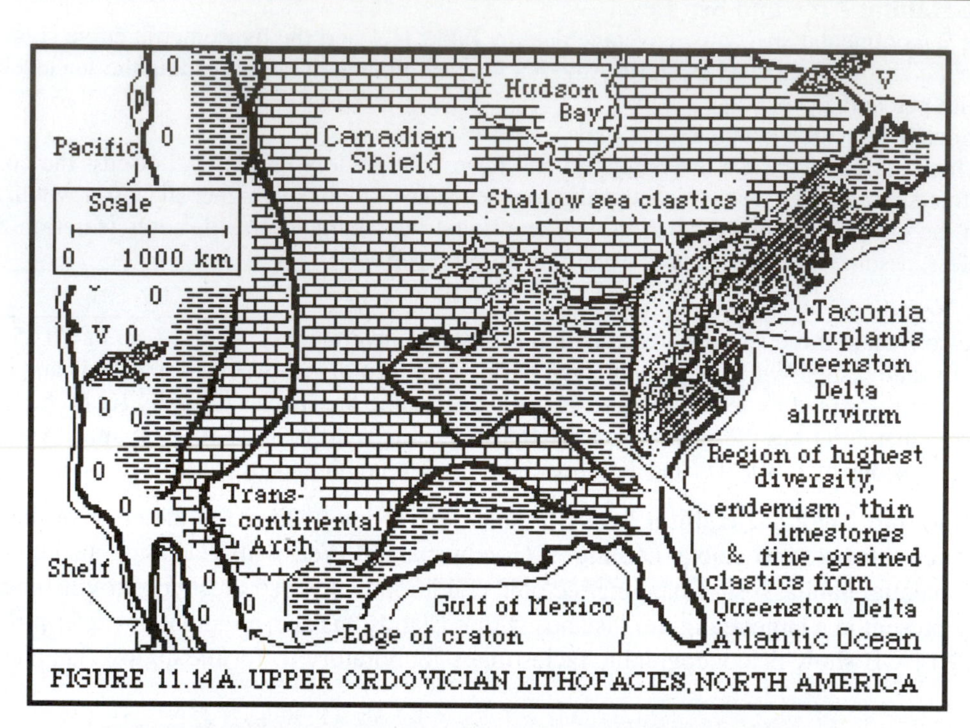

FIGURE 11.14A. UPPER ORDOVICIAN LITHOFACIES, NORTH AMERICA

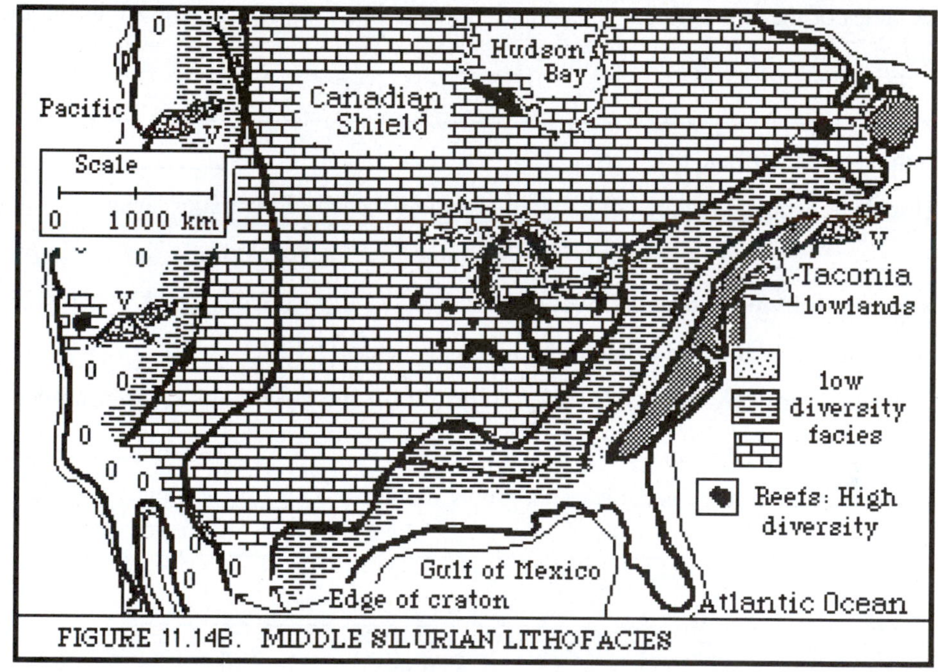

FIGURE 11.14B. MIDDLE SILURIAN LITHOFACIES

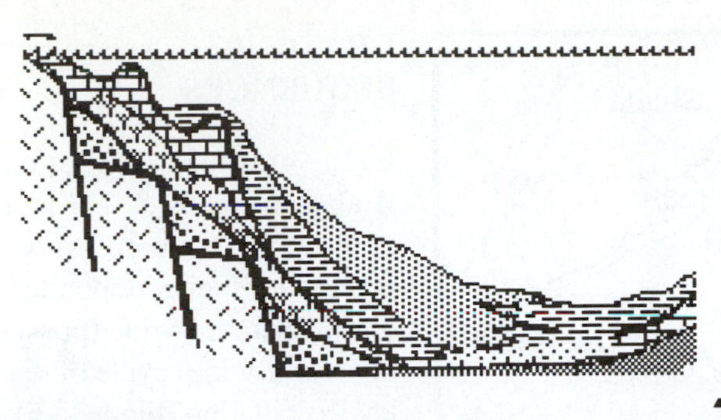

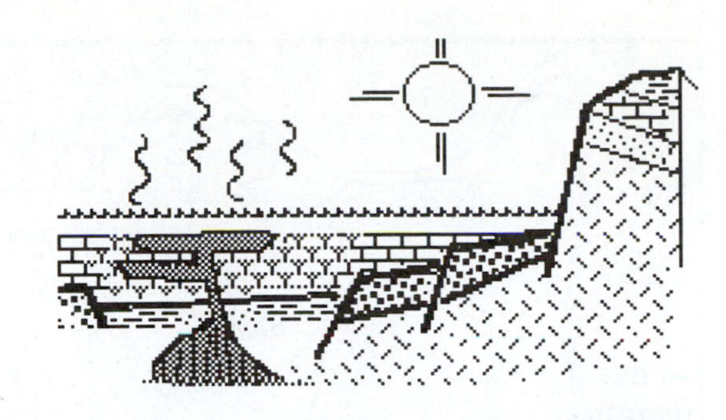

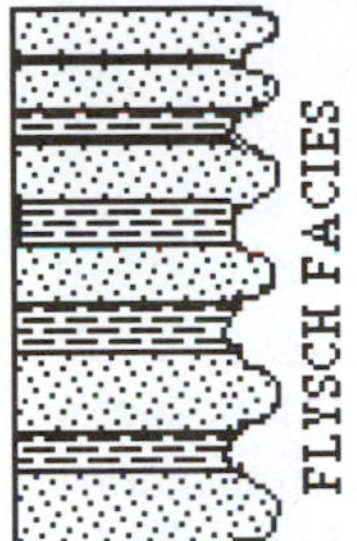

12
Plate Tectonics and Sedimentation

Objectives

To demonstrate mastery of this material you should understand: 1) how tectonic context determines sediment types; 2) why certain sedimentary sequences accumulate during tectonic evolution of continents and ocean basins; 3) how tectonic contexts of source and depositional sites affect the maturity of clastic sediments; and, given the above, 4) how to predict the sedimentary rock types and sequences that should be associated with certain tectonic settings. Colored pencils are needed.

Using This Laboratory

Familiarity with the fundamentals of plate tectonics and tectonic features (e.g., plate margin types, rift valleys, spreading ridges, etc.) is assumed. Pay close attention here to the tectonic contexts and features associated with the origins of sediment types and their sequences of accumulation.

Because plate tectonics cause continents to fragment and meld, mountains to arise and decline as continents grow about them, and ocean basins to gape, grow, shrink and close again, it must also affect the types and sequences of sediments formed in the process. In this laboratory we will see how sediment types and sequences have been affected by, and evolved with, plate tectonic movements. The exercises trace the changing character of sediments formed as ocean basins arise, expand, contract and expire, while continents conjoin and split assunder.

STABLE PLATFORMS

In the extensive stable platforms of continental interiors are sedimentary rock sequences reflecting tectonic quiescence: those portions of continents covered by "flat-lying" strata deposited in shallow inland seas little disturbed by tectonism (Figure 12.1). Widespread among these are the early Paleozoic sandstones of the North American craton lying in belts paralleling their ancient transgressive shorelines. Their exceptional maturity indicates many episodes of recycling. Widespread, shallow water limestones and dolostones also typify stable platforms. These are commonly formed far from uplifts and detrital sediments because tectonism increases erosional rates and, thus, causes deposition of voluminous detritus (Laboratory 8). Ocean currents commonly swept clays of future shales to the outer shelf and deep water continental slope and rise facies, while others accumulated within deltaic facies.

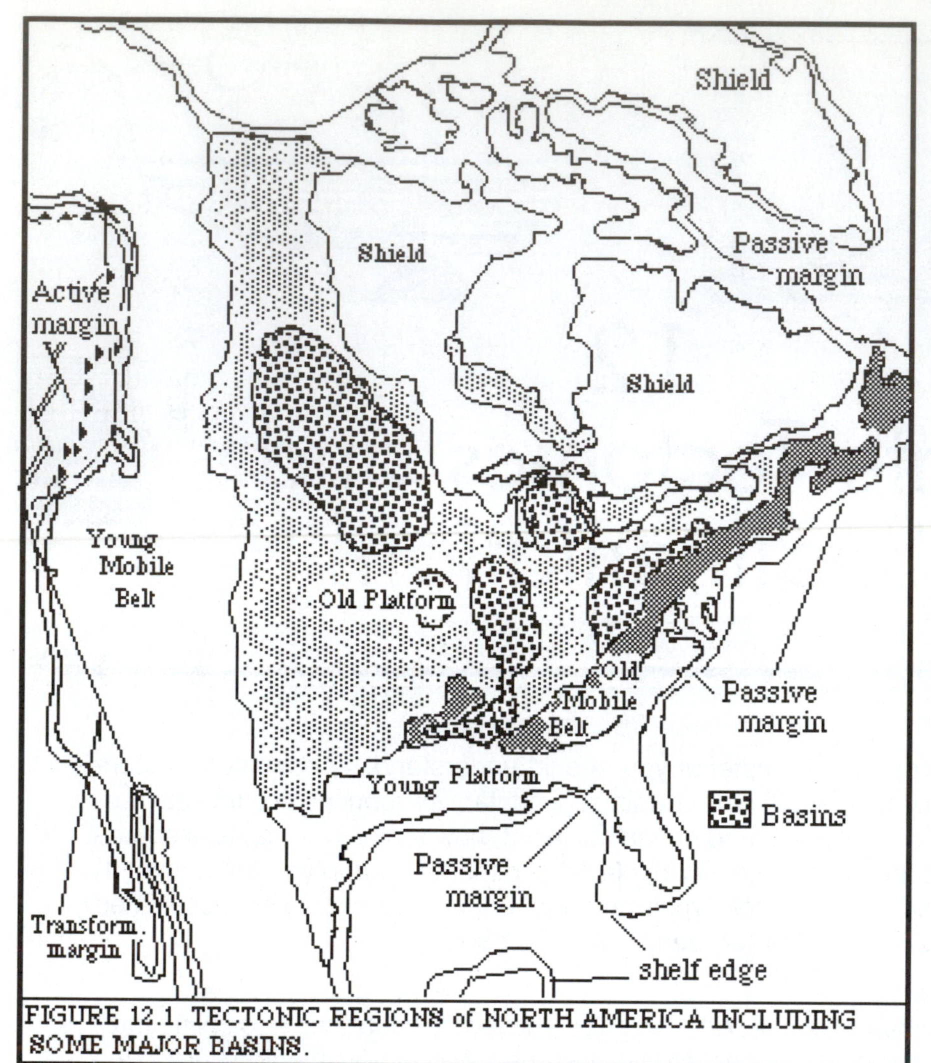

FIGURE 12.1. TECTONIC REGIONS of NORTH AMERICA INCLUDING SOME MAJOR BASINS.

CRATONIC SEQUENCES

Each cratonic sequence (Laboratories 8 & 11) is a couplet of transgressive-regressive facies deposited during a craton-wide (possibly world-wide) cycle of sea level oscillation (Figure 8.5). The cause(s) of these cycles are widely debated, and it is unlikely that a single cause explains all cycles (see text). One explanation suggests that changes in sea-floor spreading rates cause fluctuation of sea levels. When spreading rates are high, basalt extruded at spreading ridges remains hot longer than it would if spreading rates were lower (Figure 12.2A1). Hotter crust displaces more ocean basin water because its buoyancy causes the sea floor to stand higher than if the ridge had cooled rapidly. Similarly, regressions result from low spreading rates (Figure 12.2A2).

TABLE 12.1. TIME-TABLE FOR CONTINENTAL RIFTING 7 SEDIMENTATION

MY to SEPARATION	STAGE	RIFT TOPOGRAPHY	FACIES	DURATION
-30 my	Rifting	Elevated Rift Valley	Immature Clastics, Fluvial, Alluvial, Lake Black & Red Shales, other Red Beds, Flood Basalts.	10my
-20 my	Rifting-Incipient Infant Ocean Basin	Valley Floor < Sea Level	Lacustrine & Marine. Evaporites & Carbonates.	10my
-10 to +10 my	Infant Ocean Basin	2-3 km Depth 100-200 km Width	Extensive Marine Evaporites, Pillow Basalts.	20my
+10 to 30 my	Young Ocean Basin	200-600 km Width	Carbonates & Terrigenous Clastics.	20 my
+30 to 50 my	Submature-old (Late Closure)	Subsidence. Shelf, Slope & Rise	Terrig. Clastics, esp. Deltaic. Flysch, Turbidites, Pelagic.	20 my

Another tectonic mechanism, perhaps best illustrated by the low sea levels during Pangaea's assembly, is the vertical "lifting" of the continents from an accumulation of hot material insulated beneath, and by, a large "cap" of thick continental crust (Figure 12.2B).

Carbonates tend to be more widespread than detrital sediments during extensive transgressions because less craton is exposed to weathering and erosion: detrital sources are fewer, and carbonate rocks more widespread. More craton is weathered and eroded during regressions and, thus, immature clastics are more extensive than carbonates. Detrital sediments can be widespread and thick along platform margins interior to tectonically active uplands when transgressions are associated with orogenies.

CRATONIC BASINS

Minor deformation within basins has had important effects on local sedimentation by formation of two types of depressions or basins of gently folded strata: sedimentary basins and structural basins (Figure 12.3). These basins include any large, broad, low region within a continent or ocean into which sediments are shed, and thicken toward the interior (12.3A) to oval depressions of continental crust in which the strata have been warped or faulted downward to form a depression as sedimentation occurred (Figure 12.3B). Gentle downwarping of the crust and sedimentation occurred simultaneously in many basins, after faulting of deep cratonal rocks created a depression (Figure 12.4A,B). Most were shallow, persistent catchment areas which accumulated thick piles of sediment.

How can basin formation so distant from the tectonism which produces mobile belts be explained by plate tectonic theory? Most North American cratonic basins (e.g., Michigan Basin; Figure 12.1) active throughout the Phanerozoic were created by Late Proterozoic or Early Cambrian faulting accompanying early stage rifting of a proto-Pangaea super-

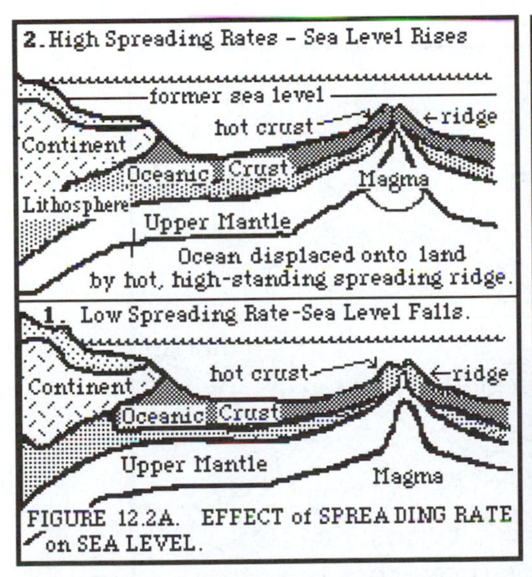

FIGURE 12.2A. EFFECT of SPREADING RATE on SEA LEVEL.

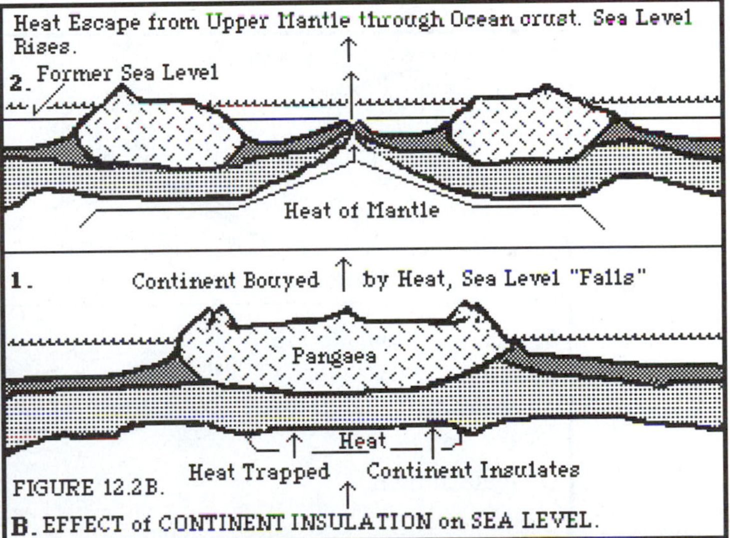

FIGURE 12.2B. B. EFFECT of CONTINENT INSULATION on SEA LEVEL.

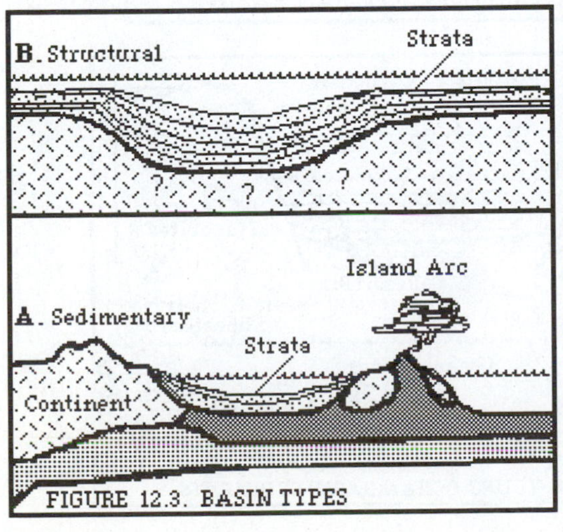

FIGURE 12.3. BASIN TYPES

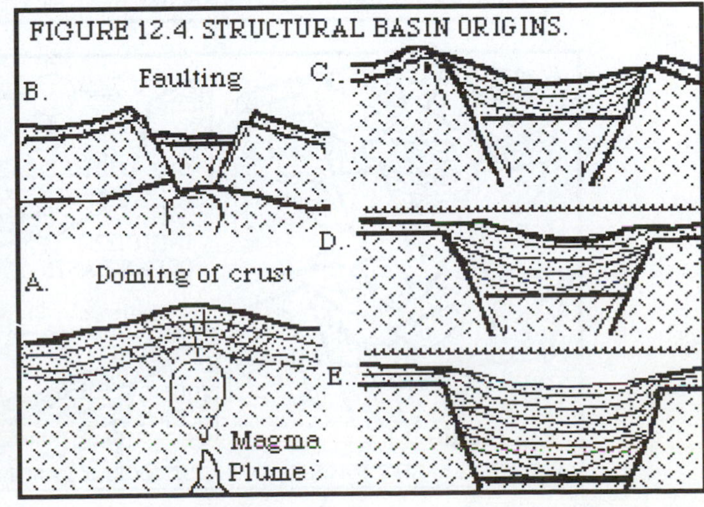

FIGURE 12.4. STRUCTURAL BASIN ORIGINS.

continent. Here rifting stopped after faulting dropped a portion of the crust downward, leaving a trough in which sediments accumulated (Figure 12.4C, D). When deformation resumed, underlying strata warped and/or faulted again, and new sediments accumulated.

RIFT ZONE SEDIMENTS

Ocean birth begins with continental rifting when large masses of hot mantle rock rise from depth toward the surface. One mechanism or cause of rifting is the development of a hot spot within continental crust. As a hot spot is born, overlying crust domes upward and radiating fractures open and extrude basalts (Figure 12.4A, 5A). When the crust domes and fractures an aulacogen or set of three (or more) centrally joined fracture zones, oriented at ≈120 ° angles, commonly form (Figure 12.5B, C). If rifting leads to sea floor spreading in each zone and thus, development of divergent plate boundaries, three plates will form. One rift commonly fails, however, and becomes a sedimentary trough while spreading begins in the active rifts. The rift margins uplift, while center falls, forming a rift valley in which sediment accumulates to a great depth (Figure 12.5D, 6). Sea floor spreading produces a new ocean basin continental margins that have irregular outlines—alternating promontories and embayments with a jig-saw puzzle match across the new ocean (Figure 12.5D, E). Many of the embayments hold buried rift valleys (Figure 12.4C-E, 12.6A,B).

Different stages in aulacogen development (Figure 12.5, 6) are seen in northeastern Africa: an enormous fracture zone, the East African Rift Zone—stretches from the Ethiopian Rift south toward Madagascar; the incipient ocean or Red Sea, and the young ocean or Gulf of Aden. The rift zone may be a failed rift, but some spreading has occurred since the Miocene.

The succession of sediment types developed in rift zones reflects their evolution from a hot spot, to rift valley, to narrow ocean basin, and finally, to a faulted continental shelf on the margin of a deep basin (Table 12.1; Figure 12.5, 6). Sediments first

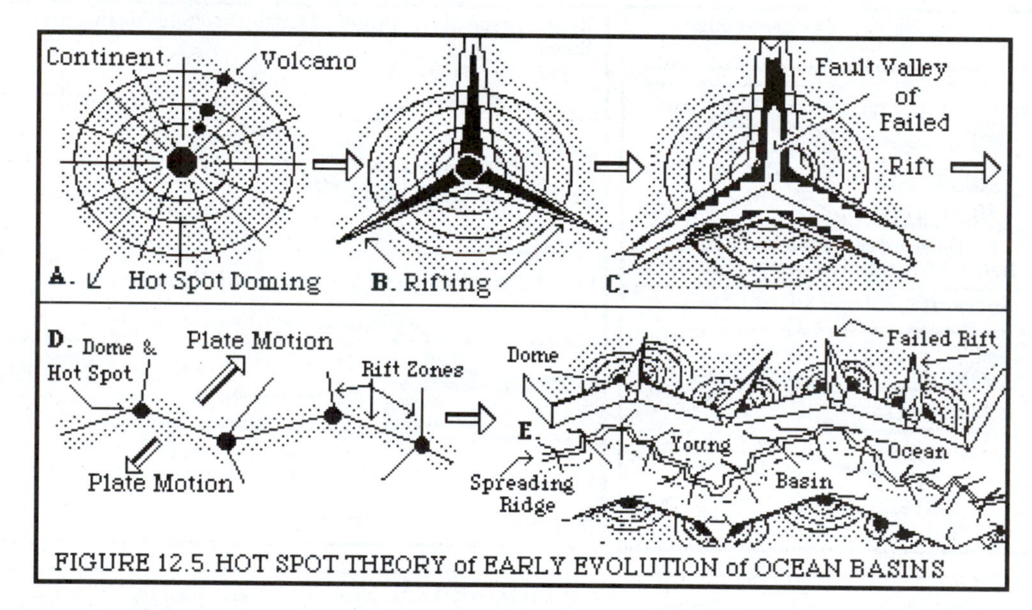

FIGURE 12.5. HOT SPOT THEORY of EARLY EVOLUTION of OCEAN BASINS

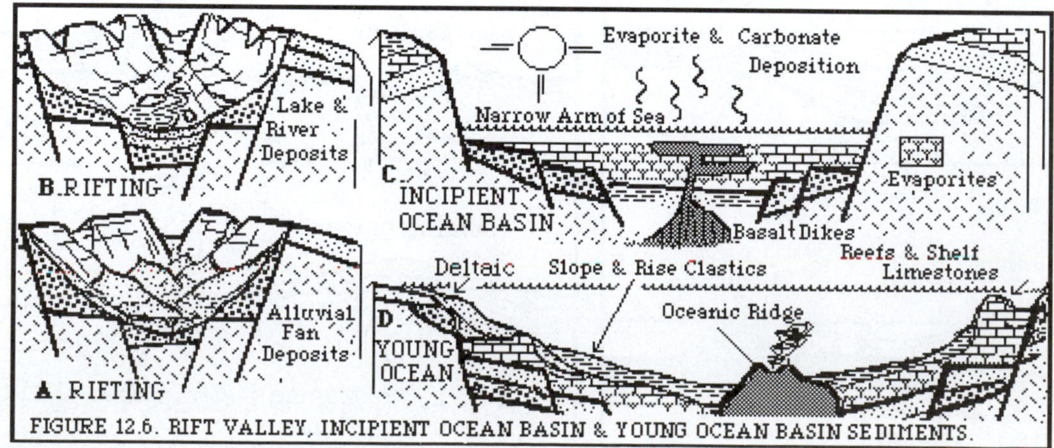

FIGURE 12.6. RIFT VALLEY, INCIPIENT OCEAN BASIN & YOUNG OCEAN BASIN SEDIMENTS.

deposited in the rift (failed or active) are terrestrial facies (red shales, immature sandstones, conglomerates and breccias of lakes, swamps, rivers, and alluvial fan facies), interbedded with flood-basalts and other volcanic rocks (Figure 12.6A). Organic carbon accumulated in lake facies becomes a source of large oil reserves. With further rifting, the margins and valley floor cool and subside . The valley floor sinks beneath sea level (e.g., Dead Sea; Figure 12.6B). As the continents separate by sea floor spreading, the rift is flooded and a large landlocked sea or infant ocean basin, 100 to 200 km (60-120 mi) wide develops (e.g., the Red Sea). Here thick evaporites form over the next 20 MY (Figure 12.6C). After about 30 MY, the new continental margins cool and subside (Figure 12.6D). Due to subsidence, the direction of drainage reverses and rivers flow toward the young continental margin. Failed rifts oriented at a high angle to the new coastline develop into the axes of major drainage systems that terminate in deltas (Figure 12.5E).

PASSIVE MARGINS of CONTINENTS

Some 10 to 30 MY after separation, the young margins of continent's margins—now some 200-600 km (120-360 mi) distant from the divergent boundary—sink below sea level, becoming tectonically inactive continental shelves or passive margins (Figure 12.7). In this phase the basin's floor sediments, clastics and evaporites of the rift valley and proto-ocean stages are buried by detrital sediments and/or reef and other carbonate sediments.

From 30 to ≥50 MY after separation the sedimentary sequence resembles that of the present Atlantic Ocean. With further subsidence, sedimentation begins on the developing shelf, slope and rise (Figure 12.7). Large deltas develop where major river systems meet coasts, and detrital coastal plain facies (Laboratory 8) form on the inner shelves. Without tectonism, little detrital sediment is shed from the low, gently sloping coast, and carbonates may be deposited in their stead. Thick carbonate deposits may form if accumulation keeps pace with sea-floor subsidence.

From the outer shelf margin toward the spreading center lay continental slope, rise and abyssal plain facies (Figure 12.8). These sediments originate as fine grained material swept to the outer shelf and farther by currents from the inner shelf; as turbidites (the detrital muds, silts and sands, and biological sediments deposited by turbidity currents) from the continental rise (Laboratory 2)); and as pelagic sediments (below) including red wind-blown dust and protistan tests (Appendix I) that slowly settle through the water column. Pelagic sediments alternating with turbidites comprise the flysch facies typical of the continental slope and rise (Figure 12.8, 12.11).

PELAGIC SEDIMENTS

Pelagic sediments include eolian desert, volcanic, and cosmic dust, carbonate muds, and calcite and silica protistan tests deposited in (usually) deep water lacking other terrigenous clastics. These commonly accumulate in deep water on abyssal plains and mid-oceanic ridges, far from continental shelves (Figure 12.9) and, thus, are most widespread in fully mature oceans. Because much pelagic sediment is of biological origin, its distribution is controlled by depth-related factors influencing calcite accumulation and the locations of nutrient-rich upwelling water.

Carbonate distribution in the deep ocean is controlled by the carbonate compensation depth (CCD).

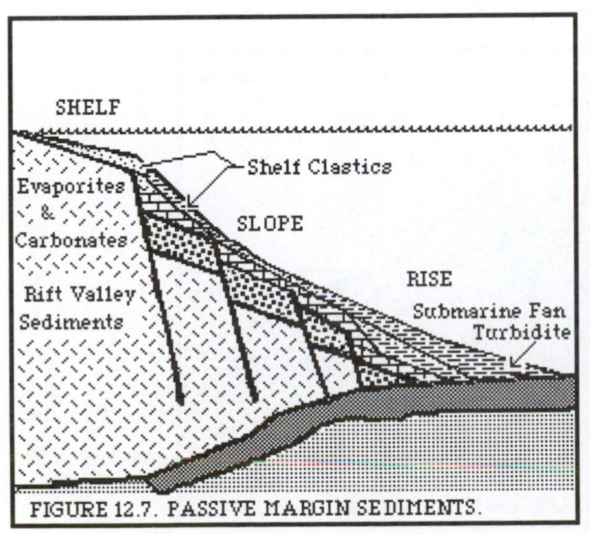

FIGURE 12.7. PASSIVE MARGIN SEDIMENTS.

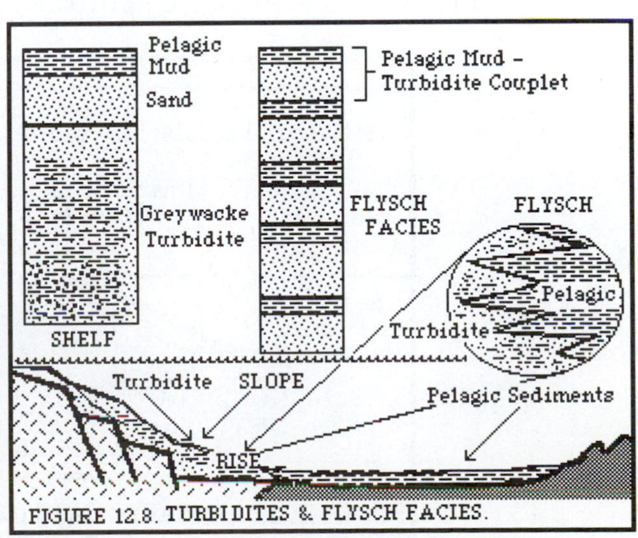

FIGURE 12.8. TURBIDITES & FLYSCH FACIES.

The CCD is the level at which calcite/aragonite is dissolved by CO_2 as fast as it is supplied by the sinking tests of protists. The CCD ranges from 3.5 km to 5.5 km (2.1-3.3 mi) in depth (Figure 12.10). Carbonate-rich clay and calcareous oozes are most abundant above the CCD, while pelagic clays and siliceous oozes are most abundant at below the CCD or where carbonates are otherwise lacking.

Pelagic sediments show latitudinal zonation reflecting biological productivity, sedimentation and climate (Figure 12.9). Calcareous oozes are most abundant in equatorial waters where biological productivity is high. Siliceous sediments are most common within 30 °N/S of the equator, along coasts where upwelling occurs, and in subpolar seas. These become beds of chert when lithified. Red desert eolian, volcanic and cosmic dust are trademarks of deep ocean facies and accumulate where biological sediments are lacking. Glacial sediments and ice-rafted till are found in deep polar seas.

The sequence of sediment accumulation resulting from sea-floor spreading and its effect on the ocean crust depth with increasing age and distance from the spreading ridge is predictable (Figure 12.11). The crust extruded at ridges consists of an upper basaltic- and a lower ultramafic layer (peridotite and serpentinite (see text)). The upper portion consists of pillow basalts and is underlain by basalt lavas, sheet-like walls of basalt dikes, and, at the base, gabbro (Figure 12.10). Some portions of the midoceanic ridges are without much sediment cover, while others are partially buried. The ridge system everywhere lies at a depth of 2.6 km ± 200 m, well above the CCD and, thus, within the range of limestone deposition. Thick deposits of iron-rich clay form near active hydrothermal vents.

As spreading proceeds, newly created ocean crust

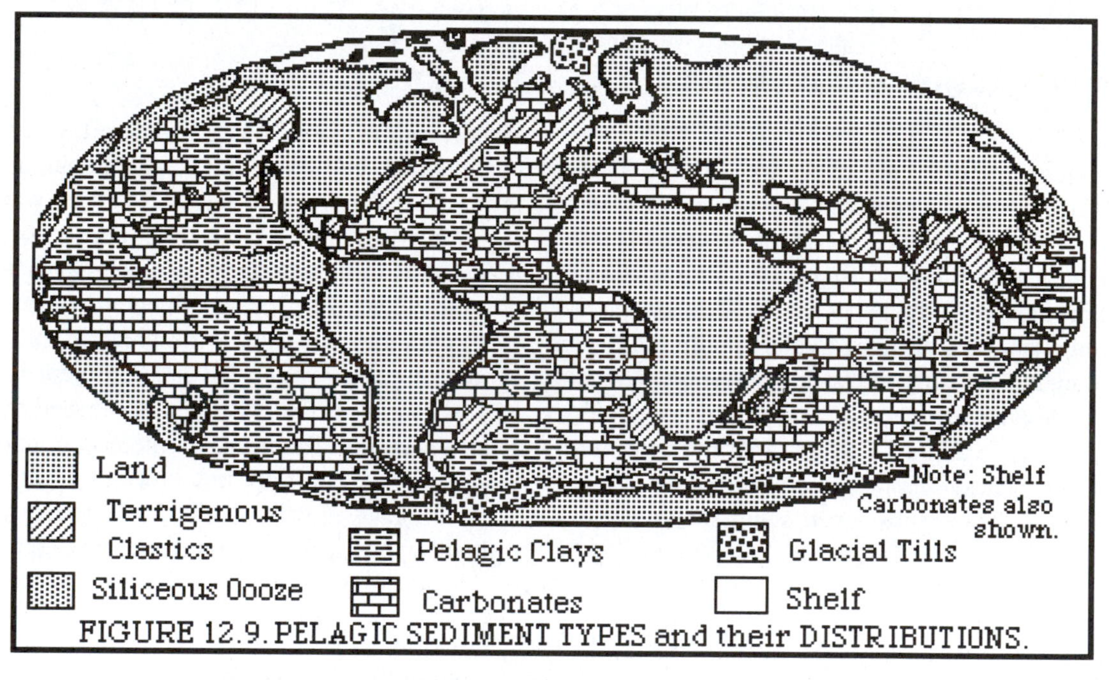

FIGURE 12.9. PELAGIC SEDIMENT TYPES and their DISTRIBUTIONS.

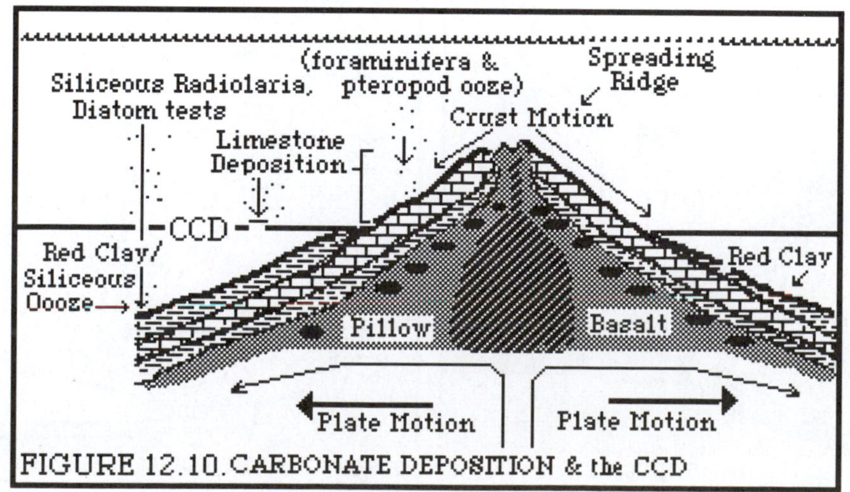

FIGURE 12.10. CARBONATE DEPOSITION & the CCD

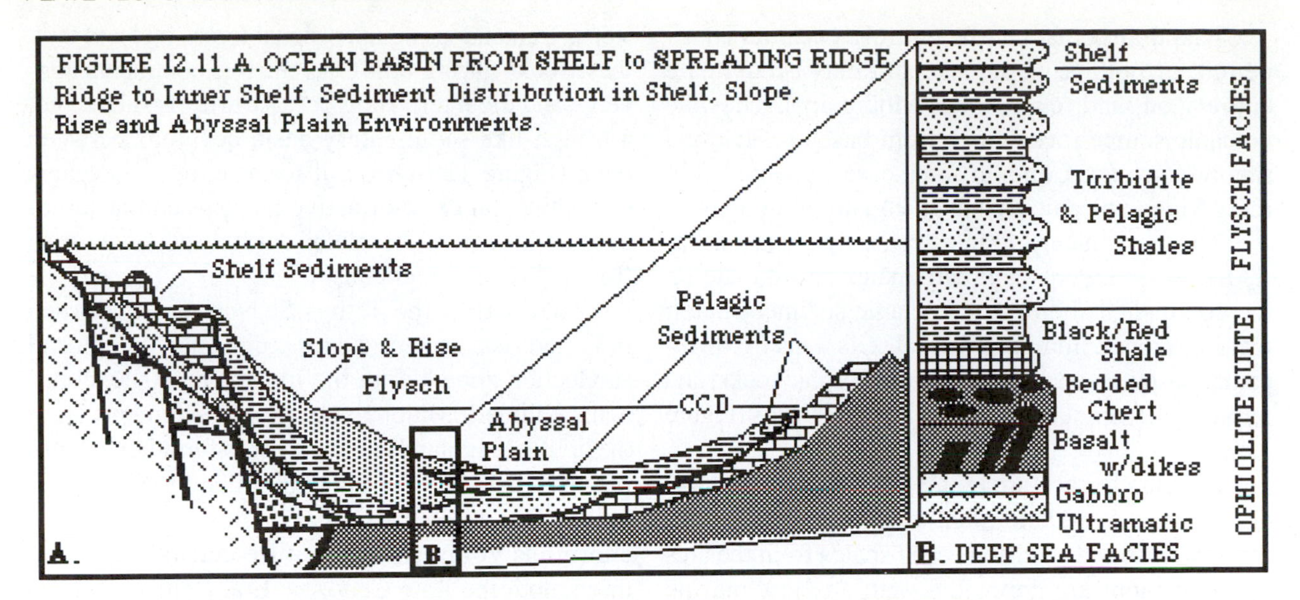

FIGURE 12.11. A. OCEAN BASIN FROM SHELF to SPREADING RIDGE. Ridge to Inner Shelf. Sediment Distribution in Shelf, Slope, Rise and Abyssal Plain Environments.

moves away from the ridge, cools, contracts and subsides, creating an ever-deepening basin. As the crust sinks, the sea floor enters increasingly deep water and descends below the CCD (Figure 12.10, 12.11A). There deep water red clays, siliceous oozes and black clays accumulate above the shallower water sediments accumulated earlier. Pelagic sediments are overlain by turbidites in regions near the continental rise where ocean crust is oldest (Figure 12.11B).

The complete ridge-rise sedimentary sequence (base-top/youngest-oldest) is: ocean crust, Fe-rich clay and limestones (above CCD), red and black shale and bedded chert (pelagic, below CCD), and graywacke of turbidites. The lower sequence from crust to pelagic facies is known as an ophiolite suite (Figure 12.11B). Notice that the vertical succession of facies in an ophiolite suite is not equivalent to the lateral distribution of facies. It does not follow Walther's Law (Laboratory 8) because strata accumulate on the ocean crust as it spreads, ages and descends, rather than on the continental crust as sea level rises or falls.

ACTIVE MARGIN SEDIMENTS

Continental margins bounded by subduction zones, island arcs and back-arc basins, (i.e., along convergent plate boundaries) or, zones where one plate moves laterally past another are termed active margins. We will consider only active continental margins of convergent plate boundaries involving continent-continent collision, continental-oceanic plate subduction and island arc formation. Sediments produced in these settings belong to mobile belts: elongate tracts of deformed, faulted rock characterized by frequent and intense earthquake and volcanic activity, and mountain building. First let us consider sedimentation within ocean trenches and regions adjacent to island arcs formed by plate convergence.

Island Arc Basins—Trenches form the deepest parts of the ocean (e.g., Marianas Trench east of Philippine Islands, ≥ 11 km (6.6 mi) depth), and are the sites of subduction (Figure 12.12). Between trenches and continent platforms lie arcuate volcanic mountain chains called island arcs formed from the partial melting of subducted crust.

Between island arcs and continent lie back-arc basins (e.g., South China Sea, Sea of Japan; Figure 12.12). Sediments here come from volcanic islands and the continental margin, and, thus, are most variable in composition and textural maturity.

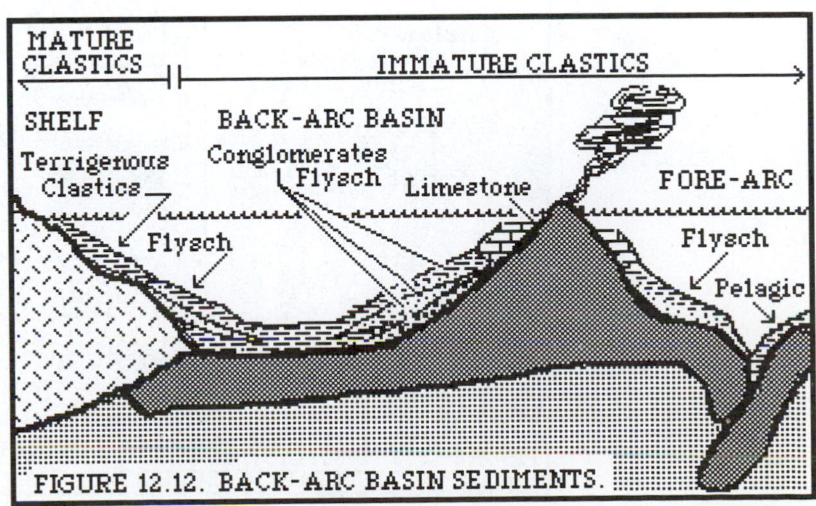

FIGURE 12.12. BACK-ARC BASIN SEDIMENTS.

Island arc crust can be entirely oceanic, continental, or both (e.g., Japan), and may accumulate entirely on land (e.g., Andes Mountains). Thus the volcanic source-rocks range from basalt to andesite, rhyolite (see text), and, in some cases, granite. Atoll reefs fringe volcanic islands in equatorial regions.

The more mature, granitic coastal plain and deltaic detrital sediments and carbonates accumulate on the continental shore, while pelagic sediments, atoll reef limestones, immature detrital sediments from volcanic (basaltic-andesitic) and metamorphic rocks, and turbidites accumulate on the island arcs' flanks (Figure 12.12).

Eventually back-arc basins fill with sediment and are uplifted with their island arcs (Figure 12.13A). Now, well sorted sandstones and shales from the arcs and continent are deposited with shallow marine carbonates, followed by nonmarine facies including coals, fresh-water shales (often red), coarse cross-bedded sandstones, and conglomerates. This shallow marine to nonmarine deltaic and alluvial fan facies is called molasse (Figure 12.13B). The entire ocean basin sequence then proceeds from ocean crust and mature deep-water sediments of an ophiolite suite, through immature turbidites and graywackes of flysch facies, to mature shallow marine and finally, immature nonmarine sediments known as a molasse facies (Figure 12.13B).

Intercontinental Basins—These basins form as an ocean is narrowed, its crust subducted and continents verge on collision. Because continental crust cannot be entirely subducted, only a portion of the advancing plate thrusts beneath the other and subduction stops. Tremendous compressional forces fold and fault these rocks during uplift. As the two continents converge and the interposing oceanic crust is subducted, a trough-like sedimentary basin develops between them (Figure 12.14A-C). It forms either on oceanic or continental crust of active margins and accumulates an immense volume of detrital sediments shed from the rising mountains.

That part of the Tethys Seaway once between India and Asia became a deep sedimentary trough and subduction zone before the Indian and Asian continents collided. Major rivers carried sediments from the uplifted highlands to the coast to form large deltas (Figure 12.14B). Continental collisions forced the uplift of so much crust, that the highlands shed an enormous volume of immature detrital sediment throughout the Late Cenozoic Era. Deltaic deposits and other molasse facies today reach the outer shelf where they grade into the fan turbidites and flysch facies of the continental slope and rise. These grade laterally into pelagic sediments of the abyssal plain. (Figure 12.14B). As the continents converge and the trough shallows, molasse facies are deposited over deeper water flysch facies. Deposition continues in a nonmarine trough adjacent to uplift (Figure 12.14C).

VOCABULARY

Active continental margin, Aulacogen, Back-arc basin, Basin, Carbonate compensation depth, Convergent plate boundary, Craton, Divergent plate boundary, Flysch, Hot spot, Island arc, Mobile belt, Molasse, Ophiolite suite, Passive margin, Pelagic sediments, Pillow basalts, Rift valley, Rift Zone, Shield, Spreading zone, Spreading ridge, Stable platform, Subduction zone, Trench.

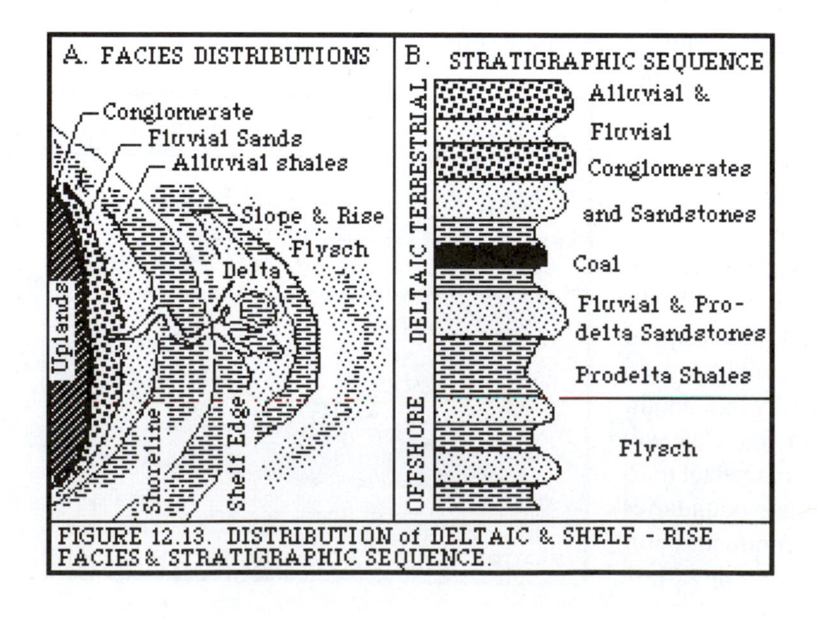

FIGURE 12.13. DISTRIBUTION of DELTAIC & SHELF - RISE FACIES & STRATIGRAPHIC SEQUENCE.

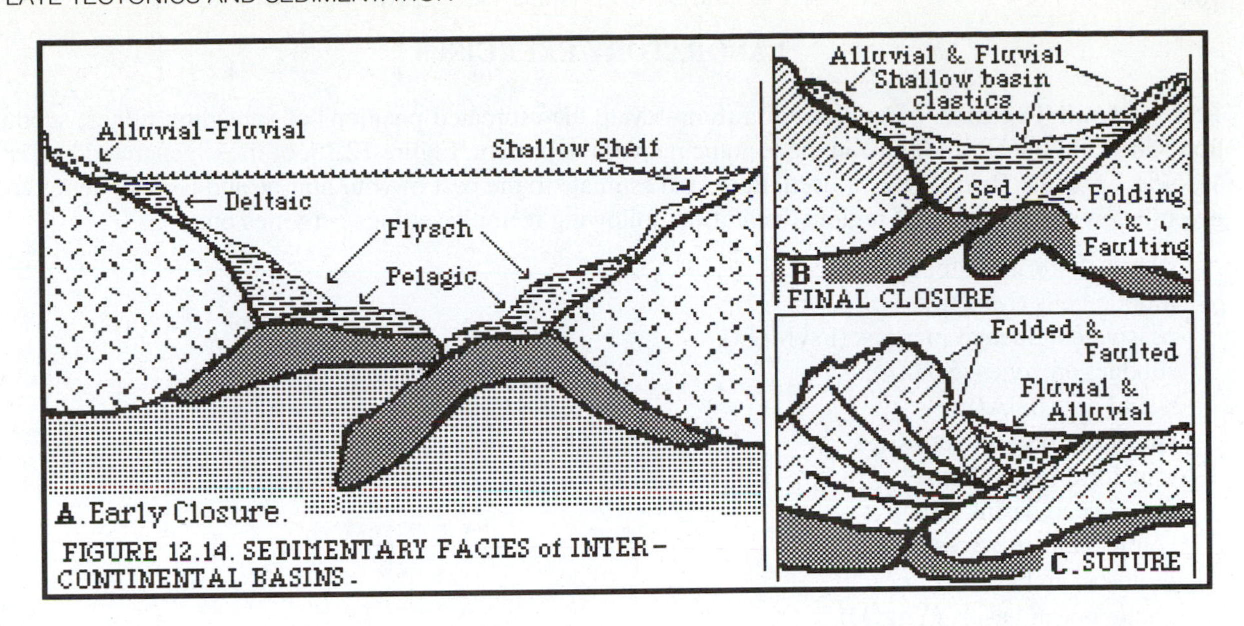

FIGURE 12.14. SEDIMENTARY FACIES of INTER-CONTINENTAL BASINS.

LABORATORY EXERCISES

12.1. Tectonic Features. Map Interpretation—Map the estimated positions of spreading ridges, subduction zones and transform faults on the tectonic map of the present, Figure 12.15. or those generated by *Terra Mobilis*, or on the *Drift Globe*, if used. Infer, and estimate to the best of your ability, and label by letter code given below the locations of regions with/of the following tectonic settings/activities on:

stable platforms (Stbl Pltf);
mobile belts (MblBlt);
passive continental margins (PsvMrgn);
subduction zones (SbdZn);
island arcs (IslAr);
spreading ridges (SprdRdg);
transform faults (TrnsFlt);
deep ocean basins (DpOc);
continental rift zones (CR);
proto-ocean basins (PrtoOc);
young ocean basins (YngOc).

A) Map the estimated positions of spreading ridges, subduction zones and transform faults on each paleogeographic map (Figure 12.15A-E; present, Eocene Epoch, Late Cretaceous, Late Jurassic, Permo-Trassic, respectively). Your instructor may have you work in groups, each responsible for one or two of the four maps of the past, and/or may have you complete the exercise using the *Drift Globe*.

B) First label the tectonic features described above on the map of the present. You may use your textbook to begin this exercises. Secondly, determine the distributions of these features in the past (B-E) by using the consistency of plate motion directions and tectonic features between periods as a guide to your reconstruction, and working backward from the present. Note that inference of spreading ridge locations becomes more speculative with age. Keep your inferences as simple as possible. Note that spreading ridges and subduction zones may be linked by transform faults.

R = Reefs, Rb = Red beds, C = Carbonate rocks, E = Evaporites, ● = Coal

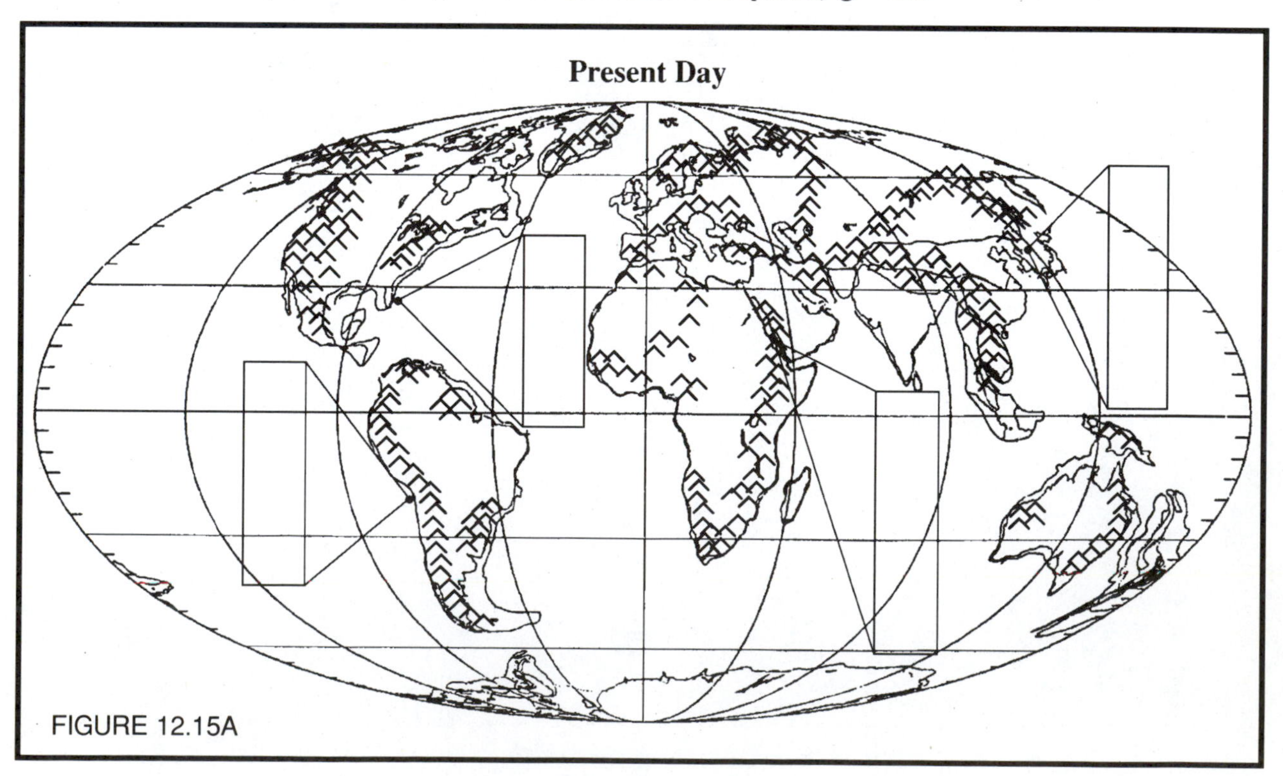

FIGURE 12.15A

Uplands **Lowlands** **Shallow Seas**

R = Reefs, Rb = Red beds, C = Carbonate rocks, E = Evaporites, ● = Coal

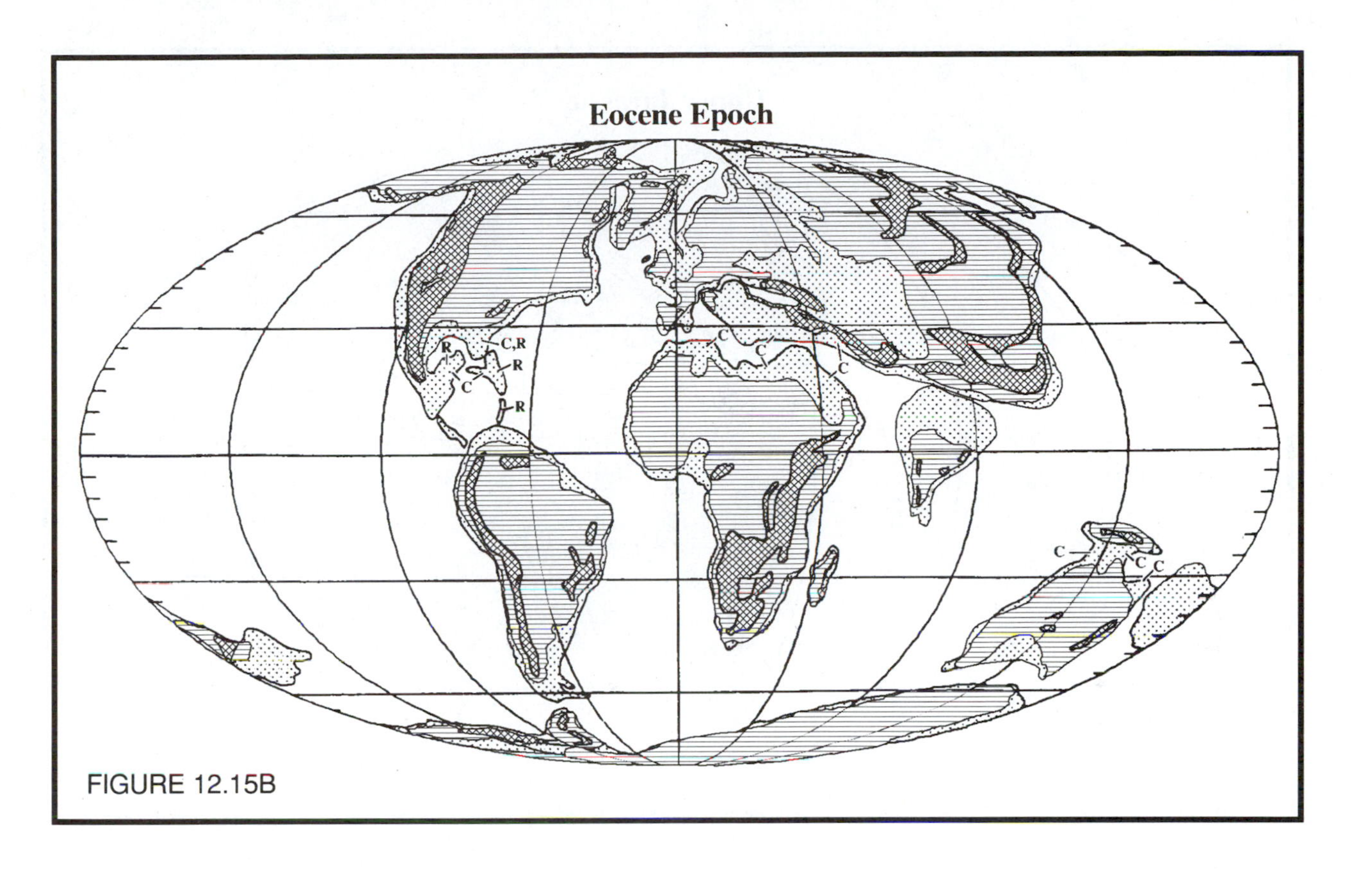

FIGURE 12.15B

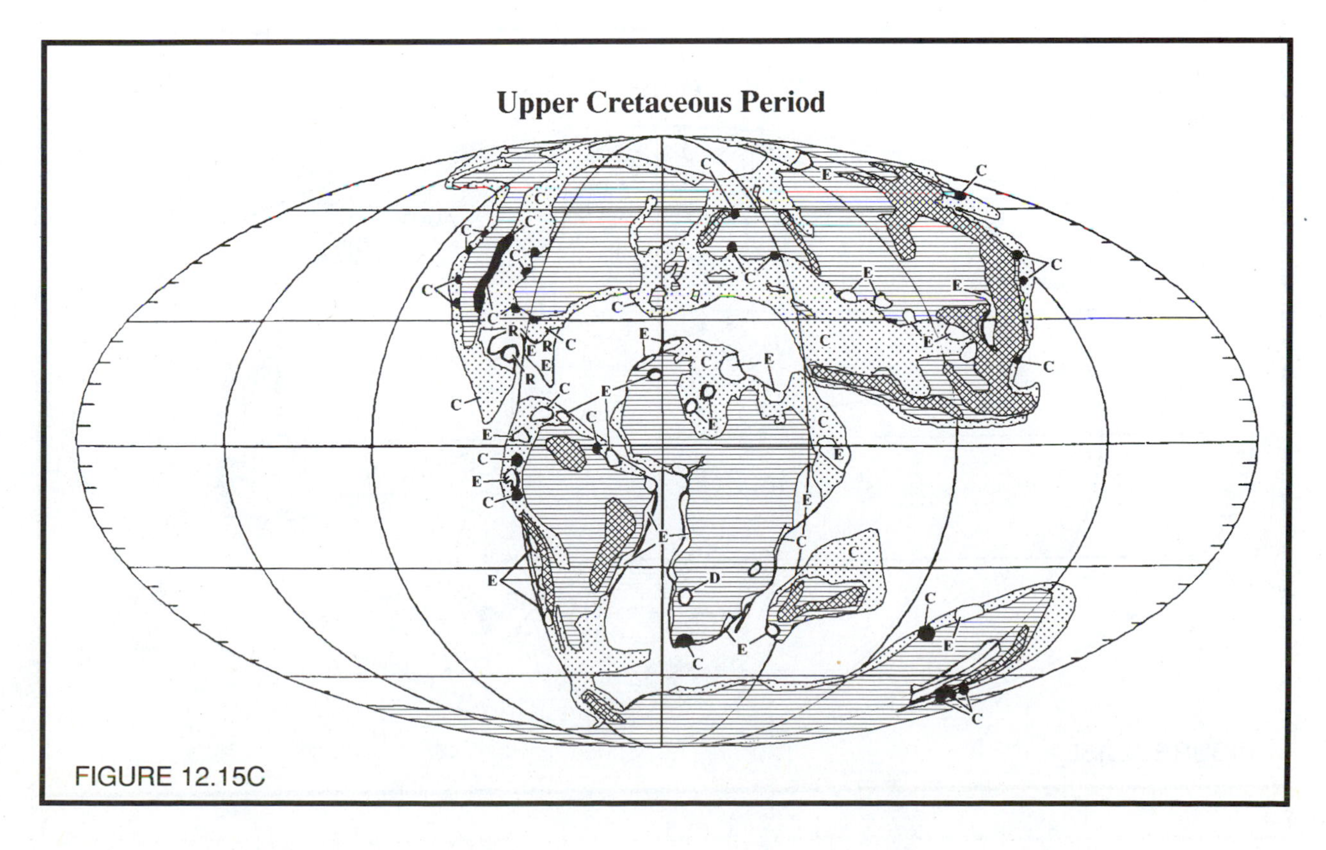

FIGURE 12.15C

Uplands Lowlands Shallow Seas

R = Reefs, Rb = Red beds, C = Carbonate rocks, E = Evaporites, ● = Coal

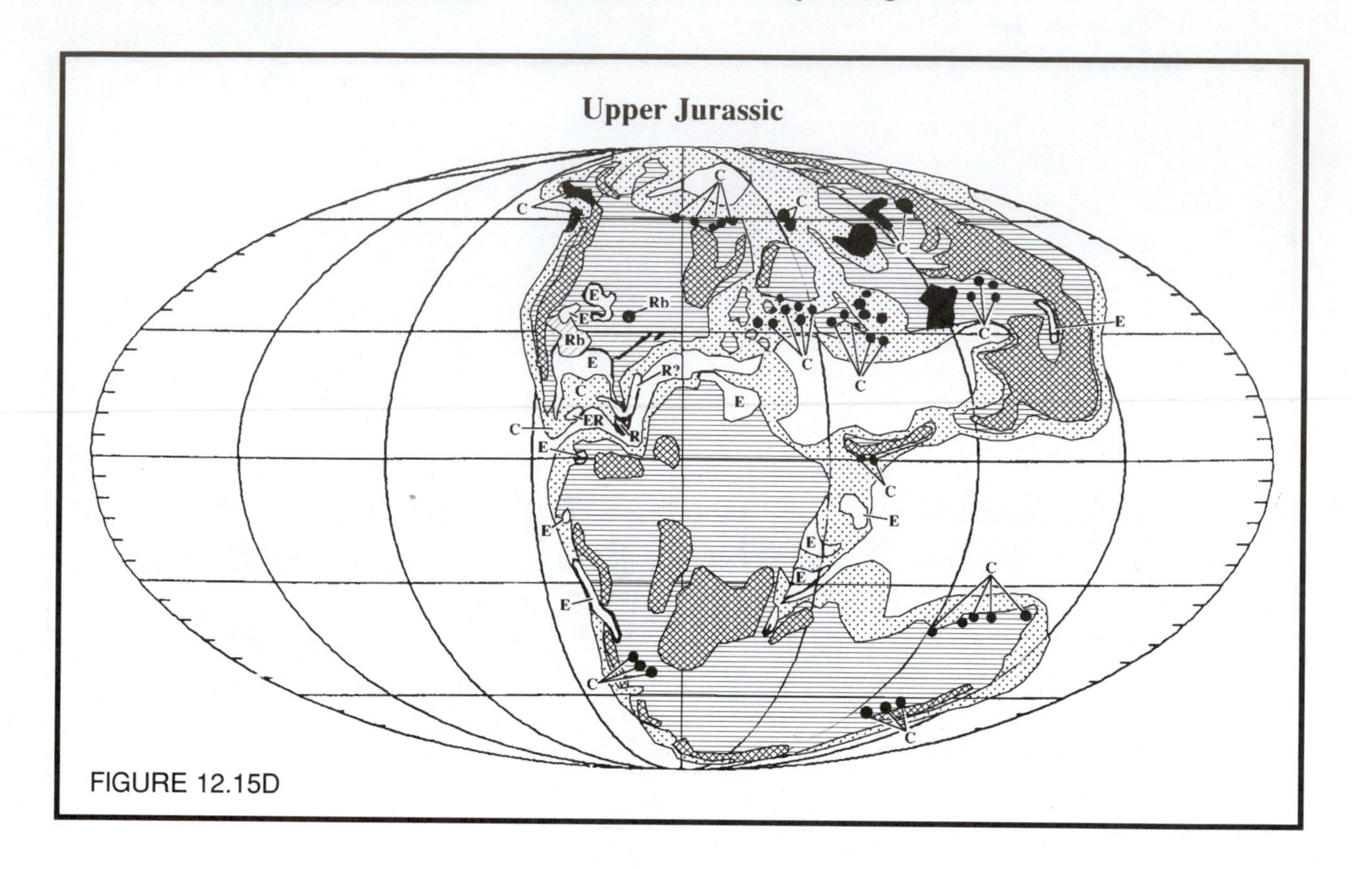

FIGURE 12.15D

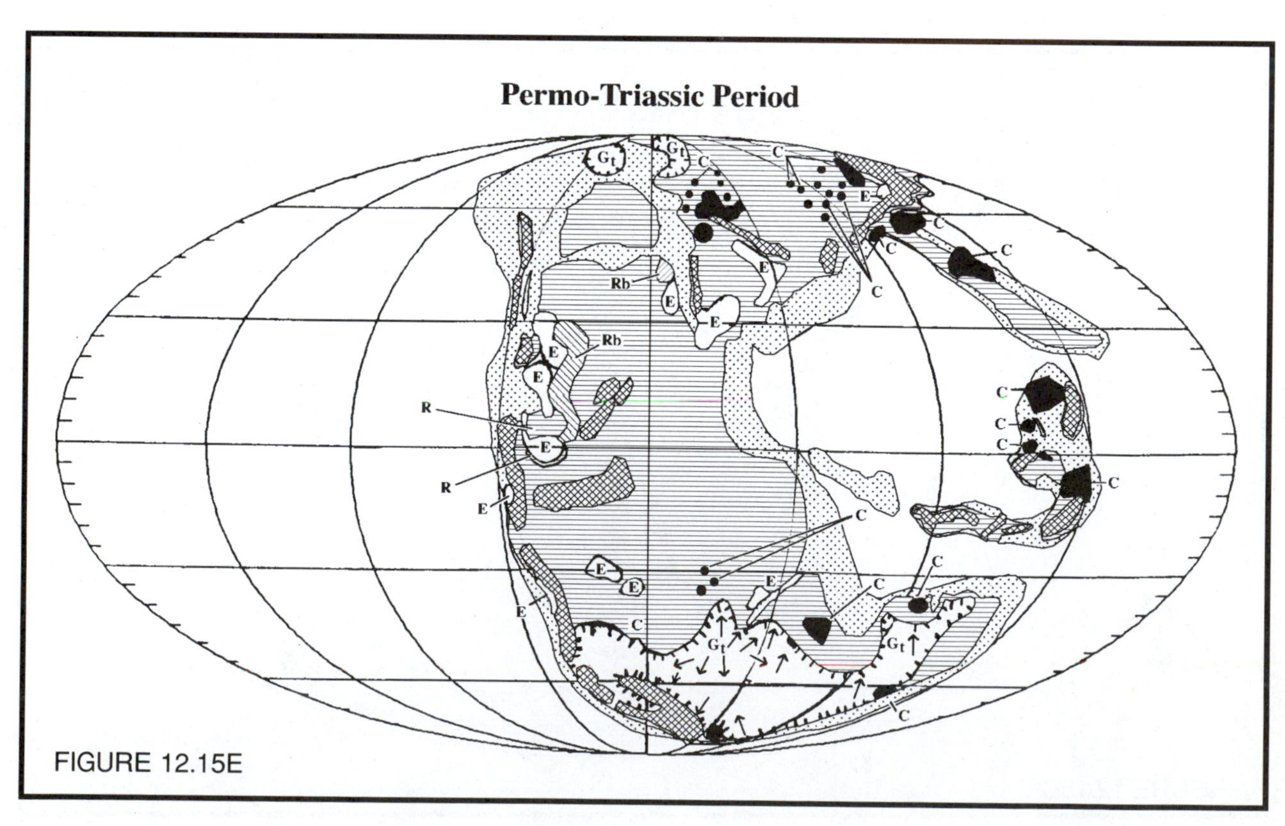

FIGURE 12.15E

12.2. Tectonic Control of Present Sedimentation—Code by letter as shown below & color-code according to your instructor's direction, the shelf and deep ocean basin regions on the map of the present, Figure 12.15A, the regions in which the following tectonically controlled facies sequences should be found:

Ophiolite (Op);
Pelagic carbonate (Pc), siliceous (Ps), clay (Pcly) & till (Pt);
Flysch (F);
Molasse; Shallow marine carbonate/clastic (Sc/c); deltaic (D);
Terrestrial alluvial (Talv), lacustrine (Tlcst), fluvial (Tflv);
Evaporites (E);
Volcanics (V).

Uplands, lowlands, shallow platform and shelf seas and ocean basins are delineated by keyed patterns.

12.3. Tectonic Control of Past Sedimentation—Code by letter as shown below & color-code according to your instructor's direction, the shelf and deep ocean basin regions on the paleogeographic maps (Figure 12.14B-E) the regions in which the following tectonically controlled facies sequences should be found:

Ophiolite (Op);
Pelagic carbonate (Pc), siliceous (Ps), clay (Pcly) & till (Pt);
Flysch (F);
Molasse; Shallow marine carbonate/clastic (Sc/c);
Deltaic (D);
Terrestrial alluvial, (Talv) lacustrine (Tlcst), fluvial (Tflv);
Evaporites (E);
Volcanics (V).

Uplands, lowlands, shallow platform and shelf seas and ocean basins are delineated by keyed patterns.

LABORATORY EXERCISES

12.4. Locating Rift Zones—Triple junctions link a network of over 100 rifts and aulacogens around the margins of the Atlantic Ocean alone; all formed during Pangaea's break-up. Attempt to predict the locations some of these aulacogens. Some lie on the continental shelves and may be inferred from a map of the ocean floor (see textbook). Recall the discussion of aulacogen formation and rifting, and see Figures 12.14A-E for locations of associated features and regions of post-Jurassic sedimentation. Paleogeographic reconstructions of Triassic—Eocene, and/or *Drift Globe* or *Terra Mobilis* simulations of plate motions will be of aid in determining the sites and order of rifting, and possible aulacogen locations.

A) Locate probable aulacogen sites on the reconstruction of Pangaea, Figure 12.16.

B) List below the change in sediment type, sediment properties, and facies formed along/within the rifting/spreading zone from rifting to 30 MY after continental crust separation.

C) Where would you look for the following and why?

1) Subsurface deposits of evaporites? ______________________________

____

2) Organic carbon in shales? ______________________________

____

3) Petroleum? ______________________________

____

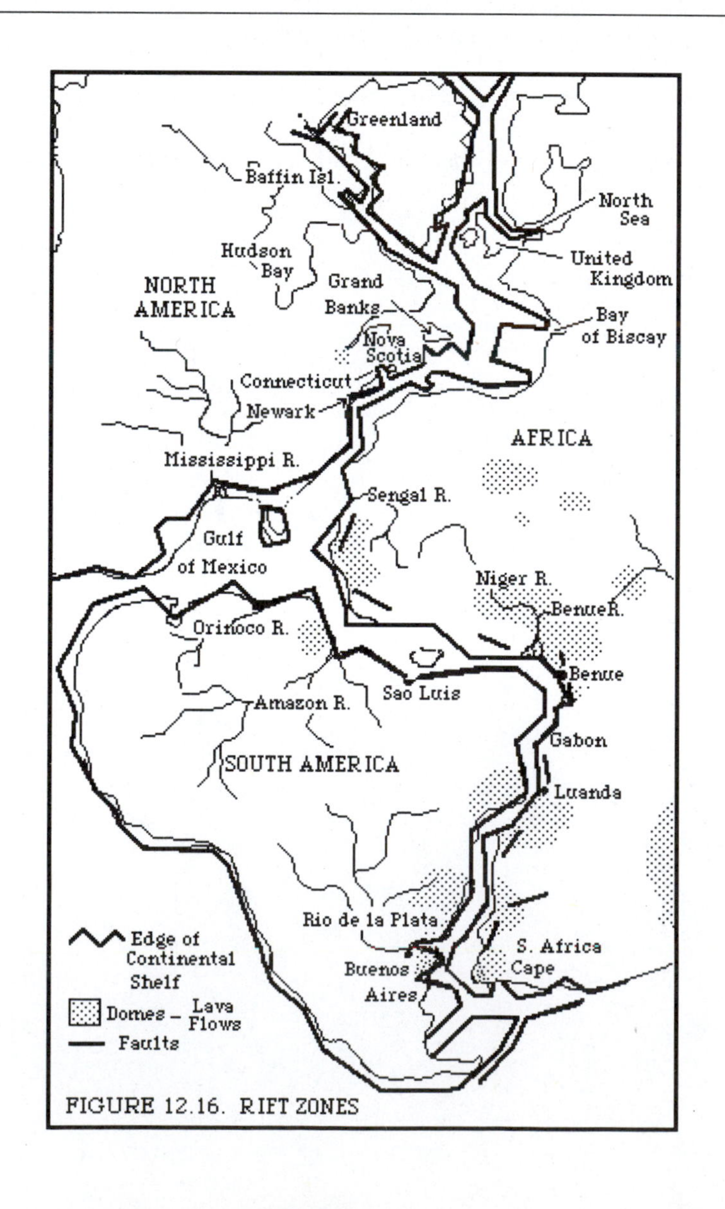

FIGURE 12.16. RIFT ZONES

12.5.A. Specimens:Tectonic Settings of Sedimentation—Your instructor will provide sets of sedimentary and igneous rock specimens that, each set taken together, comprise one or more hypothetical sedimentary sequences characteristic of sedimentation on stable platforms, rift zones, passive margins, open ocean basins, back-arc basins, and intercontinental basins. Describe each sample and determine which of the six tectonic depositional settings (above, Exercise 12.1-3.) are represented by each of the three sets. Record your observations on worksheet 12.5A and cite the bases for your conclusions.

12.5.B. Specimens: Sedimentary Sequences—Your instructor will provide sets of specimens that belong to rift, flysch, molasse, and pelagic sequences. Describe each sample in the four sets and determine which sedimentary sequence is represented by each set. Record your observations on worksheet 12.5B, and cite the reasons for your conclusions. The specimens belonging to each set are numbered arbitrarily. Arrange the specimens belonging to each sample set in the correct stratigraphic sequence and record the sequence of sample numbers and rock types on worksheet 12.5B. Your instructor may have you draw a stratigraphic column representing each sample set.

A) In what tectonic setting(s) would you expect to find rift, flysch, molasse and pelagic sequences?

____

B) In what order would you expect the four sets of samples to occur in a stratigraphic column that records a formation and subsequent closure of an ocean basin ? ______________________________

C) Record the sequence of sample sets in your stratigraphic column.

12.6. Tectonics & Past Sedimentary Sequences—Draw stratigraphic columns, labeled by letter code as in Exercise 12.3, in the three boxed inserts on the world map of the present (Figure 12.15A), the types and sequences of tectonically controlled facies you predict should be present at the sites indicated.

Use the following procedure and refer to the example column partially completed:

1) locate (or use, if completed) the tectonic settings listed in Exercise 12.1;
2) determine from this information, the type(s) of sediments which should be accumulating;
3) indicate these at the tops of the appropriate boxed stratigraphic columns; and,
4) then work up the stratigraphic column to younger facies from the Triassic Period to the Eocene Epoch by referring to your maps made for Exercise 12.3, OR by prediction from the changes in tectonic settings; and,
5) complete each sequence by labeling the facies which probably accumulated during each past age.

Simulating the breakup of Pangaea with the *Drift Globe* or *Terra Mobilis* software could provide insight and aid in your predictions.

WORKSHEET EXERCISE 12.5A.

SAMPLE	DESCRIPTION	ROCK TYPE	TECTONIC SETTING
1			
2			
3			

WORKSHEET EXERCISE 12.5B.

SAMPLE	DESCRIPTION	ROCK TYPE	FACIES -TECTONIC SETTING
1			
2			
3			
4			

12.7. Sedimentation in Cratonic Sequences—Figure 12.17 shows the cratonic sequences of North America in latitudinal (E-W; front face) and longitudinal (N-S; side faces) cross-section. The central shaded portion represents the changing area of land. Complete the diagram as follows:

A) in the portions left blank, draw the transgressive and regressive facies patterns of the onshore to offshore sequence of rock types (conglomerates, sandstone, carbonates, and shale) as they varied from the Sauk through the Tejas sequences.

B) use Walther's Law as a guide to drawing the facies relationships;

C) vary the lateral extent (latitudinal) and proportional thickness of carbonate and clastic facies according to their position in transgressive and regressive portions of each sequence;

D) vary the proportions, thicknesses and lateral extents of facies on the western and eastern coasts according to the effects of the orogenies shown to the left; and,

E) indicate levels where uncomformities would be located. Use standard lithologic symbols for each facies. __

____

F) During which periods were clastics more abundant than carbonates? ____________________

____

G) During which periods were carbonates most widespread? ____________________

____

H) Which sediment types increase in volume and extent during orogenies, regressions and transgressions?

____

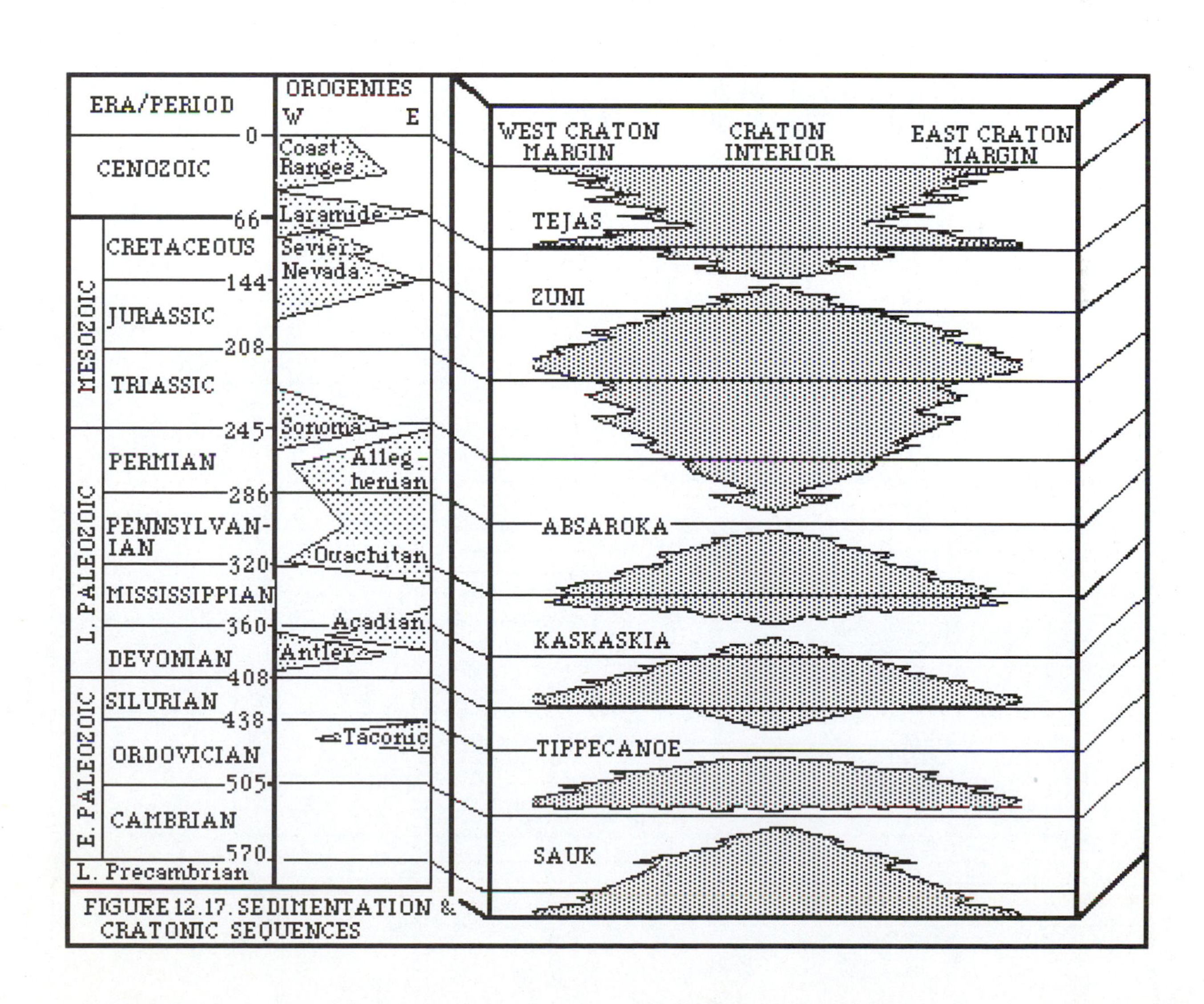

FIGURE 12.17. SEDIMENTATION & CRATONIC SEQUENCES

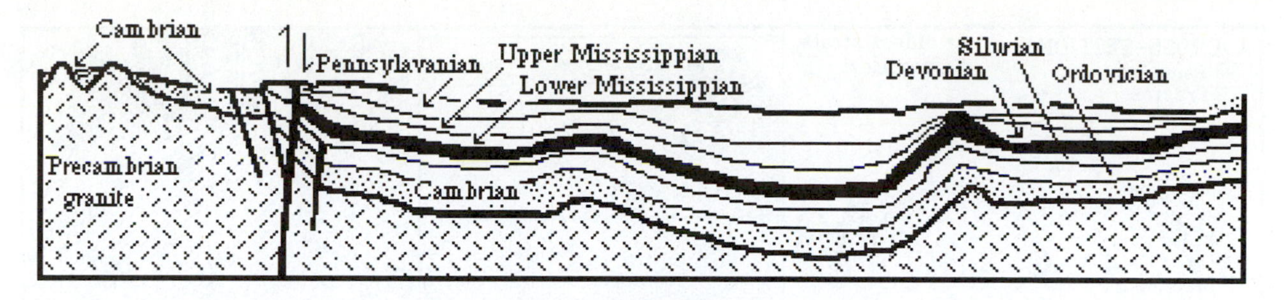

13 Geologic Maps and Cross-Sections

Objectives

To demonstrate mastery of this material you should: 1) understand the three-dimensional geometry of geologic structures; 2) recognize structures represented on geologic maps; and, 3) be able to construct geologic cross-sections. Colored pencils and a compass are needed for Exercises 13.1–13.3.

In this laboratory we review the skills needed to construct and interpret geological maps and cross-sections, and the nature of major geological structures. Geological maps hold a wealth of information: sequences of geological events, geographic distributions of rock types, and the nature and age of geological structures. Such maps are both summaries of geological history and data as well as fundamental tools of geological investigation. Map interpretation within a plate tectonic context is discussed in the final laboratory.

GEOLOGIC STRUCTURES

The physical history of the Earth is revealed by rock types and their sequence of deposition, conditions of formation and, the ways in which rocks have been deformed by tectonic activity. Folds and faults are the principal classes of structures created by tectonism. Folds are grouped into two categories: those which are areally extensive and roughly circular to ellipsoidal in outline (domes and basins), and those which are smaller, elongate to linear structures (anticlines, synclines and monoclines).

Domes—Domes are upwardly-warped folds with semi-circular to ellipsoidal outlines, and many cover thousands of square kilometers (Figure 13.1A). Beds in the center of domes have been raised higher than those of the margins, but truncated by erosion during uplift. Thus, the oldest beds are exposed in a dome's center, and the youngest on its margins (Figure 13.1B). The differing ages of the strata reflect their positions and inclinations (slope or dip) toward the dome's periphery. All beds dip gradually away from the center in all directions. Traveling in any direction from the center to periphery along a road of a constant elevation, progressively younger formations are encountered approaching the periphery, just as if you were traveling forward in time (Figure 13.1C).

Basins—Structural basins are usually large (≥thousands of sq. km) downwardly-warped folds with semi-circular to ellipsoidal outlines (Figure 13.2A). When erosion truncates a basin, the oldest formations are exposed on the periphery and the youngest toward the center (Figure 13.2B). All strata dip toward the basin's center (Figure 13.2C). A cross-section or cut-away view through the basin's center shows that the dips of the younger beds become increasingly gradual toward

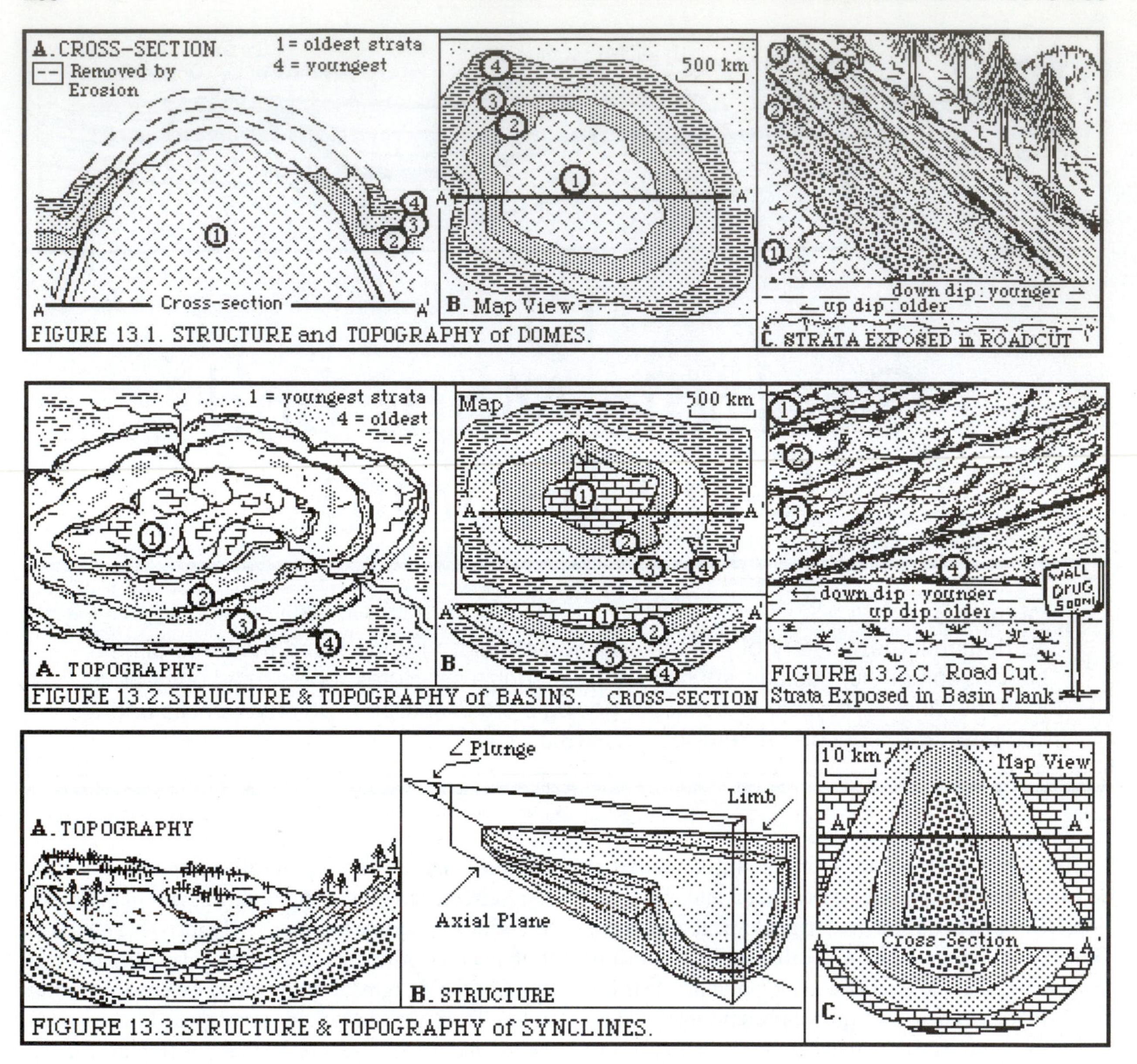

FIGURE 13.1. STRUCTURE and TOPOGRAPHY of DOMES.

FIGURE 13.2. STRUCTURE & TOPOGRAPHY of BASINS.

FIGURE 13.2.C. Road Cut. Strata Exposed in Basin Flank

FIGURE 13.3. STRUCTURE & TOPOGRAPHY of SYNCLINES.

the center. If travelling from the center to any peripheral point, and on a road of constant elevation, progressively older beds are encountered as the basin's periphery is approached, as if traveling backward in time or descending a stratigraphic section from top to bottom.

Folds—A syncline (*syn* = same or together; *cline* = slope) is an elongate structure, smaller than a basin, in which strata fold downward as in a trough. When erosion truncates a fold, the youngest beds crop-out in the structure's interior and the oldest along its margins (Figure 13.3A,C). The beds of the flanks, or limbs, dip toward each other. Their intersections form an axial plane which has a compass orientation set by its intersection with the ("horizontal") surface (Figure 13.3B). The fold axis' inclination to the horizontal is called the plunge, the axis and, thus, the entire structure, "plunges" beneath the surface at this point.

An anticline (*anti* = opposite) is an elongate fold with limbs dipping away from one another (Figure 13.4A). When truncated by erosion the oldest beds crop-out at the center, and younger beds along the margins (Figure 13.4B). In map view, anticline limbs converge in the direction of plunge, while syncline limbs open in the direction of plunge. These folds are often paired in a series of plunging folds. When truncated by erosion, such a series forms a zigzag outcrop pattern (Figure 13.5). Very gentle, regional anticlinal folds are known as arches.

Among other types of folds is the monocline (*mono* = one). Monoclines have only one steeply dipping limb (Figure 13.6A) and are commonly overlie unexposed faults.

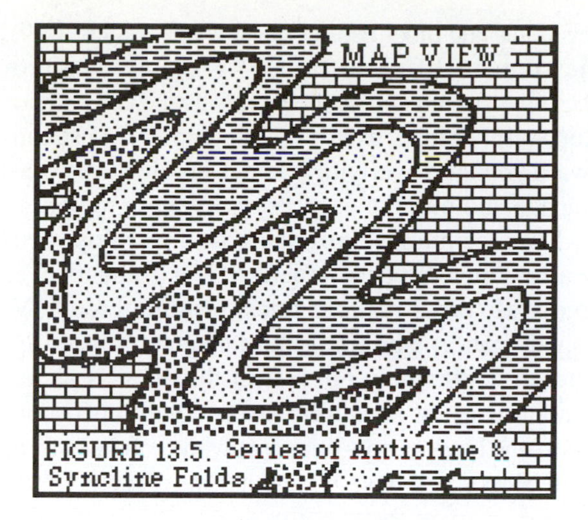

FIGURE 13.5. Series of Anticline & Syncline Folds

Faults—A fault is the plane along which strata have fractured and moved (Figure 13.7A). Faults are classified according to the relative directions of movement of the two blocks along the fault plane. The surface of fault movement (if exposed), is called a fault scarp (Figure 13.7A). If the fault plane has a slope >90° to the horizontal of the down-thrown block, that block's fault plane forms the foot wall: it has a slope up which a geologist could "walk", or on which miners, in a tunnel along the fault plane, could stand (Figure 13.7A). The block with the fault plane lying at < 90° to the horizontal of the down-thrown block, is known as the hanging wall: a wall which a geologist could only "hang from", not walk-up, or which "hung over" the heads of miners (Figure 13.7B). A normal fault results from upward motion of the foot wall block relative to that of the hanging wall block (Figure 13.7A), and a reverse fault, from upward motion of the hanging wall block relative to the foot wall block (Figure 13.7B). A reverse fault with a fault plane lying at a very low angle to the horizontal is known as a thrust fault (Figure 13.7C), because one block has been thrust over another.

Lateral faults are produced by the horizontal movement of one block past the other and parallel to the fault's trace at the surface (Figure 13.8A), and are classified according to the asymmetry of block movement. If standing on one of the fault blocks and looking across the fault zone, you see that the opposite block has moved to the left (e.g., a road crossing the fault is offset to the left), the fault is left-lateral. If the movement on the opposite block is to your right, the fault is right-lateral (Figure 13.8B).

INTERPRETING GEOLOGIC MAPS

Geologic maps portray the geographic distributions of formations or other lithostratigraphic unit

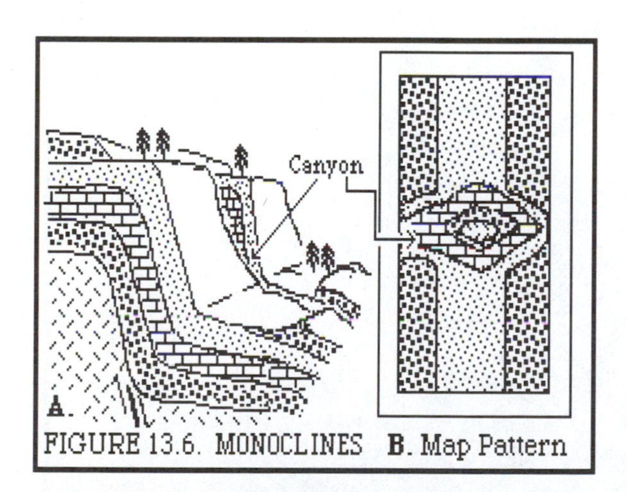

FIGURE 13.6. MONOCLINES B. Map Pattern

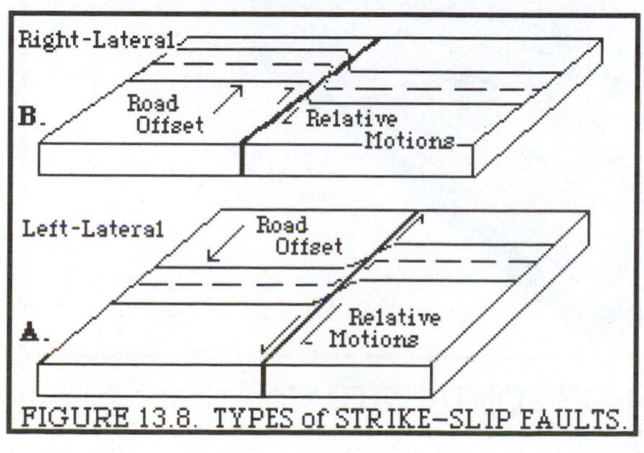

FIGURE 13.8. TYPES of STRIKE-SLIP FAULTS.

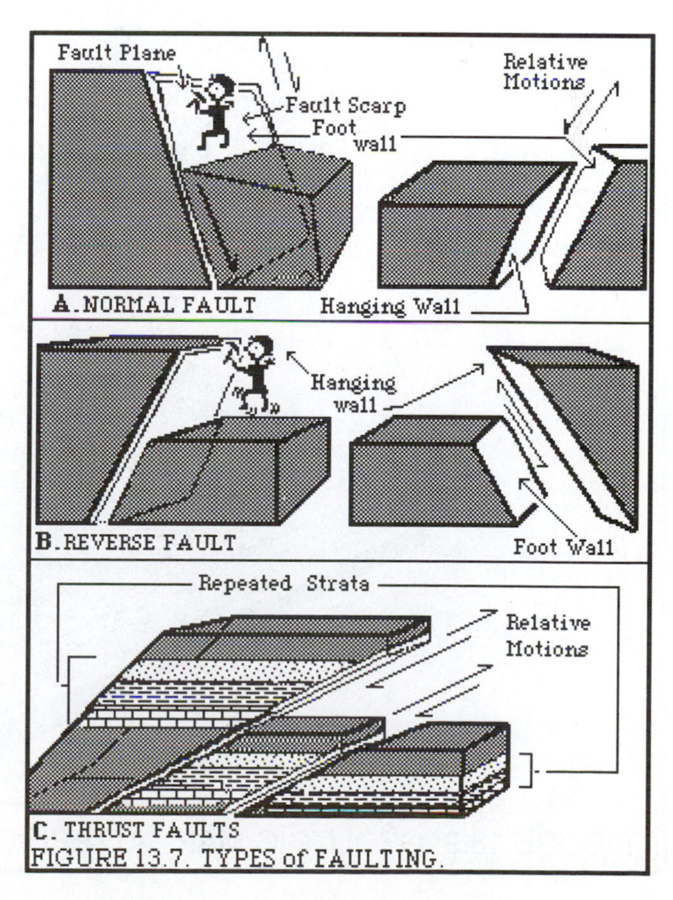

FIGURE 13.7. TYPES of FAULTING.

outcroppings, usually bedrock, but sometimes sediment, at the Earth's surface—the strata commonly, and largely hidden by a thin cover of soil and sediment. In most regions natural exposures where bedrock reaches the surface are few, and their outcrop distributions are determined by roadcuts, quarries and wells. In the arid western U.S., however, bedrock is commonly exposed and the extent of a formation's outcropping is easily realized.

To read geologic maps and determine the sequence of events and structures present, one must have a basic understanding of the relationships between geologic structures, their map patterns, and surface topography. Guidelines for reading geologic maps are discussed below.

Patterns of Horizontal Strata—Horizontal strata's outcrop patterns are most easily understood (Figure 13.9). When horizontal, or nearly so, the contacts or boundary surfaces between formations parallel the topographic contours (lines which connect points of equal elevation; (Laboratory 3)). This pattern is most easily seen in regions of stream erosion; bedrock cut by a stream will expose younger beds at higher elevations, and older beds at lower levels (Figure 13.9). The contacts of each form a "V-shaped" pattern where the streams cross them. Just as the contours of a topographic map, each "V's" apex points upstream toward younger, overlying strata. Dendritic (tree-like) drainage and outcrop patterns typify these regions.

Even seemingly horizontal beds dip or slope gently a few meters per kilometer. Their contacts over large areas cross contour lines, forming broad "V-shaped" patterns pointing in their direction of dip. Outcrop widths may not reflect thicknesses alone; weather-resistant beds form steep slopes, cliffs, and ridges, and have narrow outcrops, while soft beds form gentle slopes and have wide outcrops (Figure 13.9).

DIPPING STRATA of GEOLOGIC STRUCTURES

Orientation of Strata—Dip and Strike—Throughout our discussions are references to the inclination or dip of strata, data that are essential to constructing and understanding geologic structures and maps. Dip is the angle of a bed's inclination to the surface, measured perpendicular to its line of intersection with the

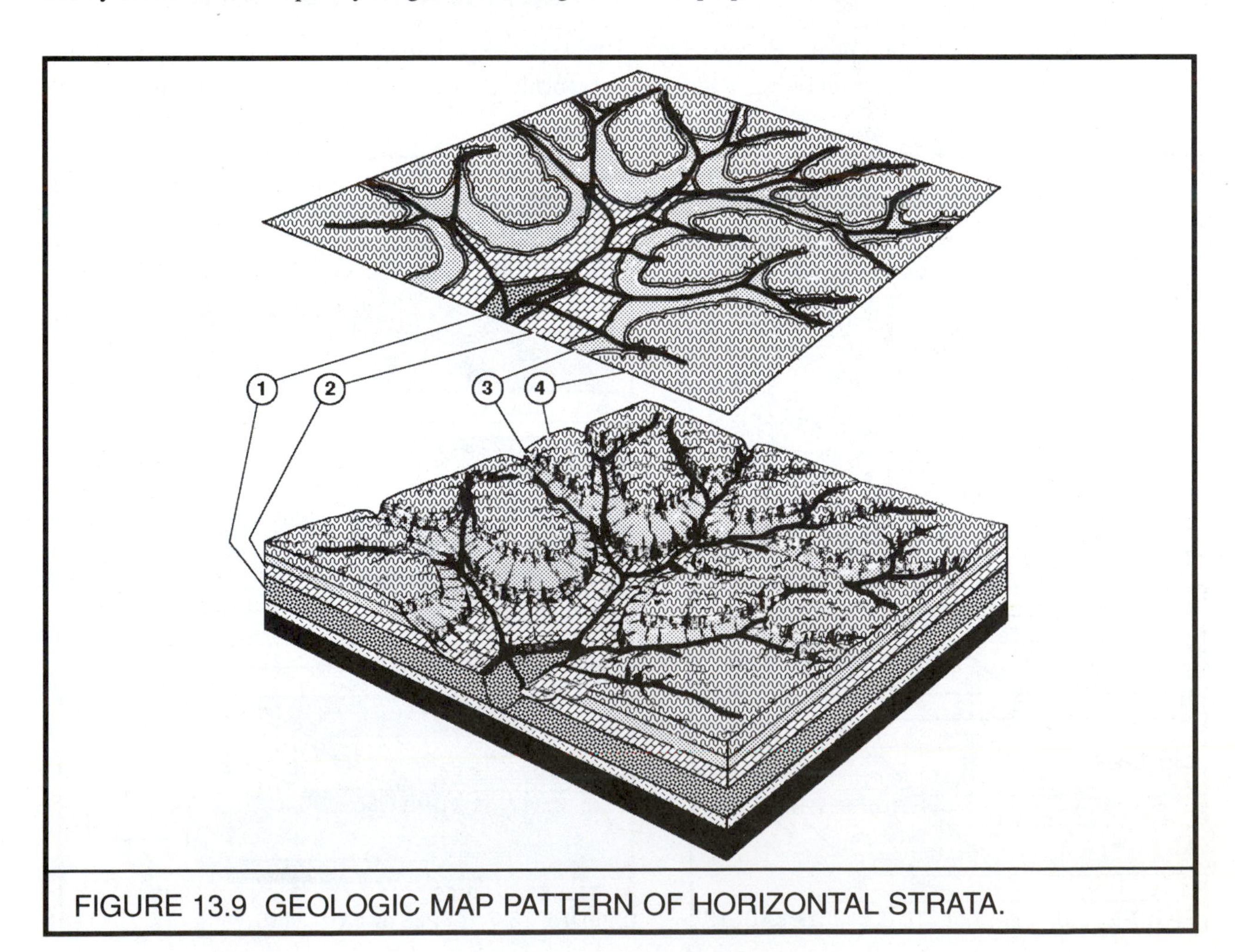

FIGURE 13.9 GEOLOGIC MAP PATTERN OF HORIZONTAL STRATA.

surface known as the strike (Figure 13.10). Any bed inclined to the surface appears to dip at some angle and in some *direction*. But the dip direction and angle casually observed is usually apparent, not actual (i.e., maximal; Figure 13.11).

An apparent dip differs from true dip in that the intersection of the exposure's surface with a dipping bed can result in a range of dips, all less than the true dip: i.e., the maximal angle of the bed's intersection with the surface. A surface exposure can cut across a bed at any angle to its surface, but only at one angle is the bed's (true) dip, and angle of intersection with the surface, identical. The maximum angle of intersection and thus, true dip, is that angle measured between the horizontal and the bed's surface along a line perpendicular to the bed's strike (Figure 13.11A). The latter, is measured and defined as the compass direction of the line formed by the bed's intersection with the surface when measured perpendicular to the maximum (true) dip. That line's direction depends on, and is defined by, the bed's direction of dip, and is always *perpendicular* to the true dip's direction (Figure 13.10A, 11A,B). A horizontal bed has no dip and every strike. A vertical bed has only one strike, so all apparent dips equal the true dip.

To illustrate strike, apparent and true dip, stack some books and prop them at an angle. Let the binding's trace be the strike and the cover be the (bed's) dip surface (13.10A). Take a pencil and orient it on the dip surface (cover) so that it is parallel to strike (the binding). Now slowly rotate the pencil so its eraser end remains fixed on the dip surface and its point moves away from the dip surface in a horizontal plane, until it lies perpendicular to the strike (binding edge). Notice that the distance between the pencil point and the dip surface (book cover) increases from zero (when the the pencil is parallel to strike) to a maximum value, when it is perpendicular to the strike. So also will

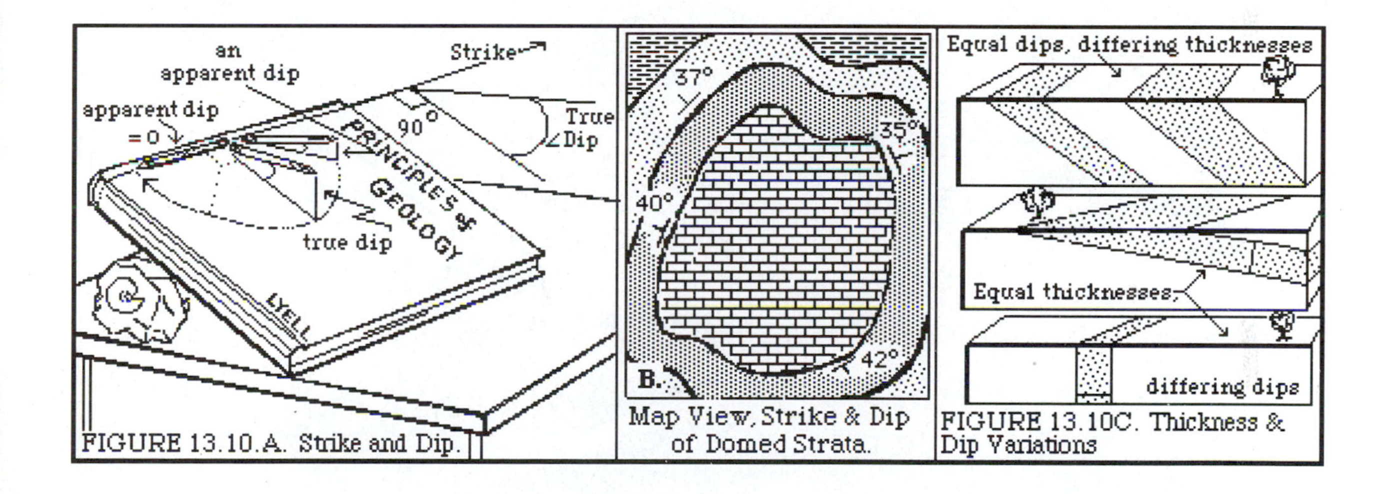

FIGURE 13.10.A. Strike and Dip. Map View, Strike & Dip of Domed Strata. FIGURE 13.10C. Thickness & Dip Variations

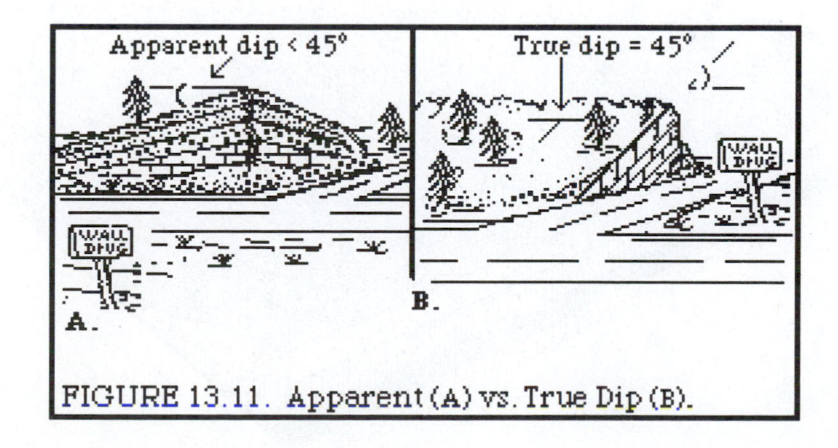

FIGURE 13.11. Apparent (A) vs. True Dip (B).

the angle between the eraser and the pencil's projection (or its shadow from an overhead light) on the cover (dashed line in Figure 13.10A) increase from zero to some maximum value. These are dip or inclination angles, and the maximum value is that of the true dip.

For two beds of the same thickness but differing dips, the bed with the gentler dip will have the wider outcrop band (Figure 13.10C). For two beds of differing thickness and the same dip, the thicker bed will have the widest outcrop band. It is easiest to envision this by holding the dip angle constant and varying the angle at which the surface cuts the bed (Figure 13.11C). Now, it is seen that the true thickness of a bed can be obtained only when measured perpendicular to the true dip and along the strike. Since only a vertical bed can have a dip which is perpendicular to the surface, the outcrop width of a formation with any other dip will be greater than its true thickness.

MAP PATTERNS of DIPPING STRATA

When formations dip, they form an outcrop pattern of parallel bands cutting across contours and physical features such as streams (Figure 13.12). A "V-shaped" outcrop pattern with the "V" pointing in the dip direction results where streams cross outcrop bands. If the dip direction is opposite the stream's gradient, the outcrop "V"'s apex points downstream: the direction of dip (Figure 13.12). For beds of equal thickness, the lower the dip, the wider the outcrop band. No "V" is present when the bed is vertical, nor when the stream's gradient is less than the bed's dip. Outcrop width of beds also depends on topographic

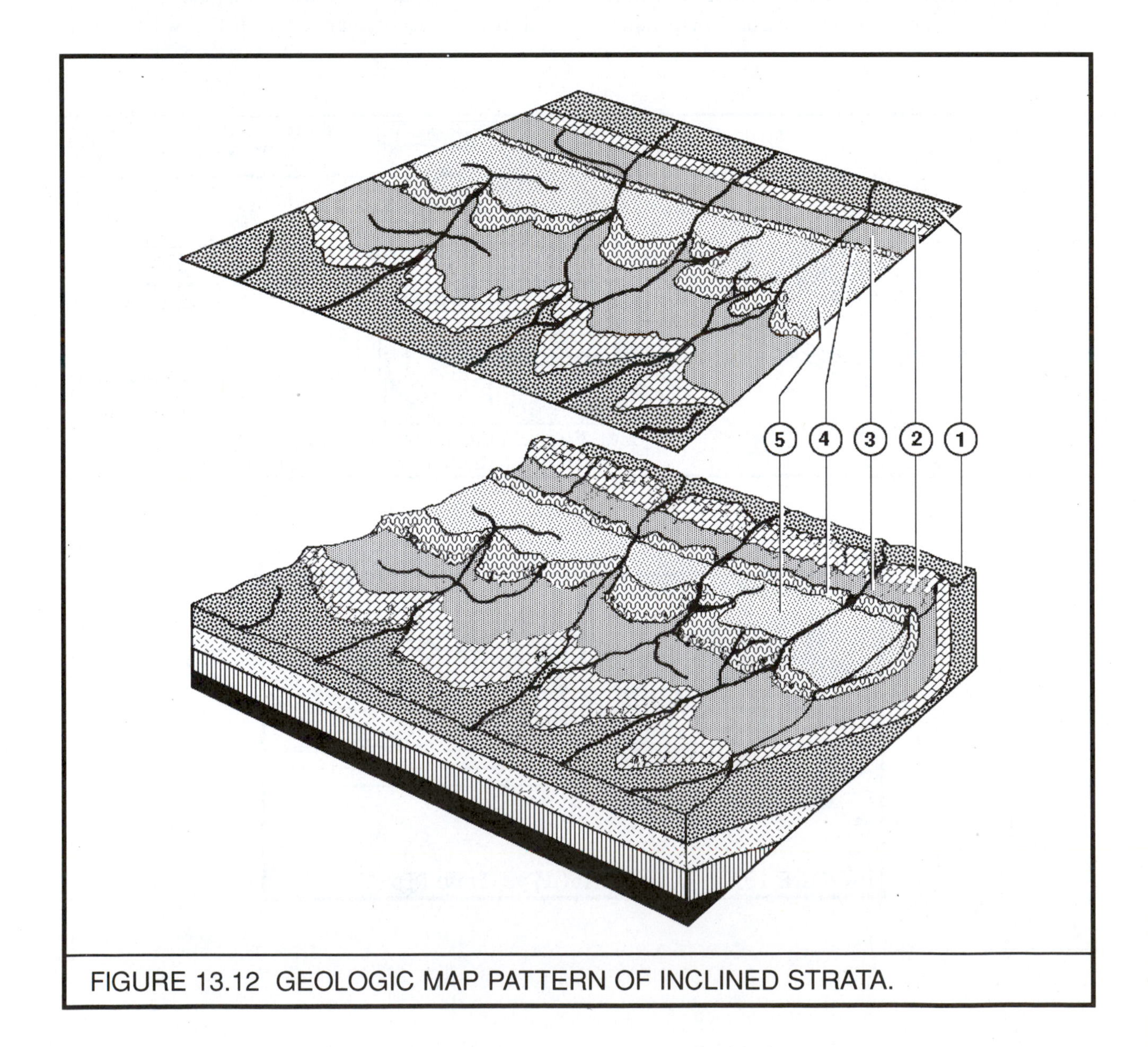

FIGURE 13.12 GEOLOGIC MAP PATTERN OF INCLINED STRATA.

relief: the steeper the slope, the narrower the outcrop band. Again the "V" points toward younger beds (unless overturned).

Dome and Basin Map Patterns—Dome outcrop map patterns form concentric bands (Figure 13.13), of decreasing age from the center (unless younger igneous rocks intrude), to the periphery. Streams running from the center to the periphery cut "V" patterns pointing in the dip direction: downstream and away from the dome's interior. A basin's outcrop pattern is also one of concentric bands, with the youngest beds, central and older beds, peripheral (Figure 13.14). The "V"'s of outcrop bands point toward the basin's center.

Fold Map Patterns—Outcrop patterns of anticlines have many dome-like features, but differ in shape and size (Figure 13.15). A stream crossing an anticline incises "V"'s pointing in opposite directions on each limb and in each limb's direction of dip. The tilt or plunge of an anticline is the angle of intersection between the fold axis and the horizontal surface. The structure's plunge direction is the compass direction of the apex or "nose" formed by the most tightly folded portion of the structure. If it plunges in two directions, there are two "noses", and the outcrop pattern is ellipsoidal (Figure 13.15). The beds of a syncline are youngest in its interior and oldest on its periphery. Streams crossing its limbs incise "V" patterns pointing toward younger, interior beds.

Unconformity Map Patterns—If an angular unconformity is present, the structural pattern of the older, underlying beds lies, not parallel with, but at an angle to, the younger, overlying beds (Figure 13.16). The two patterns "clash": contacts with the underlying structure end abruptly against the angular

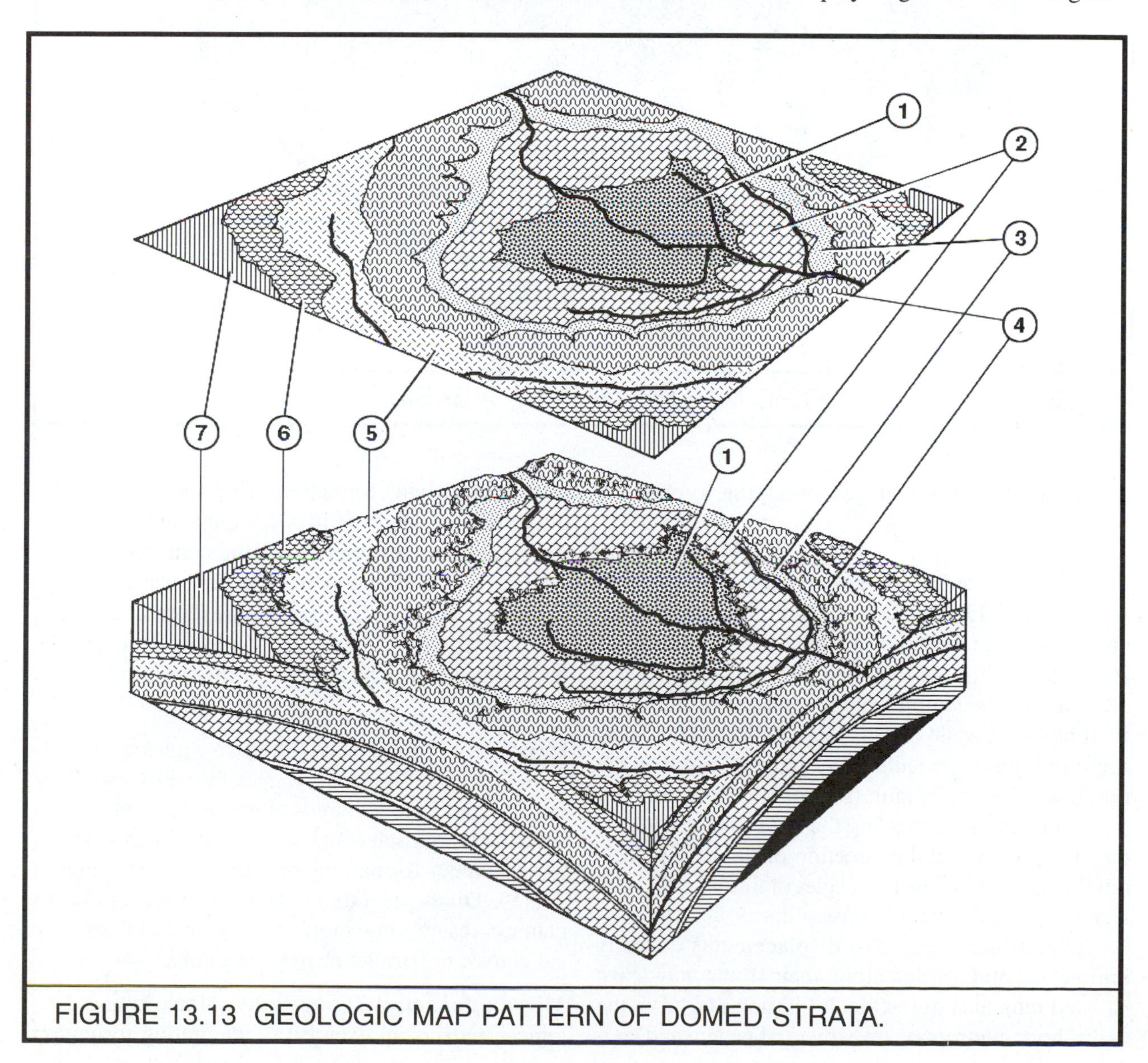

FIGURE 13.13 GEOLOGIC MAP PATTERN OF DOMED STRATA.

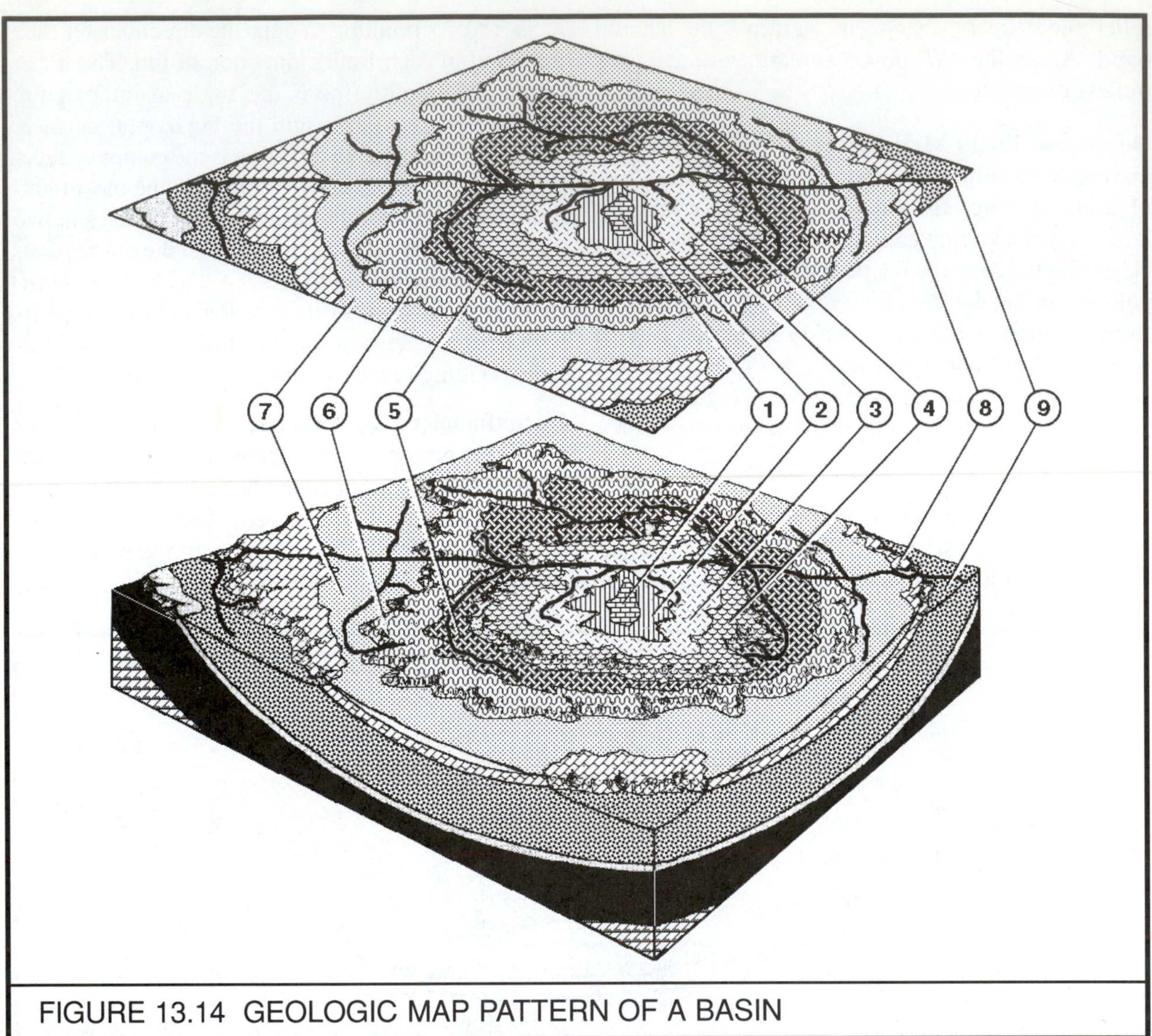

FIGURE 13.14 GEOLOGIC MAP PATTERN OF A BASIN

unconformity at the younger, overlying formation's base.

Fault Map Patterns—Faults are represented by bold lines (Figure 13.17). Because faulting causes vertical and/or lateral displacement, formation contacts terminate against the fault line and structures are offset along faults. Normal and reverse faults are usually of high angle and map as straight lines. The oldest rocks of high-angle faults are usually exposed on the upthrown side of the fault (Figure 13.17). If the fault plane dips, streams crossing the fault form slight "V" patterns pointing in the direction of the fault plane's dip (Figure 13.17). The low angles of thrust fault planes result in irregular fault line map traces.

Thrust faults can have displacements of many kilometers, and erosion along their fronts may leave isolated remnants of the block (outliers) far from the main sheet, interrupting the structural pattern and contacts of underlying formations. Erosion may also produce holes or windows in a sheet, revealing underlying formations. In both cases, the contacts between underlying formation(s) and the fault plane (overlying formation's base) will resemble those of an angular unconformity (Figure 13.16). Lateral faults produce offset formation contacts, structures, and stream courses.

Intrusion Map Patterns—Intrusive igneous masses have circular to irregular shapes in map view. They may cover tens to hundreds of square kilometers with irregular margins at contacts and cutting across structures of older formations or igneous bodies (Figure 13.18). Dikes are tabular structures with outcrop patterns resembling short, straight bars. Dikes may cut across, or parallel nearby structures.

Map Pattern and Topography—Interpreting geological maps is an exercise in integrating formation

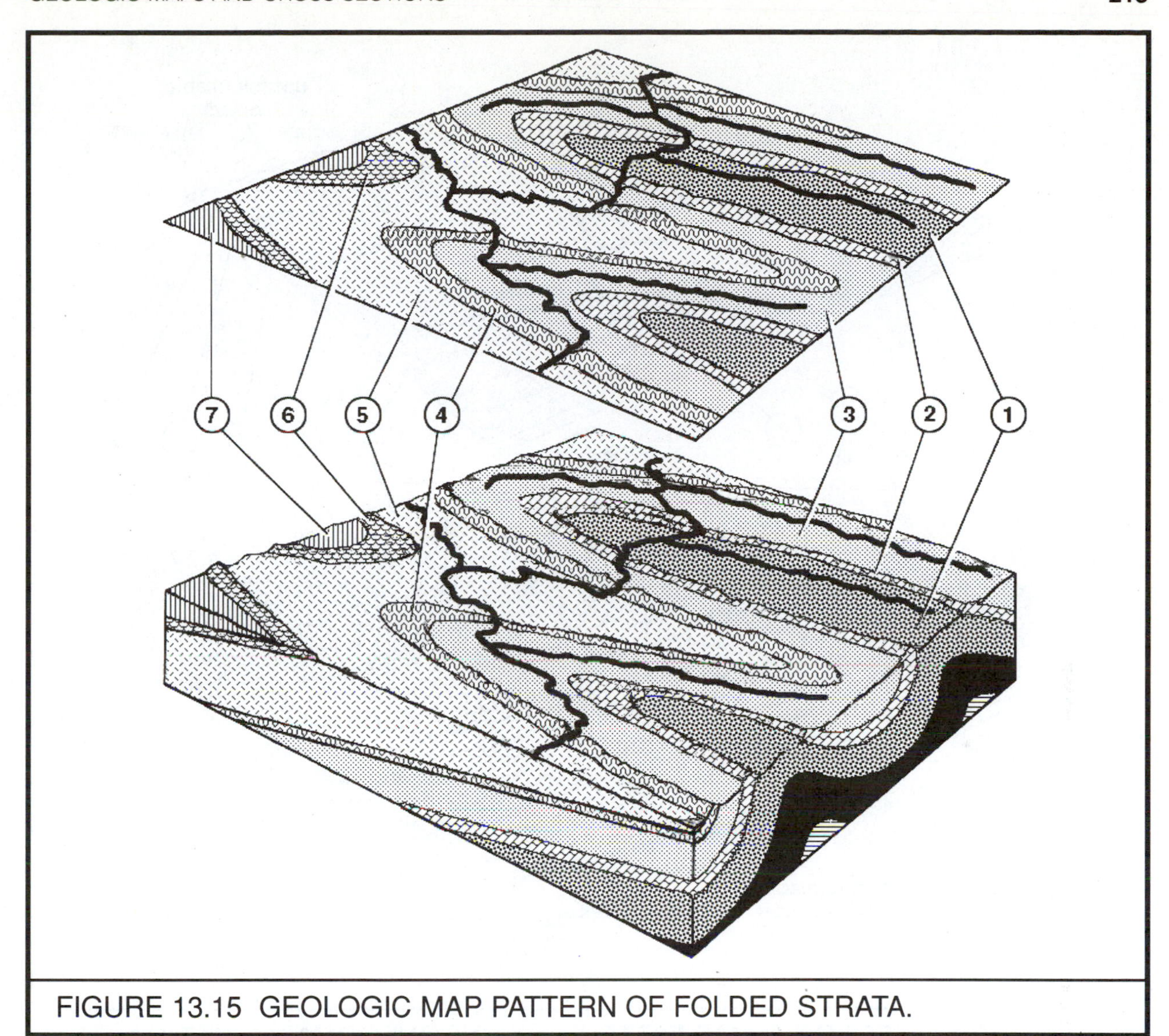

FIGURE 13.15 GEOLOGIC MAP PATTERN OF FOLDED STRATA.

data and topographic data (e.g., "V" patterns, above). The thickness of an outcrop band reflects its width and the region's topography. Generally, the wider an outcrop band, the lower the formation's dip and topographic relief. The greater the dip and (commonly) surface relief, the narrower is the outcrop band.

If the ages of strata are known, it is usually a simple matter to determine the direction and approximate angle of dip from an outcrop pattern and the general character of the topography. The compass orientation of two formation's contacts (indicated by dark lines; dashed where approximated) is their strike at a certain location. The dip direction lies at 90° (left or right) to the strike, and is shown by the outcrop "V" patterns. Formation thickness can be visually estimated by using the type and size of the structure and topography to determine whether the dip angle is high or low. Likewise, the topography of a region and a formation's outcrop width provide clues to dip direction and degree and, thus, the type of structure present.

Become familiar with the common symbols needed to read geologic maps (see geologic map keys). Strike and dip directions are indicated on a map by a single "T"-like symbol: the longer strike line has the proper compass orientation, dip direction lies at 90° to the strike and in the direction of the shorter line of the symbol, and the dip angle is noted in one corner (e.g., Figure 13.12). The ages of strata are given by letter symbols and are color-coded. Upper-case letters denote the geological system, lower-case letters denote the lithostratigraphic subdivisions (e.g., formations).

CONSTRUCTING CROSS-SECTIONS

Much of the structure we wish to understand lies beneath the outcrop patterns of geologic maps, and can be visualized by constructing a cross-section:

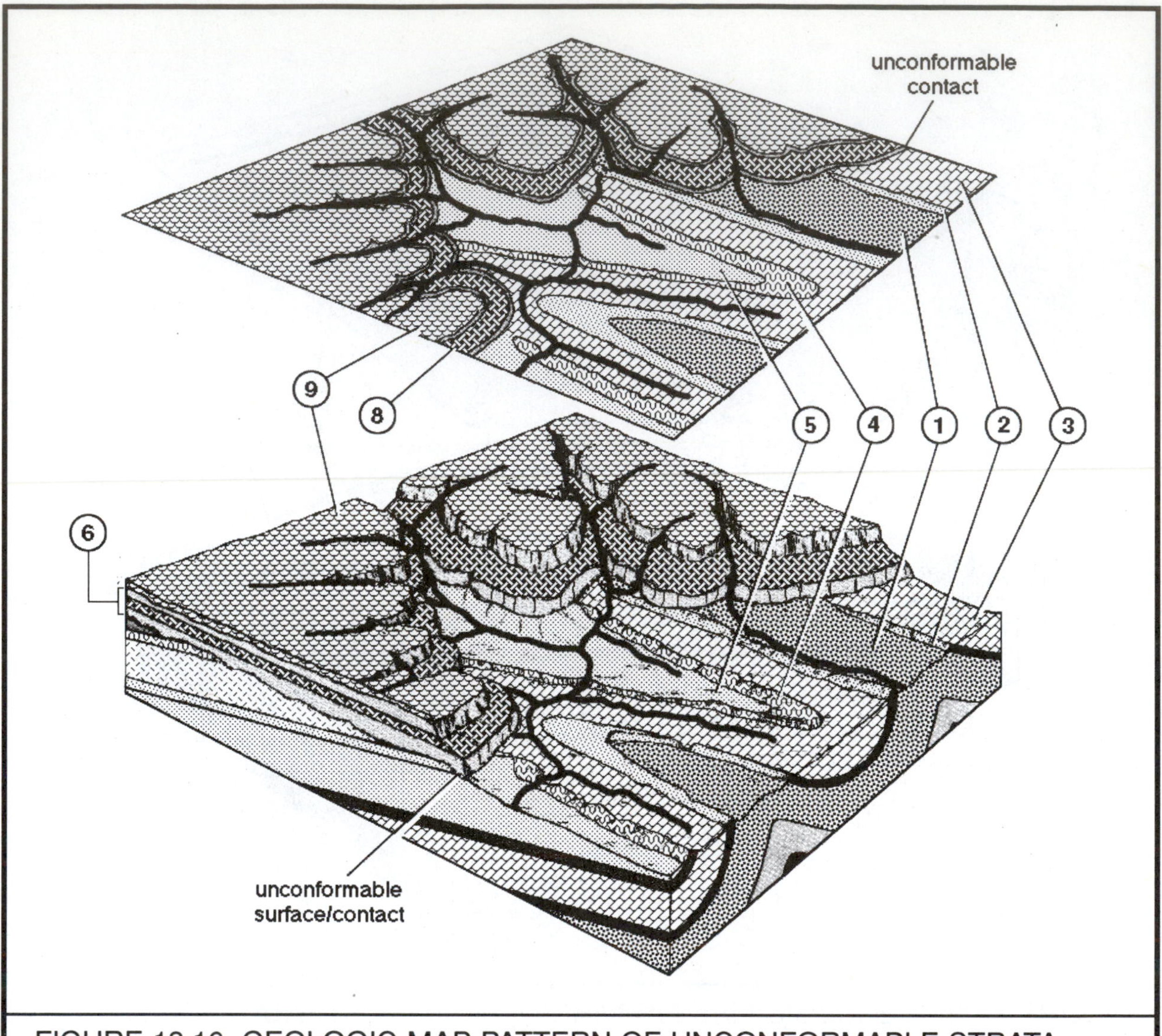

FIGURE 13.16 GEOLOGIC MAP PATTERN OF UNCONFORMABLE STRATA.

the plane of view perpendicular to the surface (below surface) we would have if the crust could be sliced along a chosen line and the blocks pulled apart .

Cross-sections are constructed as follows:

1) choose and draw (on the geologic map) the line along which the cross-section is to be constructed;
2) place a piece of paper just above this line, and mark the line's beginning and end "A" and "B" ;
3) mark each formation contact along the length of the strip (this gives the map-widths of formations);
4) record the formation symbols, dips and, elevations of each contact (if given; estimate them if not);
5) draw the horizontal line of your cross-section from the paper strip onto graph paper;
6) determine the scale of the map (e.g., a 1:250,000 scale means one inch measured on the map is equivalent to 250,000 inches on the actual surface);
7) construct a vertical scale that is two to several multiples of 10 greater than the horizontal scale so that the vertical dimension is exaggerated and detail can be seen . Mark divisions of elevation equal to 5 or 10x that of the contour interval starting with an elevation at or below the lowest point along the cross-section (the base line);
8) plot the elevations of each formation contact on the line of cross-section; and, 9) extend the formation contacts with their dip directions and (for our purposes) approximate dip angles from the base line to their elevations of exposure AND beneath the baseline in order to show the subsurface structure.

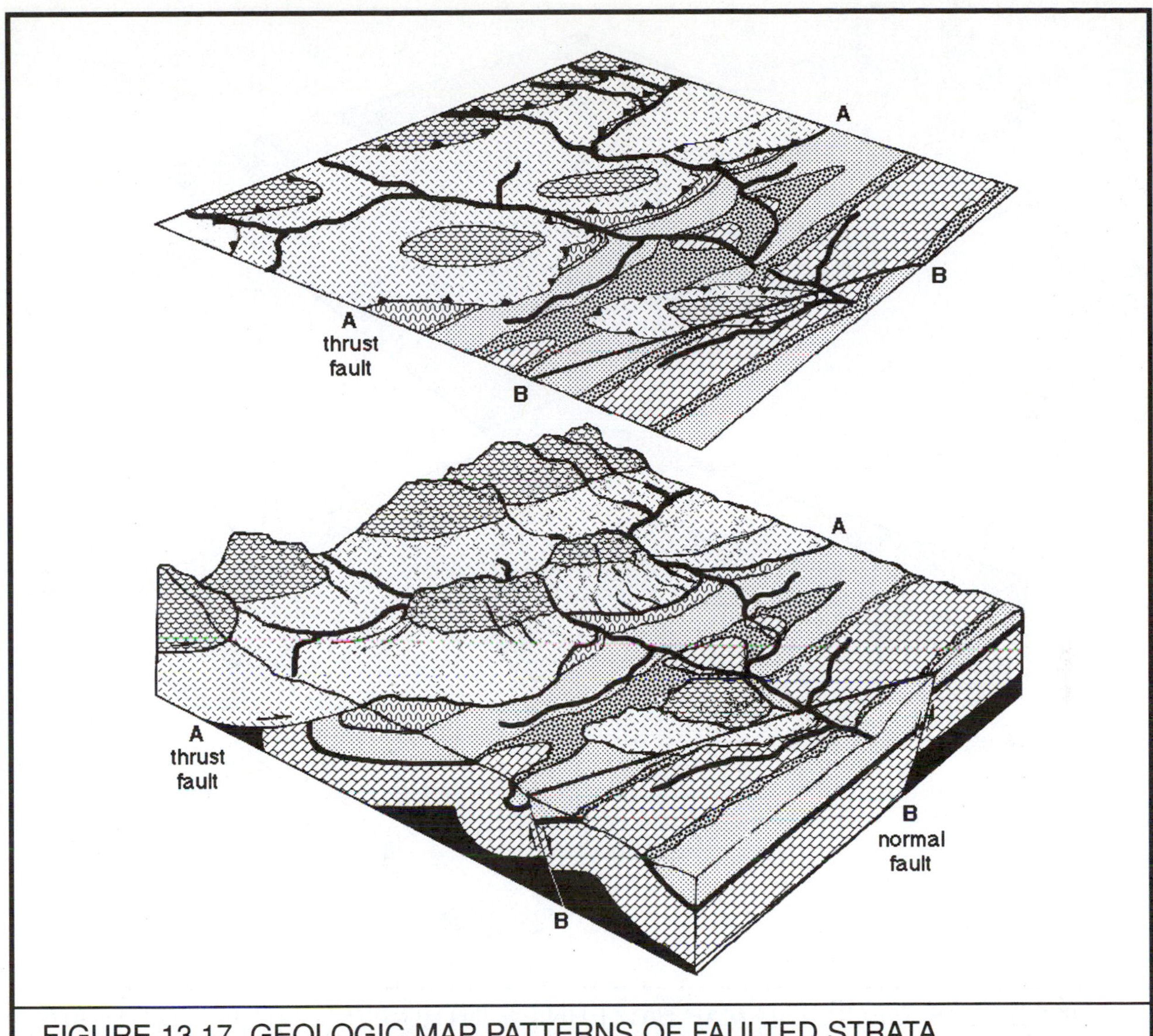

FIGURE 13.17 GEOLOGIC MAP PATTERNS OF FAULTED STRATA.

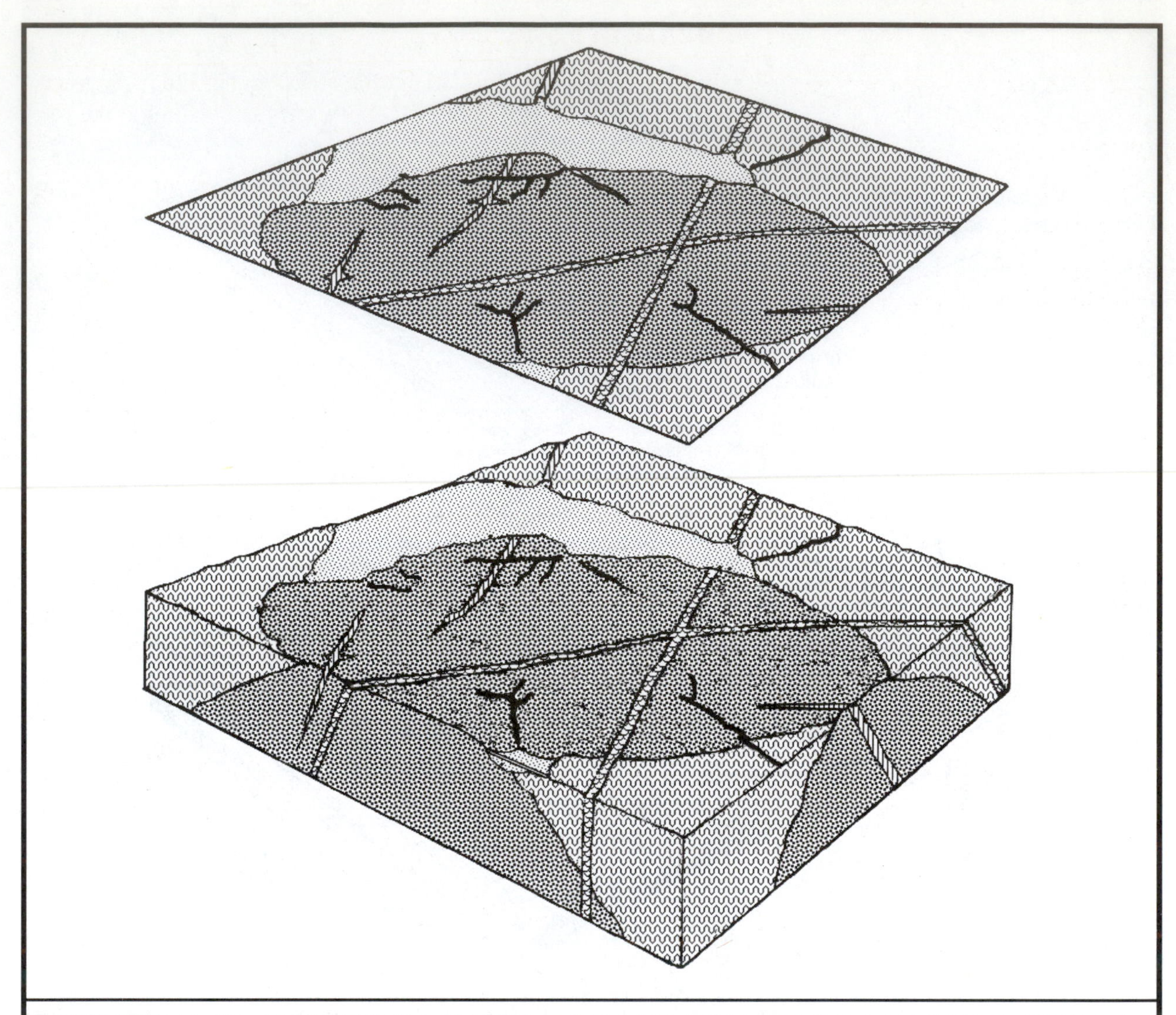

FIGURE 13.18 GEOLOGIC MAP PATTERN OF INTRUSIVES/INTRUDED STRATA.

VOCABULARY

Apparent dip, Anticline, Axial plane, Basin, Fold axis, Dip, Dome, Fault scarp, Foot wall, Hanging wall, Limb, Monocline, Normal fault, Plunge, Reverse fault, Strike, Syncline, True dip Thrust Fault.

LABORATORY EXERCISES

13.1.Strike and Dip—Determine the strikes and dips of the formations illustrated in the block diagrams and maps of Figures 13.9, 12–15 and show their orientations by drawing strike and dip symbols at the formation contacts.

13.2. Constructing a Geologic Map: Field Problems. Horizontal Strata—See Exercise 3.7 if it was not been used in Laboratory 3.

13.3. Domes and Basins—Study the geologic map of the U.S. (provided by your instructor) and locate the major domes and basins on the outline map of North America.

13.4. Geologic Map, Part of the Grand Canyon, Arizona—Plate VI.

1A) What features of topography and outcrop pattern indicate that most of the strata in this region are horizontal? ____________________

B) What are the thicknesses of the Paleozoic and Mesozoic strata? ____________________

____

2. Note the relationship between rock type and cliff/slope topography on the map and cross-section.

A) Which rock types are cliff formers? ____________________

B) Which rock types are slope formers? ____________________

3. A) What type of faulting is found in this region? ____________________

B) What is the nature of the relationship between faulting and gorge-river valley development? ________

____

____

C) One set of faults trends northwest—southeast and the other trends northeast-southwestWhich set of faults is the oldest? Why? ____________________

____

4. Why is the Tapeats Sandstone discontinuous on the north side of the Colorado River valley and along Bright Angel Creek? ____________________

5. A) What types of unconformities are present in the map region? ____________________

B) How many unconformities are present? ____________________

C) Of what age(s) are the strata immediately above the oldest unconformity? ____________________

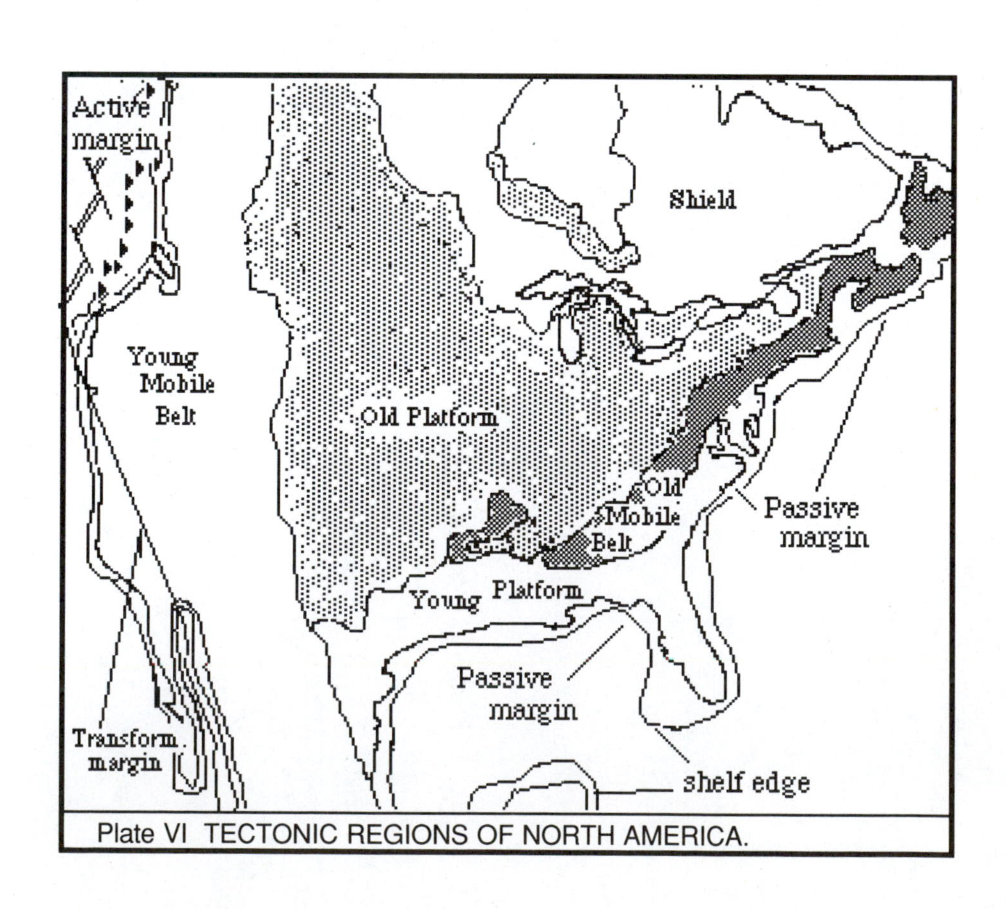

Plate VI TECTONIC REGIONS OF NORTH AMERICA.

13.5. Geologic Map of the Michigan Area: Plate VII.

1) What linear structure trends W/NW—E/SE across northern Indiana, and involves outcrops of Silurian and Devonian rocks? ______________________________

2. A) What structure is centered on Michigan and extends to adjacent areas? ____________________

B) What are the age of the oldest and youngest beds of this structure? ____________________

C) Locate sites within this structure where the "rule of v's" can be used to determine the direction of dip.

3. A) Indicate the strike and dip of the beds in this structure along North-South and East-West lines, include formations Cml to Cc. Assume that the dips are low and construct a North-South cross-section through Michigan from the Straits of Mackinac to Ft. Wayne, Indiana.

4. A) What type of fault lies to the east of Lansing, Michigan? ____________________

B) Which side of the fault moved up? Explain. ____________________

C) What is the greatest possible age that one could assign to this fault? ____________________

13.6. Geologic Map of the Wyoming—Black Hills Area.—Plate VIII.

1. The Black Hills are located in the Northeast corner of the map area (western SD and Northeastern Wyoming).

A) What structure comprises the Black Hills? Explain. ____________________

____

B) What are the ages of the beds at the center of the structure and on its margins? ____________________

____

2. A) What structure is located to the west of the Black Hills and to the east of the Bighorn Mountains (the latter is indicated by brown (no pattern), North-Central Wyoming)? ____________________

____

B) What are the ages of the oldest and youngest beds in this structure? ____________________

____

3. A) What is the small structure which lies directly west of Sheridan, Wyoming, and between the Bighorn and Greybull Rivers? Explain. ____________________

____

B) Within what larger structure is it located? ____________________

C) Of what structure are the Tertiary beds of the Wyoming area a part? ____________________

D) What is the structure associated with the rivers of the area? ____________________

4. A) What type of fault is found on the eastern flank of the Bighorn Mountains? Explain. ____________________

____

B) Of what age are the oldest rocks associated with the faults? ____________________

C) When did the first period of uplift occur in this region? ____________________

D) Speculate about the nature and sequence of events that might explain these two features of this region.

____

13.7.—Geologic Map of the South-Central U.S.—Plate IX.

1. A) What type of structure is located in the vicinity of Nashville, Tennessee? ____________________

B) What age strata are missing from the region? ____________________

C) What map pattern is produced by the streams and strata in the area? ____________________

D) What type of structure (very small) is located just north of Erin, Tennessee? ____________________

E) What type of structure (small) is located to the west of Waverly and Erin, Tennessee? ____________________

____

2. A) What is the nature of the contact between strata of the Nashville Dome (also the region to the South and the complexly deformed strata to the east) and the strata in the Mississippi Valley (western Tennessee) and trending NW-SE across South-Central Alabama?

B) In what direction do these strata dip? ____________________

C) What does this contact indicate about the geography of the area when the Cretaceous strata were ____ deposited? ____________________

3. A) What are the two dominant structures present in the deformed region stretching from East-Central Alabama to NE Tennessee? ____________________

____

B) What type of structure stretches from the NE corner of the map above Harlan, Kentucky, South past Huntsville, Tennessee and ends SW of Wartburg, Tennessee.? ____________________

C) What is the regional strike of these structures? ____________________

Exercise 13.8.—Cross-Sections.

A) Construct a geologic cross-section from Isis Temple on the north side of the canyon to El Tovar Hotel on the south edge of the canyon (Figure 13.20). Use a contour interval of 50 ft for plotting elevations. Discuss the geologic history of the region revealed by the cross-section and the map key.

B) Construct a geologic cross-section from north to South (St. Ignace —just north of the Straits of Mackinac—to Ft. Wayne Indiana) across the Michigan Basin (Plate VII). Discuss the geologic history of the Michigan Basin revealed by the cross-section and map key.

C) Construct a cross-section from the NW to SE across the Nashville Dome from Clarksville (NW Tenn.) to just SE of Dunlap, Tennessee (Plate IX). What type of structure does the cross-section cut just SE of Dunlap?

13.9. Specimens: A "Table-Top" Geologic Structure—Several slabs of rock are set out on the laboratory table tops. Each rock type represents a single "formation", and each stack of slabs represents "exposures" in a region (the lab). Your instructor has oriented the formations at each exposure with a particular strike and dip. Construct a geologic map of the lab room. Use the appropriate symbols for strikes and dips of the formations and other geological structures. Color-code your map and provide a lithologic key. Your instructor will assist you in approximating the strike and dip of each bed.

13.10. Field Problems: Discover the Structure—Figure 13.20A & B shows the general topographic features (relative elevations may be inferred from stream patterns) and geologic field data for a small region in North-Central Illinois. Using strikes and dips, exposure distributions (shown as rectangular rock type-exposure locality symbols (13.20B)), and formation contacts plotted as overlapping or adjacent rock type symbols construct:

1) a stratigraphic column in the space provided in 13.20B;
2) a geologic map of the region on 13.20A; and,
3) an East-West cross-section paralleling the Illinois River on graph paper.

Construct the map by indicating the locations of formation contacts as inferred from the exposure symbols and the directions of strike indicated. Indicate inferred contacts with dashed lines. Color each formation and make a map key.

A) What type of structure is present? ______________________

B) On which side of the structure do the beds dip most steeply? ______________________

C) Approximately when was the structure first activated? ______________________

D) What age strata are missing? ______________________

E) When did deformation begin and when did it resume? ______________________

____

____

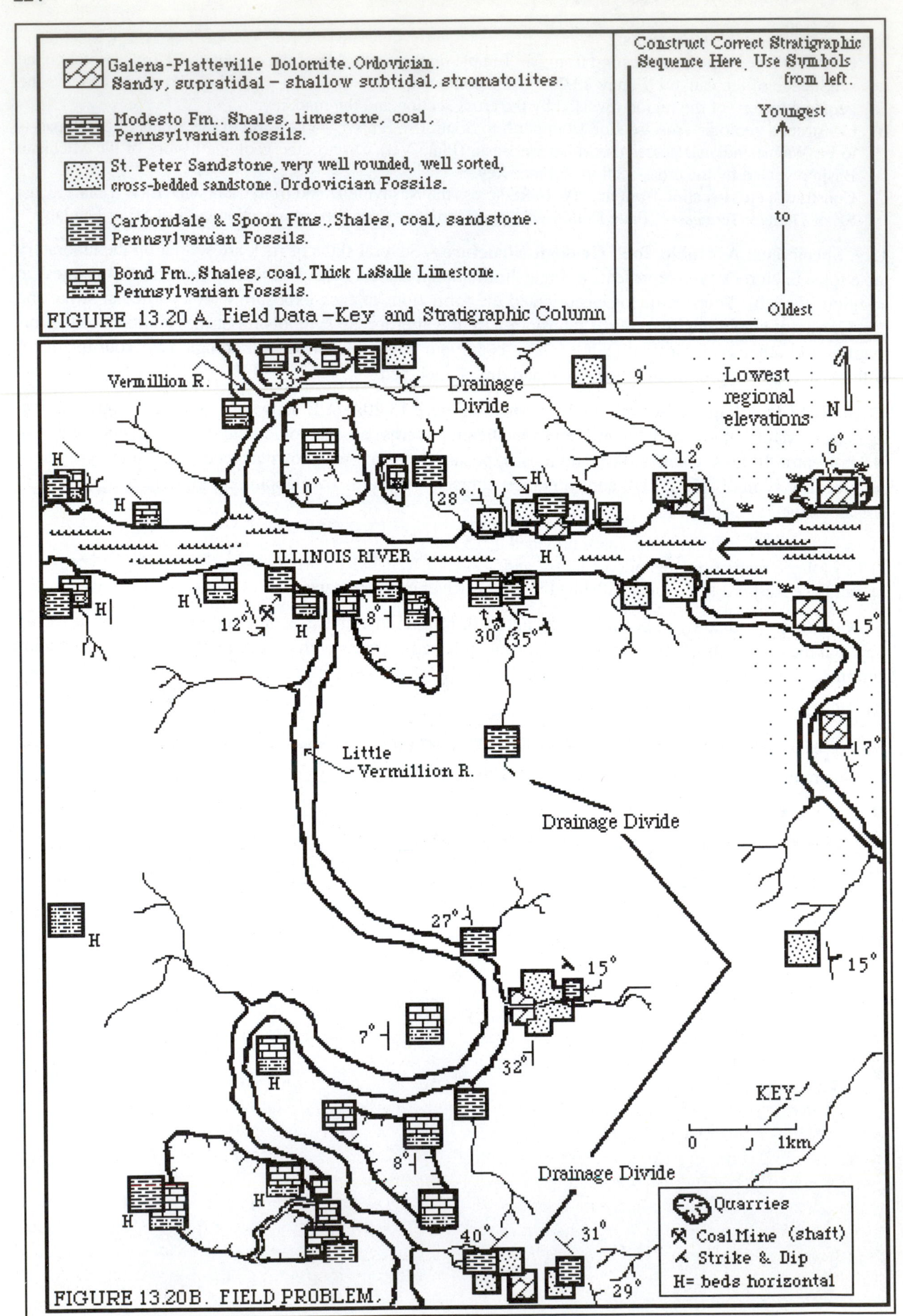

FIGURE 13.20 A. Field Data – Key and Stratigraphic Column

FIGURE 13.20B. FIELD PROBLEM.

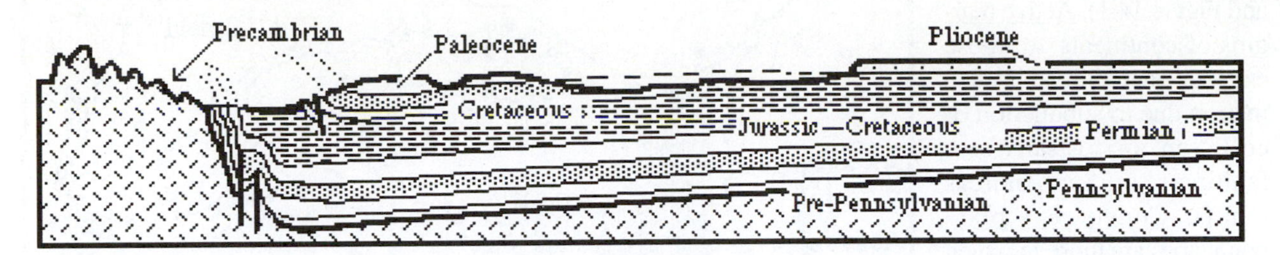

14

Geologic Maps: Tectonics, Structures and Landforms

Objectives

To demonstrate mastery of this material, you should be able to: 1) understand the types of forces producing folds and faults; 2) identify divergent, convergent, and transform plate boundaries, active and passive continental margins, orogenic belts and spreading ridges; 3) infer their approximate past positions on paleogeographic maps; 3) interpret geological events from a geologic map; and, 4) relate the structures and geologic history of a region to the major landform provinces of the United States. Colored pencils and a compass are needed.

This laboratory is an overview of the geological history of North America through the geological map interpretation and the global tectonic history of the Phanerozoic. Such maps summarize not only basic structural and stratigraphic data, but are also used to infer the origins of geological structures and how their distributions reflect plate tectonic activity. With these map-reading skills gained, you will be able to apply your knowledge of historical geology to any region of the country, and read from a simple map, the course of a continent's history.

In this Laboratory we investigate the forces and tectonic contexts which have produced the structures you studied in Laboratory 13. The exercises encompass the tectonic and geological history of North America and the principal tectonic features of the continents throughout the Phanerozoic. The final exercise relates geological history and landform provinces of the United States.

TECTONIC ORIGINS of GEOLOGIC STRUCTURES

Every continent is divisible into regions of present or past, and relatively active or inactive, tectonism. Zones of present or past activity and deformation are known as mobile belts. Orogenic belts enclose regions mantled by relatively undeformed strata (stable platforms) or long-inactive crust (shields) which together comprise the tectonically stable craton or old core (themselves comprised of smaller former plates) of a continent (Figure 14.1). Although platform strata are relatively flat-lying, they are not without structure, for within them are large, gently deformed basins and domes.

Passive margins of continents are those which, after rifting along a newly formed divergent plate boundary has ceased and sea floor spreading has created an intervening ocean basin, have become tectonically inactive (see text and Figure 14.1). Active margins of continents are those experiencing uplift and deformation due to subduction or collision, or to transform faulting (e.g., San Andreas Fault). Convergent plate boundaries are those in which two plates (with each leading margin having continental and/or oceanic crust) are moving toward one another, closing an ocean basin and/or causing uplift. Transform plate boundaries are those in which two plates are sliding past one another along a transform fault.

FIGURE 14.1. TECTONIC REGIONS of NORTH AMERICA

Basins, Domes and Normal Faulting—Basins are mildly deformed regions typical of stable platforms (Figure 14.2). They are structures formed as a result of tensional forces acting from beneath the basement rocks, causing them to move up, or drop down locally, and "forcing apart" the overlying crust. Broad synclines, arches, anticlines and monoclines are often associated with domes and basins.

Many cratonic basins originated as "failed arms" of aulacogens, and thus, are underlain by ancient rift valleys (see Laboratory 12 and text); linear zones of normal faulting (resulting from tensional forces) created when crust overlying a hot spot (see text) domed and then began, but failed, to rift completely (Figure 14.2A). Normal faulting produces a linear region of crustal blocks that have dropped down (grabens) relative to upthrown blocks (horsts) on basin flanks (often deepy buried; Figure 14.2B). Thick accumulations of

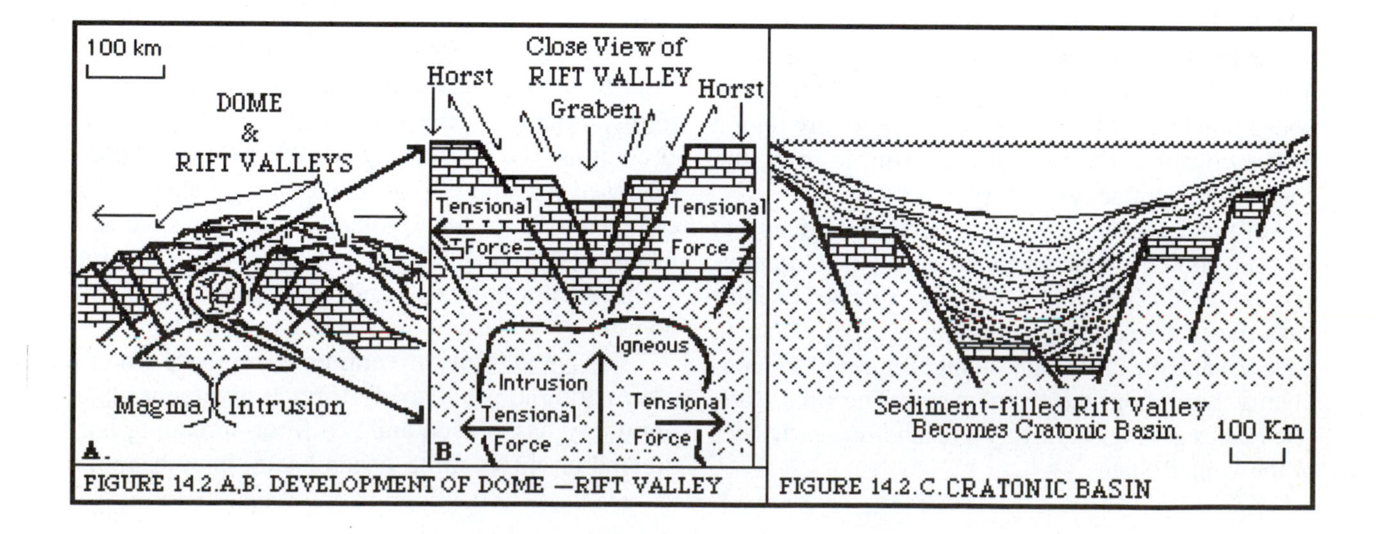

FIGURE 14.2.A,B. DEVELOPMENT OF DOME —RIFT VALLEY

FIGURE 14.2.C. CRATONIC BASIN

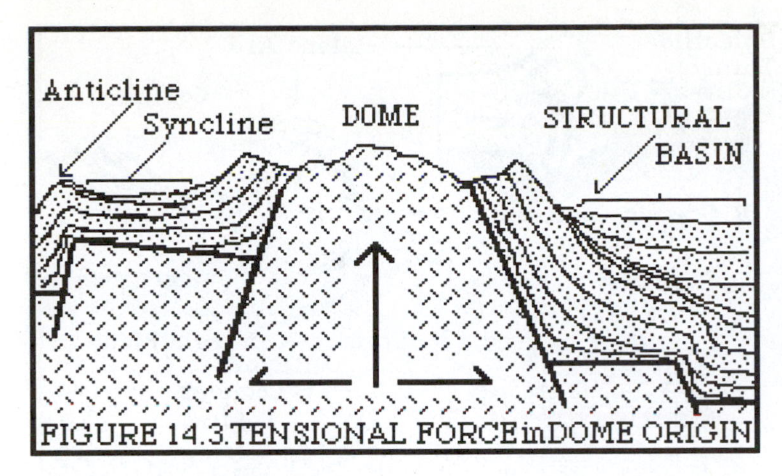

FIGURE 14.3.TENSIONAL FORCE in DOME ORIGIN

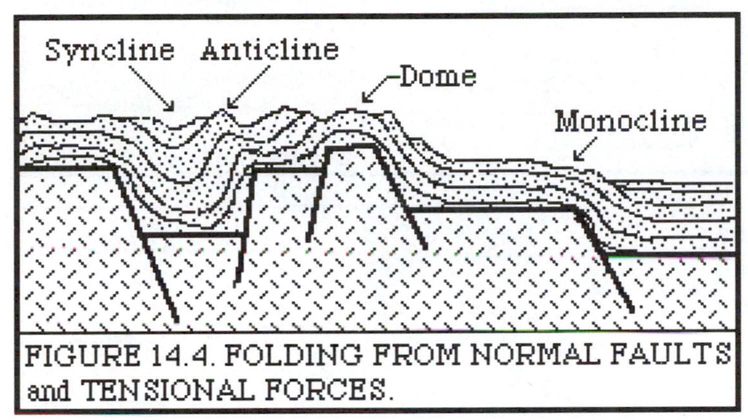

FIGURE 14.4. FOLDING FROM NORMAL FAULTS and TENSIONAL FORCES.

sediments form in the fault-bounded valleys created because they continue to subside and receive sediment long after rifting (Figure 14.2C; also Laboratory 12).

Domes are also associated with tensional forces (Figure 14.3). Commonly exposed at their centers or at depth are Precambrian basement rocks which were punched upward as if by a piston. The tensional forces causing doming produce normal faults within, and on, dome margins. Not all cratonic domes are currently explicable by plate tectonic theory. Those that are, have formed from magma plume intrusions as plates passed over hot spots (e.g., Yellowstone Caldera, WY., White Mountains of New Hampshire). Others consist of regions of Precambrian basement weakened by vulcanism or intrusion during the Precambrian and uplifted during later orogenies (e.g., St. Francois Moun-tains, southeastern Missouri; Black Hills, South Dakota; Figure 14.12).

Normal faulting is the dominant style of Late Cenozoic deformation in the Basin and Range Province (Figure 14.12); a region of horst mountains and graben valleys that stretches from New Mexico to southern Idaho. Tensional forces and crustal extension in this region are related to subduction of ocean crust (see textbook). Normal faults also occur throughout the Colorado Plateau and Rocky Mountains (Figure 14.12).

Folds and Thrust Faults—Anticlines, synclines and monoclines can be present in regions deformed largely by tensional forces. They can form when deep-seated faulting is expressed at the surface only by a fold in the drape of overlying strata (Figure 14.4). The intrusion of batholiths or smaller bodies into overlying strata can also produce folds. Folds are, however, more often the products of compressional forces (Figure 14.6). Compressional forces are greatest at active continental margins where tightly folded strata are formed during continental (and exotic terrane) collisions, or during the convergence of an oceanic and continental plate (Appalachian Mountains; Figure 14.13).

Thrust faults are perhaps the most spectacular creations of compressional forces. Their formation is easily envisioned when the crust is represented by two blocks of wood cut like that of Figure 14.7. If the blocks are held tightly togther while applying compressional force, the hanging wall block will be forced up and over that of the foot wall. In this way the stratigraphic succession of the underlying block is repeated in the overlying thrust sheets (Figure 14.6B,C). Thrust faults are developed along active margins where two continents have collided (e.g.,

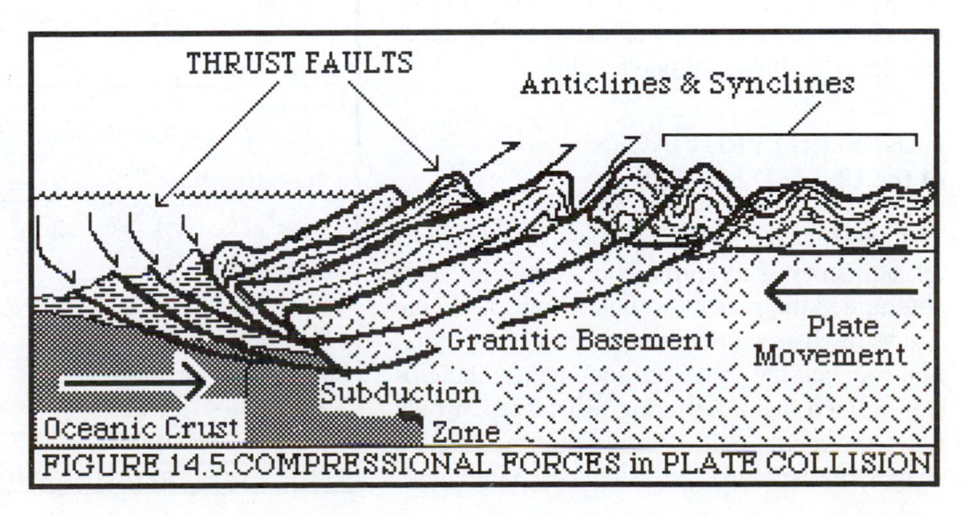

FIGURE 14.5.COMPRESSIONAL FORCES in PLATE COLLISION

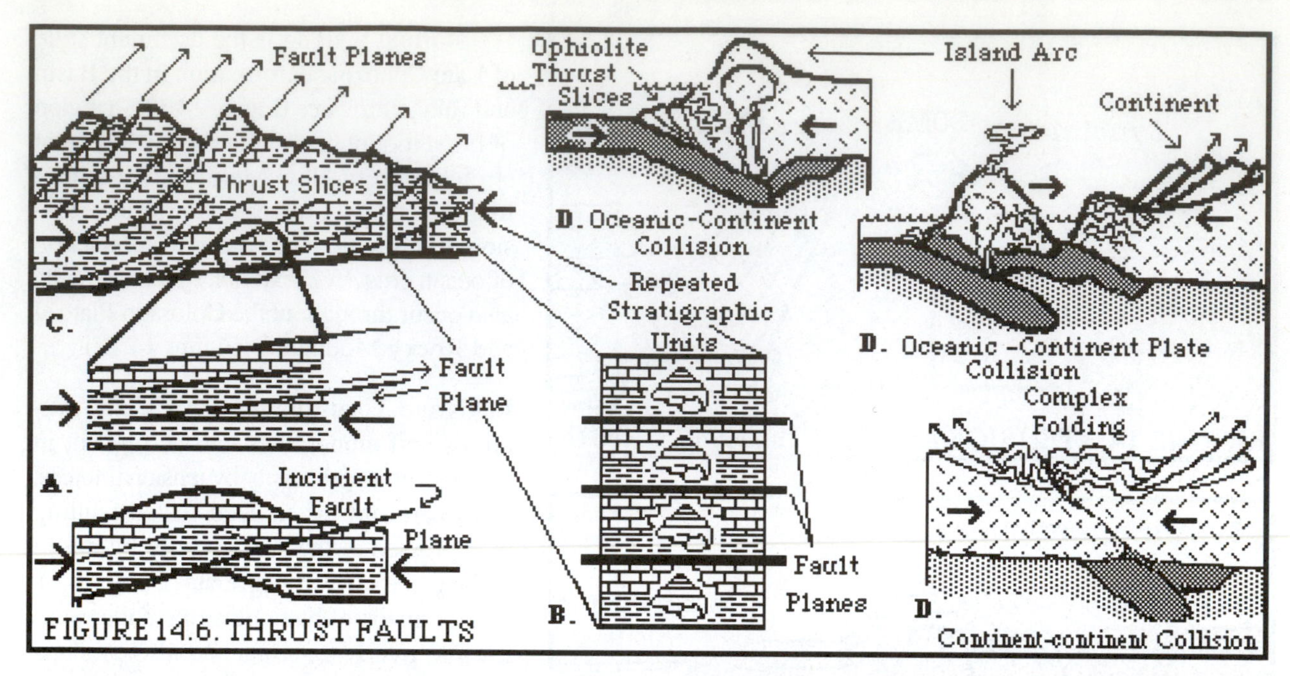

FIGURE 14.6. THRUST FAULTS

middle and Northern Rocky Mountains; Figure 14.12), where an exotic terrane has collided with a continent, or behind a zone of subduction along a continental margin, and are commonly associated with folding (Figure 14.6D).

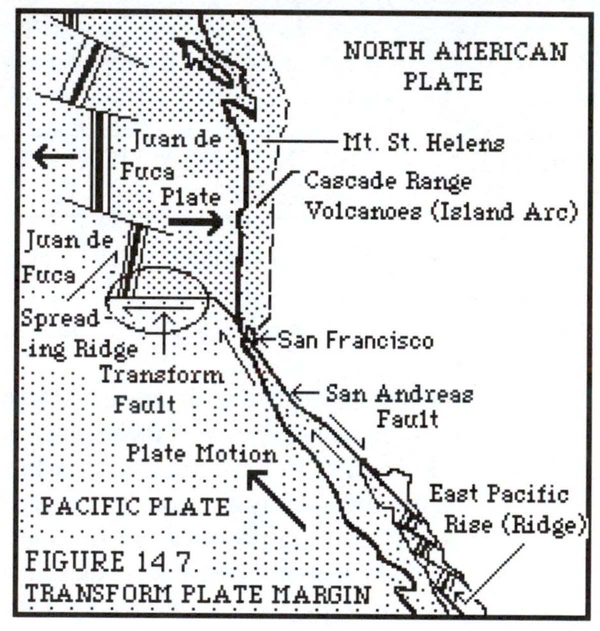

FIGURE 14.7. TRANSFORM PLATE MARGIN

Transform Faulting—These faults are not common within the continental crust, but are ubiquitous along spreading ridges. In a transform plate boundary the fault separates two plates which slide past one another (Figure 14.7). Because horizontal portions of oceanic crust are spreading along the midoceanic ridges and the crust as a whole forms a curved surface, spreading ridges must be offset at intervals by transform faults. The active portion of a transform fault lies between two offset portions of the spreading ridge (Figure 14.8). Transform faults that are also plate boundaries may link two spreading ridges, two subduction zones, or one of each, but subduction zones and spreading ridges can never be linked to one another.

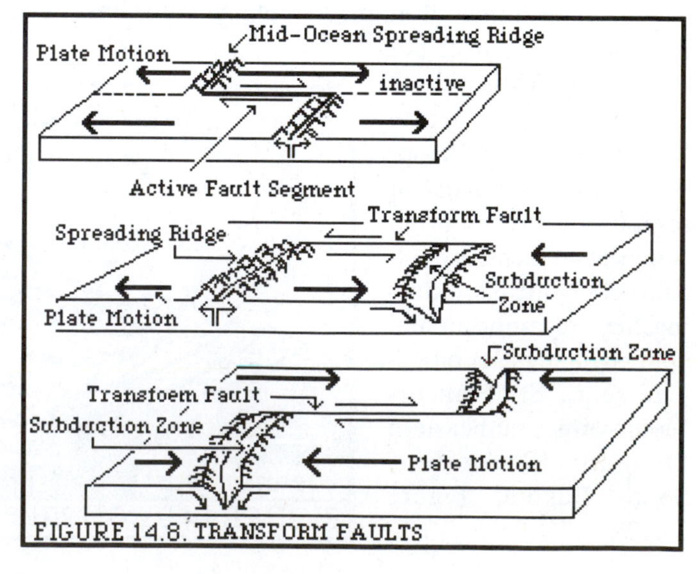

FIGURE 14.8. TRANSFORM FAULTS

LANDFORM PROVINCES of the UNITED STATES

The present face of the continent is the sum of all its history. The journeys of continents have determined the sites of mountain ranges and the course of river valleys, while styles of deformation roughed the shapes of mountains, and erosion refined them. Differing resistances of rock types to erosion considerably alter the original shapes of structures, with the resistant ridge-formers remaining long after softer beds have weathered away. Because erosion wears away mountains rather rapidly, those most striking surface features are commonly fairly young, having been born in the Tertiary, or younger. Cenozoic deformation has overprinted many Mesozoic and Paleozoic structures, while mountainous Precambrian terrains have long been worn to their roots.

Thus, some portions of any region's history are revealed simply by the types of landforms present. By integrating your knowledge of historical geology, rock types, and geological maps with a knowledge of landforms, you will be able to read some of that history directly from a landform map. If you are practiced at this exercise you may find that you can identify a geologic province or part of the country from a photograph that has "no" landmarks (to the uninitiated, that is!)!

Throughout your text and lab manual we have traced the major geological events which have shaped this and other continents. In the final exercise you will bring those past events to the present by relating geological events to the individual physical characters that define regions of the United States known as landform provinces (regions typified by particular land forms) (Figure 14.12).

After completing the geologic map exercises, use the geologic map of the US and tectonic map of North America (Plate X) to study the general geological relationships, the major styles of deformation, the ages of orogenies, and any other geological features which may characterize different regions of the United States. Then, as an exercise, relate what you know of the geological history to each of the landform provinces in Figure 14.12.

As a personal exercise, you might also want to try your hand at a "geological flash card" game by attempting to identify the landform province and region of the country to which photographs in your text belong (without using maps or figure captions).

TO OBTAIN GEOLOGIC MAPS

Geologic maps are easily and affordably obtained from several different sources: the United States Geological Survey, state geological surveys or departments of natural resources, the American Association of Petroleum Geologists, and commercial dealers.

For states east of the Mississippi:
The United States Geological Survey
Map Distribution
Reston, Virginia 22092

For western states:
Map Distribution
The United States Geological Survey
Box 25286, Federal Center, Building 41
Denver, Colorado 80225

The American Association of Petroleum Geologists publishes a geological highway map series ideal for the traveler: inexpensive combination geological highway maps of the U.S., divided into twelve regions, and containing cross-sections, stratigraphic columns, summary geological histories, brief summaries of the geology and paleontology of sites of interest and national monuments, and locations for collecting fossils, rocks and minerals.

Write to:
AAPG Bookstore
P.O. Box 979
Tulsa, OK 74101

VOCABULARY

Active continental margin, Compressional forces, Convergent plate boundary, Divergent plate boundary, Extensional force, Passive continental margin, Landform Province, Transform fault/plate boundary.

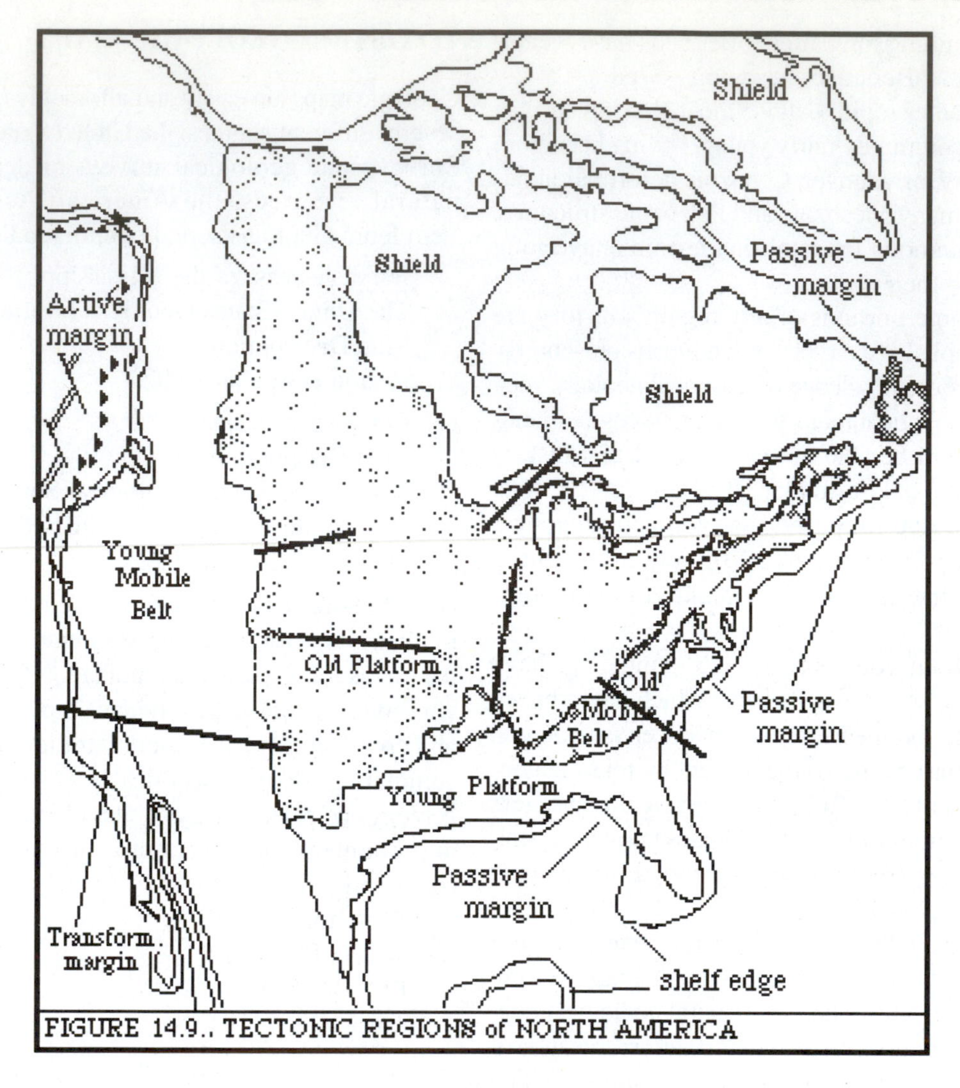

FIGURE 14.9.. TECTONIC REGIONS of NORTH AMERICA

LABORATORY EXERCISES

14.1. Tectonic Features of North America—Locate the major examples of the following tectonic features on the Geological Map of the U.S., and outline and label each on 14.9: domes, basins, arches, plateaus, tensional orogenic belts, and compressional orogenic belts.

14..2. Geologic Map of the United States—Your instructor will provide the map. You may wish to use your text in answering some of the questions in this and other map exercises.

1. Locate the major structural basins and domes (apparent from surface outcrops) of the stable platform.

A) Explain the origins of the basins in terms of plate tectonic theory. ______________________

____

____

____

B) What type of tectonic force accompanies dome and basin formations? ______________________

____

2. A) What are the major structures, tectonic forces and rock types (sedimentary, igneous and metamorphic) located in the Valley and Ridge, Blue Ridge, Piedmont and New England Uplands provinces of the ___ Appalachian Mountains? ______________________

____

B) When, by the map key, did deformation in these regions take place? ______________________

____

3. A) Describe the contact between the Mesozoic/Cenozoic strata and Paleozoic strata on the outer margin of the Appalachian Mountains from North Carolina to Alabama, and westward to the Gulf of Mexico?

____

____

B) In what region did the Cenozoic coastline of the Gulf of Mexico extend farthest inland? ______________________

____

C) What large structure lies buried beneath the sediments of this region? (see also the tectonic map of North America). ______________________

4. A) What type of faulting prevails in the southern Appalachian Mountains? ______________________

B) What type of force creates this type of fault? ______________________

5. A) What structure is represented by the outcrop pattern in central Tennessee and the tri-state region of Kentucky-Indiana-Ohio? ______________________

6. A) What is the dominant structure in the Ouachita Mountains of Southern Arkansas and eastern Oklahoma?

____

B) What type of force and tectonic event created these structures? ______________________

____

7. A) What structure is centered about the St. Francois Mountains of southeastern Missouri? ______________________

____

B) What type of faulting is present on the northern and eastern margins of this structure? ______________________

8. A) What is/are the age(s) of the intrusive igneous bodies in the Appalachian Mountain belt? (see map key).

____

B) What type of faulting is associated with the Triassic deposits found from South Carolina to Connecticut?

____

C) With what tectonic event are these intrusions and faulting associated? ______________________

9. A) What is/are the age(s) of the intrusions of batholiths within the Sierra Nevada Mountains and central Idaho? (see map key). ______________________

B) Explain their tectonic origins. ______________________

____

C) How does their origin differ from that of intrusions on the eastern coast of the US? ______________________

____

14.3. Major Tectonic Features, Mesozoic-Present Global Paleogeography—Map the estimated positions of spreading ridges, subduction zones and transform faults on the tectonic maps of Figure14.10A-E, maps generated by *Terra Mobilis* or by inspection on the *Drift Globe* , if used.

A) Infer, estimate to the best of your ability, and label by letter, the locations of regions with/of the following tectonic settings/activities on the map of the present (Figure 14.10A):

stable platforms (StblPltf);
mobile belts (MblBlt);
passive continental margins (PsvMrgn);
subduction zones (SbdZn);
island arcs (IslAr);
spreading ridges (SprdRdg);
transform faults (TrnsFlt);
deep ocean basins (DpOc);
continental rift zones (CR);
proto-ocean basins (PrtoOc);
young ocean basins (YngOc).

Use the consistency of plate motion directions and tectonic features between geologic periods as a guide to your reconstruction, and work backward from the present. Note that inference of spreading ridge locations becomes more speculative with age. Keep your inferences as simple as possible. Your instructor may have you work in groups, each responsible for one or two of the four maps, and/or may have you complete the exercise using the *Drift Globe* .

B) Indicate by letter code the regions of compressional forces (CF), tensional forces (TF), and shear forces (SF).

C) Follow the directions above for paleogeographic maps of the past (Figure 14.10B-E).

Uplands **Lowlands** **Shallow Seas**

R = Reefs, Rb = Red beds, C = Carbonate rocks, E = Evaporites, ● = Coal

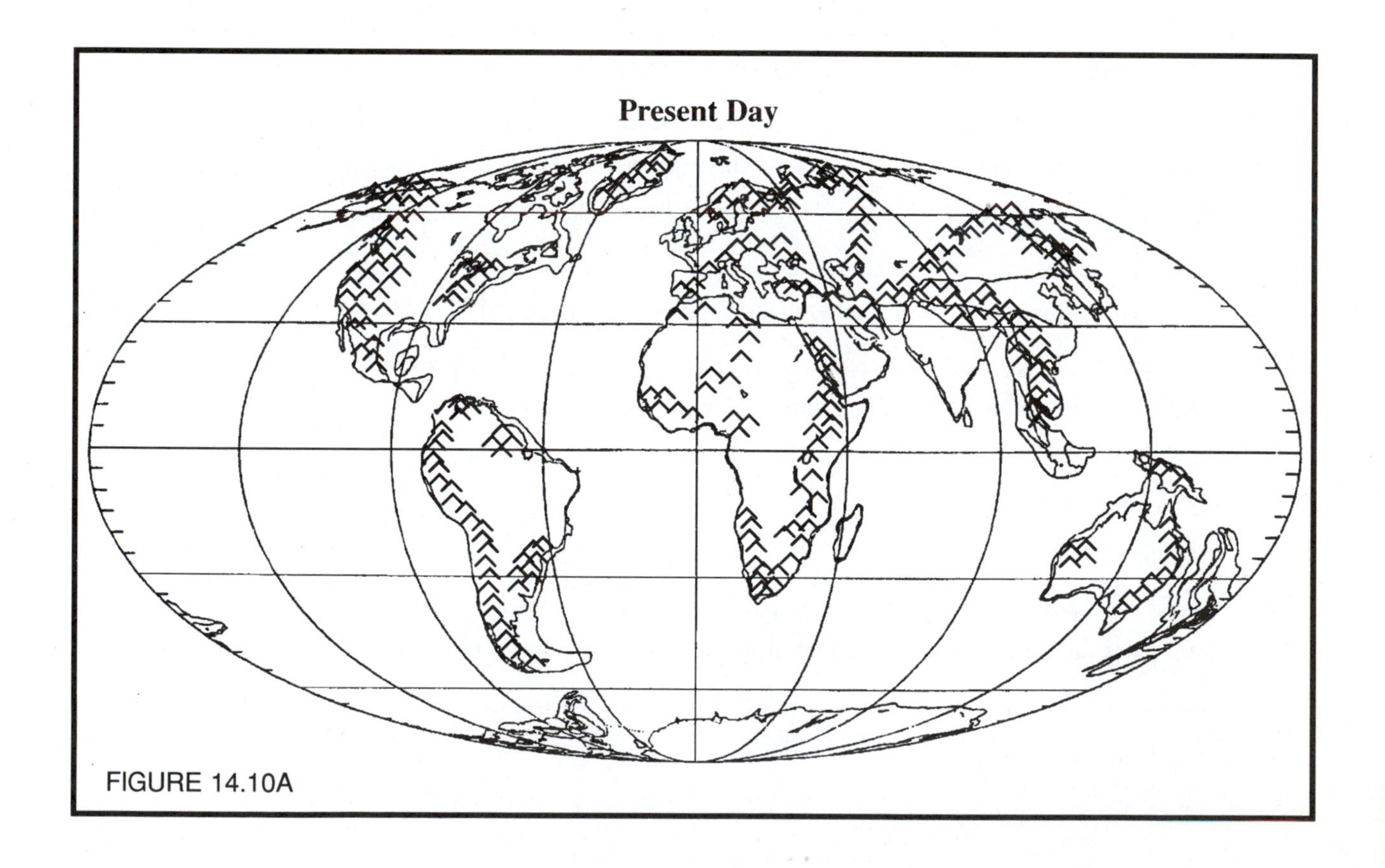

FIGURE 14.10A

Uplands **Lowlands** **Shallow Seas**

R = Reefs, Rb = Red beds, C = Carbonate rocks, E = Evaporites, ● = Coal

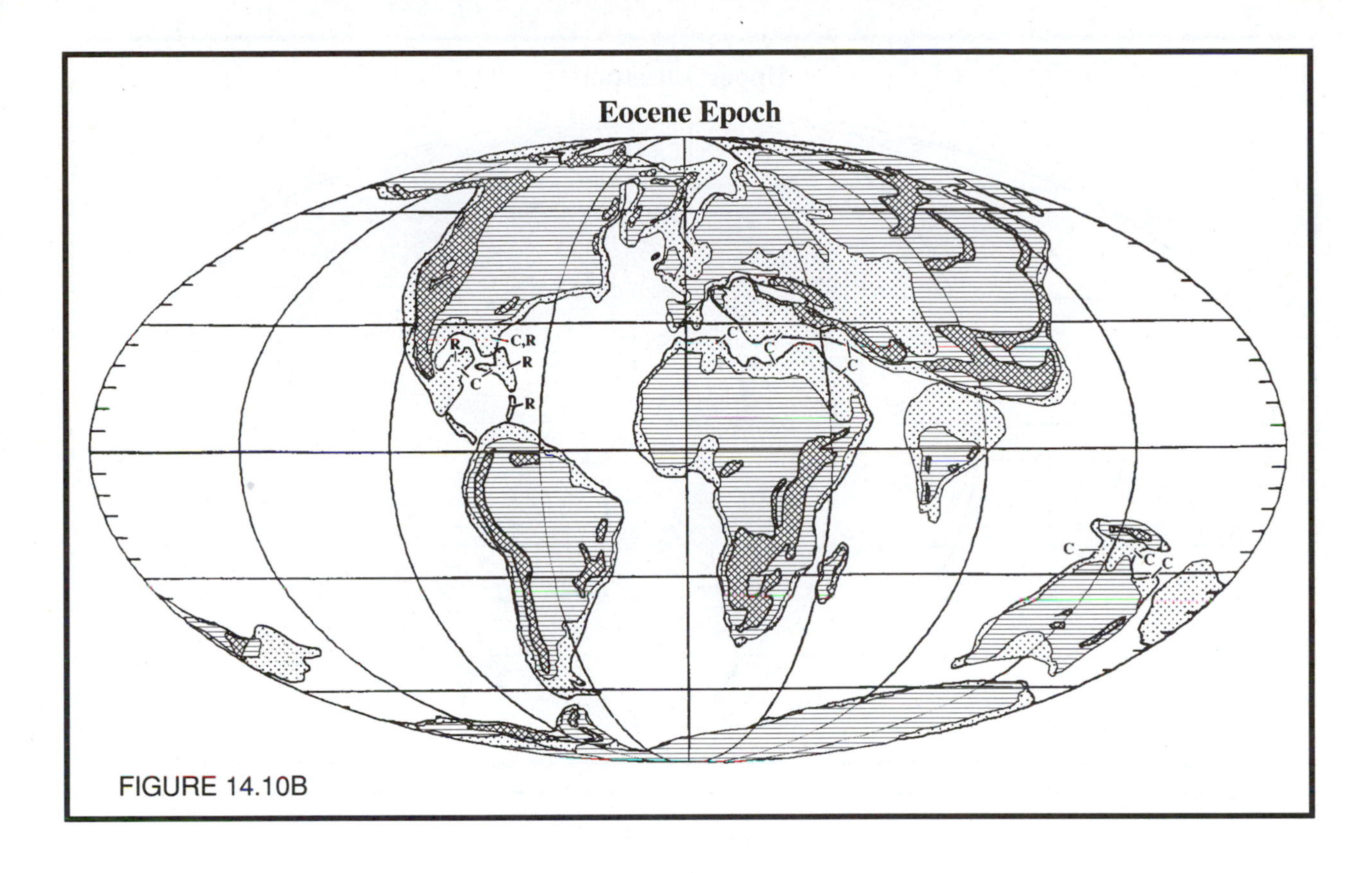

FIGURE 14.10B

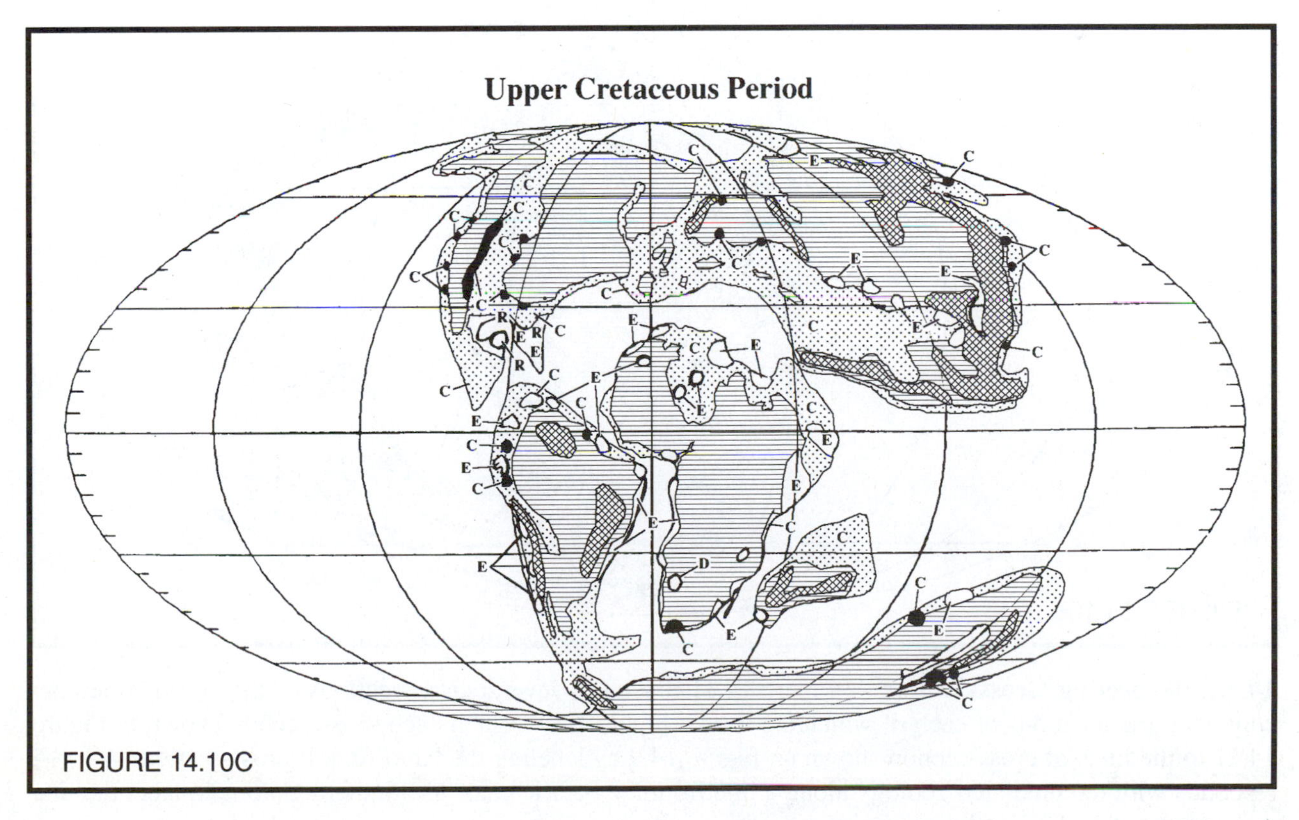

FIGURE 14.10C

Uplands Lowlands Shallow Seas

R = Reefs, Rb = Red beds, C = Carbonate rocks, E = Evaporites, ● = Coal

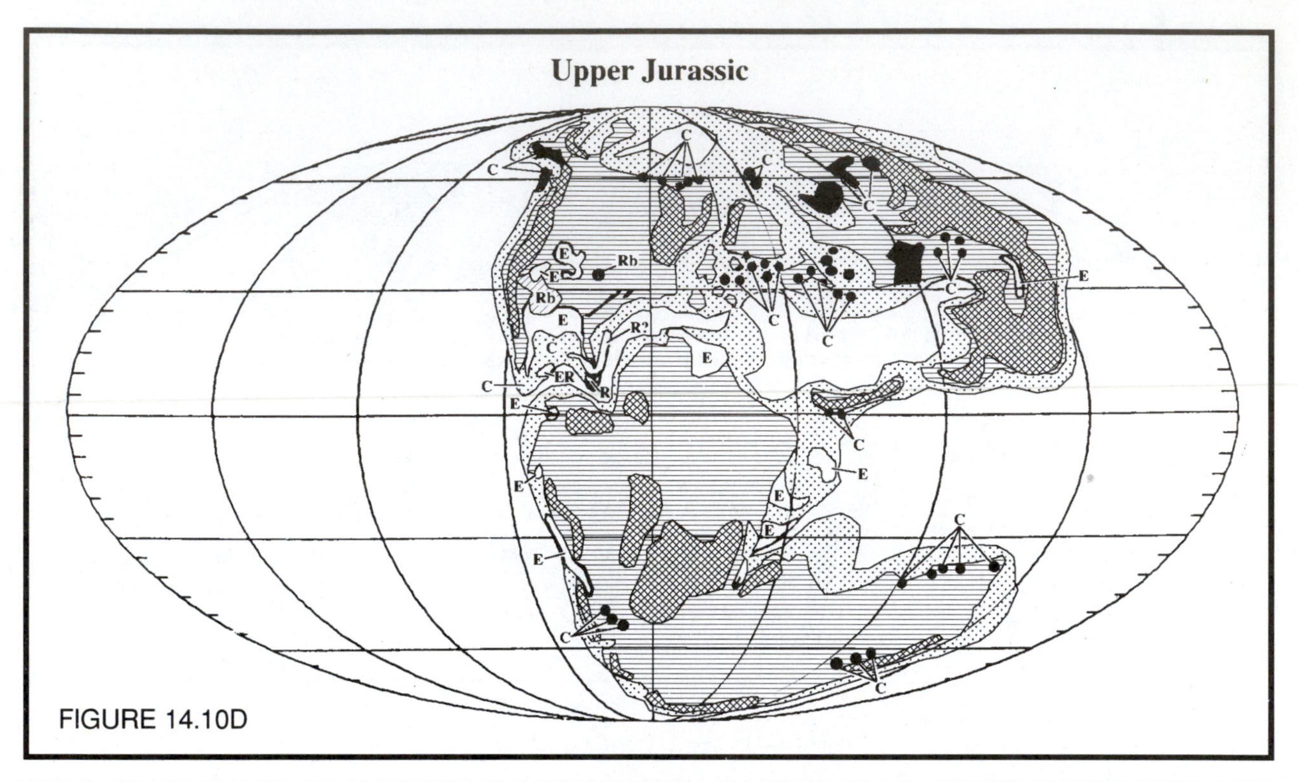

FIGURE 14.10D

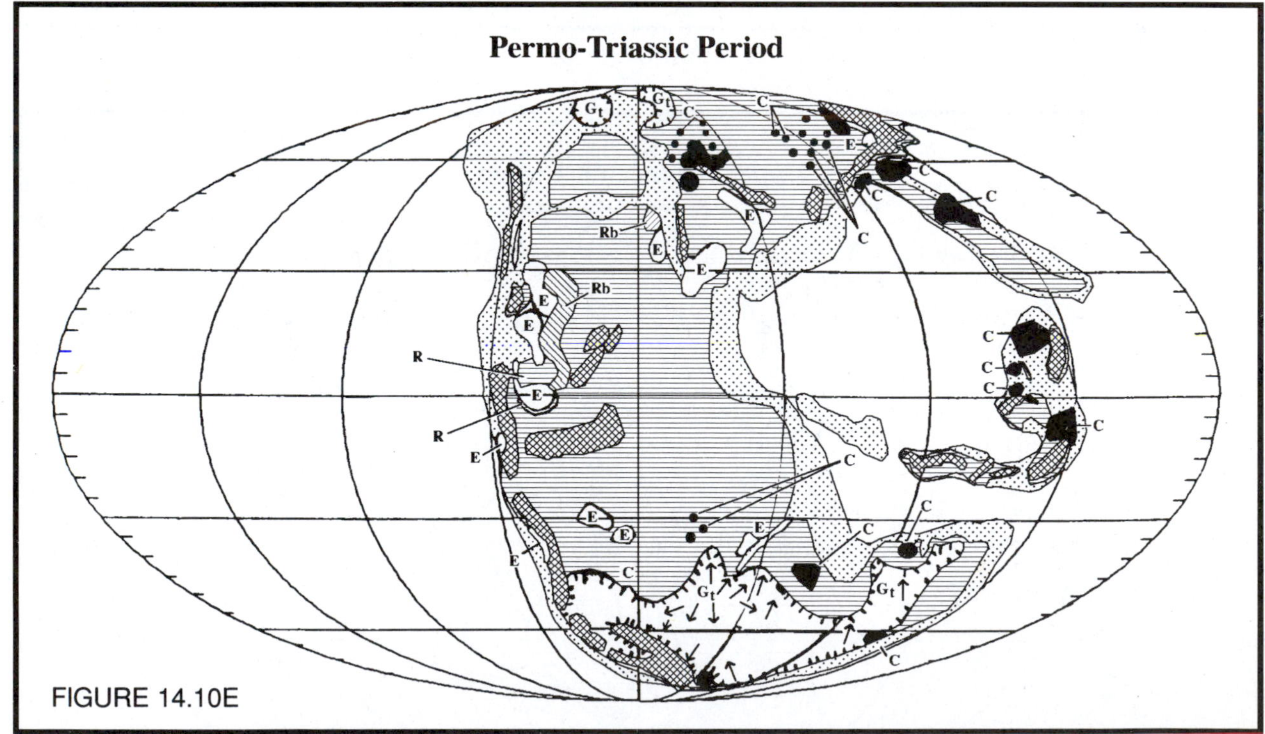

FIGURE 14.10E

14.4. Interpreting Cross-Sections—Figure 14.11 illustrates several cross-sections of the North American continent, the locations of each of which are shown Figure 14.9. Match each cross-section shown in Figure 14.11 to the lines of cross-section shown on Figure 14.9 by labeling the latter (e.g. if cross-section "A" corresponds with the predicted geology along a line from the Pacific coast to Southern Colorado, label the line so located as "A-A'"), and determine for each:

A) the tectonic context (shield, platform, and active, passive or transform margin);
B) the type(s) of force which produced the deformation in each; and,
C) the type(s) of geological structures present, and match the cross-section labels (e.g., A-A') of each cross-section in Figure14.11 with the appropriate line of cross-section lines on Figure 14.9.

Section A-A':

1) How many episodes of faulting occurred in this region and with what events were they associated? ______

2) During which geological periods did these events occur? ______

3) During which orogenic episodes were the Triassic fault basins formed? ______

4) What is a "suture zone" (right of center) and during which event and geological period was it formed?

______

5) What suite of rock types could be expected within this zone? ______

6) Contrast the early and late tectonic continental margin types evinced in this region. ______

Section B-B':

1) Explain the origin of the continental crust rocks at the far left. [Hint: what is an ophiolite (see Laboratory 12) and within which of the continental tectonic settings are such rocks formed?] ______

2) What type of faulting is associated with the ophiolites and these turbidites?

3) **A)** What major fault system (shown by the dark line) cuts across the right of the section and separates the Pacific and North American Plates? ______

B) What type of continental margin is represented by this fault system? ______

C) What type of plate boundary is associated with this type of fault? ______

4) Cite two explanations of the origin of the faulted region right of center. ______

5) How many orogenic episodes are evidenced in this region? ______

6) Contrast the crustal thickness in the right of center region with that in the other cross-sections. ______

7) Did the sediments now metamorphosed in the region of granitic mountains most likely originate in shallow or deep water, and why? ______

8) **A)** What structure is associated with the plateau region? ______

B) How might it be linked to the volcanic rocks of the region? ______

Section C-C':

1) **A)** Was the structure at the far left present before, or formed after, the Cambrian sedimentary rocks found within? ______

B) What is its age? ______________________________

____

2) When was the region at the far left uplifted and with what structure is it associated? ______________

____

3) A) What cryptic structure might explain the presence of the sedimentary deposits between the fault zone and the far right? ______________________________

B) When must such a hypothetical structure have been formed? ______________

____

4) A) What structure is indicated by the arrow and "?" at the right? ______________

B) When, according to the information given, was it formed, and why? ______________

____

Section D-D':

1) How many orogenic episodes are evidenced in this region, with what types of structures is each associated, and in what order and when did they occur? ______________________________

____

2) What is the stratigraphic and structural (age relation) evidence for the most recent episode of deformation?

____

____

3) With what tectonic type of continental margin are the structures at the far left and center associated? _____

____

____

Section E-E':

1) When did the orogenic event of this region occur? ______________________________

2) What age strata are missing? ______________________________

3) Is there evidence of two orogenic events in this region? Explain. ______________

____

____

Section F-F':

1) In what tectonic region of the continent is this section located? ______________

2) Determine the sequence of events in this region. ______________________________

____

____

3) With what tectonic structure, event, and of what age, are the Keweenawan Supergroup, Keewatin volcanics, and Duluth Gabbro associated? [Hint: how can the presence of (metamorphosed) conglomerates and other immature sedimentary rocks and basalts and gabbros within a shield area be explained?] ______________

____

4) Explain the tectonic associations of this structure with the origin of the Michigan Basin. ______________

____

5) What major global event is inferred from rocks such as the banded iron formations of the Keweenawan Supergroup? ______

14. 5. Tectonic Map of North America.—Plate X.

1. A) What type of faulting occurs on the Gulf of Mexico coastal plain and the continental shelf of the Gulf of Mexico? ______

B) What tectonic event(s) caused this faulting and during what period(s) of time did it occur? ______

2. A) What is the principal type of deformation in the Precambrian rocks of the Canadian Shield? ______

B) What type of force produced the deformation? ______

C) According to the map key, how many deformation events can be recognized? ______

3. A) Discuss the origin of the Baja Penninsula. ______

B) What will happen to southern California and the Baja Penninsula with continued sea-floor spreading and plate movement? ______

C) What is the dominant type of deformation in the west-central and southwestern coasts of California? ______

4. A) What is the dominant type of faulting in the Northern Rocky Mountains and Canadian Rocky Mountains as contrasted with the Middle and Southern Rocky Mountains? ______

B) How many orogenic episodes were there, what was the dominant type of deformation, force, and tectonic context of each, and which areas were affected by each episode? ______

5. A) Where are the volcanoes of North and Central America located? Note the spreading ridge off the coasts of Oregon and Washington. ______

B) Explain the steep submarine contours off the coasts of Alaska, Central America and Mexico. ______

C) Discuss the tectonic origins of the volcanoes of the Cascade Mountains and those of Alaska, Central America and Mexico. ______

6. Locate the Colorado Plateau.

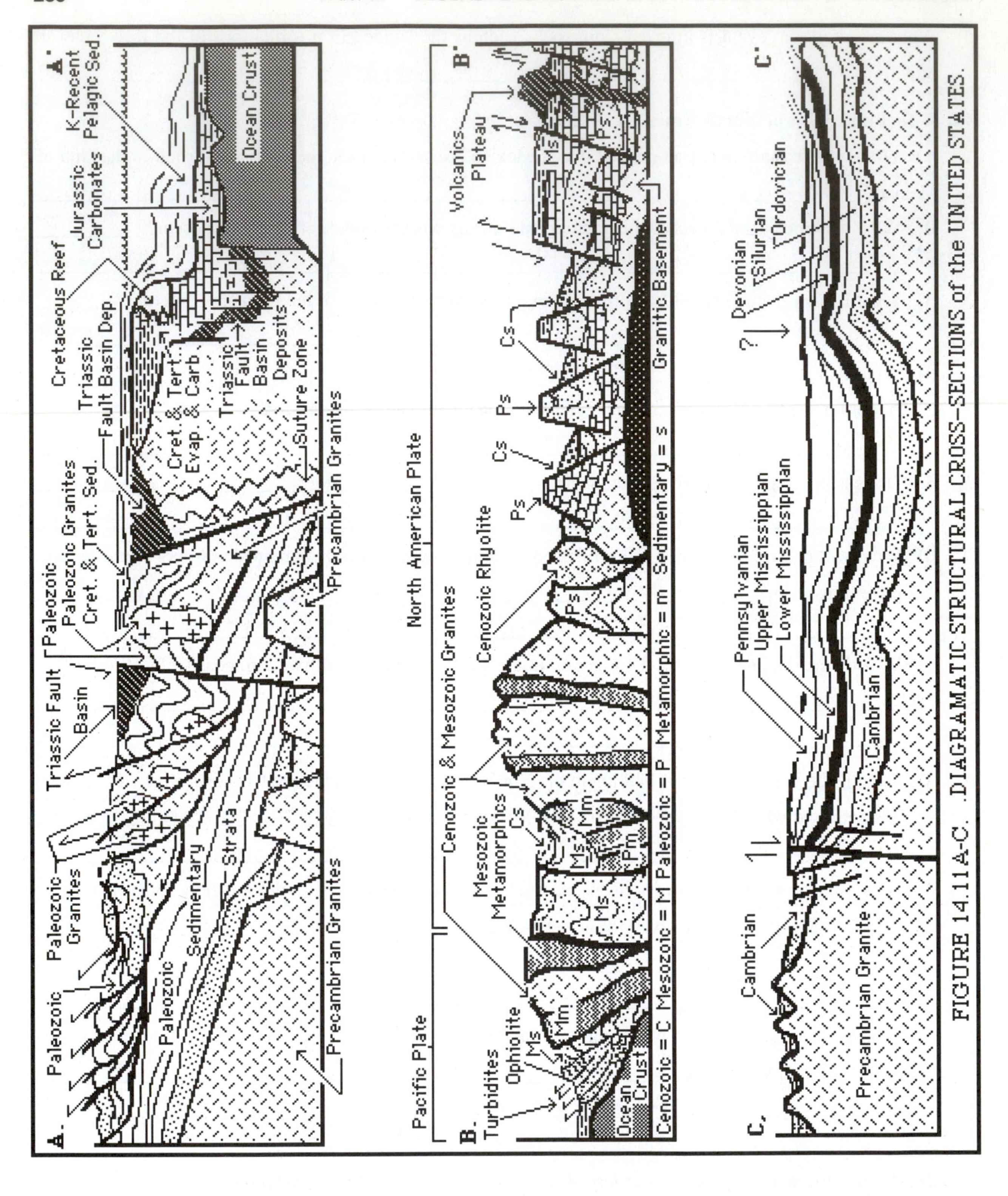

FIGURE 14.11 A-C. DIAGRAMATIC STRUCTURAL CROSS-SECTIONS of the UNITED STATES.

A) What type of faulting is found on the margins of, and within the Colorado Plateau? ______________

____

B) Is there any evidence of the cause of the uplift? ______________

7. A) Of what age are the Columbia Plateau basalts? Explain their origin in terms of plate tectonics. ______

____

____

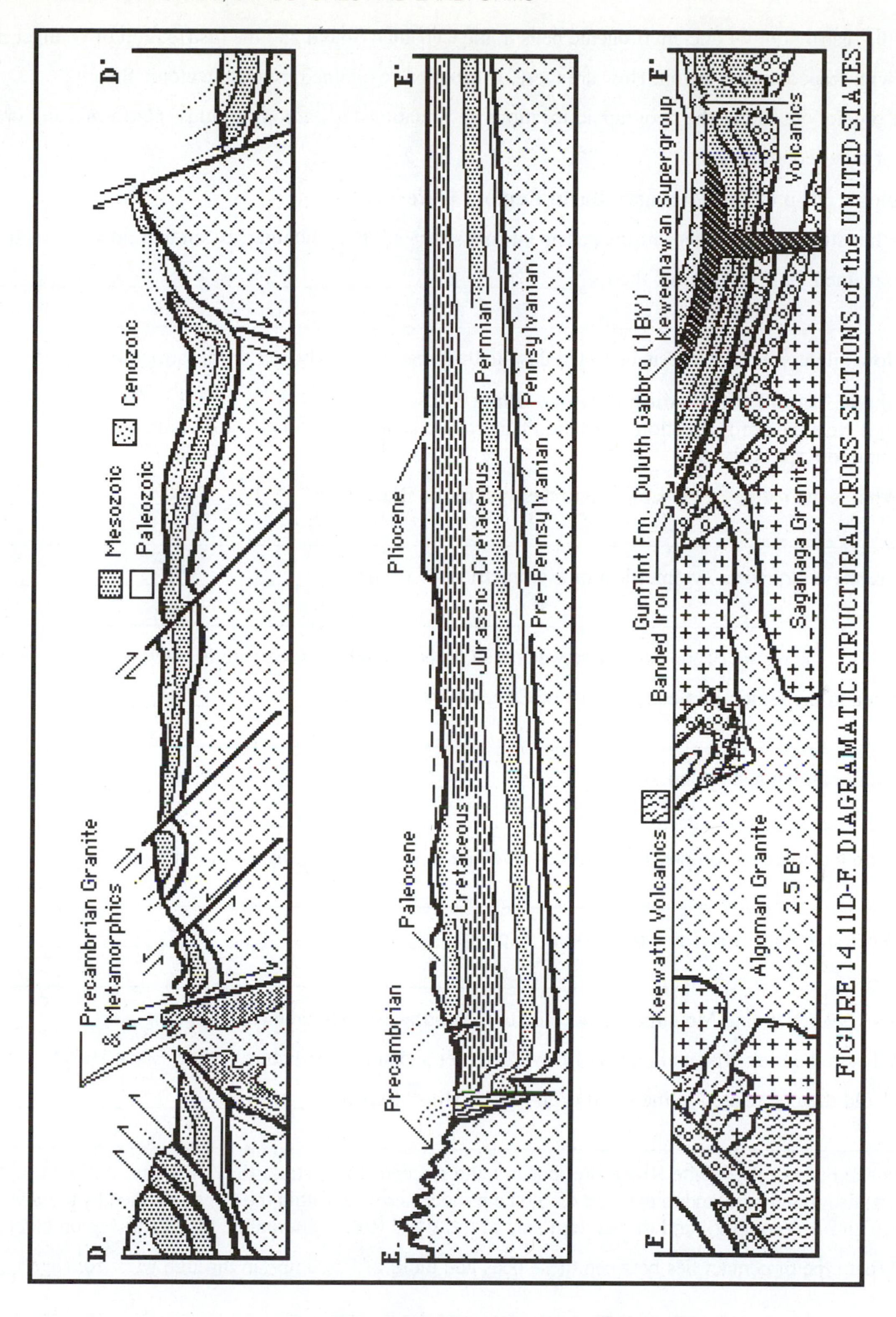

FIGURE 14.11D-F. DIAGRAMATIC STRUCTURAL CROSS-SECTIONS of the UNITED STATES.

B) Notice the chain of extinct volcanoes and laval flows that extend from the Columbia Plateau, across the Snake River Plain, to Yellowstone. How might the origin of the volcanic rocks in the vicinity of Yellowstone National Park be explained? ____________________

____

____

8. Note the distribution of ancient orogenic belts in the Canadian Shield and the distribution of younger mobile belts with respect to the craton. How are these distributions explained by plate tectonic theory?

9. Make predictions about the geographic locations and relationships among shields, platforms, and orogenic belts of other continents.

14.6. Geologic Map of the Michigan Basin Region. Plate VII.

1. A) What structure is present in the basement rocks beneath the Michigan Basin and how are such basins explained by plate tectonic theory? ______________________________

____

B) How might your hypothesis be tested despite the great depth of burial of this structure's base?

14.7. Geologic Map of the Wyoming Area.—Plate VII.

1. Sketch a west-east cross-section from the Phanerozoic sedimentary rocks just south and west of Buffalo, Wyoming to Rapid City, SD.

A) What is the nature of the structure to the west of the Black Hills? ______________________________

____

B) Are these structures symmetrical or asymmetrical? Explain. ______________________________

____

2. A) What is the nature of the contact between the Oligocene White River Group (Ow) and the surrounding strata in the region immediately east of Custer, SD? ______________________________

____

B) Strata of what age are missing? ______________________________

____

C) Were tensional or compressional forces involved in the origin of this structure? Explain. ______________________________

____

D) When did the Black Hills uplift occur? Explain. ______________________________

____

3. Devil's Tower National Monument is located to the northwest of Sundance, Wyoming.

A) What is the rock type and origin of Devil's Tower and similar structures in the Black Hills? ______________________________

B) Could they be related to the uplift of the Black Hills? Explain. ______________________________

____

4. Oligocene beds of the White River Group (Ow) and Eocene (Ews) strata that are found in the basins of the area are also found near, and to the west of, the Pathfinder Resevoir within the region drained by the Sweetwater River (central Wyoming), and the western side of the Wind River Mountains. Study this region carefully.

A) What type of contact lies between these beds and those of Precambrian through Mesozoic age? ______________________________

____

B) Of what age are the beds, and what type of structure may be buried beneath Ow sediments in this region? ______________________________

Notice that the Wind River flows northward and cuts directly through complexly deformed strata near Thermopolis. Rivers normally flow "down the slope" of a mountain, but the Wind River here does not. The distribution of Ow and Eocene strata and the unusual course of the Wind River are related.

C) What is the nature of the contact between the Precambrian rocks of the Wind River Mountains and the strata on its western flank? ______________________________

How does this outcrop pattern differ from that of the Eastern flank? Explain this pattern. ___________

D) When did the first period of uplift occur in this region? ___________

E) When did the second period of uplift begin? Speculate about the nature and sequence of events that might explain these two features of this region. ___________

14.8. Geologic Map, Part of the Grand Canyon, Arizona.—Plate VI.

1. A) What types of unconformities are present in the map region? ___________

B) How many unconformities are present? ___________

C) Of what age(s) are the strata immediately above the oldest unconformity? ___________

14.9. Geologic Map of the South-Central US.—Plate IX.

1) During what period(s) was this region deformed? ___________

2) What caused the deformtion? Contrast present and past tectonic contexts of the eastern U.S. ___________

3) What types of forces caused the deformation and what types of structures produced? ___________

4) Give one tectonic explanation for the unconformity, faulting and narrow basin in the region of the Lower Mississippi Valley (also see the Geologic Map of the U.S. and Tectonic Map of North America) ___________

14.10. Physiographic and Geological Provinces of the US.—Figure14.12.

After completing the geologic map exercises above, study the Geologic Map of the U.S., and determine the general:

A) structural features;

B) tectonic styles;

C) age ranges of strata;

D) the geologic cross-sections of Figure 14.11; and,

E) the structural features of Figure 14.9 in each of the landform provinces outlined in Figure 14.12. Complete Exercise worksheet 14.10 if so instructed. Consult your textbook or instructor if needed.

1) What principal geologic structures are found in the Central Stable Region (#1) and the Interior low plateaus (#2)? ___________

2) **A)** When were the rocks of the Atlantic Coastal Plain Region (#13) deposited? Notice the long low terraces that parallel the inner and outer margins of the province. ______________________________

____

B) What might be the origin of these terraces? ______________________________

____

C) Notice that the southeastern margin of the Ozark uplift and Ouachita Mountains (#5) and southern Appalachian Mountains (S. part #4) are abruptly truncated by the coastal plain sediments. What structures lie buried beneath the Gulf Coastal Plain (S. part #13) sediments? ______________________________

____

D) Is there any similarity in structural style or age of deformation between two of these regions? ______

____

3. What is the principal structure underlying the sediments offshore of the eastern portion of region #13? ____

____

4. A) When did the western portion of region #2 attain its present slope? ______________________________

____

B) When was the last great epeiric sea present throughout this region? ______________________________

5) Of what age are the rocks in the Superior Upland (S. portion #1)? ______________________________

6. A) Explain the origin of the mountain ranges and valleys of the Basin and Range Province. (#10) ______

____

B) What is/are the trend(s) of the ranges in Nevada and southern Arizona? ______________________________

____

7. A) What type of faulting dominates regions #6 and #7 and the southern portion of #12? ______________

____

B) What is the trend of the low coastal mountains in southern California (#12)? ______________________

____

C) To what other important feature might this trend be related? ______________________________

____

D) Speculate why this could be so. ______________________________

____

8. A) Notice the meandering course and deep canyon of the Colorado River in the Colorado Plateau Province (#9). Meandering rivers normally develop on very gradual slopes and deep canyons normally develop o steep mountain slopes. Why does the Colorado River meander in the region of the Grand Canyon? ____

____

____

9) The Columbia Plateau (#8) is surrounded by mountainous regions, but there is very little relief within the plateau. Why? ______________________________

____

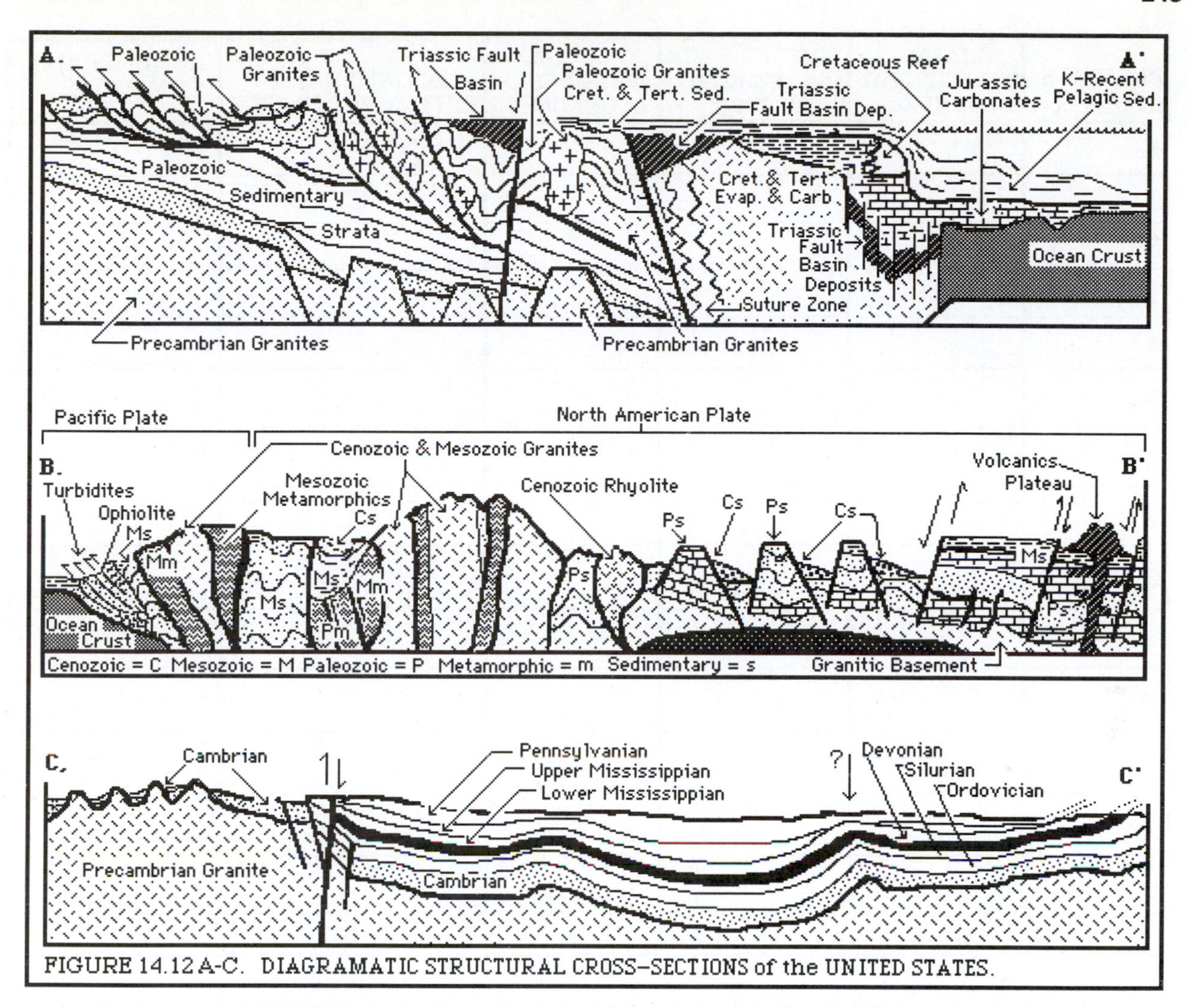

FIGURE 14.12 A-C. DIAGRAMATIC STRUCTURAL CROSS-SECTIONS of the UNITED STATES.

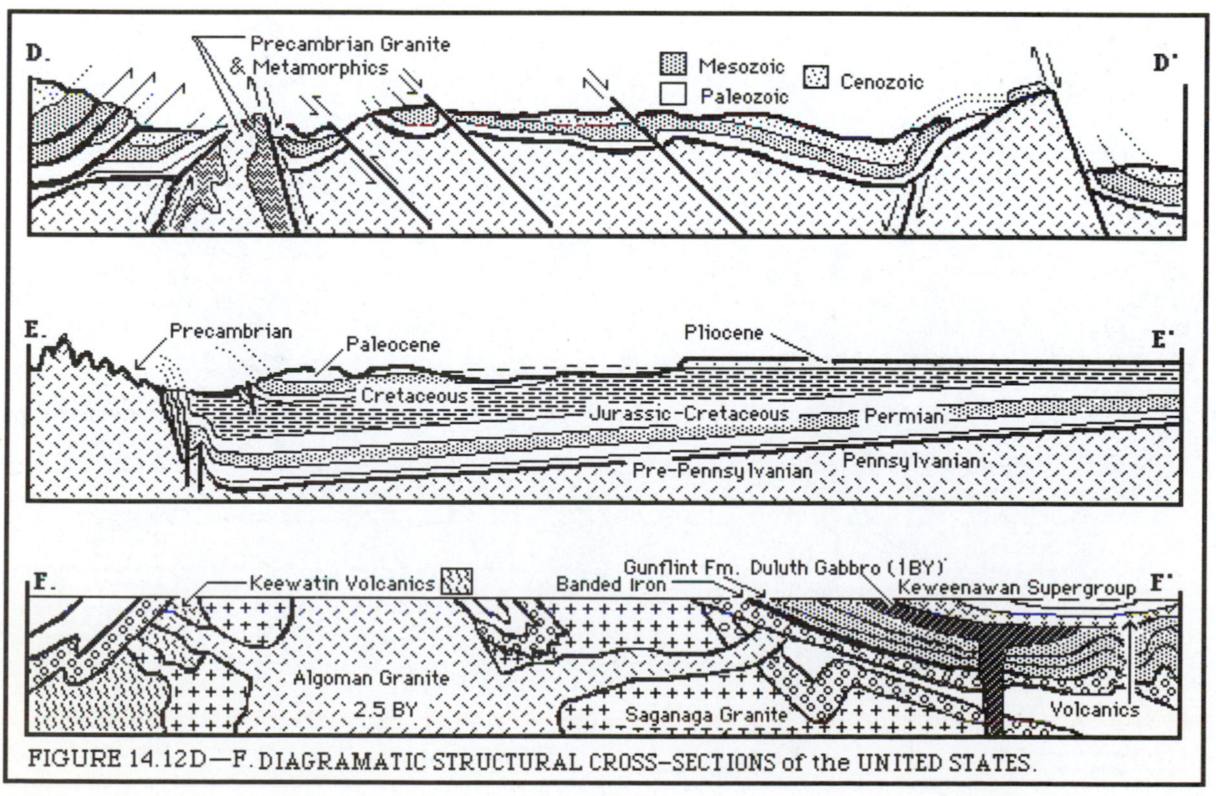

FIGURE 14.12D—F. DIAGRAMATIC STRUCTURAL CROSS-SECTIONS of the UNITED STATES.

PROVINCE see MAP Key	MAJOR STRUCTURAL FEATURES	IGNEOUS ROCK AGE, INTRUSIVE/ EXTRUSIVE	TECTONIC SETTING(S) or ORIGIN(S)	AGE(S) (ERA) of EXTENSIVE DEPOSITS	AGE(S) (ERA) of MISSING STRATA
1.					
2.					
3a.					
3b.					
4.					
5.					
6.					
7.					
8.					
9.					
10.					
11.					
12.					
13.					

WORKSHEET EXERCISE 14.10. GEOLOGIC CHARACTER of PHYSIOGRAPHIC PROVINCES.

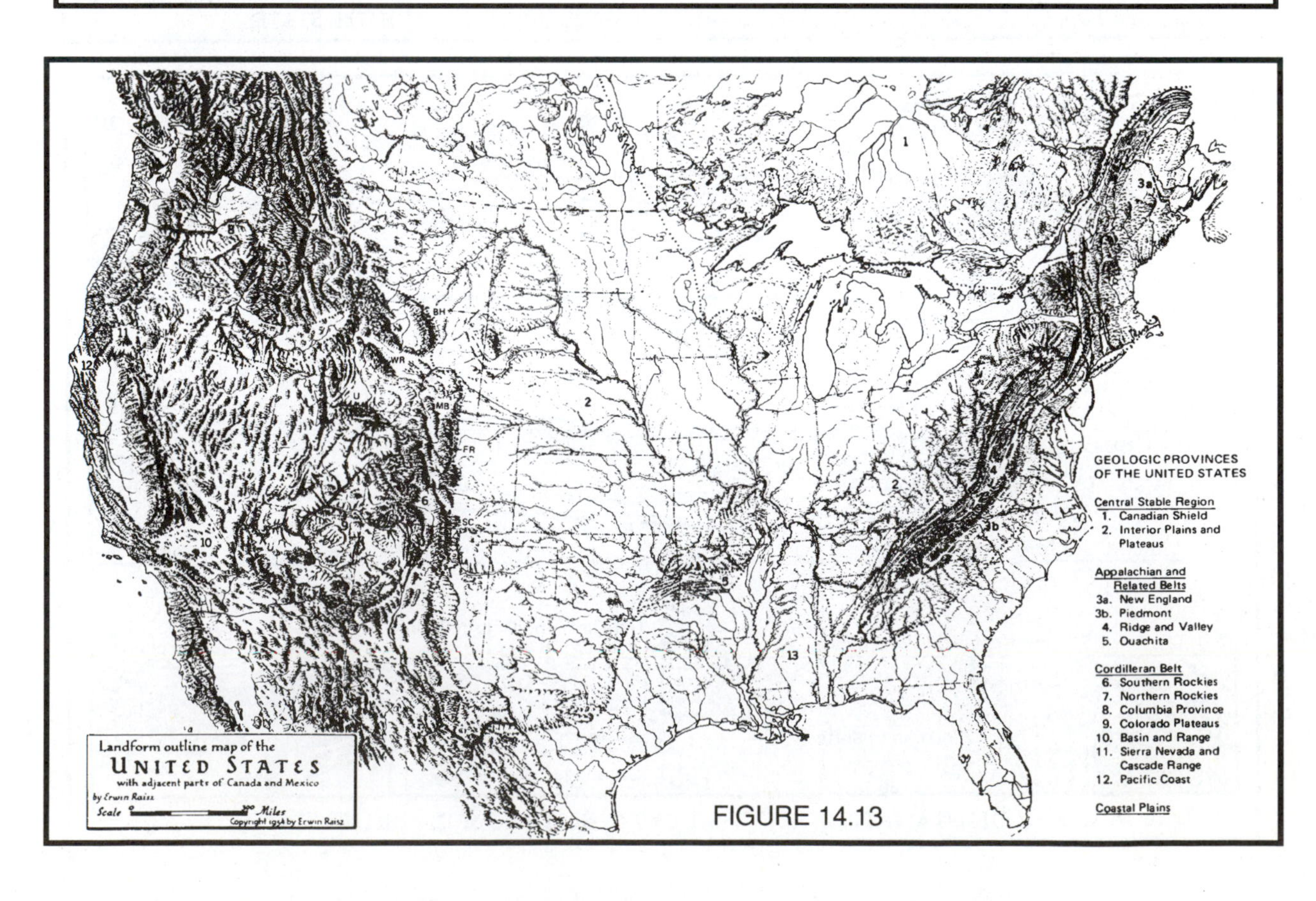

FIGURE 14.13

APPENDIX I
The Tiny and the Overlooked: Microfossils & Other Groups

The invertebrate and protist phyla discussed in this chapter are included because they are important members of the fossil record, particularly because they are widely used in biostratigraphy (Laboratory 7) and paleoecological studies (Laboratory 6).

DESCRIPTIVE PALEONTOLOGY PHYLUM ARTHROPODA

Superclass Crustacea, Class Ostracoda—The Crustacea includes crabs, crayfish, lobsters, shrimp, barnacles, sow bugs and ostracodes. Ostracoda (*ostraca* = shelled) are the most paleontologically important and common crustacean group in the fossil record. They are an unusual group of tiny animals that have a bi-valved shell or carapace composed of calcite (Figure AI.1A). The two valves are hinged and held together by a flexible ligament.

Although most ostracodes are small (<1 mm long), they have complex organ systems like other arthropods (Figure AI.1B). The head and thorax are fused and bear 7 pairs of appendages used in feeding, locomotion and sensing. The shell interiors shows numerous attachment scars that mark the sites of muscles and antennae that are unlike those of brachiopods and bivalves.

The bilaterally symmetrical carapace may be smooth or ornamented by rounded pits, polygonal depressions, ridges or spines (Figure AI.2). Although bi-valved, there are no growth lines because most ostracodes, like other arthropods, grow by molting, not by shell accretion.

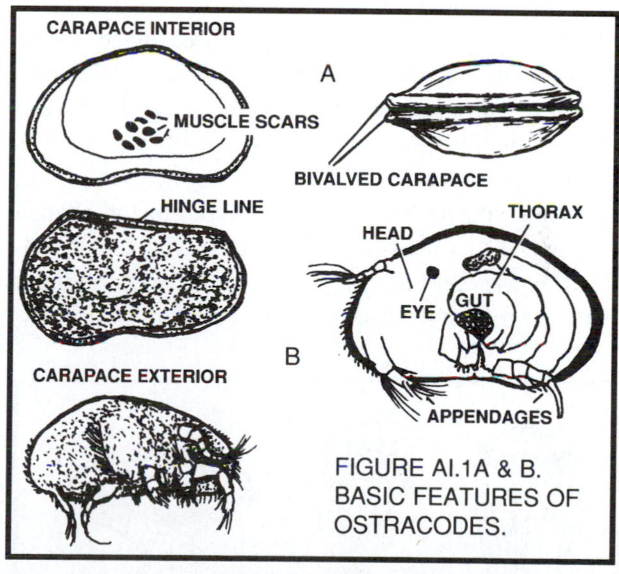

FIGURE AI.1A & B. BASIC FEATURES OF OSTRACODES.

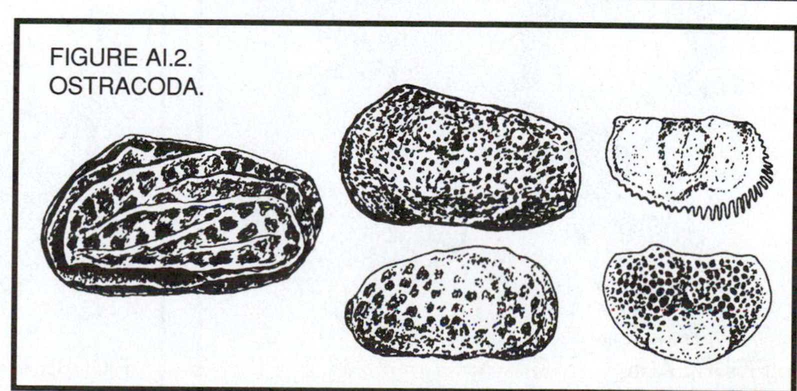

FIGURE AI.2. OSTRACODA.

PHYLUM ARCHAEOCYATHA ('ANCIENT CUP")

The Archaeocyatha (*anchaeo* = ancient; *cyatho* = cup) is an extinct phylum that appeared in the Early Cambrian, spread worldwide and became extinct by the Cambrian Period's close (Figure AI.3). Because no living animals are similar, it has been difficult to determine the group to which they were most closely related. They have been classified with corals, sponges, and calcareous algae, but are now given phylum status.

The typical skeleton or cup is a subcylindrical tube with a conical base (Figure AI.3). The tapered end is closed and was sunk in the sediment or supported upright by root-like holdfasts. The cup may consist of a single wall and the central cavity, or of two walls separated by a space called the intervallum. The intervallum is subdivided by radiating septa that join the inner and outer walls. The walls and the septa are pierced by pores and the lower intervallum spaces are subdivided by curved walls called tabulae and dissepiments. No trace of the soft tissues is known, but from the skeleton it appears that there was only a cellular level of organization, no tissue layers or organs. Most of the physiology probably took place within the intervallum, where food-bearing water flowed passively inward from the central cavity and outward through wall pores.

Most archaeocyathids have solitary vase-like cups, but some species' cones are discoidal or fan-shaped (Figure AI.4). Other species were colonial and developed by asexual budding of new cups from the outer wall. Branching colonies were common, but massive and chain-like colonies are also know. Solitary and colonial species grew in large patches and so became the first reef-builders.

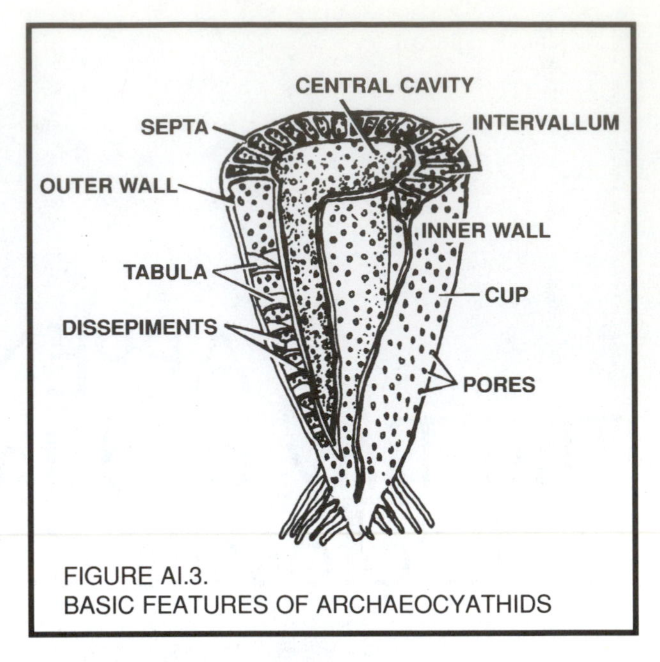

FIGURE AI.3.
BASIC FEATURES OF ARCHAEOCYATHIDS

PHYLUM CONODONTA ("CONE-TOOTHED")

Conodonts (*cono* = cone-shaped; *donta* = tooth) are an extinct, very important group of organisms known only from microscopic tooth- or jaw-like hard parts of controversial affinities (Figure AI.5). Teeth or elements usually occur isolated (Figure AI.6), but each was originally part of a natural assemblage of several elements. Like the teeth of a jaw, a natural assemblage consists of two identical, parallel "rows" each

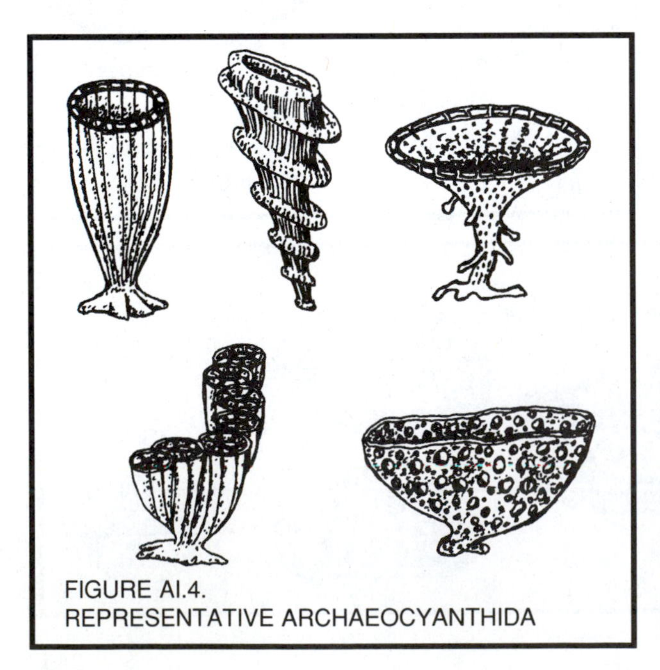
FIGURE AI.4.
REPRESENTATIVE ARCHAEOCYANTHIDA

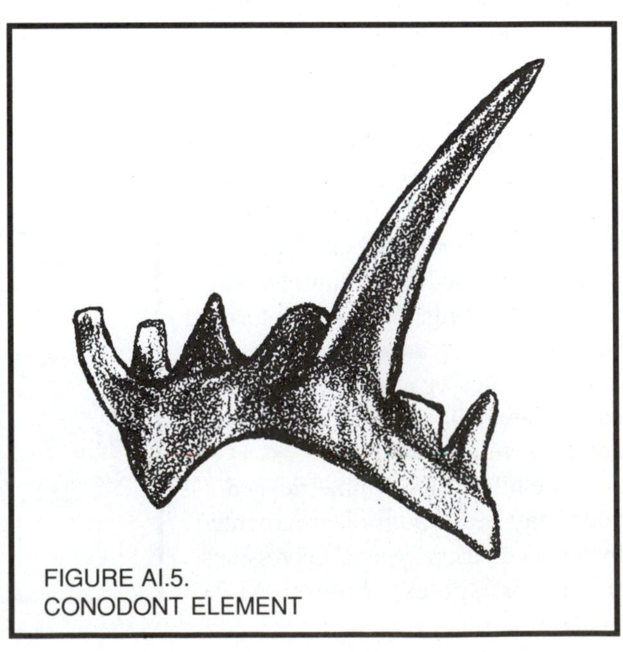
FIGURE AI.5.
CONODONT ELEMENT

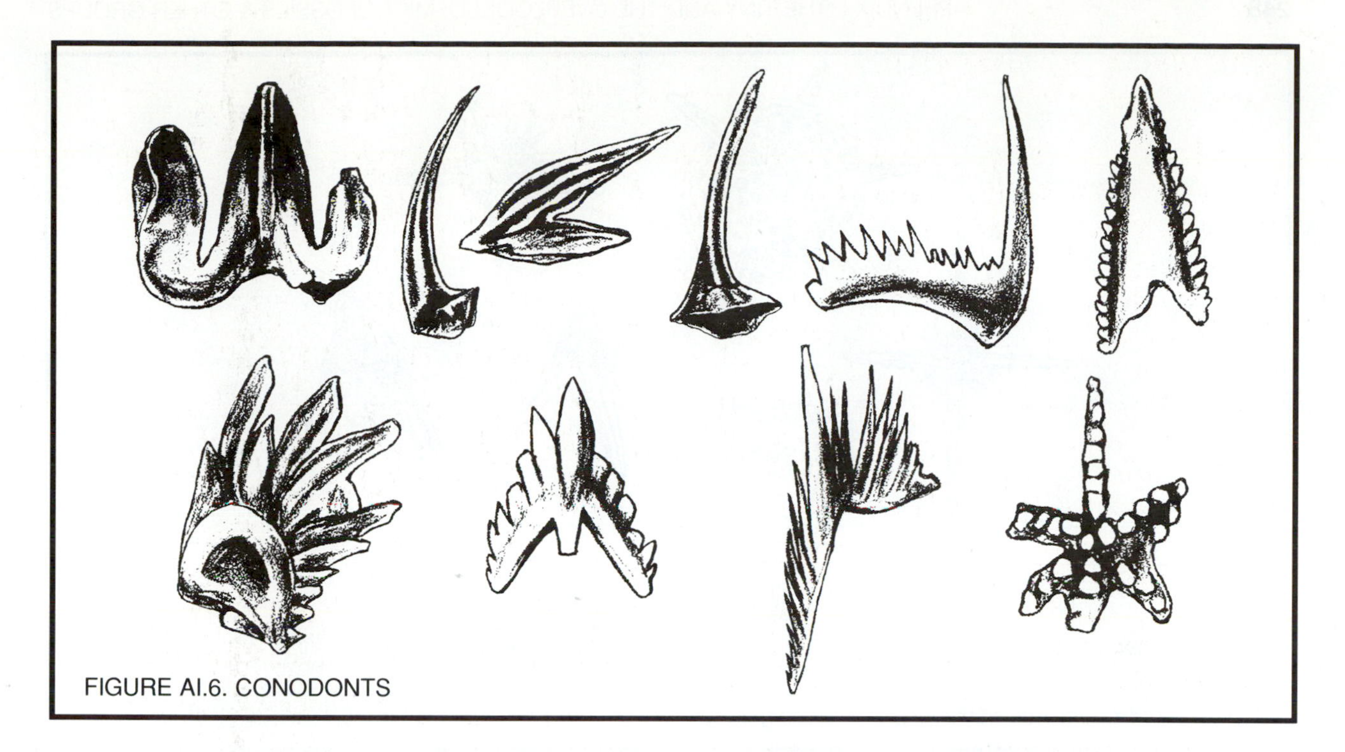

FIGURE AI.6. CONODONTS

with a distinct series of element types. Each "row" contains one of an element pair, and thus, the structure has bilateral symmetry. An assemblage may have functioned as a grasping organ or an internal supports for a ring of tentacles. Over the years these enigmatic fossils have been classified as parts of extinct cnidarians, nematodes and annelid worms, gastropods and cephalopods, crustaceans, and even extinct fish. But none of these classifications were satisfactory and the nature of these unusual fossils remained unexplained until recently.

The elusive, softbodied animal's recent discovery shows that the tooth-like structures belonged to a group very similar to chordates (animals with a spinal chord; e.g., vertebrates). A natural assemblage of elements was found in the mouth region of a small, bilaterally symmetrical worm-like animal that has a head, "tail" structure, and distinct midline. The softbodied impression, bilateral symmetry of the assemblages, and the carbonate apatite mineralogy of the elements are consistent with a chordate affinity. They are considered to be long-lost cousins of chordates but are placed in the phylum Conodonta.

KINGDOM PROTISTA

The kingdom Protista includes all single-celled eukaryotic organisms (those having a nucleus with strands of DNA (chromosomes) enclosed in a nuclear membrane). Some protists photosynthesize and others absorb organic matter for food. Most are microscopic, but despite small size, they are very common and important in the fossil record. Of the protists, only Foraminifera, Radiolaria, Diatoms and Coccolithophora are of great interest to paleontologists because these groups secreted "shells" (called tests) composed of calcium carbonate, or silica, or of cemented sediment grains and organic matter. Protists are so phenomenally abundant that thick limestone and chert deposits are formed largely of their tests. They are important also because they are at the global food chain's base. Diatoms, coccolithophores and other algae also produce most of the atmosphere's renewable oxygen and help maintain the carbon dioxide balance.

Radiolaria—Radiolaria are protists that extend cytoplasm through holes in their silica tests to envelope food particles and to aid in locomotion. All radiolaria are microscopic. Their tests form beautiful lacy spheres, helmets and bells (Figure AI.7). Because their tests are made of silica, they do not dissolve in deep, calcium deficient water as do foraminifera . Thus, they are the major source of deep-water chert.

Foraminifera—Foraminifera are the most well studied protists of the fossil record (Figure AI.9). They too collect food and move about with the aid of cytoplasm extended from the test. The tests are composed of calcite or cemented sand particles. The former type comprise a large volume of open ocean sediments. Large foraminifera are classified by internal features of the tests, and small taxa by external features. The tests of many are perforated by small holes (*foramin* = small hole); hence, foraminifera.

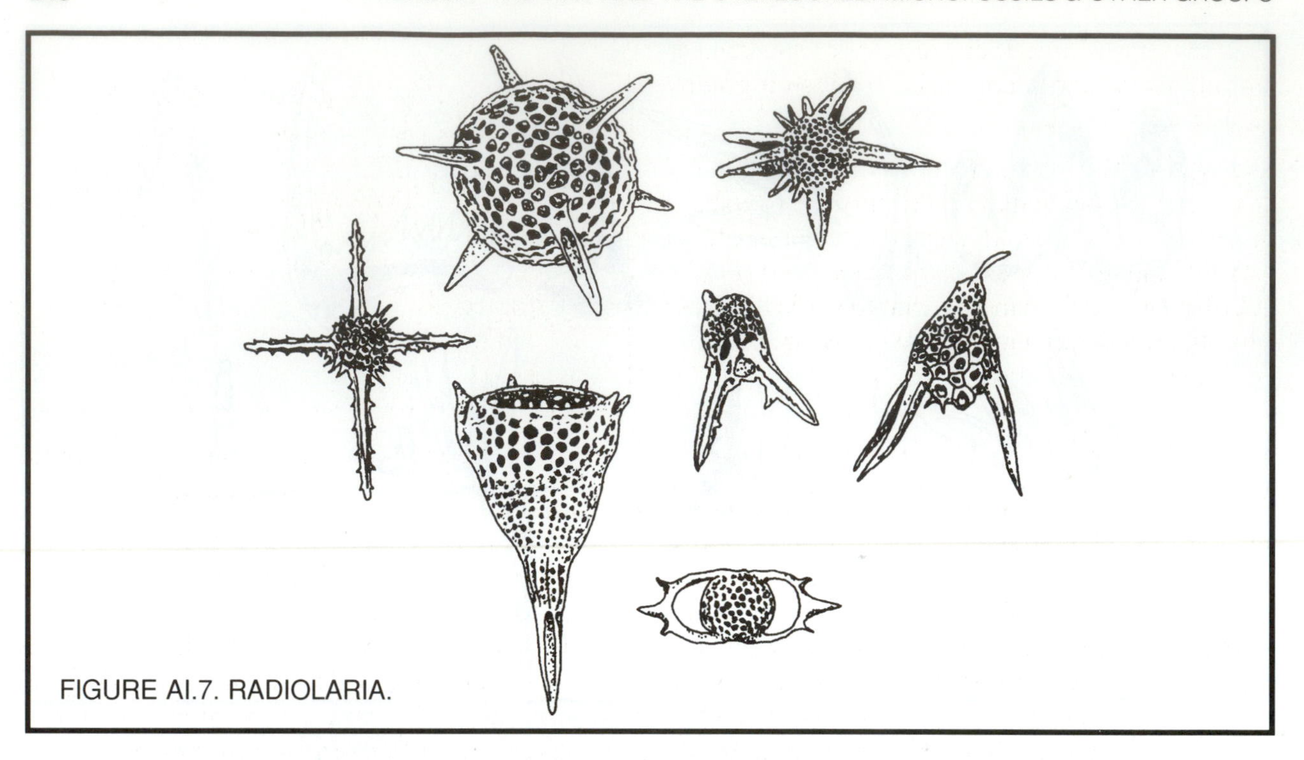

FIGURE AI.7. RADIOLARIA.

Foraminifera have evolved a great variety of test shapes and range of sizes (Figure AI.8). One species grew as large as a quarter; a phenomenal size for a single-celled creature! The species of a common Paleozoic group, the fusulinids are shaped like, and as large as, grains of rice. Others took on coiled shapes resembling tiny gastropod and cephalopod shells.

Coccolithophora—Coccolithophores, or nannofossils, are a class of golden-brown algae. Their subspherical to ellipsoidal tests (coccospheres) consist of small interlocking circular shields or coccoliths (*coccos* = seed; *lithos* = rock; Figure AI.9). There are many types of coccoliths, more than one of which may be found within a coccosphere. Coccolithophores are very important sources of deep ocean carbonate sediment. They are astronomically abundant. As many as a trillion tests may be found in one cubic centimeter of oceanic sediment!

They had not been extensively studied until the advent of the scanning electron microscope, because

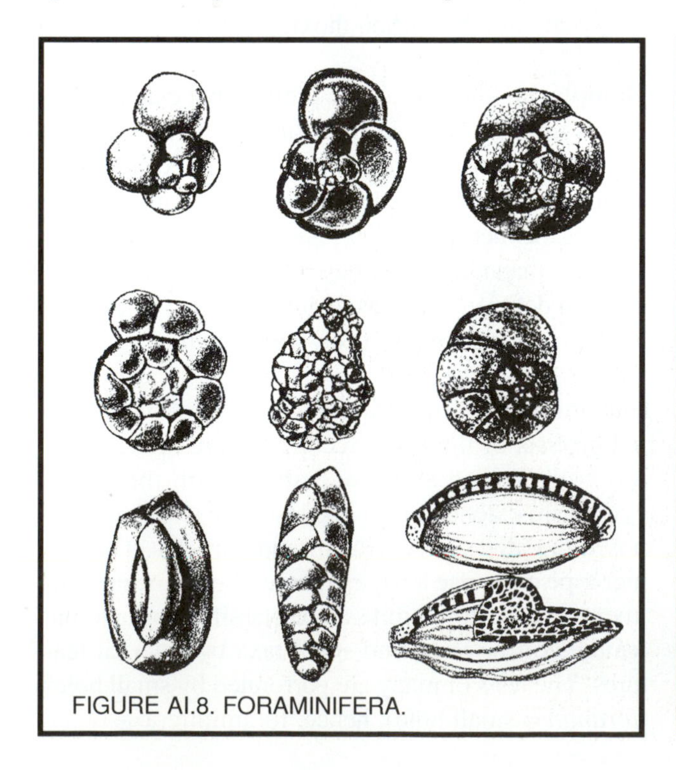

FIGURE AI.8. FORAMINIFERA.

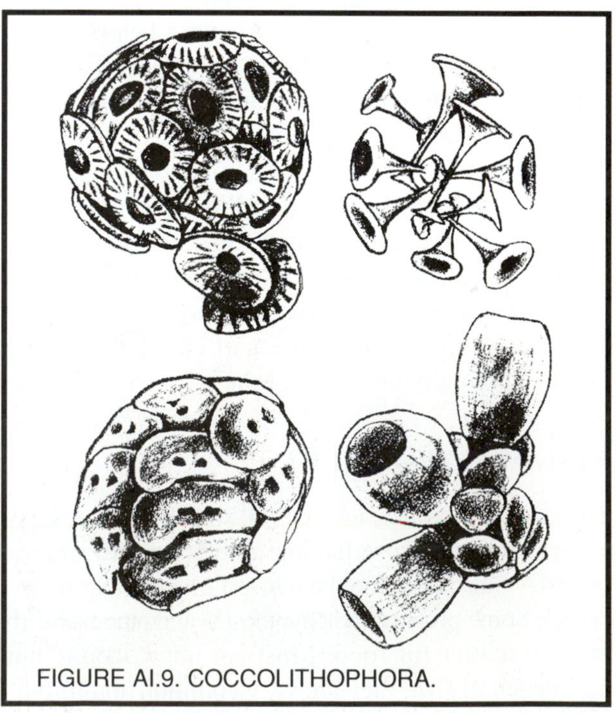

FIGURE AI.9. COCCOLITHOPHORA.

their tests are so tiny. Today they are as important as Foraminifera and Radiolaria in biostratigraphy (Laboratory 7).

Diatoms—Diatoms are algae with bivalved "shells" (*dia* = two; *toma* = cutting) called frustules of opalline silica. One circular test fits over the other like a hat box lid, and tests are circular or elongated (Figure AI.10). They photosynthesize their food and like other protists, are so exceedingly abundant that millions can be found in a gram of sediment.

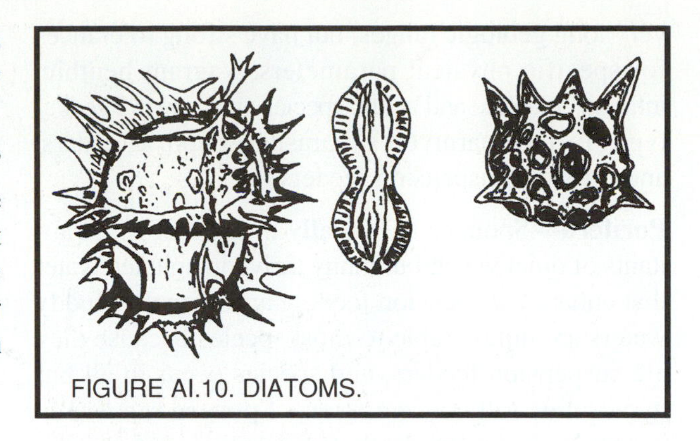

FIGURE AI.10. DIATOMS.

PALEOECOLOGY

Graptolithina—Graptolites were most abundant and diverse in nutrient-rich outer continental shelf-edge and slope environments of the early Paleozoic. These filter feeders are often the only fossils found in organic-rich black shales of poorly circulated, oxygen-poor water usually from outer shelf and slope regions. However, some did inhabit well-oxygenated, shallow marine environments. Many were sessile and benthic, but most were planktic (Figure AI.11). Their abundance in deep water black shales is partly due to better preservation of their fragile skeletons and their planktic habit. Planktic graptolites appear to have lived at different depths within the water column.

Ostracoda—Ostracodes inhabit environments as varied as those of gastropods. They occur in all aquatic environments, from lakes and rivers, to hypersaline lagoons, and abyssal depths, as well as in mosses, and caves. Thus, the groups physical tolerances span the range of depths, temperature, salinities, and other parameters. Particular species, however, have specific tolerances and can provide clues about ancient environments when those of other organisms are equivocal. This is possible because a number of genera have

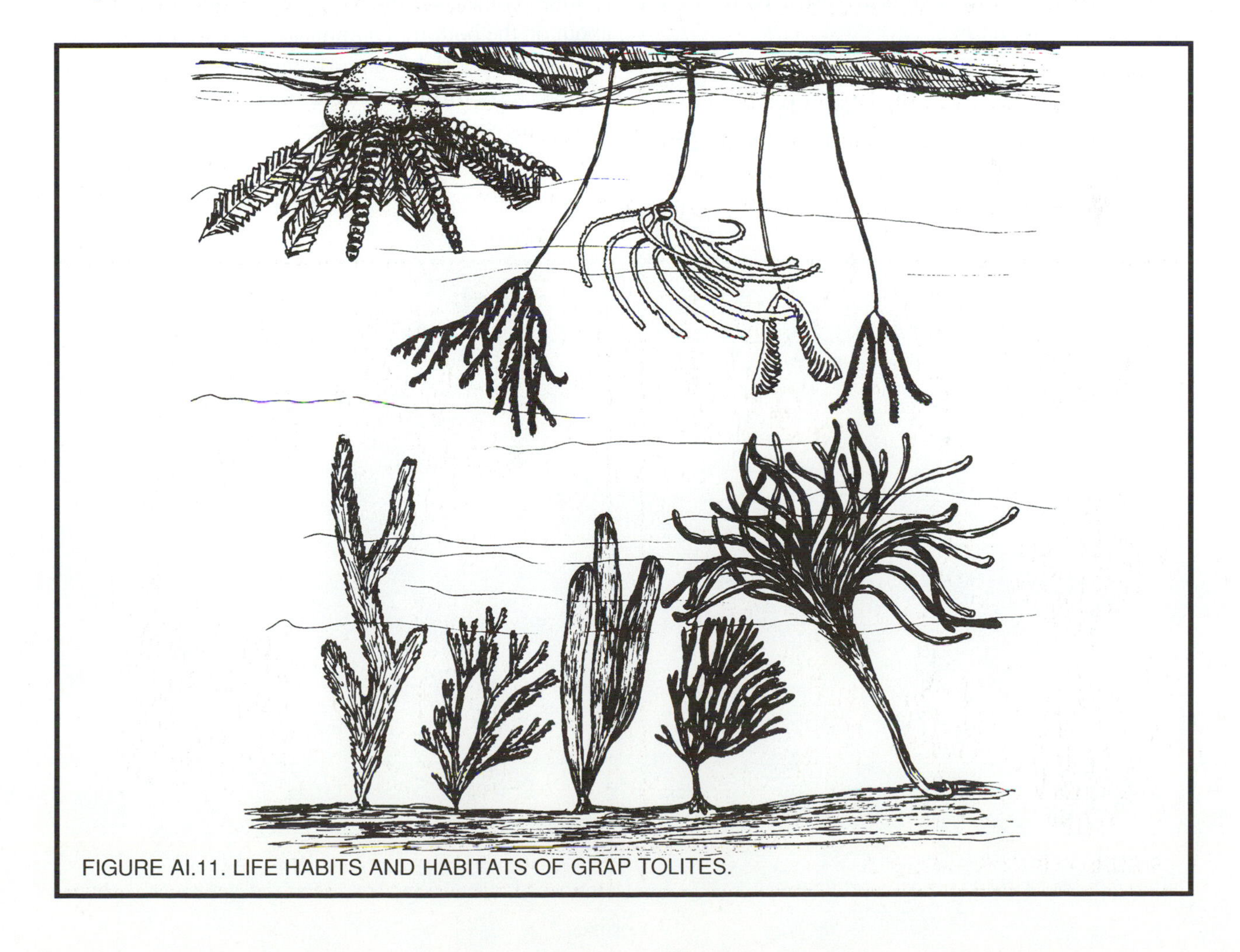

FIGURE AI.11. LIFE HABITS AND HABITATS OF GRAP TOLITES.

very long geologic ranges, but have strong tolerances for specific physical parameters. Vagrant benthic, infaunal, planktic and nektic species are known. Feeding types include carnivores, parasites, deposit feeders, and possibly suspension feeders.

Porifera—Sponges are usually thought of as inhabitants of quiet water, but many thrive in agitated water that enhance suspension feeding and growth. Muddy waters are unfavorable to most species because they are suspension feeders, and species occur in all but hypersaline salt concentrations. Sponges can live in waters of almost any depth, from shallow waters within the photic zone (200 m or 650 ft.) to abyssal depths of (5000 m or 16,400 ft.). During the Paleozoic and Mesozoic, some sponges were important framework builders in reef communities. Stromatoporoids and other calcareous sponges were particularly abundant in Paleozoic reefs and siliceous sponges were reef builders during the Jurassic.

Conodonta—Most conodonts are believed to have been planktic or nektic filter feeders. They were abundant in shallow, normal marine waters but rare in near shore and deep-sea environments. Many conodont species also appear to have been restricted to certain depths and temperatures.

Foraminifera—Most small foraminifera are benthic and epifaunal or infaunal. Small planktic species occupy all marine and brackish water habitats. Those inhabiting marginal marine environments also tolerate wide ranges of temperatures, oxygen, turbulence and turbidity. Living benthic species distributions are strongly correlated with depth and temperature, and are used extensively in reconstructing ancient climates. Large foraminifera are restricted to depths of ≤100 m (328 ft.) due to dependence on phytoplankton food or algal symbionts. Like ostracodes, they can often provide definitive environmental information when other fossils cannot.

Coccolithophora—Coccolithophores are phytoplankton and thus, restricted to the photic zone. A few species occur in brackish and fresh water, but most are marine. Particular species have little tolerance to variable physical factors and, thus, can be very useful in environmental reconstruction. A strong depth and temperature dependence makes them very useful in studies of oceanography and ancient climates. Coccolithophores are part of the food chain base and contribute a great volumes of shallow carbonate sediments.

Diatoms—Diatoms are photosynthetic phytoplankton that inhabit fresh and marine waters of the photic zone. Some diatoms were benthic and could move about on the bottom. Diatoms can be extremely abundant in nutrient rich waters, regardless of temperature. Diatom tests form extensive beds of "diatomaceous earth," the siliceous equivalent of chalk.

APPENDIX II
Fossil Identification Key

*Conodontophora, Foraminifera, Coccolithophora, Radiolaria and Diatoms not included

Key Category	Description	Organic Group*
1A.	Colonial	Go to 2
B.	Not colonial	Go to 3
2.A	Openings smaller than 1 mm	Go to 13
B.	Openings larger than 1 mm	Go to 14
C.	Openings not visible, no macroscopic internal structure	Porifera (sponges)
3.A.	Coiled	Go to 4
B.	Not coiled	Go to 6
4.A.	Chambered	Go to 5
B.	Not chambered, hollow	Gastropoda-Mollusca
5.A	Septa straight or slightly curved	Nautiloids-Mollusca
B.	Septa highly convoluted	Ammonoids-Mollusca
6.A.	Bilateral symmetry	Go to 7
B.	Pentameral symmetry	Go to 9
C.	Radial symmetry	Go to 11
7.A.	Two paired valves	Go to 8
B.	Segmented, three main segments	Trilobita-Arthropoda
8.A.	Paired valves are exact mirror images	Bivalvia-Mollusca
B.	Bilateral symmetry in each valve	Go to 17
9.A.	Globular or vaselike; polygonal plates; stem or stem attachment site present	Go to 10
	Globular or disc-like, stem absent; Secondary bilateral symmetry	Echinoidea-Echinodermata
10.A.	Five radiating grill-like grooves (ambulacral plates)	Crinozoa-Echinodermata
B.	No ambulacral plates	Blastozoa Echinodermata
11.A	Conical or tapering shell	Go to 12
B.	Cylindrical, disc-like segments	Crinozoa or Blastozoa stems

Key Category	Description	Organic Group*
12.A.	Septa forming chambers	Go to 5
	Radial septa	Rugosa (solitary coral) Anthozoa, Cnidaria
B.	Solid shell, cigar-like, radial crystals	Belemnites (Coleoidea) Mollusca
13.A.	Massive with thin laminations, each layer having many box-like chambers	Stromatoporata Porifera
B.	Not thinly laminated	Go to 15
14.A.	Vertical septa radiating from center of each opening	Rugosa (colonial coral) Anthozoa, Cnidaria
	Horizontal partitions (tabulae) forming chambers	Tabulata (colonial coral) Anthozoa, Cnidaria
15.A.	Skeleton of calcite or aragonite	Go to 16
B.	Skeleton of organic fibers	Graptolithina (Hemichordata (?))
16.A.	Living chambers tube-like	"Stony" Bryozoa (Stenolaemata)
B.	Living chambers box-like	Gymnolaemata Bryozoa
17.A .	Macroscopic	Go to 18
B.	Microscopic	Ostracoda (Arthropoda)
18.A.	Calcite shell, shell hinged	Articulata (Brachiopoda)
B.	Calcium phosphate (apatite) shell, Lacking hinges	Inarticulata (Brachiopoda)

APPENDIX III
References

The first listing is of those materials used as general references throughout the text. The second list provides a chapter breakdown of the principal source(s) of specific information and exercise data.

GENERAL

Dott, R.H., Jr. and R. L. Batten. 1988. Evolution of the Earth. Fourth Edition. McGraw-Hill, Inc. NY., NY.
Hamblin, W. K. 1978. The Earth's Dynamic Systems. Second Edition. Burgess Publishing Company, Minneapolis, MN.
Levin, H. L. 1988. The Earth Through Time. Third Edition. Saunders College Pub., NY., NY.
Press, F. and R. Siever. 1974. Earth. First Edition. W. H. Freeman and Company, San Francisco.
Stanley, S. M. 1986. Earth and Life Through Time. Second Edition. Freeman & Co., San Francisco.
Wicander, R. and J. S. Monroe. 1989. Historical Geology: Evolution of the Earth and Life Through Time. West Publishing Company, St. Paul, MN.

CHAPTER 2.

Blatt, H., G. Middleton, and R. Murray. 1972. Origin of Sedimentary Rocks. Prentice-Hall, NJ.

CHAPTER 3.

Blatt, H., G. Middleton, and R. Murray. 1972. Origin of Sedimentary Rocks. Prentice-Hall, NJ.
Krumbein, W. C. and L. L. Sloss. 1963. Stratigraphy and Sedimentation. W. H. Freeman & Co., San Francisco.

CHAPTER 5.

Clarkson, E. N. K. 1979. Invertebrate Paleontology and Evolution. Allen and Unwin, London.
Boardman, R. S., A. H. Cheetham, and A. J. Rowell (eds.). 1987. Fossil Invertebrates. Blackwell Scientific Publications, London.

CHAPTER 6.

Dodd, R. J. and Robert J. Stanton, Jr. 1981. Paleoecology, Concepts and Applications. John Wiley and Sons, New York.

Valentine, J. W. 1973. Evolutionary Paleoecology of the Marine Biosphere. Prentice-Hall, Inc., NY, NY

CHAPTER 7.

Boardman, R. S., A. H. Cheetham, and A. J. Rowell. 1987. Fossil Invertebrates. Blackwell Scientific Publications, London.

Krumbein, W. C. and L. L. Sloss. 1963. Stratigraphy and Sedimentation. W. H. Freeman & Co., San Francisco.

CHAPTER 8.

Reading, H. G. ed.). 1978. Sedimentary Environments and Facies. Elsevier, New York.

CHAPTER 9.

Facies and Paleogeographic map data from:

Dott, R. H., Jr. and R. L. Batten. 1988. Evolution of the Earth. Fourth Edition. McGraw-Hill, Inc. NY., NY.

Schopf, T. J. M. 1980. Paleoceanography. Harvard University Press, Cambridge, Massachusetts.

CHAPTER 10.

Dott, R. H., Jr. and R. L. Batten. 1988. Evolution of the Earth. Fourth Edition. McGraw-Hill, Inc. NY., NY.

Schneider, S. H. and R. Londer. 1984. The Coevolution of Climate and Life. Sierra Club Books, San Francisco.

Schopf, T. J. M. 1980. Paleoceanography. Harvard University Press, Cambridge, Massachusetts.

Windely, B. F. 1977. The Evolving Continents. Second Edition. John Wiley and Sons, NY., NY.

CHAPTER 11.

Schopf, T. J. M. 1980. Paleoceanography. Harvard University Press, Cambridge, Massachusetts.

CHAPTER 12.

Reading, H. G. ed.). 1978. Sedimentary Environments and Facies. Elsevier, New York.

Windely, B. F. 1977. The Evolving Continents. Second Edition. John Wiley and Sons, NY., NY.

PALEOGEOGRAPHIC MAPS (WORLD)

Scotese, C. R. 1986. Phanerozoic Reconstructions: A New Look at the Assembly of Asia. University of Texas Institute for Geophysics Technical Report No. 66.

Winn, K. and C. R. Scotese. 1986. Phanerozoic Paleogeography Maps. Paleoceanographic Mapping Project Progress Report No. 33-1287.

Paleoclimatological Data From:

Scotese, C. R., R. K. Bambach, Barton, C., Van der Voo, R., and Zeigler, A. M. 1979. Paleozoic Base Maps. Journal of Geology 87:217-277.

APPENDIX IV
Color Plates

FIGURE 1. Conglomerate

FIGURE 2. Breccia

FIGURE 3. Arkose

FIGURE 4. Graywacke

FIGURE 5. Quartz Sandstone

FIGURE 6. Siltstone

FIGURE 7. Shale

FIGURE 8. Fossiliferous Limestone

FIGURE 9. Unfossiliferous Limestone

PLATE I—SEDIMENTARY ROCKS.
Courtesy Sue Monroe, except Fig. 4., courtesy R.V. Dietrich, Central Michigan University

FIGURE 1. Coquina

FIGURE 2. Chalk

FIGURE 3. Crystalline Limestone

FIGURE 4. Dolostone

FIGURE 5. Chert

FIGURE 6. Coal

FIGURE 7. Halite

FIGURE 8. Gypsum

FIGURE 9. Anhydrite

PLATE II— SEDIMENTARY ROCKS
Courtesy Sue Monroe

PLATE III—FIGURE 1. DIORAMA OF A MIDDLE DEVONIAN COMMUNITY OF CENTRAL NEW YORK.

PLATE III—FIGURE 2. DIORAMA OF THE BURGESS SHALE FLORA AND FAUNA.

PLATE IV—FIGURE 1. DIORAMA OF AN UPPER ORDOVICIAN COMMUNITY FROM THE REGION OF CINCINNATI, OHIO.

PLATE IV—FIGURE 2. DIORAMA OF A SILURIAN COMMUNITY.

PLATE IV—FIGURE 3. DIORAMA OF A SILURIAN BRACKISH WATER/LAGOONAL COMMUNITY.

PLATE IV—FIGURE 4. DIORAMA OF A MISSISSIPPIAN COMMUNITY.

PLATE IV—FIGURE 1. DIORAMA OF A PENNSYLVANIAN COMMUNITY.

PLATE V—FIGURE 2. DIORAMA OF A PERMIAN COMMUNITY

PLATE V—FIGURE 3. DIORAMA OF A CRETACEOUS COMMUNITY.

PLATE V—FIGURE 4. DIORAMA OF A CENOZOIC COMMUNITY

Geologic Map of Part of the Grand Canyon, Arizona

Legend

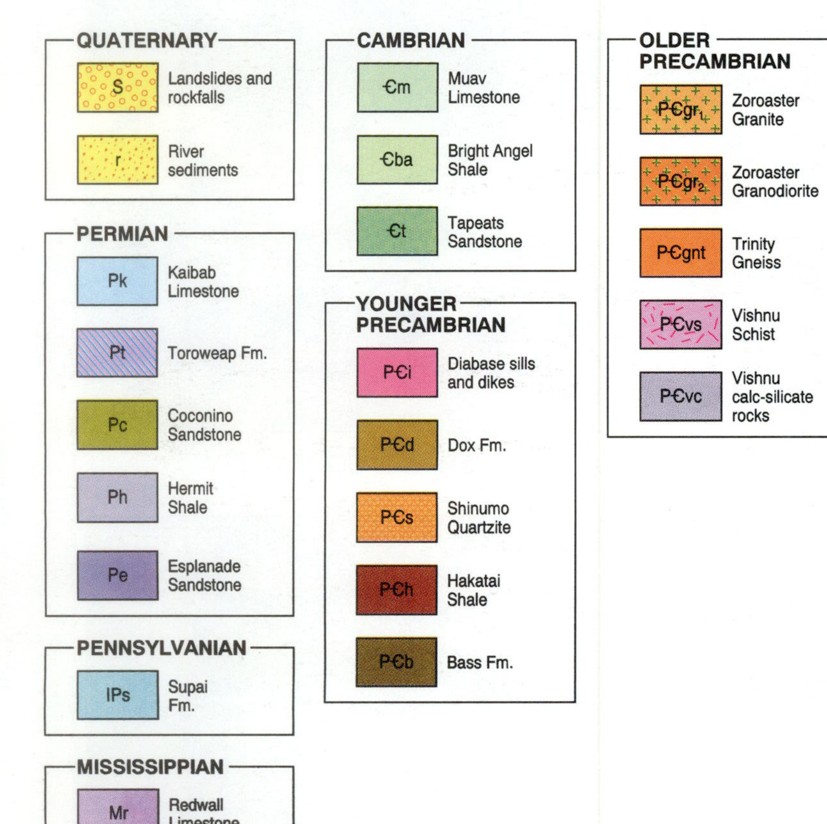

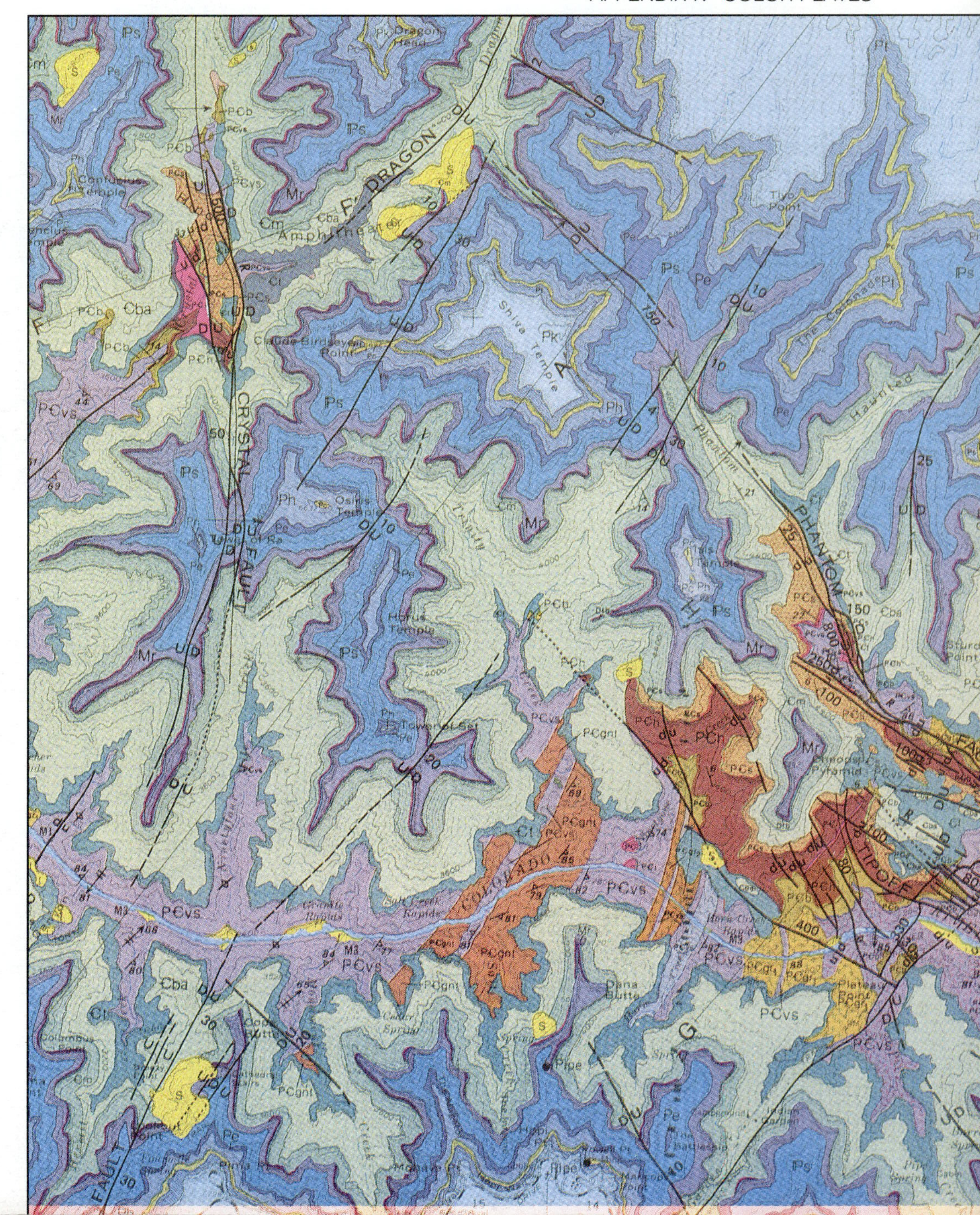

Scale 1:62,500

1 .5 0 1 miles

C.I. 50 feet

Geologic Map of Michigan and the Surrounding Area

Legend

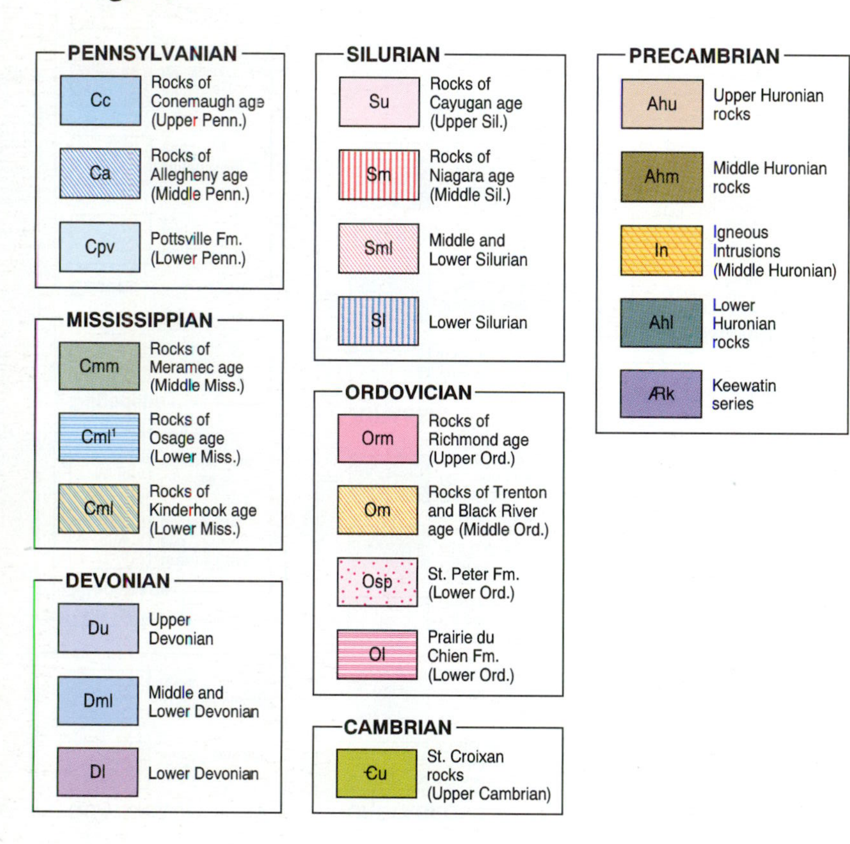

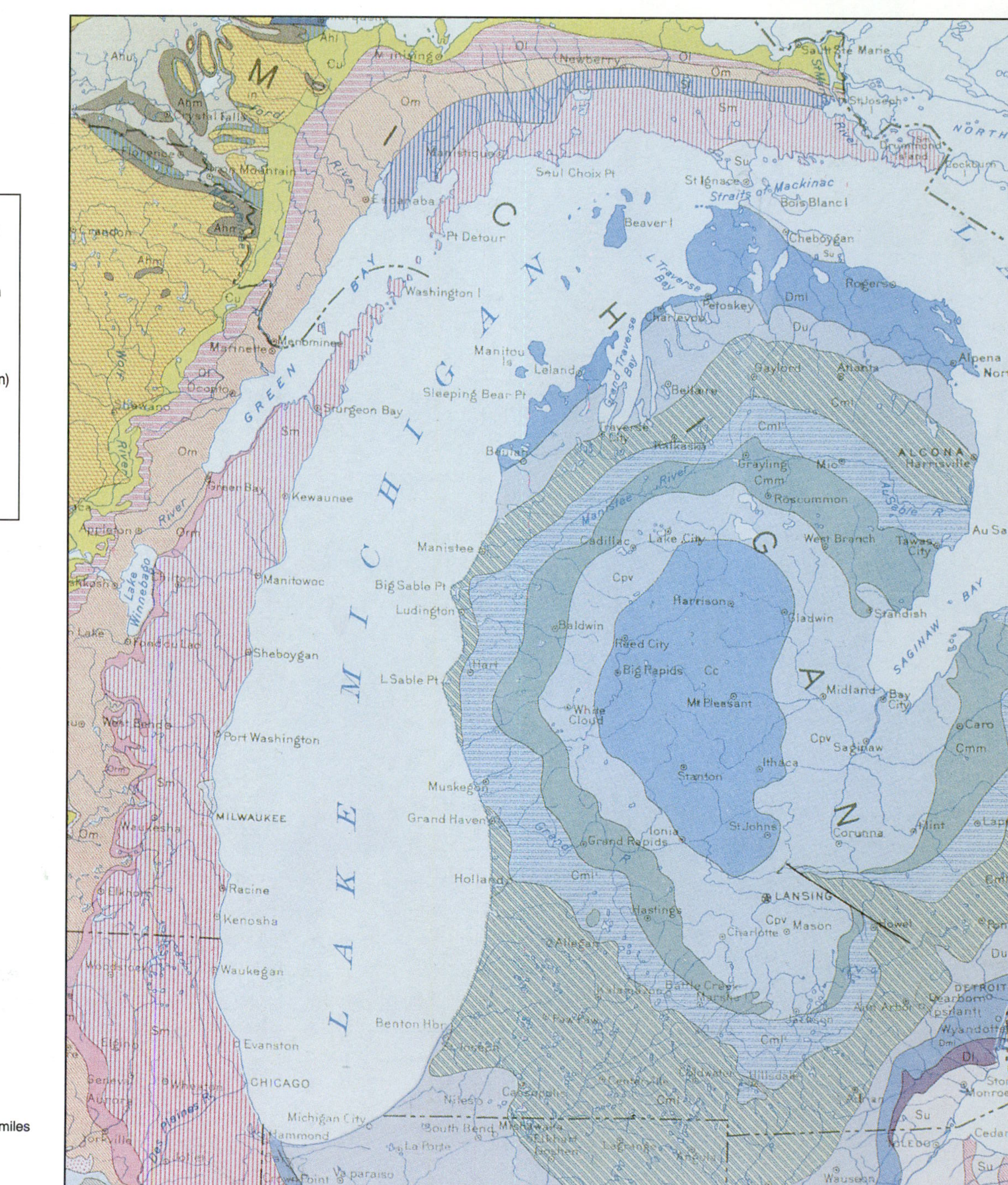

Geologic Map of the Wyoming Area

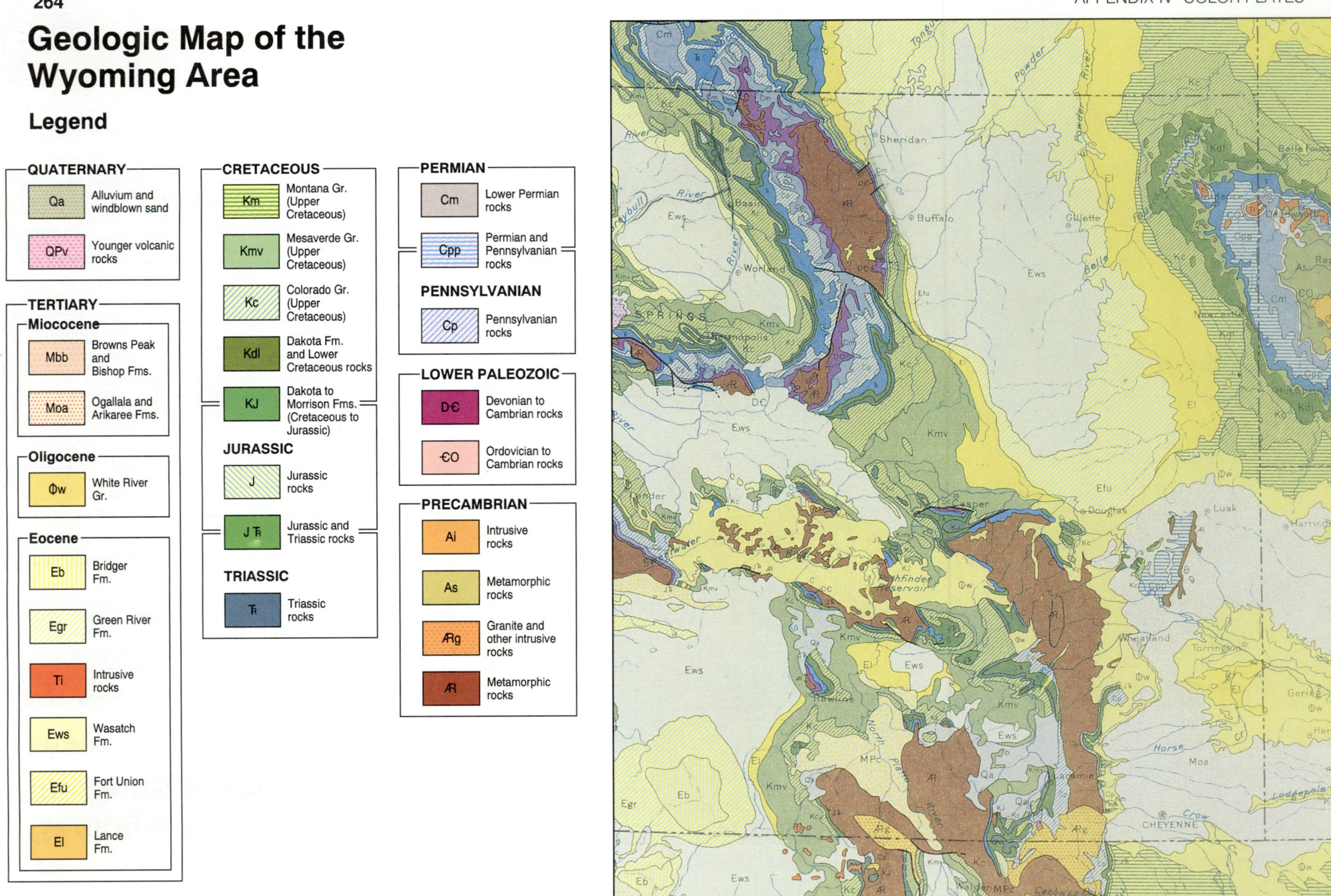

Geologic Map of the Southern States

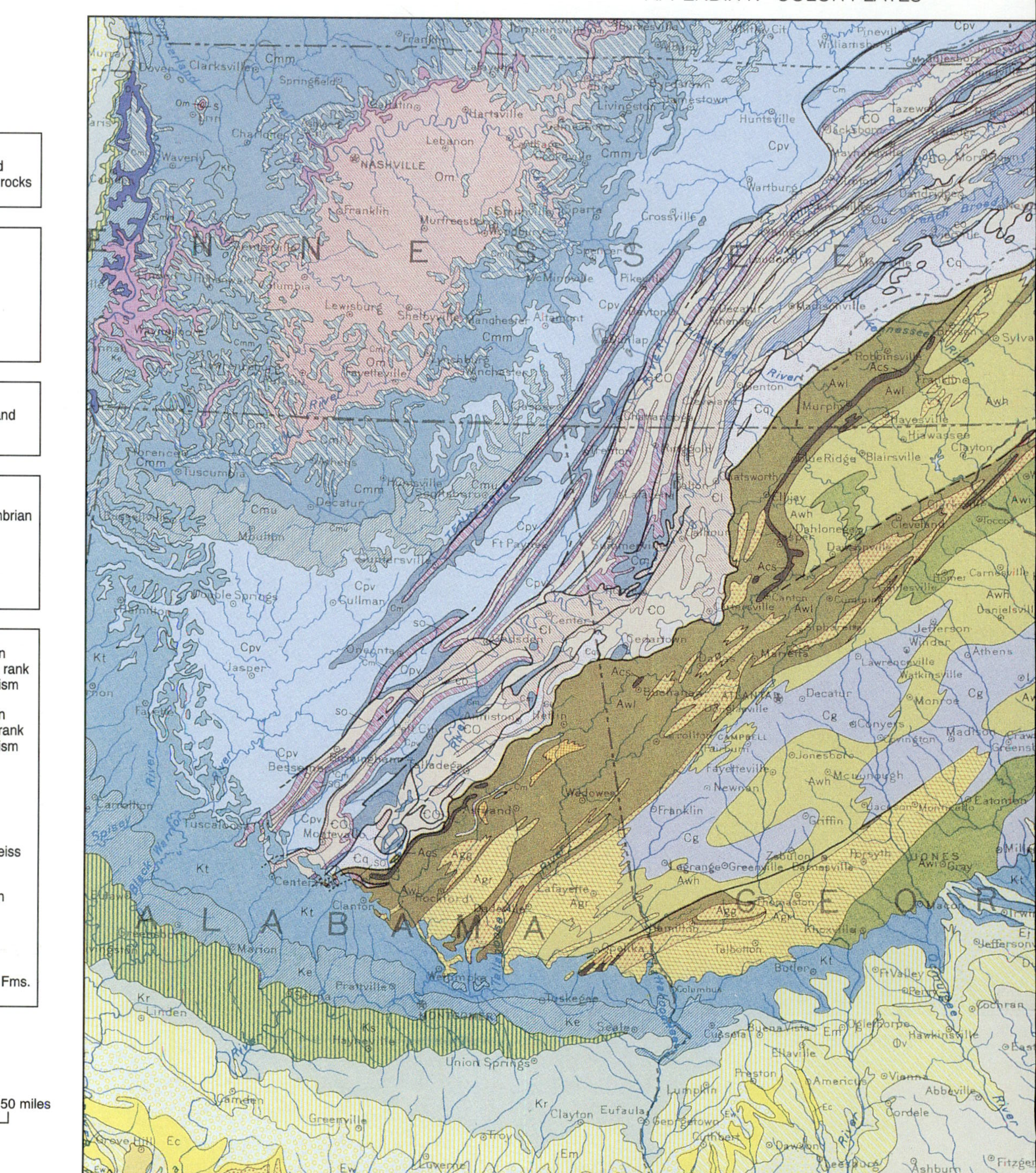

Legend

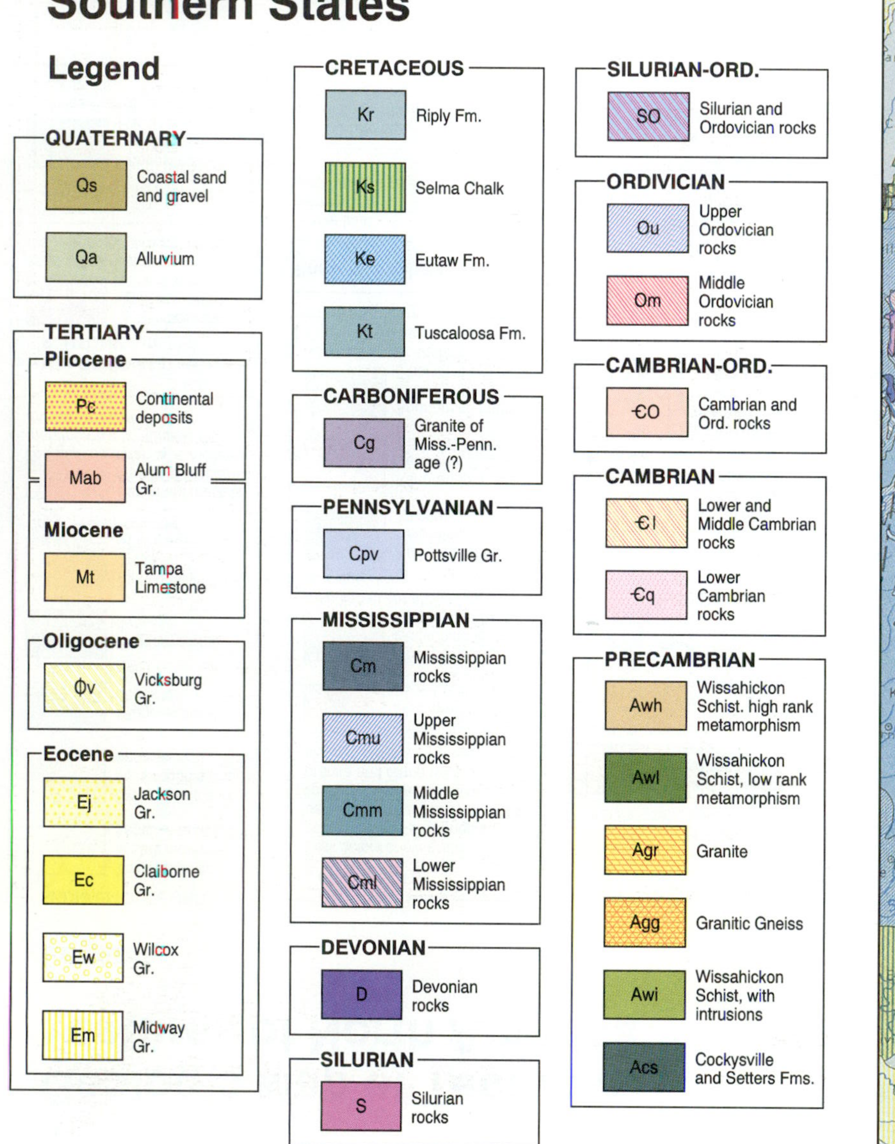

Geologic Map of Tectonic Features of North America

Legend

SEDIMENTARY UNITS

- Thick deposits in structurally negative areas
- Synorogenic and postorogenic deposits
- Miogesynclinal deposits
- Eugeosynclinal deposits
- Early geosynclinal deposits *Of Middle and Upper Proterozoic ages*
- Basement massifs *Mainly of Precambrian age. Includes metamorphic complexes that involve younger rocks.*

VOLCANIC AND PLUTONIC UNIT

- Granitic plutons *Ages are generally within the span of the tectonic cycle of the foldbelts in which they lie.*

SPECIAL UNIT

- Eugeosynclinal deposits of the Pacific border *Includes Franciscan Formation of California.*

PRECAMBRIAN FOLDBELTS

Dark colors show areas of paraschist and paragneiss derived from supracrustal rocks; light colors show areas of granite and othogneiss of plutonic origin.

- Greenville foldbelt *Deformed 880–1,000 m.y. ago*
- Rocks of the Hudsonian foldbelt *Overprinted by Elsonian event about 1,370 m.y. ago*
- Hudsonian foldbelts *Deformed 1,640–1,820 m.y. ago*
- Kenoran foldbelts *Deformed 2,390–2,600 m.y. ago*
- Anorthosite bodies *In Greenville and Elsonian belts or, alternatively, in eastern Canadian Shield*

PLATFORM AREAS

- Ice cap of Quaternary age *On Precambrian and Paleozoic basement*
- Plateau basalts and associated rocks *In North Atlantic province*
- Platform deposits on Mesozoic basement *In Arctic Coastal Plain*
- Platform deposits on Paleozoic basement *In Atlantic and Gulf Coastal Plains*
- Platform deposits on Precambrian basement *In central-craton*
- Platform deposits within the Precambrian *Mainly in the Canadian Shield*

Structural Symbols

- Normal fault *Hachures on downthrown side*
- Transcurrent fault *Arrows show relative lateral movement*
- Thrust fault *Barbs on upthrown side*
- Subsea fault *Long dashes based on topographic and geophysical evidence; short dashes, based on geophysical evidence only.*
- Axes of sea-floor spreading
- Flexure *Arrows on depressed side*
- Salt domes and salt diapirs *In Gulf Coastal Plain and Gulf of Mexico*
- ★ Volcano
- +2000 / 0 / 2000 Contours on basement surfaces beneath platform areas *All contours are below sea level except where marked with plus symbols. Interval 2,000 meters.*

Scale 1:50,000

0 1 2 miles

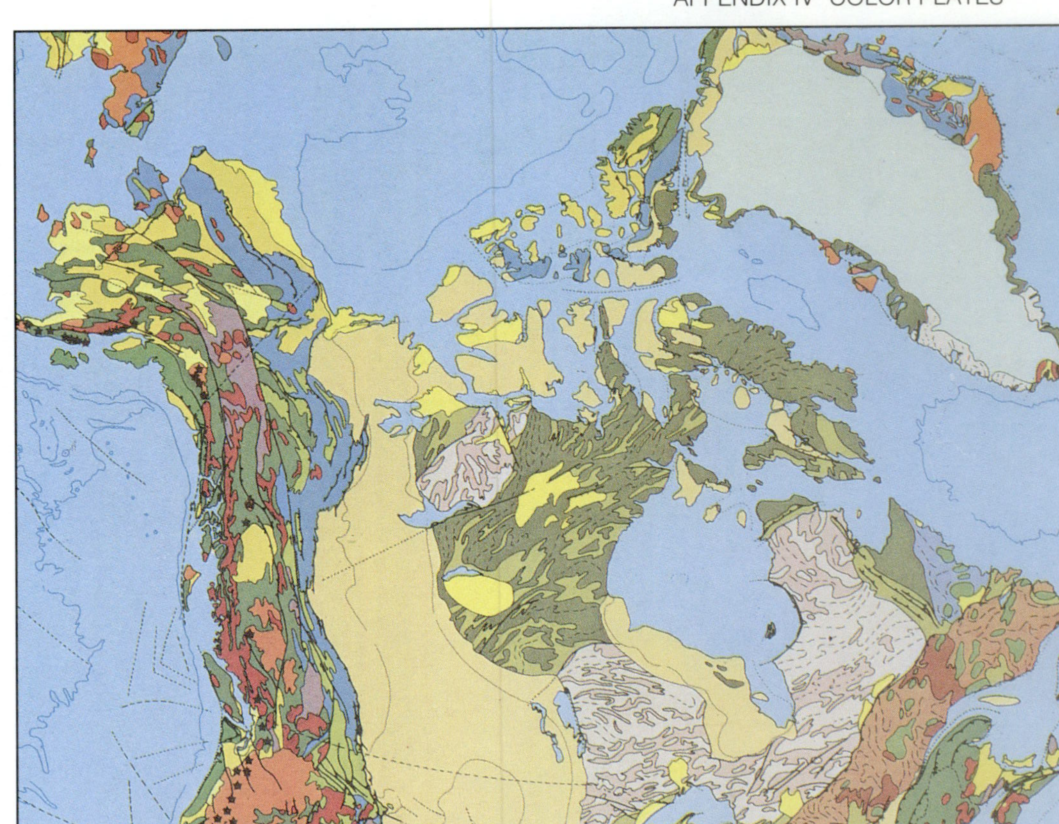

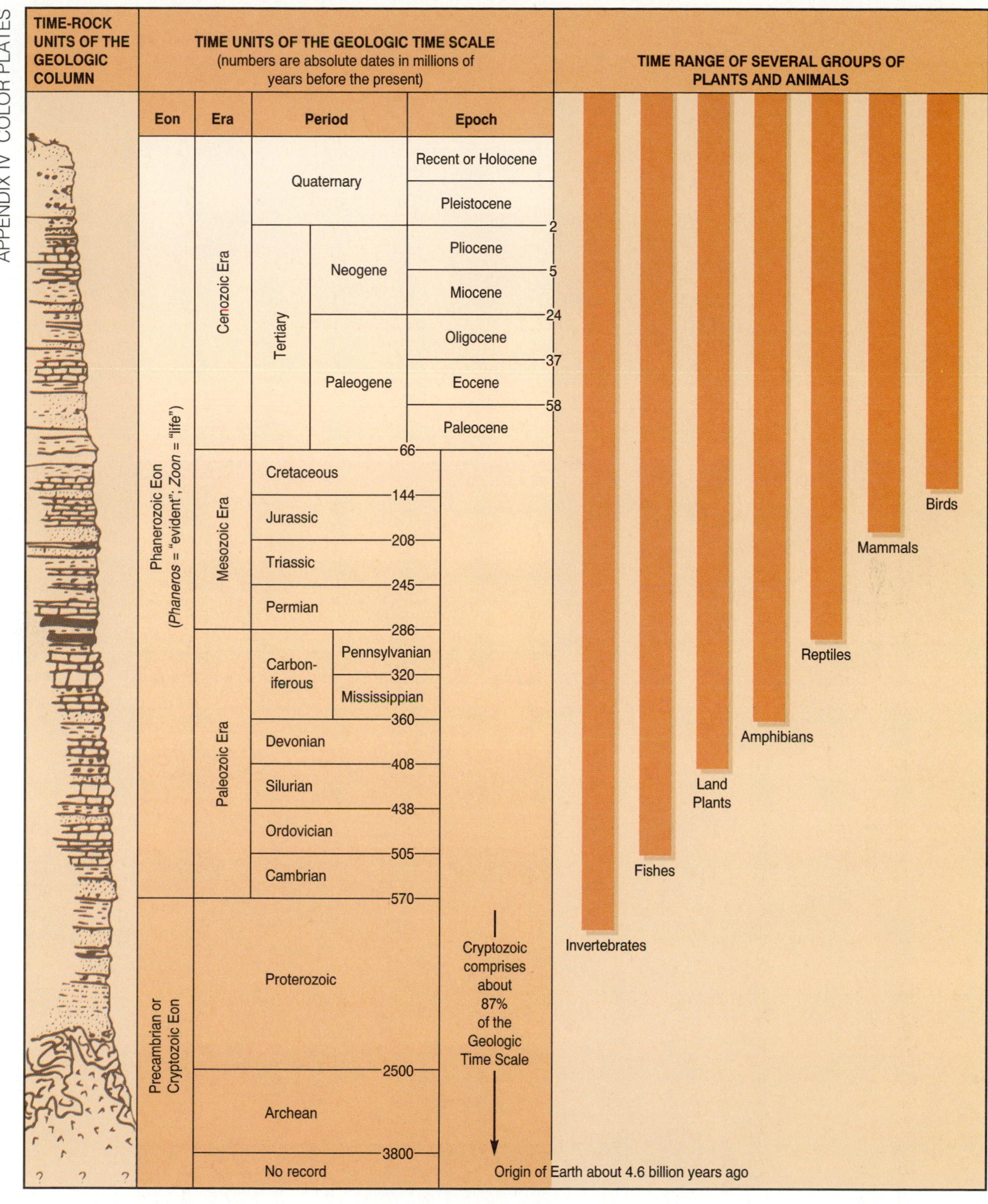
TIME-ROCK UNITS OF THE GEOLOGIC COLUMN
TIME UNITS OF THE GEOLOGIC TIME SCALE
(numbers are absolute dates in millions of years before the present)
TIME RANGE OF SEVERAL GROUPS OF PLANTS AND ANIMALS
Eon
Era
Period
Epoch
Recent or Holocene
Quaternary
Pleistocene
2
Pliocene
Neogene
5
Miocene
24
Cenozoic Era
Tertiary
Oligocene
37
Paleogene
Eocene
58
Paleocene
66
Phanerozoic Eon
(Phaneros = "evident"; Zoon = "life")
Cretaceous
144
Jurassic
Mesozoic Era
208
Triassic
245
Permian
286
Pennsylvanian
Carbon-iferous
320
Mississippian
360
Devonian
Paleozoic Era
408
Silurian
438
Ordovician
505
Cambrian
570
Proterozoic
Cryptozoic comprises about 87% of the Geologic Time Scale
Precambrian or Cryptozoic Eon
2500
Archean
3800
No record
Origin of Earth about 4.6 billion years ago
Invertebrates
Fishes
Land Plants
Amphibians
Reptiles
Mammals
Birds